Itinerant Electron Magnetism:
Fluctuation Effects

NATO Science Series

A Series presenting the results of activities sponsored by the NATO Science Committee. The Series is published by IOS Press and Kluwer Academic Publishers, in conjunction with the NATO Scientific Affairs Division.

General Sub-Series

A. Life Sciences	IOS Press
B. Physics	Kluwer Academic Publishers
C. Mathematical and Physical Sciences	Kluwer Academic Publishers
D. Behavioural and Social Sciences	Kluwer Academic Publishers
E. Applied Sciences	Kluwer Academic Publishers
F. Computer and Systems Sciences	IOS Press

Partnership Sub-Series

1. Disarmament Technologies	Kluwer Academic Publishers
2. Environmental Security	Kluwer Academic Publishers
3. High Technology	Kluwer Academic Publishers
4. Science and Technology Policy	IOS Press
5. Computer Networking	IOS Press

The Partnership Sub-Series incorporates activities undertaken in collaboration with NATO's Partners in the Euro-Atlantic Partnership Council – countries of the CIS and Central and Eastern Europe – in Priority Areas of concern to those countries.

NATO-PCO-DATA BASE

The NATO Science Series continues the series of books published formerly in the NATO ASI Series. An electronic index to the NATO ASI Series provides full bibliographical references (with keywords and/or abstracts) to more than 50000 contributions from international scientists published in all sections of the NATO ASI Series.
Access to the NATO-PCO-DATA BASE is possible via CD-ROM "NATO-PCO-DATA BASE" with user-friendly retrieval software in English, French and German (© WTV GmbH and DATAWARE Technologies Inc. 1989).

The CD-ROM of the NATO ASI Series can be ordered from: PCO, Overijse, Belgium.

3. High Technology – Volume 55

Itinerant Electron Magnetism: Fluctuation Effects

edited by

D. Wagner
W. Brauneck

Theoretische Physik III,
Ruhr-Universität Bochum,
Bochum, Germany

and

A. Solontsov

A.A. Bochvar Institute for Inorganic Materials
and State Center for Condensed Matter Physics,
Moscow, Russia

Kluwer Academic Publishers

Dordrecht / Boston / London

Published in cooperation with NATO Scientific Affairs Division

Proceedings of the NATO Advanced Research Workshop on
Itinerant Electron Magnetism: Fluctuation Effects & Critical Phenomena
Moscow, Russia
September 15–19, 1997

A C.I.P. Catalogue record for this book is available from the Library of Congress.

ISBN 0-7923-5202-5

Published by Kluwer Academic Publishers,
P.O. Box 17, 3300 AA Dordrecht, The Netherlands.

Sold and distributed in North, Central and South America
by Kluwer Academic Publishers,
101 Philip Drive, Norwell, MA 02061, U.S.A.

In all other countries, sold and distributed
by Kluwer Academic Publishers,
P.O. Box 322, 3300 AH Dordrecht, The Netherlands.

Printed on acid-free paper

All Rights Reserved
© 1998 Kluwer Academic Publishers
No part of the material protected by this copyright notice may be reproduced or utilized in any form or by any means, electronic or mechanical, including photocopying, recording or by any information storage and retrieval system, without written permission from the copyright owner.

Printed in the Netherlands

Dedicated to the Memory of
Andrei S. Borovik-Romanov

Moscow, November, 1977

Table of Contents

Dedication	v
Table of contents	vii
Preface	xi
Spin fluctuations in itinerant frustrated systems M. Shiga and H. Nakamura	1
Neutron scattering and μSR studies of spin fluctuations in frustrated itinerant magnets B.D. Rainford, S.J. Dakin, J.R. Stewart and R. Cywinski	15
Critical fluctuations and the nature of the Néel transition near the triple point in chromium alloys E. Fawcett and D.R. Noakes	27
UPd_2Al_3: An analysis of the inelastic neutron scattering spectra N. Bernhoeft, A. Hiess, B. Roessli, N. Sato, N. Aso, Y. Endoh, G.H. Lander and T. Komatsubara	43
Anisotropy of the generalized susceptibility in Mn (38% Ni) alloy in the magnetic phase transition region J.J. Milczarek, J. Jankowska-Kisielińska, K. Mikke and B. Hennion	61
Pseudodipolar interaction and antiferromagnetism of R_2CuO_4 compounds (R= Pr, Nd, Sm and Eu) S.V. Maleyev, D. Petitgrand, Ph. Bourges and A.S. Ivanov	67
Soft-mode spin fluctuations in itinerant electron magnets. A. Solontsov, A. Vasil'ev and D. Wagner	89
Effects of spin fluctuations in transition metals and alloys M. Shimizu	123
The temperature dependence of the enhanced paramagnetic susceptibility at finite magnetic field E. Pamyatnykh, A. Poltavets and M. Shabalin	151
First-principles study of itinerant-electron magnets: ground state and thermal properties. L.M. Sandratskii, M. Uhl and J. Kübler	161
Molecular dynamics approach to complex magnetic structures in itinerant-electron systems Y. Kakehashi, S. Akbar and N. Kimura	193
Spin fluctuation theory versus exact calculations V. Barar, W. Brauneck and D. Wagner	229

Magnetovolume effect and longitudinal spin fluctuations in invar alloys. ... 243
A.Z. Menshikov, V.A. Kazantsev, E.Z. Valiev and S.M. Podgornykh

High temperature thermal expansion of RMn_2 intermetallic compounds with heavy rare earth elements ... 261
I.S. Dubenko, I.Yu. Gaidukova, S.A. Granovsky, R.Z. Levitin, A.S. Markosyan, A.B. Petropavlovsky, V.E. Rodimin and V.V. Snegirev.

Magnetoelasticity and isotope effect in ferromagnets. ... 269
V.M. Zverev

Spin-flop and metamagnetic transitions in itinerant ferrimagnets ... 285
A. K. Zvezdin, I. A. Lubashevsky, R. Z. Levitin, G.M. Musaev, V. V. Platonov and O. M. Tatsenko

The phase diagram of the Kondo lattice ... 303
C. Lacroix

Pressure effect on the magnetic susceptibility of the $YbInCu_4$ and $GdInCu_4$ compounds ... 309
I.V. Svechkarev, A.S. Panfilov, S.N. Dolja, H. Nakamura and M. Shiga

Atomic volume effect on electronic structure and magnetic properties of UGa_3 compound ... 323
G.E. Grechnev, A.S. Panfilov, I.V. Svechkarev, A. Delin, O. Eriksson, B. Johansson and J.M. Wills

On the temperature dependence of the electrical resistivity of $Er_{0.55}Y_{0.45}Co_2$ ... 337
A.N. Pirogov, N.V. Baranov, A.A. Yermakov, C. Ritter, and J. Schweizer

Electrical resistivity and phase transitions in FeRh based compounds: influence of spin fluctuations ... 345
N.V. Baranov, S.V. Zemlyanski and K. Kamenev

Resistivity, magnetoresistance and Hall effect in $Co_{(100-x)}(CuO)_x$ ($10 \leq x \leq 70$ wt. %) composites ... 353
V. Prudnikov, A. Granovsky, M. Prudnikova and H.R. Khan

Theory of itinerant-electron spin-glass in amorphous Fe ... 363
T. Uchida and Y. Kakehashi

The formation of the magnetic properties in disordered binary alloys of metal-metalloid type ... 375
A.K. Arzhnikov and L.V. Dobysheva

Itinerant electrons and superconductivity in exotic layered systems ... 391
V. A. Ivanov, E. A. Ugolkova and M.Ye. Zhuravlev

The multiband analysis of the electron dispersion at the top of the valence band in undoped cuprates ... 433
S.G. Ovchinnikov

Spin-Peierls magnet CuGeO$_3$ ... 437
   G.A. Petrakovskii

Electron acoustic effects in metallic magnetic multilayers ... 451
   V.I. Okulov, V.V. Ustinov, E.A. Pamyatnykh and V.V. Slovikovskaya

Author Index ... 457

Subject Index ... 459

PREFACE

This volume contains the contributions delivered at the NATO Advanced Research Workshop "Itinerant Electron Magnetism: Fluctuation Effects and Critical Phenomena", which was held in the small village Krasnovidovo not far from Moscow, from 15th to 19th of September 1997. The book both summarizes recent developments and presents new experimental and theoretical results in itinerant electron magnetism with an emphasis on fluctuation phenomena and correlation effects.

Since the first application of quantum theory to the magnetism of electrons in metals in the late twenties, itinerant electron magnetism has been established as a separate discipline of condensed matter physics. For a long time understanding of magnetic phenomena was dominated by the independent particle models (e.g., the Stoner model), although shortcomings of these models became quite clear. But in the seventies things changed rapidly by the work of Moriya and his co-workers who tried to establish a more rigorous fundament for itinerant magnetism by including coupled spin fluctuations and electron correlations. Moreover, Moriya's work stimulated many new concepts and ideas put forward by other scientists, and made problems of itinerant magnetism the central topic in many discussions and conferences. However, it became obvious that it was not possible to formulate a generally accepted theory of itinerant magnetism and scientists turned to move to trendy topics.

So, we thought it might be useful to organise a workshop on itinerant magnetism to discuss novel ideas and concepts, which have been developed since that time. Special emphasis has been laid to fluctuation phenomena and electron correlation effects: softening (slowing down) of spin-fluctuation modes, anisotropy of the fluctuation spectrum, zero-point motions, strong spin anharmonicity, electron correlations in the ground state and magnetic frustrations.

We would like to dedicate this volume to the memory of Prof. A.S. Borovik-Romanov, who passed away shortly before the Workshop. We are grateful to him for his support to organise this meeting.

We are also indebted to A. Vasil'ev for his efforts in preparing this volume.

Finally, we would like to thank the NATO Science Committee for the support, which essentially affected the success of the Workshop and resulted in the present volume.

Dieter Wagner
Wolfgang Brauneck
Alexander Solontsov

SPIN FLUCTUATIONS IN ITINERANT FRUSTRATED SYSTEMS

M. SHIGA AND H. NAKAMURA
Department of Materials Science and Engineering,
Kyoto University, Sakyo-ku, Kyoto 606-01, Japan

Abstract. Frustration of magnetic interactions gives rise to various anomalies in the magnetic structure and the critical behavior at the transition temperature. Among them, the ground state of the fully frustrated (FFR) system, which has a macroscopic number of degenerate spin configurations, is attracting much attention. The concept of the quantum spin liquid (QSL) has been proposed as a ground state of FFR. However, no examples has been found so far in ionic crystals. We have studied magnetic and thermal properties of $Y(Sc)Mn_2$, where Mn sites form FFR lattice, and found that this compound exhibits really astonishing properties; the very large γ value of $150mJ/Kmol^2$, the enhanced thermal expansion coefficient of $50 \times 10^{-6}/K$ etc.

In order to explain these anomalies, we proposed that the ground state of this compound may be in the QSL state, which is characterized by the existence of giant zero point spin fluctuations with antiferromagnetic correlations. Neutron scattering, NMR and µSR experiments have been done to prove this proposal. Effects of impurities, which should be remarkable for the FFR system, were studied by static and dynamical methods. It has been found that the substitution of Al for Mn gives rise to a spin-liquid to spin-glass transition, supporting the FFR characters of $Y(Sc)Mn_2$.

Similar phenomena have been observed in β-Mn and β-Mn-Al alloys. It will be shown that Mn II sites in the β-Mn structure can be regarded as three-dimensional twisted Kagome lattice and so as a FFR system.

1. Magnetic Frustration and Spin Fluctuations

Frustration of magnetic interactions gives rise to various interesting properties in the ground state structure and the critical behavior at the transition point. Among them, the ground state of the so called fully frustrated (FFR) system, which has a macroscopic number of degenerate spin configurations and, therefore, cannot take a unique magnetic structure, are attracting much attention in connection with the mechanism of high T_c superconductors. A typical example of the FFR system is two-dimensional triangular Ising spin system with nearest neighbor antiferromagnetic interaction. Anderson[1] and coworkers [2] introduced the concept of a quantum spin-liquid (QSL) as a nonmagnetic ground state of the FFR system, which is characterized by strong spin fluctuations with antiferromagnetic pair correlations.

So far, candidates for QSL have been searched in ionic compounds with quasi two-dimensional lattices such as the triangular Ising lattice, the Heisenberg Kagome lattice. Recently, attempts are also extended to three-dimensional lattice, in particular, pyrochlore compounds [3]. However, in most cases, a magnetically ordered phase is formed at low temperatures.

On the other hand, metallic magnetic systems have not been intensively studied from the viewpoint of frustration. The understanding of magnetism of metallic systems, in particular, of highly correlated itinerant electron systems, has been remarkably improved by the

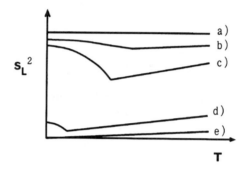

Figure 1. Types of spin fluctuations. SL2 is local amplitude of spin fluctuations.
a) local moment, b) strongly correlated system, c) Invar type, d) weakly itinerant (anti-) ferromagnet. e) enhanced Pauli paramagnet

development of the theory of spin fluctuations, which has unified two mutually opposite pictures for metallic magnetism, namely, the local moment model and the collective electron model[4]. The type of a system can be characterized by the temperature dependence of the local amplitude of spin fluctuations, which is schematically shown in Figure 1. The weakly correlated itinerant magnet is well described by the SCR theory (lines (d) and (e)). In a purely local moment system $<S_L^2>$ is equal to the Hund value of $S(S+1)$ and remains constant over the whole temperature range(line (a) in Figure 1). For strongly correlated itinerant electron systems, the temperature dependence of $<S_L^2>$ is almost constant as shown by line(b). In this case, usually a magnetically ordered state occurs below a certain critical temperature. It is interesting to study the behavior of spin fluctuations in frustrated itinerant systems with strong electron correlations where the formation of the ordered state is obstructed. In this article, we review the spin fluctuations in YMn$_2$ and β-Mn alloys, which are considered to be frustrated itinerant systems.

2. Y(Sc)Mn$_2$ and Y(Sc)(Mn$_{1-x}$Al$_x$)$_2$

YMn$_2$ has the C15 Laves phase structure, where Mn atoms form a network of corner sharing tetrahedrons. Mn-Mn interaction is antiferromagnetic and, therefore, it can be regarded as a typical fully frustrated system. At ambient pressure, it becomes antiferromagnetic below 100K accompanied with a huge volume expansion and tetragonal distortion. The spin structure is not a simple collinear one but a helically modulated one with an extremely long period of 400A [5,6] This complex spin structure may be due to frustration. The antiferromagnetic state is unstable and highly sensitive to pressure or to impurity. By substituting 3% Sc for Y, which reduces its lattice constant a little, the paramagnetic state is stabilized down to the lowest temperature [7]. The paramagnetic state is not a simple Pauli paramagnet but exhibits several anomalies such as an extraordinary large low temperature specific heat of $\gamma=150$mJ/K^{2}mol, being comparable to a heavy fermion system [8]. The enhancement of the γ value implies the existence of a large amplitude of spin fluctuations even at very low temperatures. The antiferromagnetic state also disappears by applying rather small pressure, say only 2kBar[9]. Again, the γ value is strongly enhanced by more than 10 times compared with that of an antiferromagnetic state. The γ values decreases with increasing pressure. The temperature dependence of electrical resistivity concaves upward. The coefficient of the T^2 term is also extremely large in accordance with the enhancement of the γ value [10].

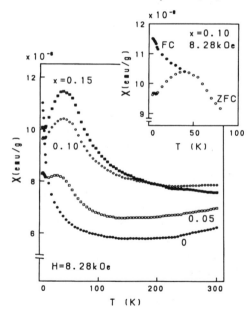

Figure 2. Temperature dependence of the susceptibility of $Y_{0.97}Sc_{0.03}(Mn_{1-x}Al_x)_2$.
Inset shows field cooled and zero-field cooled curves for x=0.1

The frustrated system is highly sensitive to impurity because the ground state degeneracy can be removed by introducing nonmagnetic sites, giving rise to a spin glass state. We have prepared $Y_{0.97}Sc_{0.03}(Mn_{1-x}Al_x)_2$ pseudo-binary compounds and studied their magnetic properties. Figure 2 shows the temperature dependence of the susceptibility. The susceptibility increases with increasing x and approaches a Curie-Weiss type at high temperatures. For x >0.05, a maximum is observed in χ- curves at low temperatures for zero field cooled samples but not for field cooled samples, suggesting spin glass freezing. These results confirm the frustrated nature of the present system and indicate the spin liquid to spin glass transition by introducing an impurity. Some microscopic measurements such as NMR[11] and µSR[12] support this picture.

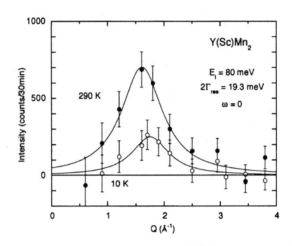

Figure 3. Magnetic neutron scattering rate of Y(Sc)Mn2 as a function of wave vector Q at zero energy transfer with energy resolution of $2\Gamma\text{res}= 19.3\text{meV}$ at 10K(○) and 290K (●)

The direct evidence for the spin liquid to spin glass transition is obtained by performing paramagnetic neutron scattering experiments to observe spin fluctuations [13]. Figure 3 shows quasi-elastic Q-scan spectra with the energy window of $2\Gamma_{res} = 19.3\text{meV}$ for Y(Sc)Mn$_2$ obtained at 10K and 290K. A large scattering amplitude was observed centered around 1.6A^{-1}. The corresponding wavelength is approximately twice of the interatomic distance, indicating antiferromagnetic correlations of spin fluctuations. The amplitude decreases with the decreasing temperature. However, scattering was still observed at 10K, suggesting the existence of strong zero-point spin fluctuations. This interpretation is confirmed by the energy spectrum of fluctuations, which is shown in Figure 4. At 10K, scattering was observed only on the neutron energy-loss side ($\omega > 0$), implying the absence of thermally excited fluctuations. The width of the spectrum is approximately 20meV (200K), indicating the characteristic frequency of zero point fluctuations of $5 10^{12}$Hz. At high temperatures, the scattering amplitude of the peak increases on both sides, indicating an increase of thermally excited spin fluctuations.

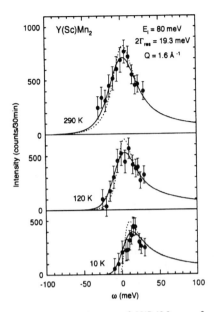

Figure 4. Magnetic neutron scattering rate of Y(Sc)Mn$_2$ as a function of energy transfer measured for a momentum transfer of $Q=1.6\text{Å}^{-1}$

These observations support the view that the magnetic state of paramagnetic Y(Sc)Mn$_2$ can be regarded as a spin liquid. The results for Y(Sc)(Mn$_{0.9}$Al$_{0.1}$)$_2$ are different. Figure 5 shows the result of the Q-scan with an energy window of 5.8meV obtained at 10, 120 and 290K. The spectrum at 290K is similar to that for Y(Sc)Mn$_2$ with a broad peak at around $Q = 1.6\text{Å}^{-1}$, although the intensity is much smaller due to the narrower energy window. In contrast to the result for Y(Sc)Mn$_2$, the scattering amplitude increases with decreasing temperature. At 10K, a double peak structure similar to that of the elastic neutron diffraction pattern was observed. Figure 6 shows the energy spectra for Y(Sc)(Mn$_{0.9}$Al$_{0.1}$)$_2$ measured at $Q =1.6\text{Å}^{-1}$. The profile of the spectrum at 290K is again similar to that for Y(Sc)Mn$_2$ with nearly the same line width. With decreasing temperature, the line width decreases rapidly and approaches the energy resolution limit of the spectrometer at 10K. This result shows that the spin fluctuations in the Al-substituted compound are strongly suppressed at low temperatures, as expected for the spin glass freezing at low temperatures. The scattering amplitude with decreasing temperature in the quasi-elastic Q-scan measurements can be simply explained in terms of narrowing of the energy spectra.

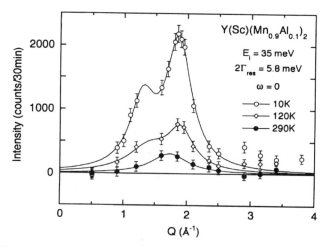

Figure 5. Magnetic neutron scattering rate of Y(Sc)(Mn$_{0.9}$Al$_{0.1}$)$_2$ as a function of wave vector Q at zero energy transfer with energy resolution of $2\Gamma_{res}$= 19.3meV at 290K(●) 120K() and 290K (O).

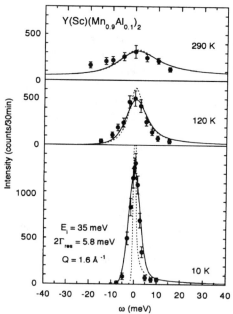

Figure 6. Magnetic neutron scattering rate of Y(Sc)(Mn$_{0.9}$Al$_{0.1}$)$_2$ as a function of energy transfer measured for a momentum transfer of Q=1.6A^{-1}

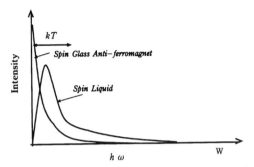

Figure 7. Schematic diagram of energy spectra of spin fluctuations for spin liquid and spin glass systems.

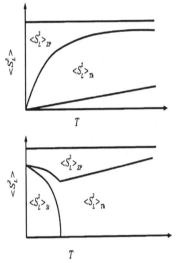

Figure 8. Schematic diagram of the temperature dependence of local amplitude of spin fluctuations for (a) spin-liquid and (b) spin glass systems.

From these spectra, we can draw the characteristic feature of spin fluctuation spectra for Y(Sc)Mn$_2$ and Y(Sc)(Mn$_{0.9}$Al$_{0.1}$)$_2$ as shown in Figure 7. For Y(Sc)Mn$_2$, there are no static fluctuations and the spectrum distributes most densely in the rather low energy range compared with the band width, W, at around thermal energy of room temperature. Therefore, spin fluctuations at 0K are totally of quantum origine, implying the quantum spin liquid nature of the ground state. With increasing temperature, low lying components of fluctuations are easily excited by thermal energy, giving rise to rapid growth of thermal spin fluctuations, which accompany the increase of magnetic entropy, resulting in the enhancement of low temperature specific heat as schematically shown in Figure 8. Here, we put $<S_L^2>_{tot}$ being constant according to Takahashi's assumption [14]. However, a recent theory, where the spin anharmonicity is taken into consideration, predicts an increase of $<S_L^2>$ at low temperatures and a decrease at high temperatures [15]. The contributions of zero-point fluctuations to thermodynamic properties are open for questions. Anyhow, the schematic picture of spin fluctuations can qualitatively explain the thermodynamic and microscopic properties of the Y(Sc)(Mn$_{1-x}$Al$_x$)$_2$ system.

Figure 7 shows only the energy spectrum of fluctuations but not the distribution in the q-space. The particular geometry of the C15 structure may give rise to anisotropic nature of fluctuations because the strong antiferromagnetic coupling is along <110> directions. In fact, the anisotropic and localized distribution of fluctuations in the q-space was found by Ballou et al.[16]. They observed strong amplitude of spin fluctuations along the Brilluion zone boundary for the (110) reciprocal lattice. This distribution can be expressed by the generalized susceptibility as

$$\chi^{-1}(Q+q,\omega) = \chi^{-1}{}_Q + a_1(q_x^2 + q_y^2) + a_2 q_z^4 - i\omega/\Gamma \quad (1)$$

In other words, the dispersion of the fluctuation spectrum has a low dimensional character. Lacroix et al. [15] calculated the temperature dependence of spin-lattice relaxation time of nuclear spins, T_1 for this special type of fluctuations and obtained an expression $1/T_1$ for the soft mode regime as

$$1/T_1 \propto T \cdot \chi(Q)^{3/4} \quad (2)$$

Using the Curie-Weiss relation for the staggered susceptibility, we have

$$1/T_1 \propto T/(T+\Theta)^{3/4} \quad (3)$$

Except for very low temperatures, the spin-spin relaxation time, T_2, is also mediated by spin fluctuations, then it is plausible that T_2 is proportional to T_1. We have measured the temperature dependence of T_1 and T_2 for YMn$_2$ in the paramagnetic phase and T_2 for paramagnetic Y$_{0.96}$Lu$_{0.04}$Mn$_2$. The results are shown in Figures 9 and 10. As seen in the figures, we have fairly good fitting by (3), supporting the low dimensional character of spin fluctuations.

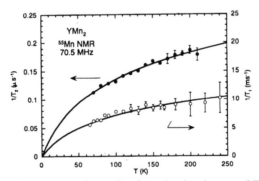

Figure 9. Temperature dependence of nuclear spin relaxation rates, $1/T_1$ and $1/T_2$ of YMn$_2$ in the paramagnetic phase.

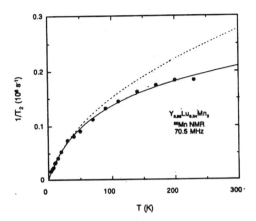

Figure 10. Temperature dependence of the nuclear spin-spin relaxation rate $1/T_2$ of Y(Lu)Mn$_2$. The solid curve indicates fitting by (3), the dotted line by $T/(T+\theta)^{1/2}$.

3. β-Mn and β-Mn$_{1-x}$Al$_x$

As well known, the maximum cohesive energy of transition metals is found at the center of the periodic table for 4d and 5d series, which is explained by the contribution of d-bands to cohesion. However, only for 3d series, the cohesive energy shows a dip for Mn. This trend may be explained if 3d bands for each spin are polarized. However, the metallic Mn is not strongly magnetic. In particular, β-Mn is believed to be nonmagnetic down to the lowest temperature and the susceptibility is of the Pauli paramagnetic type. However, β-Mn is not a simple Pauli paramagnet but exhibits anomalies similar to those observed in Y(Sc)Mn$_2$ such as a highly enhanced γ value.

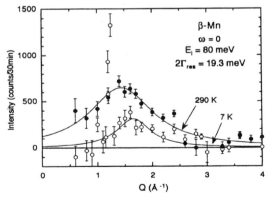

Figure 11. Magnetic neutron scattering rate of β-Mn as a function of wave vector Q at zero energy transfer with energy resolution of $2\Gamma_{res}$= 19.3meV at 7K(○) and 290K (●).

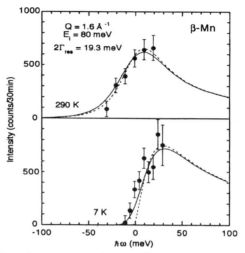

Figure 12. Magnetic neutron scattering rate of β- Mn as a function of energy transfer measured for a momentum transfer of $Q=1.6 Å^{-1}$

We performed the same experiments for β-phase Mn-Al alloys and found almost the same behaviors including neutron scattering experiments as shown in Figures 11 and 12 [17]. These results strongly suggest that anomalous behavior of β-Mn can be understood in terms of the itinerant frustrated system with strong electron correlations and that the ground state can be regarded as a spin liquid state.

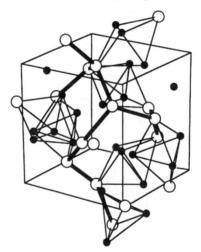

Figure 13. Crystal structure of β-Mn. Open and closed circles represent site I and site II Mn.

One may ask whether β-Mn has a structure which gives frustration or not. Figure 13 represents the crystal structure of β-Mn. At the first glance, it is difficult to see frustrated geometry. There are two different crystal sites, 8c (site I) and 12d (site II). For these two sites, Mn atoms have different magnetic characters. One (site I) is magnetically weak and the other (site II) strong. The two different sites are distinguished by NMR. Figure 14 shows the zero field NMR spectra of β-$Mn_{0.7}Al_{0.3}$, which is in a spin glass state. Two different signals are observed, one is at low frequencies with sharp line width and the other at high frequencies with a broad line width.

Figure 15 shows the concentration dependence of resonance frequencies, which is proportional to internal fields. The concentration dependence of the low frequency line, which was assigned to the site I signal, is linearly proportional to x, indicating that the internal field is originated in the transferred hyperfine field and the site I Mn is non-magnetic.

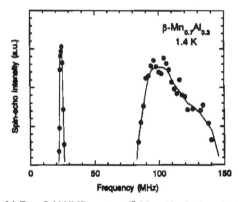

Figure 14. Zero-field NMR spectra of β-$Mn_{0.7}Al_{0.3}$ in the spin glass state.

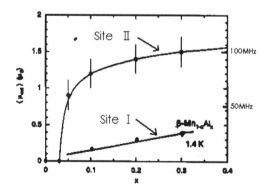

Figure 15. Concentration dependence of resonance frequencies for two sites of β-Mn alloys.

On the other hand, the high frequency signal, which was ascribed to the site II Mn, is almost constant except the paramagnetic region (x< 0.05), indicating that the site II Mn has its own magnetic moment. Figure 16 shows only one site II Mn. It is worth noting that the site II Mn atoms form a network of corner-shearing regular triangles. We believe that this is the reason why β-Mn behaves as a frustrated system.

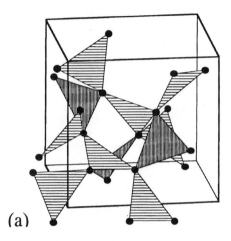

Figure 16. A schematic diagram of the site II Mn sublattice in β-Mn, which is a network of corner sharing regular triangles.

4. Summary

1. The ground state of paramagnetic YMn$_2$ under (chemical) pressure and β-Mn can be regarded as a quantum spin liquid characterized by strong zero-point spin fluctuations with antiferromagnetic correlations.
2. Thermal spin fluctuations grow rapidly with increasing temperature, which gives rise to an enhancement of the low temperature specific heat.
3. Substitution of Al for Mn causes a spin-liquid to spin-glass transition, implying frustrated characters of the system.
4. Spin fluctuations in Y(Sc)Mn$_2$ are strongly anisotoropic and has a low dimensional character, which was directly observed by neutron scattering and reflected in the temperature dependence of ^{55}Mn nuclear magnetic relaxation rate. It is likely that frustration causes lowering of the dimensionality.
5. In β-Mn, only site II Mn atoms are magnetic and they form a network of corner shearing triangles similar to the Kagome lattice.

Acknowledgements

The authors would like to thank many their collaborators, Dr. H. Wada, Mr. R. Iehara of Kyoto University, Prof. K. Kakurai, Dr. M. Nishi of the University of Tokyo, and Prof. M. Mekata of Fukui University.

References

1. Anderson, P.W. (1973) *Mater. Res. Bull.* **8,** 153.
2. Fazekas, P. and Anderson, P.W. (1974) On the ground state properties of the anisotropic antiferromagnet, *Phil.Mag* **30**, 423-440.
3. Reimer, J.N., Greedan, J.E., Kremer, R.K., Gmelin, E., and Subramanian, M.A. (1991) Short-range magnetic ordering in the highly frustrated pyrochlore $Y_2Mn_2O_7$, *Phys.Rev.* **B 43**, 3387-3394.
4. Moriya, T. (1979) Recent progress in the theory of itinerant electron magnetism, *J.Magn. Magn. Mat.* **14**, 1-46.
5. Ballou, R., Deportes J., Lemaire, R., Nakamura,Y., and Ouladdiaf, B. (1987) Helimagnetism in the cubic Laves phase YMn_2, *J.Magn.Magn.Mat.* **70**, 129-133.
6. Cywinsky, R., Kilcoyne, S.H., and Scott, C.A. (1991) Magnetic order and moment instability in YMn_2, (1991) *J.Phys.:Condens. Mater.* **3**, 6473-6488.
7. Nakamura, H, Wada, H. Yoshimura, K., Shiga, M, Nakamura, Y. Sakurai, J and Komura,Y.(1988) Effect of chemical pressure on the magnetism of YMn_2, *J.Phys.* **F18**, 981-991.
8. Wada, H., Shiga, M., and Nakamura,Y. (1989) Low temperature specific heat of nearly ferro- and antiferromagnetic compounds, *Physica* **B 161**, 197-202.
9. Fisher, R.A. Ballou,R. Emerson, J.P., Lelievre-Berna, E., and Phillips, N.E.(1992) Low temperature specific heat of YMn_2 in the paramagnetic and antiferromagnetic phases, *Proc. Int. Conf. Physics of Transition Metals, Darmstadt*, pp.830-833.
10. Bauer, E., Dubenko, I.S., Gratz, E., Hauser, R., Markosyan, A., and Payer, K (1992) Suppression of spin fluctuations in YMn_2 by hydrostatic pressure, *Proc. Int. Conf. Physics of Transition Metals, Darmstadt*, World Scientific, pp.826-829.
11. Shiga, M. (1995) Spin dynamics in frustrated itinerant systems *Proc. XXX Zakopane School of Physics, Inst. Phys.* Jagiellonian Univ. pp.57-68.
12. Mekata, M. Asano, T., Sugino, T. Nakamura, H. Asai, N. Shiga, M. Keren, A. Kojima, K. Luke, G.M., Wu, W.D. Uemura, Y.J., Dunsinger, S., and Gingras, M. (1995) Is $Y(Sc)Mn_2$ really a quantum spin liquid? *J.Magn.Magn.Mat.* **140-144**, 1767-1768.
13. Shiga, M. Wada, H. Nakamura,Y. Deportes, J. Ouladdiaf, B and Ziebeck, K.R.A. (1988) Giant spin fluctuations in $Y_{0.97}Sc_{0.03}Mn_2$, *J. Phys. Soc. Jpn.* **57**, 3141-3145.
14. Takahashi, Y. (1990) Magneto-volume effects in weakly ferromagnetic metals, *J.Phys. Condens. Matter* **2**, 8405-8415.
15. Lacroix, C, Solontsov, A., and Ballou, R. (1996) Spin fluctuations in itinerant electron antiferromagntism and anomalous properties of $Y(Sc)Mn_2$ *Phys.Rev.* **B54**, 15178-15184.
16. Ballou, R. Lelievre-Berna, E., and Fak, B. (1996) Spin fluctuations in $(Y_{0.97}Sc_{0.03})Mn_2$: A geometrically frustrated, nearly antiferromagfnetic, itinerant electron system, *Phys.Rev.Letters* **67**, 2125-2128.
17. Nakamura, H., Yoshimoto, K., Shiga, M. Nishi, M., and Kakurai, K. (1997) Strong antiferromagnetic spin fluctuations and the quantum spin-liquid state in geometrically frustrated β–Mn, and the transition to a spin-glass state caused by non-magnetic impurity, *J. Phys. Condens. Mater.* **9**, 4701-4728.

NEUTRON SCATTERING AND µSR STUDIES OF SPIN FLUCTUATIONS IN FRUSTRATED ITINERANT MAGNETS

B.D.RAINFORD AND S.J.DAKIN
*Department of Physics and Astronomy,
University of Southampton SO17 1BJ, England,*
J.R.STEWART AND R.CYWINSKI
Department of Physics and Astronomy, University of St. Andrews, St. Andrews, KY16 9SS, Scotland.

Abstract. Extensive measurements have been made of the spin dynamics of YMn_2 and its alloys, and of β-Mn and β-Mn-Al alloys, using a combination of inelastic neutron scattering and muon spin relaxation. The dynamical response measured by time-of flight neutron spectroscopy in YMn_2 alloys is strongly peaked at the characteristic wavevectors of the antiferromagnetic spin fluctuations. It appears that the wavevectors found in recent single crystal studies of YMn_2 are similar to those found in Monte Carlo simulations of localised moments in the antiferromagnetic pyrochlore lattice. In YMn_2 alloys the neutron linewidths are strongly correlated with the unit cell volume: substitution of Mn by Al expands the cell and decreases the linewidth, while substitutions of Sc for Y and Fe for Mn contract the cell and increase the linewidth. µSR measurements on β-Mn-Al alloys show a crossover from the Lorentzian to stretched exponential damping above 9% Al, the concentration at which the spin glass temperature begins to rise rapidly. This suggests a regime dominated by spin fluctuations below 9% Al and a regime with stable Mn moments above this composition.

1. Introduction

The extraordinary magnetic properties of YMn_2 and β-Mn result from a rather unique combination of circumstances, namely, the proximity of the itinerant electron system to a moment instability (YMn_2) or moment formation (β-Mn), together with the topological frustration arising from antiferromagnetic correlations on lattices of corner sharing tetrahedra (YMn_2) or triangles (β-Mn). Both of these systems show large amplitude spin fluctuations and a great sensitivity of the magnetic properties to the unit cell volume. YMn_2 displays a dramatic first order phase transition to an antiferromagnetically (AF) ordered state at 100 K, which is accompanied by a 5% increase in the cell volume [1]. This AF transition may be suppressed completely by a pressure of only 0.4 Gpa, or by the substitution of 2% of Y by Sc or of 2.5% of Mn by Fe. In these cases the ground state appears to be very similar to the spin liquid ground state of β-Mn, which shows no sign of magnetic order down to low temperatures. Substitution of Mn by Al in both β-Mn and YMn$_2$ tends to stabilise a local moment on the Mn atoms, and causes a reduction of the magnetovolume effects. In studies of these systems microscopic probes like NMR, μSR and neutron scattering have much to offer. The subject owes a great deal to the extensive work on both systems carried out by Prof. Shiga's group [1,2]. In the this paper we present a contribution to the systematic study of YMn$_2$, β-Mn and their alloys using inelastic neutron scattering and μSR techniques.

2. YMn$_2$ and Related Alloys

2.1 SPIN CORRELATIONS FROM INELASTIC NEUTRON SCATTERING

Neutron scattering studies of YMn$_2$ have been carried out by a number of groups, mostly using neutron polarisation analysis (PA) to separate the spin fluctuation response from the phonon scattering [3,4]. The PA technique has the disadvantage that it is necessary to scan point by point in the (**K**, ω) space and count rates are low compared to conventional spectrometers. We chose instead to use time-of-flight spectroscopy with unpolarised neutrons, allowing spectra to be collected over a wide range of **K** and ω simultaneously [5]. The separation of the magnetic response from the phonon scattering was effected by extrapolating the measured signal from high **K** values, where the magnetic form factor $f^2(\mathbf{K})$ has dropped to a low value, to lower **K**. Extensive Monte Carlo simulations of the multiple phonon scattering were made to establish the form of this extrapolation. Details of the procedure are given in [5]. The method is particularly well suited to systems with antiferromagnetic correlations where the integrated magnetic response $\chi(\mathbf{K})f^2(\mathbf{K})$ peaks strongly at intermediate values of **K**. Constant K cuts through the (**K**,ω) scattering surface were well represented by a Lorentzian lineshape function:

$$\text{Im } \chi(\mathbf{K}, \omega)/\omega = \chi(\mathbf{K}) \Gamma / (\Gamma^2 + \omega^2) \qquad (1)$$

For the incident energies used (60 to100 meV) energy transfers could be measured out to 40-50 meV in the **K** range between 1 Å^{-1} and 3 Å^{-1}, where the magnetic scattering peaks. In all the samples studied, namely pure YMn$_2$ and four alloys with 4% Fe, 3% Al and 10% Al substitution for Mn, and 3% Sc substitution for Y, there was still significant scattering out to the edge of the accessible (**K**, ω) domain [6]. It follows that there is a component of the spin fluctuation response extending out to high frequencies. It should be noted that measurements on polycrystalline samples involve a spherical (powder) average over the direction of the scattering vector **K**, so that the simple lineshape (1) masks the detailed dependence of the linewidth Γ on the wavevector **K**. However the wavevector - dependent susceptibility χ(**K**) can be derived directly from the lineshape (1) using the Kramers-Kronig relations. Fig.1 shows χ(**K**) as a function of the temperature derived in this way for the alloy Y(Mn$_{0.96}$Fe$_{0.04}$)$_2$. The susceptibility was placed on an absolute scale by the normalisation of the scattering against the incoherent scattering of a standard vanadium sample. The substitution of 4% Mn by Fe contracts the unit cell of YMn$_2$ and suppresses the first order transition. This allows measurements of the spin fluctuation response to be made down to low temperatures. It

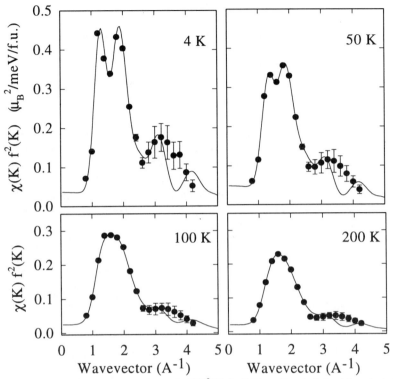

Figure 1. Temperature dependence of $\chi(K)f^2(K)$ for $Y(Mn_{0.96}Fe_{0.04})_2$ derived from inelastic neutron spectra. The solid lines are a fit to a model of the powder average of $\chi(q)$ as discussed in the text.

can be seen that there is substantial temperature dependence to $\chi(K)$: the peak response doubles between 200 K and 4 K, and at the lower temperatures a double - peaked structure develops. In our earlier paper [5] we analysed this data assuming that the characteristic wavevectors of the spin fluctuations were close to the reciprocal lattice vectors of the antiferromagnetically ordered phase of YMn_2, namely τ_{110}, τ_{210}, τ_{211}, etc. Recent important measurements on single crystals of YMn_2 have shown that this is not the case [7]. Ballou et al., using a triple axis spectrometer, showed that the magnetic response peaks in kidney - shaped blobs displaced from τ_{110} towards the corner of the Brillouin zone at $(Q, Q, 0)$ with $Q \approx 1.25$ (Fig.2, left). We have now reanalysed the $\chi(K)$ data in the light of this new information. The solid lines in Fig.1 are fits to a model for the spherically averaged form of $\chi(q)$ in which the form of the susceptibility was taken, for simplicity, to be a Gaussian function of q centred on $\tau = (Q,Q,0)$ type positions ($q = K - \tau$):

$$\chi(q) = A \exp[-|q|^2/2\sigma^2] \qquad (2)$$

The equivalent positions were derived from a Brillouin zone doubled in each direction, namely, zone centres at (2,2,2), (4,0,0), (4,4,0) etc., plus all equivalent vectors {**Q**,**Q**,0}. The doubling of the Brillouin zone required to describe the characteristic wavevectors found in the single crystal measurements implies that the antiferromagnetic correlations are strongest within a subunit of the unit cell, most likely a single tetrahedron of four nearest neighbour Mn atoms. The form of the spherical average of $\chi(\mathbf{q})$ used in the fits is given in [5]. The values of **Q** found in the fits in Fig.1 ranged from 1.13 Å^{-1} at 4 K to 1.21 Å^{-1} at 200 K, somewhat smaller than the value 1.25 Å^{-1} found for YMn$_2$ in the single crystal data [7]. The Gaussian width σ of the (**Q**, **Q**, 0) peak increased smoothly from 0.19 Å^{-1} at 4 K to 0.26 Å^{-1} at 200 K, reflecting a decrease in the correlation length of the magnetic short range order with increasing the temperature. Similar quality fits were obtained for all the alloys measured, though there were some striking qualitative differences in the forms of χ (**K**). The Y$_{0.97}$Sc$_{0.03}$Mn$_2$ alloy might be expected to give results very similar to those of Fig.1, since Sc substitution contracts the unit cell by a similar amount and suppresses the first order AF transition. However, we find that $\chi(\mathbf{K})$ for this alloy is practically independent of the temperature and does not show the double peaked structure at low temperature. This suggests magnetic correlations of a shorter range than for the 4%Fe alloy. The two samples with 3 and 10% Al substituted for Mn developed double peaked structure at temperatures as high as 100 K. It is well known that Al substitution expands the cell volume, tending to stabilise the Mn moment. The χ (**K**) data are consistent with this picture. A more detailed consideration of the Y(Mn$_{1-x}$Al$_x$)$_2$ data will be presented in a future paper.

2.2 FRUSTRATION AND THE FORM OF THE SPIN CORRELATIONS

It is generally accepted that frustration must play a key role in determining the magnetic properties of YMn$_2$ and its alloys, since the main feature of magnetism is antiferromagnetic coupling between Mn atoms on a lattice of corner sharing tetrahedra. Yet it has not been apparent so far that the form of the spin correlations observed in these alloys is what one would expect for such a lattice. In recent years, however, interest in frustrated localised magnets has blossomed and there have been extensive Monte Carlo simulations for the pyrochlore lattice, which is exactly the structure of the Mn sublattice in YMn$_2$. When we compare the neutron data for YMn$_2$ and its alloys with these simulations it is immediately apparent that there are close similarities. Firstly, Reimers [8] presented powder average S(**K**) data for the pyrochlore lattice for nearest neighbour exchange which has a form remarkably similar to the high temperature data in Fig.1, corresponding to the short range order on a scale of the order of one unit cell. The inclusion of a small amount of second neighbour exchange gives a multiple peaked structure, similar to the low temperature data in Fig.1. The correlations in this case extend out to two or three unit cells. Zinkin and Harris [9] have used a local mean field approach to calculate the form of S(**q**) throughout the Brillouin zone (BZ) for the pyrochlore lattice with nearest neighbour exchange. The remarkable feature of these calculations is the presence of narrow necks in the contours of S(**q**) at particular points like (002) and (220) in the zone (Fig.2, right). Chalker [10] has shown that these necks result from having zero net spin on the basic tetrahedral unit, leading to very long range AF correlations between successive (1,1,1) planes (which in the pyrochlore lattice form Kagome nets). More relevant here, however, is that for the (1,-1,0) plane of the BZ

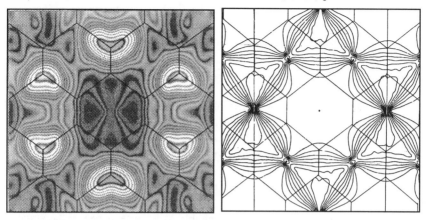

Figure 2. (left): Contours of scattering from spin fluctuations in Y(Sc)Mn$_2$ in the (1,-1,0) plane of the BZ, after [7]. (right): Contours of S(**q**) in the (1-1,0) plane of the BZ for the pyrochlore lattice with n.n. exchange, after [9]. In both plots the BZ boundaries are overlaid. The [001] direction is horizontal and the [1̄10] direction is vertical.

the highest contours of S(**q**) extend out from the narrow necks along the (h,h,0) directions towards the corners of the BZ at the {5/4,5/4,0} type positions. These positions are precisely where the scattering is found to peak in the single crystal data for YMn$_2$! The match is not perfect: when plotted over several zones the S(**q**) data has the appearance of a pattern of "bow-tie" shapes (Fig.2, right), whereas the "kidney" shapes of the YMn$_2$ data contours (Fig.2, left) are elongated along the edges of the BZ boundaries. S(**q**) calculations including further neighbour exchange interactions would be of interest to see if the match could be improved. It would appear then that the distinctive wavevectors associated with the spin fluctuations in YMn$_2$ are characteristic of the frustrated correlations on the pyrochlore lattice. There is much more to explore here through further measurements on single crystals, for example whether the form of the correlations changes significantly on alloying or with temperature.

Knowledge of the dynamical response is crucial for the understanding of the bulk magnetic properties, for example Pinette and Lacroix [11] have argued that the form of $\chi(\mathbf{q}, T)$ in Y(Sc)Mn$_2$ leads directly to an explanation of the heavy fermion behaviour found in these alloys.

2.3 SPIN DYNAMICS AND :µSR

The inelastic neutron spectra from the single crystal samples of Y$_{0.97}$Sc$_{0.03}$Mn$_2$ and YMn$_2$ [7] are well described by a Lorentzian lineshape:

$$\mathrm{Im}\chi(q,\omega)/\omega = \chi(\mathbf{q})\,\Gamma(\mathbf{q}) / (\Gamma(\mathbf{q})^2 + T^2) , \qquad (3)$$

There is not a great deal of information about the form of $\Gamma(\mathbf{q})$ or its temperature dependence, except that the linewidth varies slowly with wavevector. We have attempted to extract further information about Γ from the time-of-flight data on polycrystalline samples by fitting the constant **K** cuts through the data to a powder average of (3). In order to do this we assumed a particular form for $\Gamma(\mathbf{q})$, namely:

$$\Gamma(\mathbf{q}) = \Gamma_0 (\kappa^2 + \mathbf{q}^2) \tag{4}$$

This is the variation predicted for antiferromagnetic spin fluctuations by Moriya [12]. We further assumed an Orstein-Zernicke form for $\chi(\mathbf{q})$:

$$\chi(\mathbf{q}) = \chi(0) \, \kappa^2 / (\kappa^2 + \mathbf{q}^2) \tag{5}$$

since this makes $\chi(\mathbf{q})\Gamma(\mathbf{q})$ independent on q and leads to an analytic form for the powder average. The results for the linewidth $\Gamma(0)$ [13] derived from the fits appear to be too small compared to the single crystal data, probably as a consequence of the assumption of (5) which does not give a good account of the powder average form of $\chi(\mathbf{K})$. However, we believe that the scaling of the linewidth with temperature and with alloy composition is at least qualitatively correct. This is shown in Fig.3. The filled symbols are derived from the neutron data on YMn$_2$ (squares), and the alloys with 4%Fe (diamonds), 3%Al (inverted triangles) and 10% Al (triangles) substitution for Mn, and 3%Sc (circles) substitution for Y. It is clear that the expansion of the unit cell volume by Al substitution decreases the linewidth, while the contraction of the unit cell in Y(Sc)Mn$_2$ and Y(MnFe)$_2$ increases the linewidth.

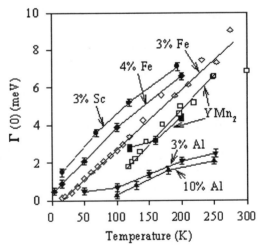

Figure 3. Linewidth parameter for YMn$_2$ alloys derived from fits to inelastic neutron spectra (solid symbols) and from scaled μSR data using equation (5) (open symbols).

μSR measurements also give direct information about the spin dynamics. In studies of YMn$_2$ and YMn$_{1.94}$Fe$_{0.06}$ [14,15] the muon relaxation rate λ increases as the temperature is lowered, diverging close to T$_N$ in the case of YMn$_2$ and following a power law dependence for the 3% Fe alloy, where the magnetic order has been suppressed. This is evidence for slowing down of the spin fluctuations. In the case of a Lorentzian lineshape function for the dynamical response (3) it can be shown that the muon relaxation rate is

$$\lambda = (BT/N) \Sigma \chi(\mathbf{q}) / \Gamma(\mathbf{q}), \qquad (6)$$

where B is a coupling constant. In practice we do not have detailed information about χ(**q**) or Γ(**q**), but it would appear from the inelastic neutron data on single crystals that the wavevector dependence of the linewidth Γ(**q**) is rather weak. In this case we may rewrite (4) as

$$\lambda = BT\chi_L / \Gamma, \qquad (7)$$

where χ$_L$ is the local susceptibility, which may be derived from χ(**K**) as described in [5]. We have shown, using this approach, that the systematics of λ(T) for YMn$_2$ and the alloys with Fe substitution agree very well with the neutron linewidths, once a scale parameter (i.e. the coupling constant B) can be determined [13]. The open symbols in figure 3 are estimates of the linewidth Γ, derived from the measured μSR relaxation rates λ using (7) for YMn$_2$ (open squares) and for an alloy with 3%Fe substitution for Mn (open triangles). The temperature dependence of χ$_L$, the local susceptibility, was taken to be Curie-Weiss like, as determined by the neutron measurements [5], and a single value of the coupling parameter B was used to scale these data to the linewidths derived from the neutron measurements (filled circles).

In the case of YMn$_{1.8}$Al$_{0.2}$ [16] the character of the muon relaxation was rather different, in that the relaxation function was fitted best by a stretched exponential form:

$$G_z(t) = a_0 \exp[-(\lambda t)^\beta] \qquad (8)$$

The exponent β was found to be temperature dependent, tending towards the value 1/3 just above the spin glass temperature T$_G$. This form has been found in a large number of concentrated spin glass systems, and in this context provides additional confirmation that substitution of Mn by Al tends to stabilise local moments on the Mn. We were able to apply (7) to these data also: the temperature dependence Γ(T) was found by Motoya et al. [4] to follow an Arrenhius law: Γ(T) =Γ$_0$ exp(-E$_a$/k$_B$T) with E$_a$/k$_B$ = 278.4 K, while our time-of-flight data showed that the local susceptibility χ$_L$ followed a Curie-Weiss law χ$_L$= C/(T+θ) with θ = 93 K. Putting these results into (7) gave a form for λ(T) that agreed remarkably well with the experiment [16].

μSR data are collected and analysed much more quickly than inelastic neutron data. Once the scaling between λ, Γ and χ$_L$ has been established, the muon technique allows the systematics of the spin dynamics to be explored as a function of temperature and alloy composition.

3. β-Mn and β-Mn-Al Alloys

β-Mn is the only stable allotrope of elemental manganese that does not display a magnetic moment at any temperature. Previous work by Prof. Shiga's group has shown convincingly that the magnetic properties of β-Mn are characterised by strong spin fluctuations associated with Mn moments on the verge of localisation [2]. The cubic A13 type crystal structure has 20 atoms in the unit cell, with two inequivalent sites. Eight Mn occupy the site with trigonal point symmetry (site I), while the remaining 12 Mn occupy the orthorhombic site II. It appears from NMR and NQR data that the Mn on site I are very weakly magnetic. But the large nuclear spin relaxation rate and high resonance frequency for the site II Mn are consistent with a strong hyperfine field at the nucleus arising from coupling with strong, slow spin fluctuations [2]. The similarity with Y(Sc)Mn$_2$ alloys suggests that geometrical frustration also plays an important role, but the origins of this frustration in the rather complex crystal structure of β-Mn have not been very clear. In a recent review however [2] Nakamura et al. suggest a description of the connectivity of the type II sites in the unit cell in terms of a network of corner-sharing regular triangles, with similarities to the Kagome lattice. It would be of great interest to perform Monte Carlo simulations of the spin correlations for localised spins on the site II sublattice of β-Mn to see whether this conjectured frustration is borne out.

3.1 INELASTIC NEUTRON SCATTERING FROM β-Mn AND β-Mn-Al.

Inelastic neutron scattering measurements on β-Mn and β-Mn$_{0.9}$Al$_{0.1}$ have been made by Nakamura et al [2] using polarisation analysis. In β-Mn the linewidth of the spectra increases from 20 to 30 meV between 7 K and 290 K; these widths are much larger than those found in YMn$_2$. For the 10% Al alloy the linewidth is very much smaller, of order 10 meV at 290 K, and decreases linearly with temperature. Our own neutron data taken using the time-of-flight method is broadly in agreement with this picture, though the analysis is not yet complete. One difference is that we used a higher incoming energy than that used in [2] for measurements on a 10%Al alloy. It appears that in this alloy there is still a component of the response with a large energy width at low temperatures, which was not reported in [2]. Initial estimates of the energy integrated response, $\chi(\mathbf{K})f^2(\mathbf{K})$, for β-Mn show a **K** dependence with a smooth fall-off in the response at wavevectors below 1.5 Å^{-1}; there is little evidence of the peaked structures seen in YMn$_2$ and its alloys (see, e.g., Fig.1). More details will be presented in a future paper.

3.2 μSR STUDIES OF MOMENT LOCALISATION IN β-Mn-Al ALLOYS

We have followed the evolution of the spin dynamics in β-Mn$_{1-x}$Al$_x$ alloys in zero field longitudinal μSR measurements [17]. For alloys with 0<x≤0.09 the form of the relaxation function of the muon polarisation could be described by:

$$P_z(t) = a_0 P_z^{KT}(t) \exp(-\lambda t), \qquad (9)$$

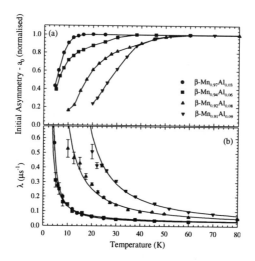

Figure 4. a) Initial asymmetry a_0 and b) relaxation rate λ *versus* temperature for β-Mn$_{1-x}$Al$_x$ alloys with concentrations x = 0.03, 0.06, 0.08, 0.09.

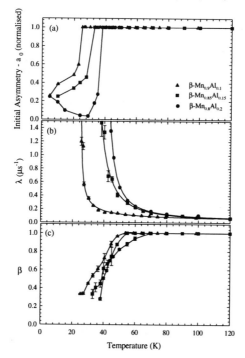

Figure 5. The temperature dependence of a) the initial asymmetry, b) the relaxation rate 8, and c) the stretch exponent β for β-Mn$_{1-x}$Al$_x$ alloys with x= 0.10, 0.15 and 0.20.

where $P_z^{KT}(t)$, the static Kubo-Toyabe (KT) function, accounts for the muon depolarisation due to a distribution of static nuclear dipolar fields. The second term arises from the rapidly fluctuating atomic spins. The KT component was found to be independent of temperature and alloy composition. For pure β-Mn the depolarisation due to the atomic spins was found to be independent of temperature also, with a value of $\lambda = 0.02\mu s^{-1}$. This is consistent with rapid motional narrowing due to the large, weakly temperature dependent linewidth of the spin fluctuation spectrum of β-Mn. As Al is substituted into β-Mn, λ develops strong temperature dependence, diverging towards low temperature (Fig.4). This too is consistent with the neutron studies where the inelastic linewidth is seen to decrease as the temperature is lowered in the Al - substituted β-Mn. The initial asymmetry begins to falls below 60 K, finally reaching 1/3 of its high temperature value (Fig.4). This behaviour indicates quasistatic correlations in the low temperature regime. The transition temperatures determined from the divergence of λ and the loss of asymmetry are shown in the magnetic phase diagram, Fig.6. For Al concentrations above x = 0.09 the form of the muon relaxation function changes from the $P_z^{KT}(t)$ exp(-λt) form to the stretched exponential $P_z^{KT}(t)$ exp[-(λt)$^\beta$]. The value of β decreases slowly from one at high temperatures to a approximately 0.3 at the transition temperature (Fig.5). As mentioned in Sec. 2.3, such a response is characteristic of concentrated spin glass systems, and reflects a broadening of the distribution of spin relaxation rates as the spin glass transition T_G is

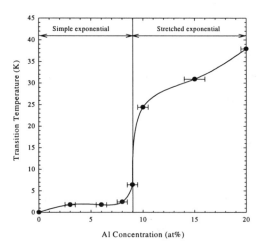

Figure 6. The magnetic phase diagram for β-Mn$_{1-x}$Al$_x$ alloys. The vertical line separates the spin liquid regime x<9% from the spin glass regime x≥9%.

approached. In the concentration regime 0.09<x≤0.20 the relaxation rate λ diverges as the temperature is decreased towards T_G (Fig.5). The spin glass transition can also be seen as a sharp drop in the asymmetry at T_G. The magnetic phase diagram (Fig.6) shows that the spin glass transition temperatures rise rapidly for concentrations above x=0.09.

This feature, taken together with the change in form of the relaxation function, suggests that the character of the magnetic ground state changes near the 9%Al substitution. The picture emerges that in the low concentration regime ($0 \leq x < 0.09$), the ground state is a spin liquid, dominated by the zero point spin fluctuations, while above $x=0.09$ a spin glass state appears, in which the Mn moments are well localised, but still topologically frustrated.

4. Conclusions

A much more complete picture of the dynamical response in the frustrated itinerant antiferromagnets YMn_2 and β-Mn has been built up by recent measurements. The great sensitivity of the dynamics to cell volume is a feature common to both systems. The first measurements on single crystals of YMn_2 by Ballou et al. have pinned down the characteristic AF wavevectors: these appear to be close to those expected for near neighbour AF correlations on the pyrochlore lattice. There is still a great deal to be done in exploring how, or whether, these wavevectors change on alloying. Computer simulations for localised spins on the β-Mn lattice would furnish useful insight into the nature of the spin correlations, but a proper treatment of frustration in the context of itinerant magnetism remains a considerable theoretical challenge.

Acknowledgements

We would like to thank Mark Harris and John Chalker for useful discussions relating to the spin correlations in the pyrochlore lattice.

References

1. Shiga, M. (1988) Magnetism and spin fluctuations of Laves phase manganese compounds, *Physica* **B 149**, 293-305, and references therein.
2. Nakamura, H., Yoshimoto, K., Shiga, M., Nishi, M. and Kakurai, K. (1997) Strong antiferromagnetic spin fluctuations and the quantum spin-liquid state in geometrically frustrated β-Mn, and the transition to a spin-glass state caused by a non-magnetic impurity, *J.Phys.: Condens. Matter* **9**, 4701-4728.
3. Déportes, J., Ouladiaff, B. and Ziebeck, K.R.A. (1987) Spin fluctuations in the paramagnetic state of YMn_2, *J. Magn. Magn. Mater.* **70**, 14-16; ibid. (1987) Thermal dependence of the longitudinal spin fluctuations in YMn_2, *J. Physique* **48**, 1029-1034.
4. Freltoft, T., Böni, P., Shirane, G. and Motoya, K. (1988) Neutron scattering study of the itinerant electron magnet YMn_2, *Phys. Rev.* **B 37**, 3454-3460; Motoya, K., Freltoft, T., Böni, P. and Shirane, G. (1988) Neutron scattering study of spin fluctuations in $Y(Mn_{1-x}Al_x)_2$, *Phys. Rev.* **B 38**, 4796-4802.
5. Rainford, B.D., Dakin, S.J. and Cywinski, R. (1992) Spin fluctuations in YMn_2 and related alloys, *J.Magn. Magn. Mater.* **104-107**, 1257-1263.
6. Dakin, S.J., (1993), Ph.D. thesis (unpublished), Southampton University.
7. Ballou, R., Lelièvre-Berna, E. and Fåk, B. (1996) Spin fluctuations in $(Y_{0.97}Sc_{0.03})Mn_2$: a geometrically frustrated, nearly antiferromagnetic, itinerant electron system., *Phys. Rev. Lett.* **76**, 2125-2128.
8. Reimers, J.N. (1992) Absence of long range order in a three dimensional geometrically frustrated antiferromagnet, *Phys. Rev.* **B 45**, 7287-7294.
9. Zinkin, M.P. and Harris, M.J. (1995) Local mean field calculation of the frustrated pyrochlore spin structure, , *J.Magn. Magn. Mater.* **140-144**, 1803-1804.

10. Chalker, J.T., private communication.
11. Pinettes, C. and Lacroix, C. (1994) A model for heavy fermion behaviour of YMn_2 –influence of frustration on the spin fluctuation spectrum, *J. Phys.: Conden. Matter* **6**, 10093-10104.
12. Moriya, T. (1985) *Spin Fluctuations in Itinerant Electron Magnetism*, Springer-Verlag, Berlin.
13. Rainford, B.D., Cywinski, R., and Dakin, S.J. (1995) Neutron and μSR studies of spin fluctuations in YMn_2 and related alloys, *J.Magn. Magn. Mater.* **140-144,** 805-806.
14. Cywinski, R., Kilcoyne, S.H. and Scott, C.A. (1991) Magnetic order and moment stability in YMn_2, *J. Phys.: Condens. Matter,* **3**, 6473-6488.
15. Cywinski, R., Kilcoyne, S.H., Cox, S.F.J, Scott, C.A. and Schaerpf, O. (1990) A μSR and neutron polarisation analysis study of Mn moment delocalisation in Fe substituted YMn_2, *Hyperfine Inter.,* **64**, 427-434.
16. Cywinski, R. and Rainford, B.D. (1994) Spin dynamics in the spin glass phase of $Y(Mn_{1-x}Al_x)_2$, *Hyperfine Interactions* **85,** 215-220.
17. Stewart, J.R., Hillier, A.D., Kilcoyne, S.H., Manuel, P., Telling, M.T.F. and Cywinski, R. (1998) Moment Localisation in β-MnAl, , *J.Magn. Magn. Mater.,* (Proceedings of ICM'97) to be published.

CRITICAL FLUCTUATIONS AND THE NATURE OF THE NÉEL TRANSITION NEAR THE TRIPLE POINT IN CHROMIUM ALLOYS

ERIC FAWCETT

Physics Department, University of Toronto
Toronto, Ont. M5S 1A7
Canada

AND

D.R. NOAKES

Physics Department, Virginia State University
Petersburg, VA 23806
USA

Abstract. Inelastic neutron scattering measurements are described in pure chromium and in dilute alloys of Cr with V and Re that show the role of critical fluctuations in determining the nature of the Néel transition near the triple point between the paramagnetic phase and incommensurate and commensurate spin-density-wave (SDW) phases. In pure chromium, field-cooled so as to produce a single-Q_x state, critical fluctuations grow rapidly at Q_y and Q_z, as well as at Q_x, as the Néel temperature T_N is approached from below, so that at the Néel transition the entry to the cubic paramagnetic phase is almost continuous. Doping with V rapidly reduces the amplitude of the critical fluctuations in the paramagnetic phase, as the system moves away from the triple point, and the weak first-order transition seen in pure chromium becomes continuous with the addition of about 0.1%V. Adding 0.18%Re moves the system towards the triple point and increases the amplitude of the critical fluctuations. Extrapolation of the parameters of the Sato-Maki model, which fits the inelastic scattering data rather well, indicates that there is a mode-softening transition that occurs at a finite value, about 0.017 rlu, of the incommensurability parameter, where the SDW becomes commensurate, with a first-order Néel transition.

1. Introduction

Chromium enters an incommensurate spin-density-wave (SDW) antiferromagnetic state at a Néel temperature of $T_N = 312$ K. The addition of small amounts of nonmagnetic impurity elements from the right of chromium in the periodic table generally increases the ordering temperature and makes the ordered state more nearly commensurate [1]. This is opposite to the effect of adding vanadium, which lies to the left of chromium in the periodic table, and which depresses T_N and increases the incommensurability parameter δ. Theoretical studies attribute these effects primarily to the shifting of the Fermi level in the band structure of pure chromium by the change in the average number of electrons per atom. Chromium alloy magnetic phase diagrams are often drawn with V on one side and Mn or Re on the other [2,3] to illustrate this effect. Factors other than electron concentration may play a role, however. V has a local moment in the paramagnetic state of chromium, as indicated by Curie-Weiss temperature dependence of the susceptibility, [4] whereas Re and (less than 10%Mn; all percentages are atomic) do not [1].

It has been known for some time that the addition of only a small amount of Mn or Re impurity sends the wavelength of the incommensurate SDW to infinity, thus creating a "commensurate SDW" having simple antiferromagnetic ordering with a continuous Néel transition at a considerably higher value of T_N. The critical behavior of the magnetic excitations observed by neutron scattering near that transition was clearly established [1]. It was found later that the addition of small amounts (less than 0.2%) V to chromium, while slightly decreasing the wavelength of the incommensurability, reduces the first-order jump in the amplitude of the SDW at the Néel transition to zero, leaving a truly continuous transition, not just a disorder-broadened first-order transition, to the incommensurate SDW phase [5].

Neutron scattering measured just above the Néel temperature in single crystals of chromium and Cr(V) alloys may be fit closely by a model for magnetic critical scattering that assumes the critical behavior to be associated with a second-order transition, even though there is a weak first-order transition at T_N in pure chromium [6]. This Sato-Maki model [2] employs an incommensurate critical susceptibility $\chi(\mathbf{Q}, \omega)$ which embodies intrinsically the sixfold symmetry around the commensurate (010) position of the magnetic Bragg peaks that appear in the SDW phase at T_N.

The Sato-Maki model is also applicable to incommensurate SDW alloys on the other side of chromium in the magnetic phase diagram, which increase δ. In Cr+0.18%Re, however, we find that, in contrast with the effect of introducing V impurity, the intensity of critical scattering is increased relative to that in pure chromium [7]. Furthermore, the magnetic stiffness associated with the critical fluctuations in the paramagnetic phase close to the Néel transition approaches zero at a finite value of the incommensurability parameter, $\delta = 0.017$ rlu, instead of zero, as usually assumed. The mode of the incommensurate SDW softens as δ approaches this critical value with Re doping, and conversely hardens as δ increases when V is added to chromium [7].

Our understanding of the physical process, whereby the symmetry change from the tetragonal incommensurate SDW phase (orthorhombic if we also take into account the polarization of the SDW) to the cubic paramagnetic phase occurs at the Néel transition in pure chromium was greatly advanced by a study of critical fluctuations at the "silent satellite" positions [8]. These are the off-x-axis positions for a single crystal of Cr field-cooled so as to have a single-wave-vector SDW along the cubic x-axis, at which no Bragg scattering is observed, in contrast with the SDW satellites centered on the simple-antiferromagnet reciprocal lattice point that appear on the x-axis at the Néel transition. As the temperature increases towards the Néel temperature T_N, the intensity of the silent satellites increases rapidly from very low values, until at T_N a discontinuous jump, corresponding to the first-order nature of the Néel transition in chromium, results in equal scattering at all six satellites required by the cubic symmetry of the paramagnetic phase.

2. Experiment

The study at AECL Chalk River of critical neutron scattering in the paramagnetic phase close to the Néel temperature T_N in pure chromium, Cr+0.2at.%V, and Cr+0.5at.%V, was described in the Refs. [6]. As a preliminary to these measurements of the dynamic susceptibility, the elastic scattering was measured to determine the precise value of T_N for each sample, where the Bragg satellite peaks appear as the SDW phase is entered [9].

The single crystal of Cr+0.18%Re, which had been previously measured by Mikke and Jankowska [10], was studied at the H8 spectrometer at Brookhaven's HFBR. The spectrometer uses a graphite monochromator and analyzer, and was operated at a fixed scattered wavelength of 2.37 with a pyrolytic graphite filter. Data were collected near the (010) position in the (001) plane. Because magnetic satellites exist just out of the spectrometer plane at $(0,1,\pm\delta)$, extra horizontal collimator blades were mounted so as to bring beam acceptance to 40 minutes of arc in the vertical as well as the horizontal plane.

The Néel transition at $T_N = 325.5 \pm 0.4$ K to the incommensurate SDW phase is first order in this sample [9,10], as it is in pure chromium. A small

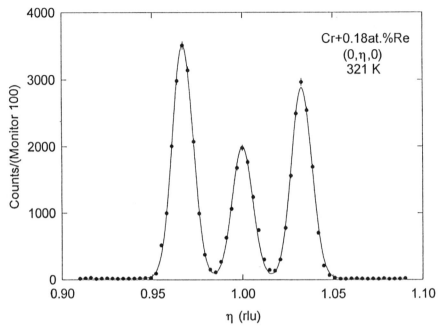

Figure 1. Radial elastic **Q**-scan in Cr+0.18at.%Re at 321 K, below T_N; the solid line shows the fit of Gaussian lines having the width of the instrumental resolution.

fraction of the sample volume executes a Néel transition to the commensurate SDW phase at some temperature above 400 K [10]. Since the composition of the sample (0.18%Re) is near the triple point at about 0.3%Re, and the Néel temperature to the commensurate-SDW phase increases rapidly for higher Re concentrations [1], local sample composition need not vary much to cause this effect. Fig. 1 shows a wave-vector **q**-scan for elastic scattering through the two incommensurate magnetic Bragg peaks flanking the commensurate peak at temperature 321 K.

The study at Brookhaven National Laboratory of the silent satellites in pure chromium was described in [8].

3. Results for Cr(Re)

The susceptibility function used to analyze critical magnetic scattering in CrRe, as in pure chromium and the CrV alloys [6] was of the form proposed by Sato and Maki [2]:

$$\chi(\mathbf{Q},\omega) = \frac{(\chi^0/r^2) A^2}{A^4 [\kappa^2(T) + R(\mathbf{q})]^2 + \omega^2}, \tag{1}$$

where χ^0 is the coefficient of the Curie law susceptibility in the non-interacting limit, r is the length scale of the magnetic interaction, A_2 is "the magnetic stiffness", κ is the inverse correlation length, and $\mathbf{q} = \mathbf{q} - a^*(0,1,0)$, with

$$R(\mathbf{q}) = \frac{1}{4\delta^2}\left[\left(|\mathbf{q}|^2 - \delta^2\right)^2 + 4\left(\{q_xq_y\}^2 + \{q_yq_z\}^2 + \{q_zq_x\}^2\right)\right]. \quad (2)$$

While this susceptibility function peaks near the six incommensurate satellite positions in any constant-energy cut, folding of the cross-section with the spectrometer resolution function extends scattering intensity into the region between the satellites around the commensurate position. The incommensurate peaks are still resolved for low energy transfers in chromium and Cr(V) alloys [6], but the smaller value of the incommensurability parameter δ in Cr+0.18%Re, and the somewhat poorer energy resolution in this experiment, render the incommensurate peaks unresolved, as seen in Fig. 2, even at the lowest energy transfer of 2 meV.

The resolution-folded Sato-Maki model fits the neutron scattering data reasonably well, as is shown by the solid lines in Fig. 2. The deduced values for the amplitude scale, χ^0/r^2 (plotted as its reciprocal r^2/χ^0) and the magnetic stiffness, A^2, are shown in comparison to these quantities for chromium and CrV alloys in Fig. 3, as a function of the (fit) incommensurability parameter δ. In this figure an interesting correlation is revealed, as indicated by the dashed lines: the magnetic stiffness appears to be tending to zero, and the amplitude scale appears to be diverging, as δ tends not to zero (the commensurate SDW), but to a finite value of about 0.017 rlu. This suggests that δ does not go smoothly to zero with increasing Re concentration, but rather that there is a range of long-wavelength critical fluctuations (and therefore spin-density wave in the ordered phase) that is inaccessible. This would mean that there is a discontinuity in δ at the triple point in the magnetic phase diagram of the alloy system, where the Néel transition lines to the incommensurate ISDW and commensurate CSDW phases meet the ISDW-to-CSDW transition line. Such behavior is indeed found in a Cr+2.8%Fe alloy, when it is taken through the triple point by applying pressure [11].

Fig. 3 shows that the amplitude scale χ^0/r^2 seems to vary as the inverse of A^2 when the alloy composition varies. This means that the amplitude diverges, while the stiffness approaches zero at the same finite value of δ. This remarkable behavior needs to be explained by any theory that claims to describe the properties of chromium alloys.

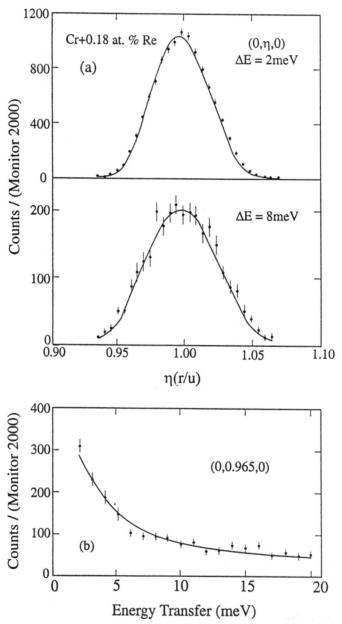

Figure 2. Inelastic neutron scattering from Cr+0.18at.%Re at 335 K, above T_N: a) constant energy scans, and b) constant **Q** scan. Solid lines are the least squares fit described in the text.

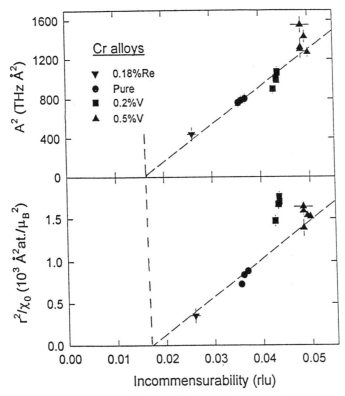

Figure 3. Correlations among Sato-Maki critical neutron scattering model parameters in chromium alloy single crystals just above T_N; A^2 is the magnetic stiffness, r^2/χ^0 is the reciprocal of the amplitude scale. The straight lines are guides for the eye [7].

4. Discussion

The first-order nature of the Néel transition in pure chromium is seen very clearly in the elastic neutron scattering, as illustrated in the upper panel of Fig. 4. When only 0.2%V is introduced, the transition becomes continuous, as shown in the lower panel of Fig. 4. We note that the Néel temperature decreases from about $T_N = 310$ K to $T_N = 289$ K for Cr+0.2%V due to the lowering of the Fermi level by doping with V [1,3], while the transition temperature is still well defined to within a fraction of a degree. Thus the change in the nature of the phase transition is not due to smearing of the first-order transition due to inhomogeneous distribution of the V impurity.

Further evidence for this effect, and therefore for the existence of a tricritical point, where the first-order Néel transition becomes continuous, somewhere along the line of phase transitions between pure chromium and the composition Cr+0.2%V, was provided by a study of the thermal ex-

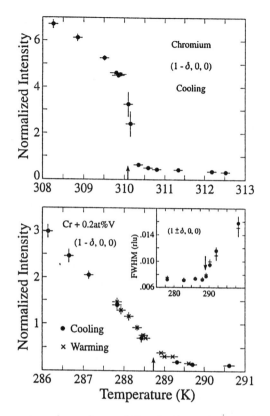

Figure 4. Temperature dependence of the elastic neutron scattering through the Néel transition in Cr and Cr+0.2at.%V [6].

pansion [5]. The results reproduced in Fig. 5 show that the tricritical point must occur quite close to the composition Cr+0.1%V. We note that the continuous transition in Cr+0.2%V is smeared in this sample, and the study by de Camargo *et al.* [5] of the better sample used for the neutron scattering study, illustrated in Fig. 4, gave a sharp transition in the thermal expansion, just as in the Bragg scattering amplitude.

The Néel transition in pure chromium is usually referred to as being "weakly first-order". Hysteresis is generally observed in a system having a strong first-order transition, with supercooling or superheating beyond the transition temperature, until some nucleation center can find the lower energy phase. The weak first-order transition in pure chromium has never been reported in the literature as being hysteretic, though there are unpublished data that appear to exhibit hysteresis [12]. The fact that the jump in the order parameter is so small probably accounts for the fact that the behavior of the inelastic neutron scattering above T_N in pure chromium is very similar to that in Cr+0.2%V, in which the phase transition is con-

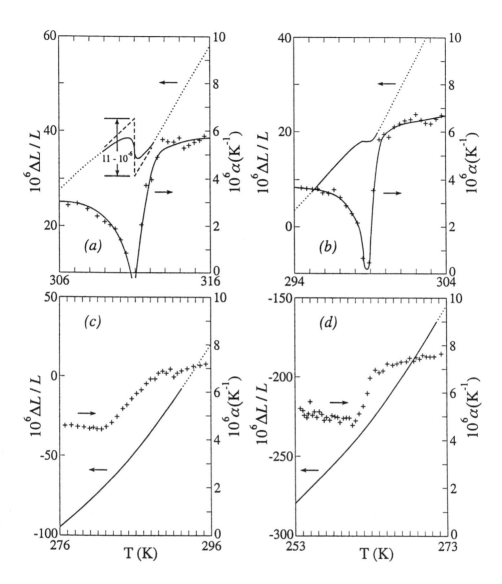

Figure 5. Temperature dependence of the thermal expansion near the Néel temperature in CrV alloys. The linear strain $\Delta L/L$ with $\Delta L=(L_{293}-L_T)$, shown by the data points (most omitted for clarity), and the fitted thermal expansivity α, shown as ++++++ with _____ being a guide to the eye: a) pure Cr; b) Cr+0.1at.%V; c) Cr+0.2at.%V; d) Cr+0.5at.%V [5].

tinuous (see Fig. 4), and both are believed to exhibit critical fluctuations. Unfortunately, there is no theory available that deals with critical fluctuations above a first-order transition, and all we can do is speculate that some ordering interaction in the system is leading to a continuous transition as the temperature decreases towards the phase transition, but some other interaction intervenes to cause a jump instead. The occurrence of a weak first-order transition in pure chromium has never been explained satisfactorily (see [5] for references to theoretical work on this problem).

We now turn our attention to the effect of V and Re doping on the critical fluctuations in the paramagnetic phase of chromium. Doping with V has been found to produce a strong effect on several physical properties [13]: the nuclear magnetic relaxation time, electrical resistivity and magnetic susceptibility; as well as inelastic neutron scattering, Bragg scattering and thermal expansion, as illustrated in Figs. 3, 4 and 5, respectively. All indicate that V doping suppresses the critical fluctuations.

The data for inelastic neutron scattering shown in Fig. 3 provide quantitative results for this effect for three Cr(V) samples having different V content; and extending to the right of the phase diagram by doping with Re, we see also in Fig. 3 that 0.18% Re introduced into chromium *increases* the amplitude of the critical fluctuations by a factor of about two, roughly the same as the decrease of the amplitude on doping with 0.2%V. It is interesting to note also that, while 0.2%V reduces to zero the magnitude of the first-order length change at the Néel transition (see Fig. 5), its size in Cr+0.18%Re is roughly the same as in pure chromium [14].

While we do not at this time have an explanation for the correlations seen in Fig. 3 between the parameters of the Sato-Maki theory fit to the dilute Cr alloy data, we shall offer here several ideas that may eventually lead to a satisfactory theoretical description:

1. The fact that, at higher Re concentrations, the ISDW-CSDW phase transition. is first order [15] is consistent with a jump in δ over a forbidden range. The occurrence of a first-order ISDW-CSDW transition is also evident in the published data CrRu [16], CrSi [17] and CrIr [18], as well as for CrMn and CrFe [1]. An ISDW-CSDW phase transition is seen also in alloys of chromium with Co, Pt, Rh, Ge and Ga, but it is not known whether or not it is first-order, since the experimental data are consistent with either a continuous transition or a smeared first-order transition (see Table IX and Table XV in Fawcett *et al.* [1]). The nature of the ISDW-CSDW phase transition has not been discussed in these previous publications, but current theoretical considerations might explain why it is generally first-order and at the same time explain the jump in δ over a forbidden range.
2. Arguments have been published that crystallographic commensurate-

to-incommensurate transitions must always be discontinuous, due to interactions between incommensurate domain walls [18]. Since little is known about ISDW domain walls, it is not clear if these arguments can in fact be applied to them.

3. Reduction of stiffness is softening, so the effect we are seeing may indicate a kind of "mode-softening" transition for $\delta \leq 0.017$ rlu. Normally, however, the mode involved is a real (underdamped) excitation with a dispersion relation that softens, whereas critical fluctuations are virtual (overdamped) excitations with a self-energy functional R($\mathbf{Q}$), which resembles but is not exactly a dispersion relation that appears to be softening. Magnetic soft-mode transitions, as usually discussed, involve magnetic interactions in a material generating a dispersion relation for a normally dispersionless (single-ion) "crystalline electric field" excitation above a non-magnetic singlet ground state [19], which then softens to cause magnetic ordering. No analogous excitation of non-magnetic origin is known in chromium alloys.

The rapid increase in the intensity of the critical fluctuations, which is indicated in Fig. 3 by the decrease towards zero of the reciprocal amplitude scale r^2/χ^0, as δ approaches the critical value, 0.017 rlu, is illustrated schematically in Fig. 6. The present data for T_N in CrV and CrRe alloys are used to construct the phase diagram shown in Fig. 7.

Sternlieb et al. [8] realized the significance for pure chromium of the existence in the paramagnetic alloy, Cr+5%V, of spin fluctuations giving rise to inelastic scattering at the incommensurate points corresponding to the nesting vector $\mathbf{Q'}$ of the Fermi surface [20]. One might expect to find a peak in the wave-vector dependent susceptibility $\chi(\mathbf{Q},\omega)$ at $\mathbf{Q'}$, which becomes a singularity in $\chi(\mathbf{Q},0)$ at low temperature, corresponding to the onset of long-range magnetic order, i.e., a SDW with $\mathbf{Q} \sim \mathbf{Q'}$, when the V content is reduced to less than about 4%V [1]. Thus, in a single-$\mathbf{Q}$ sample of chromium, with $\mathbf{Q} = a^*(1 \pm \delta, 0, 0)$, the other two pairs of off-axis satellites at $a^*(1, \pm\delta, 0)$ and $a^*(1, 0, \pm\delta)$ might be expected to give rise to peaks in $\chi(\mathbf{Q},\omega)$ below the Néel temperature T_N corresponding to modes of excitation that are termed "silent satellites", just as excitations occur at all six satellite points in the paramagnetic phase.

Fig. 8 illustrates the experimental evidence for these silent satellites. As temperature increases towards T_N, their intensity increases rapidly from very low values, until right at T_N a discontinuous jump (corresponding to the first-order nature of the Néel transition in chromium) results in equal scattering at all six satellites in the paramagnetic phase. Thus the Néel transition is driven by critical fluctuations at all six satellites, not just the two at the wave vectors $\pm\mathbf{Q}_x$ of the SDW.

While this picture of the relation between critical fluctuations of the

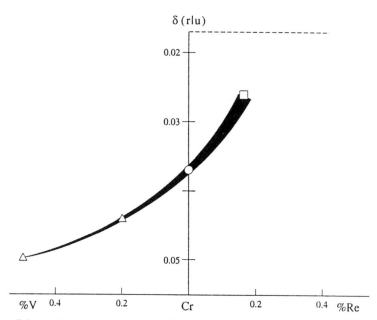

Figure 6. Schematic diagram showing the rapid increase of the incommensurability parameter δ, and of the intensity of critical fluctuations in the paramagnetic phase of chromium alloys, as the critical value, $\delta = 0.017$ rlu (indicated by the dashed line) is approached.

incommensurate SDW in the ordered phase and the paramagnetic phase is unique to chromium, similar behavior has been observed in γ-Mn, where the polarization direction of the simple antiferromagnet can be selected by stress-cooling, and the critical fluctuations in the ordered phase occur at the magnetic reciprocal lattice points corresponding to the other two polarization directions [21]. The theoretical analysis of the effect should enlarge our general understanding of phase transitions.

5. Conclusion

The work of Sternlieb *et al.* [8] on the on the development of critical fluctuations in the incommensurate SDW phase of a single-**Q** crystal of pure chromium, through the growth of the silent satellites as the weak first-order Néel transition is approached, encourages similar studies in dilute alloys of Cr. In particular, the continuous transition in Cr+0.2%V is likely to give rise to rather different behavior. The experimental problem will be to achieve a single-**Q** state by field-cooling a dilute alloy [1,22]. Indeed, the

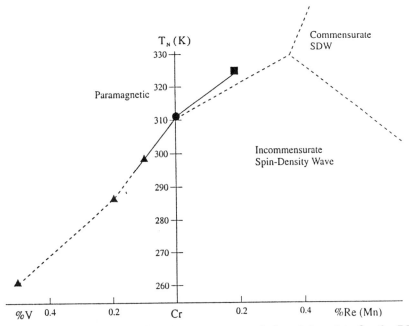

Figure 7. Phase diagram of the Cr alloy system around the triple point. On the RHS, the solid line indicates Cr(Re), while the dash line indicates Cr(Mn) through the triple point. On the LHS, the change from solid to dashed line indicates the change in transition order at the tricritical point for Cr(V).

success of field-cooling in pure chromium may be contingent on its having a weak first-order transition.

On the other hand, with Re doping the Néel transition to the incommensurate SDW phase is first-order [10,14], and the amplitude of the critical fluctuations in the paramagnetic phase increases considerably [6]. Thus one might expect that it would be possible to obtain a single-**Q** sample by field-cooling a sample of the composition about Cr+0.2%Re, and that the silent satellites woulld be even more in evidence than in pure chromium. A careful study of this system (or an incommensurate SDW CrMn sample) should provide considerable insight into the nature of the Néel transition near the soft-mode transition that seems to mark the triple point in the Cr alloy system.

Acknowledgements. This work was supported in part by US DOE grant DE-FG05-88ER45353 and the Natural Sciences and Engineering Research Council of Canada. Research at Brookhaven National Laboratory was supported by the Division of Materials Science, US DOE, contract DE-AC02-CH00016.

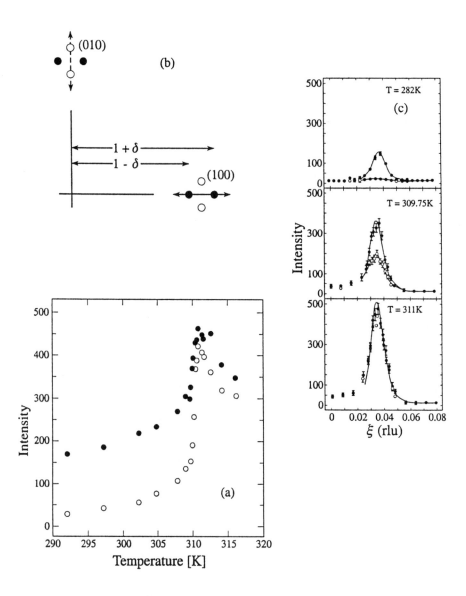

Figure 8. Silent satellites: a) temperature dependence of the peak intensity for longitudinal scans at constant energy, E= 0.5 meV, for single-**Q** Cr, where solid circles correspond to the dashed-line **Q**-scan in the inset (b). Inset (c) shows the data for a **Q**-scan through single peaks of the on- and off-axis satellites at two temperatures close to the Néel temperature, $T_N = 310.3$ K, and at a temperature 28 K below T_N [8].

References

1. For reviews, see Fawcett, E. (1988) Spin-density-wave antiferromagnetism in chromium, *Rev. Mod. Phys.* **60**, 209-283; and Fawcett, E., Alberts, H.L., Galkin, V.Y., Noakes, D.R., and Yakhmi J.V. (1994) Spin-density-wave antiferromagnetism in chromium aloys, *Rev. Mod. Phys.* **66**, 25-127.
2. Sato, H., and Maki, K. (1974) Theory of inelastic neutron scattering from Cr and its alloys near the Néel temperature, *Int. J. Magn.* **6**, 183-209.
3. Machida, K. and Fujita, M. (1984) Soliton lattice structure of incommensurate spin-density waves, *Phys. Rev. B* **30**, 5284-5299; Fishman, R.S., and Liu, S.H. (1992) Effect of impurities on the magnetic ordering in chromium, *Phys. Rev. B* **45**, 12306-12318 and (1993) Magnetic structure and paramagnetic dynamics of chromium and its alloys, *Phys. Rev. B* **47**, 11870-11882.
4. Hill, P., Ali, N., Oliviera, A.J.A., Ortiz, W.A., de Camargo, P.C., and Fawcett E. (1994) Local moments in the paramagnetic phase of dilute CrV alloys, *J. Phys. Cond. Matter* **6**, 1761-1767.
5. Fawcett, E., Roberts, R.B., Day, R., and White, G.K. (1986) Possible tricritical point in dilute chromium alloys, *Europhys. Lett.* **1**, 473-478; de Camargo, P.C., Castro, E.P., and Fawcett, E. (1988) Thermal expansion and ultrasonic velocity and attenuation at the continuous Néel transition in Cr+0.2at.%V, *J. Phys. F* **18**, L219-L222.
6. Noakes, D.R., Fawcett, E., Holden, T.M., and de Camargo, P.C. (1990) Critical scattering in Cr+0.2at.%V and in chromium, *Phys. Rev. Lett.* **65**, 369-372; Fawcett, E., Holden, T.M., and Noakes, D.R. (1991) The effect of 0.2at.%V impurity on the dynamic magnetic susceptibility of chromium, *Physica B* **174**, 18-21; Noakes, D.R., Fawcett, E., and Holden, T.M. (1997) Concentration dependence of critical scattering from Cr(V) alloys above the Néel temperature, *Phys. Rev. B* **55**, 12504-12509.
7. Noakes, D.R., Fawcett, E., Sternlieb, B.J., Shirane, G., and Jankowska J. (in press) Nature of the triple point in chromium alloys, *Physica B* [ICNS-97].
8. Sternlieb, B.J., Hill, J.P., Inami, T., Shirane, G., Lee, W.-T., Werner, S.A., and Fawcett, E. (1995) Silent satellites: critical fluctuations in chromium, *Phys. Rev. Lett.* **75**, 541-544.
9. Fawcett, E. and Noakes, D.R. (1993) The order of the Néel transition in Cr and dilute Cr alloys with V and Re, *Int. J. Mod. Phys. B* **7**, 624-629.
10. Mikke, K. and Jankowska, J. (1980) Magnetic excitations in the longitudinal and transverse spin-density-wave phases of a Cr+0.18%Re alloy, *J. Phys. F* **10**, L159-L163.
11. Fawcett, E. and Vettier, C. (1982) The phase diagram of CrFe in the pressure-temperature plane, *J. Phys. (Paris)* **43**, 1365-1369.
12. Shirane, G. (private communication).
13. Fawcett, E. (1992) Suppression of spin fluctuations in Cr by doping with V, *J. Phys. Cond. Matter* **4**, 923-928.
14. Dev Mukherjee, G., Bansal, C., and Fawcett, E. (1994) Effect of 0.18at.%Re on the Néel transition in chromium, in Srivastava, V. et al. (eds.), *AIP Conference Proceedings 286*, American Institute of Physics, New York, pp. 330-332.
15. Boshoff, A.H., Alberts, H.L., du Plessis, P. de V., and Vettier, A.M. (1993) Utrasonic and neutron diffraction studies of dilute Cr-Re and Cr-Ru alloy single crystals, *J. Phys. Cond. Matter* **5**, 5353-5370.
16. Martynova, J., Alberts, H.L., and Smit, P. (1996) A high-pressure ultrasonic investigation of the Néel transition in a Cr+0.07at.%Ir alloy single crystal, *J. Phys. Condens. Matter* **8**, 4045-4054.
17. Anderson, R.A., Alberts, H.L., and Smit, P. (1993) Magnetoelasticity of dilute Cr-Si alloy single crystals, *J. Phys. Cond. Matter* **5**, 1733-1752.
18. Lajzerowicz, J., Levanyuk, A.P., and Minyukov, S.A. (1996) Impossibility of ob-

serving a continuous commensurate-incommensurate transition, *Phys. Rev. B* **54**, 12073-12075, and refs. cited therein.

19. Youngblood, R.W., Aeppli, G., Axe, J.D., and Griffin, J.A. (1982) Spin dynamics of a model singlet ground-state system, *Phys. Rev. Lett.* **49**, 1724-1727; Kötzler, J., Neuhaus-Steinmetz, H., Froese, A., and Görlitz, D. (1988) Relaxation-coupled order-parameter oscillation in a transverse ising system, *Phys. Rev. Lett.* **60**, 647-650; Grahl, M. and Kötzler, J. (1992) Narrow central peak of the magnetization dynamics at and below T_C in the singlet-singlet ferromagnet $LiTbF_4$, *J. Magn. and Magn. Mater.* **104-107**, 219-221; Marx, C., Görlitz, D., and Kötzler, J. (1996), Soft modes and damping near the Curie temperature of $LiTbF_4$: effect of magnetic field *Phys. Lett. A* **210**, 141-145; McEwen, K.A., Stirling, W.G., and Vettier, C. (1983) Soft-mode behavior of magnetic excitations in praseodymium under uniaxial stress, *Physica B* **120**, 152-155; Zheludev, A., Tranquada, J.M., Vogt, T., and Buttrey, D.J. (1996) Magnetic excitations and soft-mode transition in the quasi-one-dimensional mixed-spin antiferromagnet Pr_2BaNiO_5, *Phys. Rev. B* **54**, 6437-6447.

20. Fawcett, E., Werner, S.A., Goldman, A., and Shirane, G. (1988) Observation of incommensurate spin-density-wave paramagnons, *Phys. Rev. Lett.* **61**, 558-561.

21. Mikke, K., Jankowska-Kisielinska, J., and Hennion, B. (in press) Inelastic neutron scattering in some γ-Mn alloys below and above T_N, *Physica B* [ICNS-97].

22. Sternlieb, B.J., Lorenzo, E., Shirane, G., Werner, S.A., and Fawcett, E. (1995) Magnetism in the spin-density-wave alloy $Cr_{1-x}Mn_x$ (x=0.007), *Phys. Rev. B* **50**, 16438-16443.

UPd_2Al_3 : AN ANALYSIS OF THE INELASTIC NEUTRON SCATTERING SPECTRA

N. BERNHOEFT AND A. HIESS
Institute Laue Langevin
F-38042 Grenoble France
B. ROESSLI
Paul Scherrer Institute
CH-5232 Villigen Switzerland
N. SATO, N. ASO AND Y. ENDOH
Physics Department, Tohoku University
Sendai 980-77 Japan
G.H. LANDER
European Commission, Inst. Transuranium Elements,
D-76125 Karlsruhe Germany
AND T. KOMATSUBARA
Low Temperature Science, Tohoku University
Sendai 980-77 Japan

Abstract. The heavy fermion superconductor UPd_2Al_3 exhibits the relatively uncommon combination of an antiferromagnetic phase transition, at $T_N = 14.3$ K, followed by a superconducting phase transition below 2 K without destruction of the ordered magnetic moment. Neutron inelastic scattering at low energy transfers in the vicinity of the antiferromagnetic ordering wave vector has been used to probe the magnetic fluctuation spectrum in the temperature range 130 mK < T < 20 K. In the normal antiferromagnetic phase the spectra may be characterized in terms of a damped spin wave interacting in a mean field manner with a quasielastic mode. By neutron polarization analysis the two modes have been shown to be polarized *transverse* to the sub-lattice magnetization vector. The spectra are used to estimate both the extrapolated low temperature linear term of the normal state heat capacity and the magnitude of T_N. Satisfactory agreement with the measured values is obtained. In the superconducting state at the lowest temperatures evidence is presented for the opening of a gap at low energies which may suggest a connection between the antiferromagnetic fluctuations and superconductivity.

1 Introduction.

The discovery of the heavy fermion superconductor UPd_2Al_3 which exhibits both an antiferromagnetic phase transition, $T_N = 14.3$ K, and a superconducting phase transition below 2 K has aroused great interest. The unusually large low temperature ordered moment of 0.85 μ_B was taken as an early sign that this might be a useful material in which to study the interplay of magnetism and superconductivity [1].

UPd$_2$Al$_3$ crystallizes in the hexagonal PrNi$_2$Al$_3$ structure (space group *P6/mmm*) with lattice constants a = 5.350Å and c = 4.185Å at room temperature. Neutron and x-ray scattering measurements have revealed that the moments are coupled in ferromagnetic sheets in the basal plane and that these sheets are then ordered antiferromagnetically along the *c*-axis with a wave vector $\mathbf{Q}_O$ = (0 0 1/2) [2-4]. The measured discontinuity in the heat capacity at the superconducting transition temperature is large, $\Delta C = 1.2 \gamma T_C$ (γ = 140 *mJ / mole K^2*) [5-6] and suggests that a significant density of thermally activated modes with energies up to $0.5 meV \approx 3 k_B T_C$ are involved in the formation of the superconducting state.

Initial neutron and x-ray scattering experiments gave contradictory results. An anomaly in the temperature dependence of the magnetic Bragg intensity below T_c was reported by Krimmel *et al* [7] and later cast in doubt by Kita *et al*. [3] and Gaulin *et al*. [8][1]. Neutron inelastic scattering experiments by Petersen *et al*. [9] revealed no changes in the spin wave spectrum around the antiferromagnetic zone center on cooling through T_c. Thus, as yet, there has been little conclusive evidence for a strong interplay between magnetism and superconductivity in this compound.

More recent experiments, carried out at higher energy resolution than the preliminary work of Petersen *et al*. by ourselves [10] and another group [11] have however discovered that there exist *two* distinct contributions to the inelastic scattering spectrum of UPd$_2$Al$_3$ at low energies for temperatures below the antiferromagnetic phase transition. The neutron spectra, as a function of energy transfer[2] in the range 0 to -3 *meV* at fixed wave vector transfer in the vicinity of $\mathbf{Q}_O$, exhibit two maxima. The higher energy feature appears to be the broad spin wave like mode found in [9] which becomes strongly damped as T approaches T_N. However, in contrast to the deductions of Petersen *et al.*[9], we find an energy gap to the spin wave at the antiferromagnetic zone center *below* which exists a second low energy mode. This lower energy mode is heavily (over) damped at all temperatures measured down to T_c. Immediately below T_c the low energy mode remains heavily overdamped with evidence for a change in spectral form to a gapped inelastic mode occurring at the lowest temperatures, 130 mK to 500 mK.

Preliminary experimental results, analyzed in terms of two independent modes, one spin wave like and one overdamped have been presented previously [10]. The aim of this work is to extend the previous analysis in the light of new data. It has been our aim to see if the passage from the normal to superconducting state may be represented in a simple manner and, additionally, if some insight into various thermodynamic properties such as the values of T_N and γ may be obtained.

---

[1] Currently it appears that a weak effect might exist at least in some samples [11].
[2] The neutron looses energy in the scattering processes considered, for ease of discussion in the text 'higher energy' refers to a larger modulus of energy transfer. In the diagrams neutron energy loss processes are negative.

To make the analysis of the measured spectra we invoke the following model: take two modes in which all internal interactions have been included apart from their mutual coupling, then add a mean field interaction. On account of the generic form of the dynamical susceptibility [12] the mutual coupling in a mean field approximation leads to a low frequency enhancement of the total susceptibility and a concomitant renormalisation of the effective low energy linewidth.

The mean field coupling yields [13, 14]

$$M_1 = \chi_1[H + \lambda M_2] \quad (1a)$$

$$M_2 = \chi_2[H + \lambda M_1] \quad (1b)$$

$$M = M_1 + M_2 \quad (2)$$

and hence

$$\chi = \frac{\chi_1 + \chi_2 + 2\lambda\chi_1\chi_2}{1 - \lambda^2\chi_1\chi_2} \quad (3)$$

where χ_1 and χ_2 are the starting susceptibilities (i.e. with all interactions *apart* from the mean field coupling between χ_1 and χ_2 switched on) and λ represents the coupling coefficient between the low, χ_1, and intermediate, χ_2, frequency modes. In the limit of low frequencies the real part of χ tends to a constant whilst the imaginary part is proportional to the frequency [12]. This, for the dissipative component of the total susceptibility as $\omega \to 0$ yields a denominator

$$1 - \lambda^2 \operatorname{Re}[\chi_1]\operatorname{Re}[\chi_2] \quad (4)$$

whilst at high frequencies the bare susceptibilities must vanish and the denominator goes to unity.

2 Experimental Results.

2.1. THE SPECTROMETER AND SAMPLES.

Two sets of inelastic neutron scattering experiments were carried out on the IN14 triple axis spectrometer situated on the cold source at the Institute Laue Langevin, Grenoble, France. The first experiments covered the temperature range 0.5 K to 20 K using a He^3 insert cryostat at zero applied field, whilst the second set of experiments were performed with improved cooling capabilities (T_{min} 130 mK) and the possibility to apply a magnetic field of up to 2.5 Tesla at the sample coupled with full polarization analysis of the neutron beam. In the first experiment the spectrometer was operated at constant final wave vector 1.3Å^{-1} with an energy resolution of 140 μeV full width at half maximum together with a resolution in wave vector of 0.06 Å^{-1} along the c*-axis, whilst in the latter experiment the resolution was improved working at a final wave vector 1.15 Å^{-1} giving an energy resolution of 90 μeV and a wave vector resolution of 0.05 Å^{-1}. The data have been set on an absolute scale using both a standard vanadium sample and the sample's nuclear incoherent cross section. The uncertainties in calibration lead to relatively important errors in estimation of the absolute values of the parameters used to characterize the spectra of the order of 20%.

The samples used in the studies were pulled from a carefully homogenized melt of high purity elements by the Czochralski method with a nominal composition of UPd$_{2.02}$Al$_{3.03}$ [15, 16]. The superconducting transition temperatures of the two crystals were estimated to be $1.90 \pm 0.07 K$ and $1.88 \pm 0.07 K$ respectively as used in the two sets of experiments outlined above.

2.2. THE TEMPERATURE DEPENDENCE OF SCATTERING; T>T$_C$ WITH UNPOLARISED BEAM.

Scans at constant wave vector in the neighborhood of the antiferromagnetic wave vector Q_o were performed at a series of temperatures in the normal antiferromagnetic and paramagnetic states and representative scans are shown in figure 1. From the temperature and wave vector dependencies of the observed scattering we infer that the scattering is antiferromagnetic in nature. This conclusion has been reinforced by the results of full polarisation analysis. As will be shown, these results identify the inelastic response as being *transverse* to the sub-lattice magnetization density vector and are described in detail in Sections 2.4 and 2.5 below.

In the analysis the spectra are taken to be proportional to the imaginary part of the dynamic magnetic susceptibility. Above and in the neighborhood of T$_N$ the response is quasielastic and may be characterized by a dynamic susceptibility of the following form

$$\chi^{-1}(\bar{q},\omega) = \chi^{-1}(\bar{q})\left[1 - \frac{i\omega}{\Gamma(\bar{q})}\right] \quad (5)$$

with

$$\bar{q} = Q_o + q, \quad (6)$$
$$\chi^{-1}(\bar{q}) = \chi^{-1}(Q_o) + cq^2 \quad (7)$$

and

$$\Gamma(\bar{q}) = u\chi^{-1}(\bar{q}). \quad (8)$$

In this formulation $\chi(Q_o)$ is the static magnetic susceptibility at the antiferromagnetic wave vector, c is a torsional stiffness coefficient, q is $|q|$ as measured from the antiferromagnetic wave vector and u is a measure of the relaxation rate of the fluctuations.

Examples of data in this region are given for $k_f = 1.15 \text{Å}^{-1}$ in the top frame of figure 1. The lines in Fig. 1 have been calculated with the spectrometer resolution taken in the energy-wave vector plane of the experiment and a dispersion surface based on the model for χ given in Eq. (5) to (8). At these temperatures there is but one quasielastic mode and so the mean field coupling parameter λ introduced in Eq. (1) plays no role. As may be seen from the figure, over the full experimental range the model calculation is in good agreement with the data. The parameters c and $\hbar u$ take the values $2.8 \ 10^4 \ \text{Å}^2$ and 1.3 µeV respectively, whilst in the paramagnetic phase $\chi(\mathbf{Q}_o)$ follows an approximate Curie-Weiss like temperature dependence as shown in Figure 4.

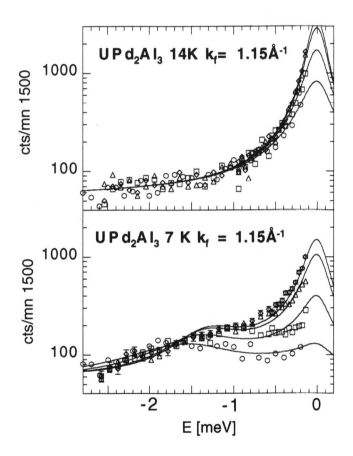

Figure 1. Upper frame: The low energy scattering from UPd$_2$Al$_3$ taken in the vicinity of T_N. The diamonds, triangles, squares and circles correspond with scattering wave vectors, given in reciprocal lattice units, of (0 0 .5), (0 0 .515), (0 0 .530) and (0 0 .545). Counting time 5 min. per point. The solid lines are calculated with a dispersion surface given by Eq. (5) - (8) convoluted with the instrumental resolution taken in the ω - q cut of the scattering plane. The parameter values are given in Figures 2 and 3. Lower frame: As in upper frame but at a temperature of 7 K. The error bars given for the diamonds are typical in all plots in both frames.

For temperatures below T_N the solid lines are fits in which the dynamical susceptibility is given by Eq. (3) with the low energy response represented as a quasielastic excitation

$$\chi_1^{-1}(\bar{q},\omega) = \chi_1^{-1}(\bar{q})\left[1 - \frac{i\omega}{\Gamma_1(\bar{q})}\right] \qquad (9)$$

together with a damped spin wave at pole $\omega_o(q)$ at higher energies. This damped pole takes the following form

$$\chi_2^{-1}(\bar{q},\omega) = \chi_2^{-1}(\bar{q})\left[1 - \frac{\omega(\omega + i\Gamma_2(\bar{q}))}{\omega^2(q) - i\Gamma_2(\bar{q})(\omega + i\Gamma_2(\bar{q}))}\right] \qquad (10)$$

$$\hbar\omega(q) = \hbar\omega_o + Dq^2 \qquad (11)$$

where $\chi_i^{-1}(\bar{q})$ and $\Gamma_i(\bar{q})$ take the same functional form as above T_N in both Eqs. (7) and (8). The spin wave pole gap at the antiferromagnetic zone center is given by $\hbar\omega_o$ and D is the stiffness. This quadratic dispersion, local to the zone center, joins to the linear dispersion at higher q as observed by Petersen et al. [9] around $q = 0.06$ Å^{-1}. In calculating the intensities account has been taken of the relative orientation of scattering wave vector **q** and the sub-lattice magnetic moment direction. In the given experimental geometry the neutron cross section is sensitive to transverse and longitudinal fluctuations in the plane of the sub-lattice magnetization and insensitive to transverse fluctuations perpendicular to this plane. It is important to the analysis that the predominant fluctuations are transverse, this point is uniquely demonstrated by our experiments with polarization analysis. For the moment we will use this fact and substantiate it in subsection 2.4 below.

In modeling the data over the range $T_c < T < T_N$ we have kept the coupling parameter λ as a real constant independent of q (local in space), ω (instantaneous in time) and temperature. The numerical value of λ which, together with the given values of $\chi_{1,2}(q,\omega)$ best describes the data is $\lambda = 200$. As with the other parameters, the error in this Figure is estimated to be around 20%. In this approach the evolution of line width and intensity over the measured energy and wave vector intervals may be consistently accounted for by a natural softening of the spin wave pole and energy gap together with a thermal renormalisation of $\chi_{1,2}$. The temperature dependencies of χ_1 and χ_2 and spin wave stiffness and gap are given in Figure 2.

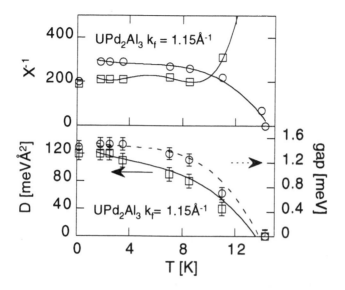

Figure 2. Upper frame: The temperature dependence of the antiferromagnetic susceptibility, the circles refer to the low frequency component and the squares to the spin wave. The lines are smooth extrapolations from above T_c through the data used to generate the temperature dependence as shown in Figure 4. Lower frame: Temperature dependencies of spin wave gap and stiffness.

The parameters $\hbar u_1$ and $\hbar u_2$, which define the energy scale of the Γ_1, the quasielastic linewidth, and Γ_2, the damping of the spin wave, are essentially temperature independent over this range with values of 1.3 μeV and 1.4 μeV, respectively. Near T_N the torsional stiffness of the quasielastic (c_1) and the spin wave mode (c_2) take similar values and join smoothly with the value of c in the paramagnetic phase, however below about 10 K the quasielastic stiffness c_1 increases by almost an order of magnitude. The microscopic origin of such behavior remains to be understood.

To explore more fully the temperature dependence of the response at low energies a scan at $\mathbf{Q}_o$ and fixed energy transfer of -0.2 meV (neutron energy loss) was made. This scan together with the calculation based on a smooth interpolation of the parameters already presented in Figure 2 is given in Figure 4, the agreement is acceptable. Thus parameters are available to give a description of the energy, wave vector and temperature dependence of the magnetic fluctuations in the vicinity of $\mathbf{Q}_o$ in the normal state. In the paramagnetic phase the susceptibility is fitted to a Curie Weiss type susceptibility, $\chi^{-1}(Q_o) = 90(T - T_N)$, with $T_N = 14.35 \pm 0.05 K$ together with the previously determined values of the microscopic parameters c and $\hbar u$.

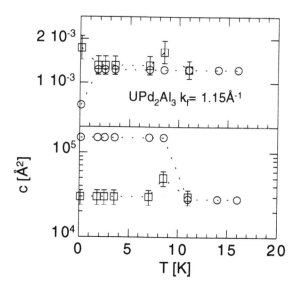

Figure 3. Upper frame: The temperature dependence of the parameter u (equation (8)). Lower frame: The temperature dependence of the parameter c (Eq. (7)). Symbols as in upper frame of Figure 2 above.

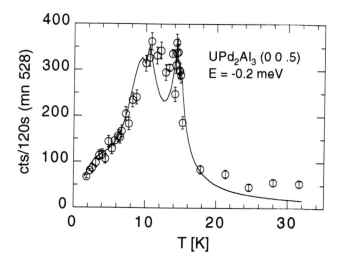

Figure 4. The temperature dependence of scattering intensity at (0 0 .5) and an energy transfer of -0.2 meV (neutron energy loss). The solid line below T_N results from the theoretical model developed in the main text with parameters taken from Figures 2 and 3. Above T_N an approximate fit to a Curie-Weiss susceptibility has been made.

2.3. THE TEMPERATURE DEPENDENCE OF SCATTERING; T<T_C WITH UNPOLARISED BEAM

On cooling below T_c no immediate changes in spectral form are evident, and the similar set of microscopic parameters which described the scattering just above T_c give a good description of the data immediately below T_c. However, on cooling to the lowest temperatures a change in spectral form does occur with the development of an *inelastic* damped pole at low energies, Figure 5. The low energy mode before coupling develops an energy gap of $0.53 \pm 0.03 meV$ which is approximately $3k_BT_c$ and has a rapid dispersion of some $7 \: 10^3$ meVÅ^3 with a relatively narrow width given by $\hbar\Gamma_{sc} = \hbar u_{sc}\chi_1^{-1}(q)$ where $\hbar u_{sc} = 0.5$ μeV. The low damping rate, corresponding with the fall in $\hbar u$ by a factor of approximately 3, from the normal state may signal a paucity of low energy decay channels in the superconducting state.

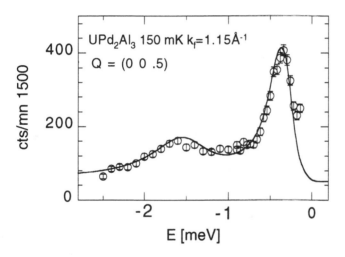

Figure 5. Energy dependence of scattering of unpolarised neutrons at Q_o and a temperature of 150 mK. The solid line incorporates an energy gap in the low energy spectrum of 0.53 *meV* together with the wave vector dependent form of λ. Note the experimental gap is of the order 0.4 meV, this reduction from the bare value of .53 *meV* is a consequence of the mode interactions.

This change in spectral lineshape is accompanied by a change of form of the mean field coupling coefficient λ; at $q = 0$ the value of λ is essentially unchanged however it falls rapidly to approximately one half this value for $q > 0.01$ Å^{-1}. The approximate wave vector dependence is of the following form:

$$\lambda = \lambda(q) = 120 + 50\exp(-q^2 / 0.005^2) \tag{12}$$

To consider what this implies in real space we return to Eqs. (1a) and (1b). Explicitly inserting the spatial dependence gives for example in (1a),

$$M_1(\mathbf{r}) = \int d\mathbf{r}' \chi_1(\mathbf{r} - \mathbf{r}')[H(\mathbf{r}') + \lambda M_2(\mathbf{r}')] \tag{13}$$

with

$$\lambda M_2(\mathbf{r}') = \int d\mathbf{r}'' \lambda(\mathbf{r}' - \mathbf{r}'') M_2(\mathbf{r}''). \tag{14}$$

Thus the molecular field from sub-system (2) at $\mathbf{r}'$ is felt by sub-system (1) at $\mathbf{r}$ through the spatial range of χ_1. This range is the correlation length $\zeta_1 = 2\pi\sqrt{\chi_1 c_1} = 140 \text{Å}$. The existence of a spatial dependence to λ implies that the molecular field generated by sub-system (2) becomes non-local, its effect is felt over some 2000 Å. The parametric forms used for $\chi_{1,2}$ are summarized in Table 1 at the end of the article.

2.4. POLARISATION ANALYSIS, T>T_C.

In order to determine the polarization and magnetic character of both the spin wave and quasielastic scattering a full polarization analysis has been carried out. The ratio of counts recorded in the flipper on to flipper off configuration at 20 K on the (0 0 1) nuclear peak was 18 in an applied vertical field of 2.5 Tesla. The sample was then field cooled into the antiferromagnetic phase and the field was lowered to 0.8 Tesla at 8 K. Under these conditions the (0 0 .5) antiferromagnetic peak had an intensity 15 times lower than that observed in the similar configuration without polarization analysis. The flipping ratio at the antiferromagnetic peak was 2.5. The reduced flipping ratio is in agreement with the magnetic phase diagram [3] in which the sub-lattice moments lie at approximately 30 degrees to the horizontal under the given conditions of temperature and applied field. In the experimental geometry the flipping ratio may be expressed as

$$R = \frac{flipperon}{flipperoff} = \frac{\sin^2(\alpha)S^{xx} + \cos^2(\alpha)S^{zz}}{\cos^2(\alpha)S^{xx} + \sin^2(\alpha)S^{zz}} \tag{15}$$

where S^{xx} and S^{zz} are the transverse and longitudinal parts of the response function respectively. At the antiferromagnetic Bragg peak this reduces to

$$R^{-1} = \tan^2(\alpha) \tag{16}$$

where α is the angle of the sub-lattice moments to the horizontal scattering plane. The observed ratio of 2.5 is consistent with an angle α of 32 degrees.

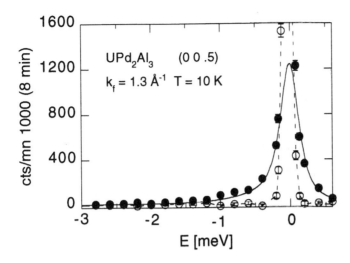

Figure 6. The transverse (filled circles) and longitudinal (open circles) response at 10 K. The solid line is a scaled plot of the fit to the unpolarised data at 10 K. The dashed line is a guide to the eye.

Working at $k_f = 1.3 Å^{-1}$ data were collected at 10 K and 2.2 K in the antiferromagnetic phase. The data at both temperatures show the scattering to be *transverse* to the magnetic moment, the further geometric constraint implied by working with the scattering wave vector perpendicular to the plane containing the sub-lattice moments means that one may identify the fluctuations as being transverse *and* lying in the basal (*a-b*) plane of the sample. Data are shown at 10 K, with a background of 6 cts/mn 1000 subtracted and corrected for the measured flipping ratio, in Figure 6.

2.5 POLARISATION ANALYSIS, $T<T_C$.

In the superconducting state at 150 mK a similar experiment and analysis at $k_f = 1.15 Å^{-1}$ indicated the scattering remains *transverse* and in the basal plane. This is shown in Figure 7 where the strong low energy peak can be seen to join smoothly with the spin wave at higher energies as in the top frame of Figure 5. The solid line is a scaled copy of the solid line in Figure 5 where data was collected without polarisation analysis (no applied field) in the same spectrometer configuration.

Below the intense transverse response, there may be a weak level of scattering in the longitudinal channel at low energies, however it is not possible at this stage to state unambiguously whether this is intrinsic or arises from the imperfect polarisation. Further experiments with a sample geometry tilted by 30 degrees in the vertical plane, enabling one to reduce the offset angle α to zero are planned to resolve this issue. Clearly this could be a key to understanding the residual heat capacity at the lowest temperatures in the superconducting phase, this point is further discussed below.

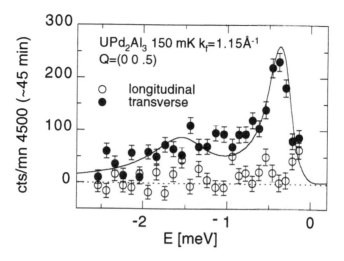

Figure 7. The transverse (filled circles) and longitudinal (open circles) response at 150 mK. The solid line is a scaled plot of the fit to the unpolarised data at 150 mK. The dashed line is the zero level.

3. Discussion.

Using Eq. (3) through (11) we have been able to represent, in a consistent manner, all the data over the temperature interval $T<T_c$ to $T>T_N$. In the normal phase the natural variables of the problem are the spin wave stiffness and energy gap and the amplitudes of the spin wave and quasielastic response; their temperature dependencies are shown in Figure 2. Over the range of the data the spin wave response is represented with a quadratic dispersion and, as the pole becomes over damped, blends smoothly with the form of response function above T_N (Eqs. (5) to (8)). For wave vectors q greater than 0.06 Å^{-1} the measured response is similar in energy to that previously reported by Petersen *et al* [9].

We find the spin wave stiffness and energy gap to renormalise smoothly with temperature through T_c whilst the spin wave damping coefficient changes little over the measured temperature interval 0.15 K to 10 K. It would appear that the spin wave dynamics are more or less unaffected by the presence of superconductivity; given the scale of the low temperature spin wave energy gap is close to T_N this is perhaps not unexpected.

More unusual is the quasielastic response. This component carries an unusually large spectral weight. Within our analysis the temperature dependence of this spectral weight in the normal phase is conventional, simply its magnitude is at least one order of magnitude greater than might be expected. This has two consequences. First, if a gap or partial gap opens in the spectrum of excitations in the superconducting phase one may expect a *measurable* renormalisation of this low energy mode, and indeed this is seen.

Second, the low relaxation frequency associated with this mode will, in the vicinity of T_c, contribute a major portion of the heat capacity and hence naturally lead to a direct association of the low energy antiferromagnetic fluctuations with the superconducting phase transition. Writing the inverse lifetime of the quasielastic mode as $u_1 = bQ_o$ we find b is directly comparable in magnitude with estimates based on the measured Fermi Liquid response in paramagnetic metals where, in the ballistic regime one has the relation $\Gamma(q) = bq\chi^{-1}(q)$ and b lies in the range 1.5 - 3 µeVÅ [17-20].

In analogy with the Stoner enhancement factor [21] one may define an enhancement of the *total* susceptibility through Eq. (1) above,

$$enh(q, \omega = 0, T) = \frac{1}{1 - \lambda^2(q)\chi_1(q)\chi_2(q)} \qquad (17)$$

which takes into account the joint thermal evolution of the coupling constant and the individual susceptibilities of the spin wave and low energy modes. This quantity is plotted for selected temperatures in the normal and superconducting phase in Figure 8. The first characteristic is that it falls towards unity at high q, this reflects that the mutual enhancement of modes to the total susceptibility is focused around the antiferromagnetic wave vector. Second, in the normal phase, the enhancement around $\mathbf{Q}_o$ out to $q = 0.02$ *rlu* falls on lowering the temperature towards T_c. This fall in the total susceptibility around $\mathbf{Q}_o$ may be imagined as a screening of χ for wave packets of typical size 100 - 300 Å.

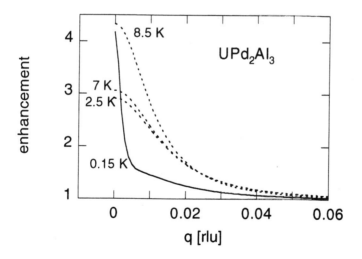

Figure 8. The experimental enhancement factor, as defined in Eq. (17), for various temperatures in both the normal antiferromagnetic and superconducting antiferromagnetic states.

Finally, on reaching the superconducting state the enhancement is strongly reduced for all wave vectors greater than 0.005 Å^{-1}. In the superconducting phase the enhancement at low q indicates the effective magnetic fields to be mutually coupled over a spatial scale of some 2000 Å. This is analogous to the estimated London penetration depth [1] and dominates the magnetic correlation length (140 Å) which appears to confine the magnetic fluctuations to within the typical superconducting coherence length (100 -200 Å) in this type of material [1].

3.1. CALCULATIONS OF THERMAL PROPERTIES.

We use respectively the high temperature magnetic fluctuation spectra to estimate T_N, and the low temperature spectra at 2.2 K, dominated over the thermal range by the quasielastic mode, to calculate the linear coefficient of the specific heat as extrapolated to T = 0 K.

3.2. CALCULATION OF T_N.

T_N is defined as the point where $\chi^{-1}(\mathbf{Q}_o) = 0$. Within the framework of the elementary mode coupling theory, in which a Gaussian approximation to the contribution from dynamic fluctuations in the first anharmonic term (quadratic term) in the free energy density renormalises the coefficient α of the leading (quadratic) term, one has, in the limit that magnetization density fluctuations are more important than Stoner (independent quasiparticle) excitations [22-25],

$$\chi^{-1}(\mathbf{Q}_o) = \alpha \left[1 - \frac{5}{3} \frac{\langle m^2 \rangle}{|M_o(\mathbf{Q}_o)|^2} \right]. \qquad (18)$$

T_N is then given by the condition

$$1 = \frac{5}{3} \frac{\langle m^2 \rangle}{|M_o(\mathbf{Q}_o)|^2} \qquad (19)$$

where $M_o(\mathbf{Q}_o)$ is the low temperature magnetization at wave vector $\mathbf{Q}_o$ and $\langle m^2 \rangle$ is the thermal variance of the local magnetization. For the purpose of calculation we take the torsional coefficient c and the characteristic diffusion frequency u determined at 16 K (as defined Eqs. (7) and 8). The fluctuations are integrated up to a thermal cut off given by $k_B T_N = \hbar \Gamma(\bar{q})$ which yields the following approximate formula for T_N,

$$k_B T_N = 2.42 c M_o^{4/3} (\hbar u)^{1/3}. \qquad (20)$$

Inserting numerical values yields the estimate $T_N = 16$ K in reasonable accord with the experimental value.

3.3. CALCULATION OF γ.

In the similar manner the contribution of the low lying modes to the entropy may be calculated [22-23]. Extrapolating from the normal state at low temperatures one has,

$$\gamma = \frac{V\pi k_B^2}{3\hbar} \sum_{\bar{q}} \frac{1}{\chi(\bar{q})} \left(\frac{\chi''(\bar{q}, \omega)}{\omega} \right)_{\omega=0}. \quad (21)$$

That is, the extrapolated low temperature value of γ is given by the spectral weight of the fluctuations around zero energy. Assuming an isotropic spectrum around $\mathbf{Q}_o$ leads to

$$\gamma = \frac{V k_B^2}{6\pi\hbar} \frac{V\kappa}{u_1 c_1} \int_0^{x_z^*} \frac{dx \cdot x^2}{1+x^2} \quad (22)$$

where V is the volume per mole of uranium ions, κ the inverse (antiferromagnetic) correlation length equal to $\sqrt{\chi^{-1} c^{-1}}$ and $x = \frac{q}{\kappa}$. The integration is up to the Brillouin zone boundary. Inserting the values from the analysis of the low temperature data one obtains the estimate $\gamma = 0.1$ $J / mole$ K^2 in reasonable agreement with the extrapolated value from the low temperature normal state.

On passing into the superconducting state a gap develops at low temperatures which will remove spectral weight from the integral. In this context it may be useful to remark that, if the *magnetic* correlation length is limited by the superconducting coherence length below T_c, then one may express γ in the following manner

$$\gamma \propto \frac{1}{u\chi^{-1}(\mathbf{Q}_o)} = \frac{1}{\Gamma(\mathbf{Q}_o)}. \quad (23)$$

Given this result, the reported intrinsic value of $\gamma = 15$ $mJ / mole$ K^2 [6] extrapolated in the superconducting phase, would correspond to a mode having a relaxation rate approximately one order of magnitude greater than that of the low energy quasielastic mode observed at 2 K (viz 2 meV as opposed to 0.2 meV). If the transverse inelastic response arises from a part of the excitation spectrum of the quasiparticle states which condenses, then a second mode, *not* involved with the condensate, having spectral weight at the origin and a characteristic frequency of $\hbar\Gamma(\mathbf{Q}_o) \approx 2 meV$ would be required to provide the necessary entropy.

Conclusion.

Two components in the low energy inelastic response of UPd$_2$Al$_3$ have been observed below T$_N$. One is dispersive with an energy gap and shows no appreciable change on passing through T$_c$ [9]. On raising the temperature this spin wave like mode softens and on approaching T$_N$ and above it is manifested as a quasielastic mode. The second component to the low energy response exists in the antiferromagnetic phase. It is a quasielastic mode with a characteristic energy below that of the spin wave energy gap. This mode, in contrast with the spin wave, is renormalised by the transition to the superconducting state. The energy scale of 0.2-0.5 meV and the major contribution to the low temperature heat capacity ($\gamma = 0.1\ J\ /\ mole\ K^2$) in the normal state hint that the low energy (antiferro)magnetic fluctuation spectrum may, in a semi-quantitative manner, be intimately connected with the transition to the superconducting state. Given that no enhancement of the quasielastic spectrum has yet been seen in the vicinity of T$_c$, it is plausible that the peaking of heat capacity at 2 K is the result of the energy gap which, having opened in the spectrum at low temperatures, collapses on approaching T$_c$. This is consistent with the data collected at low temperatures where at 1.8 K the spectral form of the scattering is, within our resolution, entirely similar to that at 2.2 K. Further experiments to explore both the anisotropy and possible longitudinal spectral response at low temperatures are planned.

TABLE 1. Parametric forms used in the analysis.

	paramagnetic	normal antiferromagnetic	Superconducting Antiferromagnetic
$\dfrac{\chi_1''}{\omega}$	Lorentzian	Lorentzian	Damped Oscillator
$\dfrac{\chi_2''}{\omega}$	Lorentzian	Damped Oscillator (spin wave)	Damped Oscillator (spin wave)
$\lambda(q)$	0	A_1	$A_2 + A_3 e^{-(q/\sigma)^2}$

Acknowledgments.

One of us, NB, gratefully acknowledges the patience and generous support of the directors and staff of the ILL, Grenoble and that of ITU Karlsruhe which has made this analysis possible.

References.

1. Steglich, F., Ahlheim, U., Bohm., A., Bredl, C.D., Caspary, R., Geibel, C., Grauel, A., Helfrich, R., Kohler, R., Lang, M., Mehner, A., Modler, R., Shank, C., Wassilew, C., Weber, G., Assmus, W., Sato,

N. and Komatsubara, T. (1991) Phase Transitions in the Heavy Fermi Liquid State: $CeCu_2Si_2$, UNi_2Al_3 and UPd_2Al_3,, *Physica* **C185**, 379-384.
2. Krimmel, A., Fischer, P., Roessli, B, Maletta, H., Geibel, C., Shank, C., Grauel, A., Loidl, A., Steglich, F. (1992) Neutron diffraction study of the heavy-fermion superconductors UM_2Al_3 ,(M = Pd, Ni), *Z. Phys.* **B86,** 161-162.
3. Kita, H., Donni, A., Endoh, Y., Kakurai, K., Sato, N., Komatsubara, T.; (1994) Single crystal neutron diffraction study of the magnetic phase diagram of the Heavy Fermion Superconductor UPd_2Al_3), *J. Phys. Soc. Japan.* **63,** 726-735.
4. Paolasini, L., Paixao, J.A., Lander, G.H., Burlet, P., Sato, N., Komatsubara, T.; (1994) Field dependence of magnetic structure of UPd_2Al_3 in the normal state, *Phys. Rev.* **B49,** 7072-7075.
5. Geibel, C., Shank, C., Thies, S., Kitazawa, H., Bredl, C.D., Bohm, A., Rau, M., Grauel, A., Caspary, R., Helfrich, R., Ahlheim, U., Weber, G., Steglich, F.; (1991) Heavy-Fermion superconductivity at Tc = 2K in the antiferromagnet Upd_2Al_3,, *Z. Phys.* **B84,**1-2.
6. Steglich, F., Gegenwart, P., Geibel, C., Helfrich, R., Hellman, P., Lang, M., Link, A., Modler, R., Sparn, G., Buttgen, N., Loidl, A.; (1996) New observations concerning magnetism and superconductivity in the heavy-fermion metals, *Physica* **B223,** 1-8.
7. Krimmel, A., Loidl, A., Fisher, P., Roessli, B., Donni, A., Kita, H., Sato, N., Endoh, Y., Komatsubara, T., Geibel, C., Steglich, F. (1993) Single crystal neutron diffraction studies of the heavy fermion superconductor UPd_2Al_3, *Solid State Comm.* **87,** 829-831.
8. Gaulin, B.D,, Gibbs, D., Isaacs, E.D., Lussier, J.G., Reimers, J.N., Schroder. A., Taillefer, L., Zschack, P. (1994) Resonant magnetic x-ray scattering study of Phase transitions in UPd_2Al_3, *Phys. Rev. Lett.* **73,** 890-893.
9. Petersen, T., Mason, T.E., Aeppli, G., Ramirez, A.P., Bucher, E., Kleiman, R.N.; (1994) Magnetic fluctuations and the superconducting transition in the heavy-fermion material UPd_2Al_3, *Physica* **B199,** 151-153.
10. Sato, N., Aso, N., Lander, G.H., Roessli, B., Komatsubara, T., Endoh, Y. (1997) Spin fluctuations in the Heavy Fermion Superconductor UPd_2Al_3 Studied by Neutron inelastic scattering, *J. Phys. Soc. Japan.* **66,** 1884-1887.
11. Metoki, N., Haga, Y., Koike. Y., Aso, N., Onuki, Y. (1997) Coupling between Magnetic and Superconducting Order Parameters and Evidence for the spin excitation gap in superconducting state of a heavy fermion superconductor UPd_2Al_3, *J Phys. Soc. Japan* **66,** 2560-2563.
12. Marshall, W. and Lovesey, S.W. (1971) *Theory of Thermal Neutron Scattering,* Oxford University Press, Oxford.
13. Bernhoeft, N. and Lonzarich, G.G. (1995) Scattering of slow neutrons from long wavelength magnetic fluctuations in UPt_3, *J. Phys. Condens. Matter* **7,** 7325-7333.
14. Bernhoeft, N. (1997) An Introduction to Magnetisation Density Fluctuations in Charged Fermi Liquids, Zuoz Summer School, (ed. A Furrer) ISSN 1019-6447, Villigen, Switzerland, p 73-87.
15. Sato, N., Sakon, T., Takeda, N., Komatsubara, T., Geibel, C., Steglich, F. (1992) Anisotropy in a Heavy Fermion Superconductor: UPd_2Al_3, *J. Phys. Soc. Japan.* **61,** 32-34 .
16. Sato, N., Inada, Y., Sakon, T., Imamura, K., Ishiguro, A., Kimura, J., Swada, A., Komatsubara, T.; (1994) Magnetic, Electrical and Thermal Properties of Heavy Fermion Superconductor UPd_2Al_3, *IEEE Trans. Magn.* **30,** 1145-1147.
17. Lonzarich, G.G., Bernhoeft, N., McK Paul, D. (1989) Spin density fluctuations in magnetic metals, *Physica* **B156-157** 699-705.
18. Bernhoeft N, Lonzarich GG, Mitchell PW, McK Paul D; (1983) Magnetic excitations in Ni_3Al at low energies and long wavelengths, *Phys. Rev.* **B28,** 422-424.
19. Bernhoeft, N., Lonzarich, G.G., McK Paul, D., Mitchell, P.W.; (1986) Magnetic fluctuation spectra of very weak itinerant ferromagnets, *Physica* **136B,** 443-446.
20. Bernhoeft, N., Hayden, S.M., Lonzarich, G.G., McK Paul, D., Lindley, E.J. (1989) Dispersive magnetic density fluctuations in Ni_3Ga, *Phys. Rev. Lett.* **62,** 657-660.
21. White, R.M. (1983) *Quantum Theory of Magnetism,* v.**32** Solid State Sciences, Springer, Berlin.
22. Edwards, D.M. and Lonzarich, G.G. (1992) The entropy of fluctuating moments at low temperatures, *Philosophical Magazine* **B65,** 1185-1189.

23. Lonzarich, G.G. (1986) The magnetic equation of state and heat capacity in weak itinerant ferromagnets, *J. Magn. Magn. Mat.* **54,** 612-616.
24. Moriya, T. and Takimoto, T. (1995) Anomalous Properties around Magnetic instabilities in heavy electron systems, *J. Phys. Soc. Japan.* **64,** 960-969.
25. Solontsov, A. and Lacroix, C. (1997) Specfic heat of soft-mode spin fluctuations in itinerant electron magnets, *Physics Lett.* **A224,** 298-302.

ANISOTROPY OF THE GENERALIZED SUSCEPTIBILITY IN Mn(38%Ni) ALLOY IN THE MAGNETIC PHASE TRANSITION REGION

J.J. MILCZAREK, J. JANKOWSKA-KISIELIŃSKA AND K. MIKKE
Institute of Atomic Energy,
05-400 Otwock, Świerk, POLAND
AND
B. HENNION
Laboratoire Leon Brillouin,
CE-Saclay, 91191 Gif sur Yvette Cedex, FRANCE

Abstract. The critical inelastic neutron scattering was investigated for the first time in the Mn(38%Ni) alloy. The intensity distribution determined in the (001) plane around the (100) rlp confirmed the large, independent of temperature, anisotropy of the transverse component of the static generalized susceptibility. The spectral width of the dynamic part of generalized susceptibility is anisotropic in q - space. The temperature dependence of the spectral width at q = 0 is characteristic for the continuous phase transitions.

The phase diagrams of γ-Mn alloys offer a variety of structural and magnetic phase transitions of both 1st and 2nd order types [1,2]. In alloys of high manganese concentration c_{Mn} the phase transition is a 1st order magneto-structural transformation but with decreasing c_{Mn} it becomes a 2nd order purely magnetic phase transition from paramagnetic to antiferromagnetic phase without any change of the fcc lattice structure. Recent theories relate all those transformations to the stability of multiple spin density waves in the itinerant electron systems [3,4].

The previous studies [5-7] of the magnetic phase transition in the fcc γ-Mn-Fe and γ-Mn-Ni alloys revealed a large directional anisotropy of the transverse spin-correlation lengths. This effect is similar to that found in USb [8] for longitudinal spin-correlation lengths. It is interesting that both effects were observed in cubic systems.

The aim of the present work was to determine the anisotropy and dynamics of the transverse and longitudinal components of the generalized susceptibility near the Néel temperature for the Mn(38%Ni) alloy.

The Mn(38%Ni) alloy undergoes the continuous P-AF1 phase transition at 400 K in the fcc lattice. The critical exponent for the sublattice magnetization is close to 0.4. The transverse fluctuations were found to increase significantly at the transition point although their correlation lengths were finite at the Néel temperature. It was found that the correlation length in the direction parallel to the [001] axis was two times larger than the ones in the perpendicular directions.

Neutron scattering allows the investigation of both the parallel $\chi^{zz}(\mathbf{q})$ and transverse $\chi^{xx}(\mathbf{q})$ (with respect to the [001] direction) components of the generalized susceptibility tensor.

The investigated sample was in the form of a cube with the volume of 1 cm^3. The neutron scattering measurements were performed on the triple axis neutron spectrometer at the ORPHÉE reactor in LLB CEN, Saclay. The measurements were done with the constant energy of scattered neutrons (E_f = 14.68 meV). Pyrolytic graphite was used as the monochromator and the analyzer. The horizontal collimations were 26'-30'-10'-10', starting from the reactor core, and vertical collimations were 70'-103'-155'-410'.

The measurements with a zero energy transfer were performed in the vicinity of the (100) and (110) rlp in the (001) and (1,-1,0) planes. The intensity distribution determined in the (001) plane around the (100) rlp confirmed large and independent of temperature anisotropy of the transverse component of the static generalized susceptibility. The intensity distribution measured in the same plane in the vicinity of the (110) rlp did not reveal any significant departures from the isotropic features i.e. the constant intensity contours remain circular. The measurements performed in the (1,-1,0) plane around both the (001) and (110) rlp have shown that the correlation length in the [001] direction is larger than the one in the [110] direction.

These results indicate that the anisotropy axis is parallel to the direction of the wave vector of the spin density wave, which according to the current theories [3,4], describes the magnetic structure of the alloy.

For the first time the critical inelastic neutron scattering was investigated in the Mn(38%Ni) alloy. The observed effect was discussed in terms of the anisotropic static generalized susceptibility and anisotropic component of the dynamic susceptibility:

$$\chi^{xx}(\mathbf{q},\omega) \propto \frac{M}{1+\left(\xi_\parallel q_z\right)^2 + \xi_\perp^2\left(q_x^2 + q_y^2\right)} \cdot \frac{\Gamma(q)}{\Gamma^2(q)+\omega^2}, \quad (1)$$

where

$$\Gamma(q) = \Gamma_o\left(1+(\xi_\parallel q_z)^2 + \xi_\perp^2(q_x^2 + q_y^2)\right) \quad (2)$$

To determine $\xi_\parallel$, $\xi_\perp$, and Γ_o the fitting procedure was used in which the scattering cross-section of the form given by Eq.(1) was convoluted with the instrument resolution function. The experimental data were fitted with two Lorentzian components of the form given by Eq. (1), one of which is dominant. The background and elastic incoherent neutron scattering were also incorporated into the fitting procedure. The examples of the fitted results are presented in Fig.1.

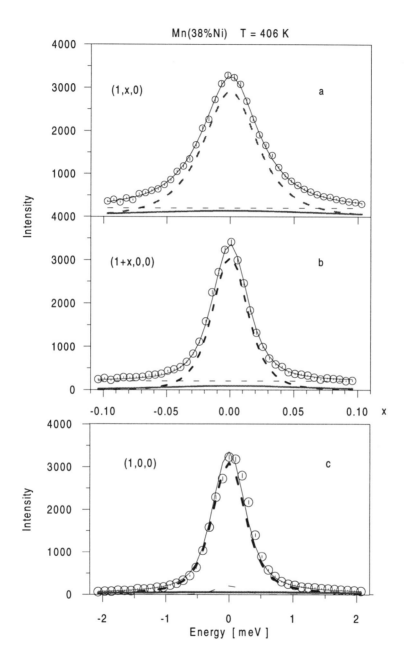

Figure 1. The neutron scattering data in the vicinity of (100) rlp at 406 K. The thick dashed line represents the main Lorentzian component, the thick solid line is the second Lorentzian component and thin dashed line represents the background and incoherent scattering. The thin solid line represents the sum of all of the components. (a) and (b) are the intensity distributions for elastic scattering in the direction perpendicular and parallel to [100], respectively. (c) is the intensity distribution vs energy transfer at (100).

The dependence of the spectral width Γ on q appears to be less anisotropic than assumed in Eq.(2). The temperature dependence of Γ_0 is characteristic for the continuous phase transitions (see Fig. 2).

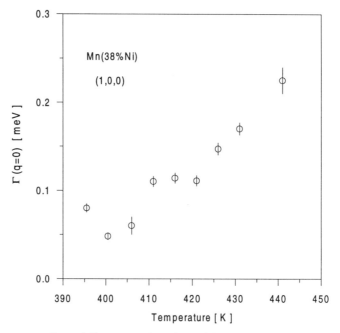

Figure 2. Temperature dependence of the spectral width $\Gamma(q=0)$.

The static parts of the generalized susceptibility are anisotropic in the way qualitatively similar to the ones calculated for 4f systems [9]. It is difficult to indicate the origin of the anisotropy in manganese alloys. The system possesses full crystal and magnetic cubic symmetry. The magnetic ordering is induced by the direct exchange interaction between 3d electrons. There is no reason for expectation of significant spin - orbit coupling or quadrupolar interaction since there is no evidence for non-zero quadrupolar moments. Nevertheless, quadrupolar interactions and magneto-elastic coupling between strains and quadrupoles were used to calculate the spin wave energy gap it in the localized model of γ- Mn [10].

Acknowledgment

This work was partly supported by the 'Human Capital and Mobility - Access to Large Scale Facilities PECO Extension' Program (Contract N° ERB CIPD CT 940080).

References

1. Vintaikin, E.Z., Udovenko, V.A., and Gogua, L.D., (1978) Low temperature fcc-fct transformation in γ Mn-Fe alloys, *Izv. Vys. Uch. Zav. Fizika* **7**, 146 (in Russian).
2. Honda, N., Tanji, Y. and Nakagawa, Y. (1976) Lattice distortion and elastic properties of antiferromagnetic γ Mn-Ni alloys, *J. Phys. Soc. Jap.* **41**, 1931-1937.
3. Jo, T. and Hirai, K. (1986) Lattice distortion and multiple spin density wave state in γ-Mn alloys, *J. Phys. Soc. Jap.* **55**, 2017-2023.
4. Long, M.W. and Yeung, W. (1987) Spin - orientation phase transition in itinerant multiple spin - density - wave systems, *J. Phys. C: Solid State Phys*. **20**, 5839-5866.
5. Ishikawa, Y., Endoh. Y., and Ikeda, S. (1973) Magnetic critical scattering from an itinerant antiferromagnet of $\gamma Fe_{0.5}Mn_{0.5}$ alloy I. Quasielastic scattering, *J. Phys. Soc. Jap.* **35**, 1616-1626.
6. Milczarek, J.J., Mikke, K., and Jaworska, E. (1988) Transverse spin fluctuations in γ - Mn(37%Fe) alloy, *J. de Physique* **49**, C8 - 183 - 184.
7. Milczarek, J.J., Jaworska, E., and Mikke, K. (1989) Magnetic critical scattering in γ-Mn alloys, *Physica* **B 156-157**, 238-240.
8. Lander, G.H., Sinha, S.K., Sparlin D.M., and Vogt, G. (1978) Spin correlations in actinide materials: a neutron study of USb, *Phys. Rev. Lett.* **40**, 523-526.
9. Hälg, B. and Furrer, A. (1986) Anisotropic exchange and spin dynamics in the type-I (-IA) antiferromagnets CeAs, CeSb, and USb: a neutron study, *Phys. Rev.* **B34**, 6258-6279.
10. Holden, T.M., Mikke, K., Fawcett, E., and Fernandez-Baca, J. (1992) The magnetism of metallic manganese alloys, *Proc. Int. Conf. Physics of Transition Metals*, World Scientific, Singapore pp. 644-650.

PSEUDODIPOLAR INTERACTION AND ANTIFERROMAGNETISM OF R_2CuO_4 COMPOUNDS (R=PR,ND, SM AND EU)

S.V. MALEYEV
Petersburg Nuclear Physics Institute
Gatchina, St.Petersburg 188350, Russia

D. PETITGRAND AND PH.BOURGES
Laboratoiré Léon Brillouin
CE-Saclay 91191 Gif-sur-Yvette, Cedex, France

AND

A.S.IVANOV
Institute Laue-Langevin
B.P. 156. F-38042 Grenoble Cédex 9, France

Abstract. It is shown that the pseudodipolar interaction allows to explain static and dynamic properties of non-collinear antiferromagnets R_2CuO_4 in the temperature range above the rare-earth (RE) ordering temperature. Experimental data for inelastic neutron scattering in Pr_2CuO_4 are presented. Parameters of the pseudodipolar interactions are determined. Comparison of the spin-wave neutron scattering in Pr_2CuO_4 and Nd_2CuO_4 reveals a strong dependence of the in-plane anisotropy on the type of the RE ion. The ground-state energy in magnetic field applied parallel to the CuO_2 planes is derived and analyzed. In particular it is shown that the spin-flop transition is of second order if the field is applied along the [1,1,0] direction.

1. Introduction

During the last decade magnetic properties of dielectric cuprates, which are parent compounds for the High-T_c superconductors, were extensively investigated. In particularly, compounds R_2CuO_4 (R=Pr,Nd,Sm and Eu) were studied [1]–[5]. In these materials as well as in other insulating cuprates La_2CuO_4 and $YBa_2Cu_3O_{6+x}$ the long-range antiferromagnetic order sta-

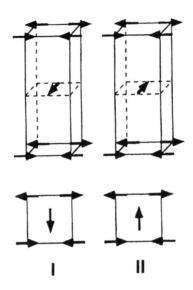

Figure 1. Two observed non-collinear magnetic structures: I (La$_2$NiO$_4$ type) left; Pr, Nd (I,III), Sm and II (La$_2$CuO$_4$ type) right; Eu, Nd (II), Sr$_2$CuO$_2$Cl$_2$ and their projections on the basal plane. For Nd$_2$CuO$_4$ phases I, II and III are observed at $T_N \simeq 250K > T > 80K$, $80K > T > 30K$ and $30K > T$, respectively.

bilizes below $T_N \sim 250 - 400K$ due to very strong in-plane exchange interaction between Cu^{2+} $S = 1/2$ spins and weak interplane coupling. The easy-plane anisotropy retains spins in CuO$_2$ planes. However, some structural peculiarities distinguish the R$_2$CuO$_4$ family as well as Sr$_2$CuO$_2$Cl$_2$ [6] from other antiferromagnetic cuprates. For the following it is important that R$_2$CuO$_4$ materials have a tetragonal structure where Cu^{2+} ions form a body-centered lattice. As a result in the antiferromagnetic state Cu^{2+} ions in the adjacent CuO$_2$ planes do not interact in the mean field approximation if one assumes conventional isotropic exchange coupling. Thus some weak interactions may manifest themselves, which are masked in other cuprates. This suggestion is confirmed by non-collinearity of the magnetic structure revealed by neutron scattering in a magnetic field [1]–[5] (see Fig.1). In Ref. [1] the non-collinearity has been attributed to biquadratic exchange. However this exchange gives equal contributions to the ground state energy of both structures shown in Fig.1. Hence, it should be some other interaction which distinguishes them. We suggest that it is the pseudodipolar interaction (PD).

The magnetic moment of the rare-earth (RE) ions is the second peculiarity of the R$_2$CuO$_4$ family. Recently in a careful review [7] it was shown that the magnetic properties of the material strongly depend on the type of the RE ion. In the case of Pr_2CuO_4 the ions Pr^{3+} have a nonmagnetic singlet ground state and a first excited state which is a doublet at 18 meV [1]. Hence, well below the Neel temperature $T_N \simeq 280K$ the Pr^{3+} momenta are slightly polarized due to the interaction with the ordered spins of the Cu^{2+} ions and remain inactivated thermally [1]. As a result at low T they give some temperature independent contribution to the effective Cu–Cu interaction.

In the case of Nd^{3+} and Sm^{3+} ions the ground state is a Kramers doublet which is splitted slightly due to the interaction with the copper spins. Neglecting this splitting one has the single-ion susceptibility in the form $X = C/T$. In the case of Nd_2CuO_4 this susceptibility is responsible for two spin reorientational transitions at $T_1 \simeq 80K$ and $T_2 \simeq 30K$ [7, 8]. There are also RE momenta ordering transitions at 1.5K and 6K for Nd_2CuO_4 and Sm_2CuO_4, respectively. Moreover in the case of Sm_2CuO_4 there is some additional transition at 20K which waits for further investigations [4, 7]. $Eu_2CuO structure. The Cu^{3+}$ ion has $J = 0$ and does not contribute to the low-temperature magnetism of the compound. Unfortunately, at $T = 140K$ there is a structural phase transition [9] and the spin ordering below this temperature has not been determined yet.

This paper is devoted to the low-temperature properties of the Cu subsystem above the RE ordering temperature. We assume that in this temperature range the influence of the RE ions on the Cu^{2+} spins may be properly taken into account by a renormalization of weak Cu–Cu interactions such as the easy-plane anisotropy and parameters of the pseudodipolar (PD) Hamiltonian. Hence, for Pr_2CuO_4 our theoretical consideration should be applicable in the whole temperature range well below $T_N \simeq 280K$ with temperature independent parameters. In the case of Nd_2CuO_4 the theory should hold for $T > 1.5K$ and according to [7] the parameters strongly depend on T. For Sm_2CuO_4 the temperature range of the theory is restricted to $T > 20K$. The theory is hardly applicable to Eu_2CuO_4 due to the above mentioned phase transition at $T = 140K$. We note also that low temperature spin-waves in Nd subsystem below 1.5K have been studied recently (see [7] and references therein).

The non-collinearity may be easily explained if one assumes a dependence of the interaction on the direction of the bounds connecting two Cu^{2+} spins (see Sec.2 and Ref.[7]). This interaction may be represented as the (PD) interaction proposed by Van-Vleck in 1937 which has the following form

$$V_{pd} = \frac{1}{2} \sum_{\ell\ell'} V(\mathbf{R}_{\ell\ell'}) S_{\ell'}^{\beta} \hat{R}_{\ell\ell'}^{\beta} \hat{R}_{\ell\ell'}^{\alpha} S_{\ell}^{\beta} , \qquad (1)$$

where ℓ and ℓ' label lattice sites and the function $V(R)$ decreases faster than R^{-3} as $R \to \infty$.

For the first time the PD interaction between Cu^{2+} and Nd^{3+} ions has been considered in [11]. Recently it has been used for the complete description of the spin structure of the R_2CuO_4 family [7]. In particular, it has been shown that it is responsible for the non-collinear structure of Cu as well as RE subsystems. It was used also for the explanation of the weak ferromagnetism of the tetragonal compound $Sr_2Cu_3O_4Cl_2$ [12]. A microscopic derivation of the PD interaction between the Cu spins in CuO_2 planes has been presented in [13, 14] in the frame of the Hubbard model using on-site Coulomb and spin-orbit interactions. It should be noted also that the PD interaction appears in metals as a result of skew scattering [15] which is a consequence of the ordinary exchange interaction between ions with $L \neq 0$ and the conducting band [16]. The result may be easily generalized for the interaction of d and f electrons via oxygen $2p$ orbitals.

In this paper we develop a spin-wave theory of the non-collinear cuprates above the RE ordering temperature taking into account both interplane and intraplane pseudodipolar interactions. We demonstrate that the former splits the in-plane spin-wave spectrum into acoustic and optical branches. The interplane PD interaction gives rise to a fourth-fold (square) anisotropy [12, 13]. As a result an in-plane spin-wave gap appears due to the interaction between spin-waves. This gap has been obtained in [14] using a phenomenological Hamiltonian. We confirm this result by microscopic calculations and determine its dependence on the sublattice orientation in the (ab) plane. This dependence is important for the determination of the spin configuration in the magnetic field.

Then we present detailed neutron scattering data for Pr_2CuO_4. We observed the acoustic and optical spin-wave branches and the in-plane and out-of-plane spin-wave gaps. From these data we determine the parameters of the PD interaction. In particular, the interplane PD interaction appears to be by on order of magnitude larger than calculated in [14]. Comparing this result with the strong T dependence of the gaps in Nd_2CuO_4 observed previously [10] we conclude that along with the PD interaction and the anisotropic exchange considered in [14] there are additional contributions to the square and uniaxial anisotropies connected with the RE ions.

The spin orientation in a magnetic field, experimentally studied in [1]–[5] is the next problem considered in the paper. We derive the ground state energy, which determines this orientation, for a magnetic field arbitrarily directed in the CuO_2 planes. In particular, we show that if the field is along the [1,1,0] direction the spin-flop transition is of second order and the following relation holds

$$g\mu H_c = \Delta_0 , \qquad (2)$$

where H_c is the critical field and Δ_0 is the in-plane spin-wave gap at $H = 0$. Using this equation we analyze the data of Ref. [1] and confirm the large value of Δ_0 obtained also in our inelastic scattering experiments. The spin-wave spectrum in a magnetic field is considered too. We show that at the spin-flop transition one of the branches becomes gapless.

This paper is a short version of a review which will be published elsewhere [17].

2. Spin Waves and Anisotropy

We represent the Hamiltonian of the Cu^{2+} spin system in the following form

$$H = H_{ex} + H_A + H_{PD} + g\mu \mathbf{H} \sum_{n\ell} \mathbf{S}_\ell , \qquad (3)$$

where H_{ex} is the isotropic exchange interaction, H_A is the uniaxial anisotropy, H_{PD} is the pseudodipolar interaction and the last term is the interaction with the external magnetic field, and $g\mu > 0$.

Below we consider the interaction between nearest neighbors only. In this case H_{ex} can be written as

$$\begin{aligned} H_{ex} &= \frac{1}{2} \sum_{\ell,\ell',n} J_{\ell\ell'} \mathbf{S}_{\ell n} \mathbf{S}_{\ell' n} - \frac{1}{2} \sum_{\ell n n'} I_{nn'} \mathbf{S}_{\ell n} \mathbf{S}_{\ell n'} \\ &+ \frac{1}{2} \sum_{\ell,\ell_1,n} M_{\ell,\ell_1} \mathbf{S}_{\ell,n} \mathbf{S}_{\ell_1,n} , \end{aligned} \qquad (4)$$

where ℓ and ℓ' denote sites in the CuO_2 planes and n denotes planes; $J_{\ell\ell'}$, $I_{nn'}$ and $M_{\ell\ell_1}$ are the nearest-neighbor exchange interactions in the CuO_2 planes along the c axis and between Cu^{2+} ions in adjacent planes, respectively. In particular, the index ℓ_1 in the last term of Eq.(4) is equal to $\ell + (\pm a, \pm b, \pm c)/2$.

The uniaxial anisotropy in Eq.(3) we consider as a part of the anisotropic exchange and represent it as

$$H_A = -\frac{1}{2} \sum_{\ell,\ell',n} A_{\ell\ell'} S^c_{\ell n} S^c_{\ell' n} , \qquad (5)$$

and the easy-plane anisotropy takes place if $A_{\ell\ell'} > 0$.

It is convenient to subdivide the PD interaction into interplane and intraplane parts. For the former we have

$$H_I = \frac{1}{2} \sum_{\ell,\ell_1,n} Q_{\ell\ell_1} (\mathbf{S}_{\ell,n} \hat{R}_{\ell\ell_1})(\mathbf{S}_{\ell_1,n} \hat{R}_{\ell\ell_1}) . \qquad (6)$$

Slightly modifying Eq.(1) we represent the intraplane PD interaction in the traceless form as in [14]:

$$H_P = \frac{1}{2} \sum_{\ell,\ell'} P_{\ell\ell'} \left[(\mathbf{S}_\ell \hat{R}_{\ell\ell'})(\mathbf{S}_{\ell'} \hat{R}_{\ell\ell'}) - \frac{1}{2}(S_\ell^a S_{\ell'}^a + S_\ell^b S_{\ell'}^b) \right], \tag{7}$$

where a and b are the coordinate axes in the CuO_2 plane. The interaction (7) may be rewritten in the form (1) by redefining both interactions $J_{\ell\ell'}$ and $A_{\ell\ell'}$.

Using conventional approximations and neglecting PD interaction between adjacent planes we obtain the following expressions for the in-plane and out-of-plane spin-wave branches

$$\epsilon_{\mathbf{k},in\,1,2}^2 = S^2(J_0 - J_\mathbf{k} + I_0 - I_\mathbf{k} + \gamma_\mathbf{k} \cos 2\varphi_{1,2})(J_0 + J_\mathbf{k}),$$
$$\epsilon_{\mathbf{k}\,out}^2 = S^2(J_0 - J_\mathbf{k} + I_0 - I_\mathbf{k} + A_\mathbf{k})(J_0 + J_\mathbf{k}), \tag{8}$$

where $\gamma_\mathbf{k} = P(\cos k_a - \cos k_b)$, φ_{12} are angles which determine the spin orientation in neighboring planes, $\mathbf{k}$ belongs to the magnetic zone, and we neglect $I_0, A_\mathbf{k}$ and P in comparison with $J_0 + J_\mathbf{k}$. These expressions coincide with those given in [14] if we put $J = J_{av}$, $A = J_{av} - J_z$, $P = -\delta J$ and neglect $I_\mathbf{k}$. We see that the in-plane branch remains gapless in spite of the fact that the PD interaction violates the rotational invariance around the c-axis. As we will show below a corresponding gap appears as a result of the interaction between spin-waves. The out-of-plane gap has the form

$$\Delta_{out}^2 = 8S^2 J_0 A = 32 S^2 J A. \tag{9}$$

We now rederive the φ-dependent contribution to the ground-state energy which gives rise to the square anisotropy considered in Ref. [14]. This anisotropy is responsible for the spin-wave gap in the in-plane mode. In [14] this gap has been calculated using an effective Hamiltonian. We demonstrate that it appears as a result of the interaction between spin-waves and determine the φ-dependence of the gap, which is important in the case of a non-zero magnetic field.

According to [13] the first correction to the ground state energy is given by zero-point motion and has the form

$$\Delta E = \frac{1}{2} \sum_\mathbf{k} \varepsilon_\mathbf{k}, \tag{10}$$

where the sum is extended over the whole Brillouin zone, $\varepsilon_\mathbf{k}$ is given by (8), and we put $I_\mathbf{k} = A_\mathbf{k} = 0$. As a result we get the φ-dependent part of this energy in the following form

$$\Delta E_\varphi = \frac{S N \Delta_0^2}{16 J_0} \sin^2 2\varphi, \tag{11}$$

where

$$\Delta_0^2 = \frac{2SP^2}{(2\pi)^2} \int_{-\pi}^{\pi} dk_a dk_b \frac{(2+\cos k_a + \cos k_b)^{1/2}}{(2-\cos k_a - \cos k_b)^{3/2}}$$
$$\times (\cos k_a - \cos k_b)^2 \simeq 1.28 \, SP^2 \, . \tag{12}$$

These expressions coincide with the results of Ref. [13].

Using Eq.(11) and taking into account the classical contribution to the ground-state energy from the interplane PD interaction we obtain the energy which determines the ground state configuration of the system

$$E_0 = S^2 N J_0 \left[\frac{\Delta_0^2}{4(2SJ_0)^2} (\sin^2 2\varphi_1 + \sin^2 2\varphi_2) \right.$$
$$\left. + \frac{8Q_0}{J_0} \sin(\varphi_1 + \varphi_2) \right] , \tag{13}$$

where $Q_0 = Q a^2/(2a^2 + C^2)$ and Q is the interplane PD interaction. For $Q > 0$ ($Q < 0$) this energy has a minimum at $\varphi_1 = 0$ and $\varphi_2 = -\pi/2$ ($\varphi_2 = \pi/2$), and we obtain the first (second) type of structure shown in Fig.1.

From Eq.(13) one sees that the PD interaction gives rise to the square anisotropy which should be responsible for the gap in the in-plane spin wave mode. In Ref. [14] this gap has been determined using a specially constructed effective Hamiltonian. It may be shown that the lowest order perturbation theory gives the same result and also allows to determine the φ dependence of the in-plane gap which is necessary for the consideration of the system in a magnetic field. The final result has the form

$$\Delta_{in}^2 = \Delta_0^2 \cos^4 \varphi \, . \tag{14}$$

For $\varphi = 0$ this expression coincides with the result of Ref. [14]. However, as we will see below the φ-dependence of the in-plane gap becomes important in a magnetic field.

The PD interaction between adjacent planes is very small and has to be taken into account along the $(0, 0, k_c)$ direction only. Otherwise it is masked by the strong dispersion in the ab plane connected with the huge intraplane exchange. As a result, we have two in-plane branches along this direction:

$$\epsilon_{ac}^2 = 2SJ_0 \left[2SI(1 - \cos k_c) + R_0 \left(1 - \cos \frac{k_c}{2}\right) \right] + \Delta_0^2,$$
$$\epsilon_{opt}^2 = 2SJ_0 \left[2SI(1 - \cos k_c) + R_0 \left(1 + \cos \frac{k_c}{2}\right) \right] + \Delta_0^2, \tag{15}$$

$$\Delta_{out}^2 = 8\langle S\rangle^2 J_0 A + 2SJ_0 R_0, \tag{16}$$

where $R_0 = 8|Q_0|$ and $\langle S\rangle$ is the average value of the Cu^{2+} spin. It should be emphasized that the φ dependence of the in-plane gap given by Eq.(14) is not a specific result of the PD interaction (7). In fact it is a direct consequence of the square symmetry of the CuO_2 planes.

3. Experimental Results

In this section we present results of the inelastic neutron scattering experiments in Pr_2CuO_4. We will determine from them principal parameters of the theory: the strength of the intra-plane (P) and inter-plane (Q) pseudodipolar interactions. Available experimental results for the inelastic scattering in Nd_2CuO_4 will be discussed too.

The Pr_2CuO_4 single crystal of very good mosaicity (less than 10') with $T_N = 252K$ was grown in air from the melt in a crucible. The sample, already utilized in higher energies spin-waves experiments [10], was a thin plate of about a volume of 0.5 cm^3. It has been mounted with the reciprocal directions (110) and (001) within the scattering plane. Inelastic neutron scattering experiments were performed using the triple-axis spectrometers at the thermal neutron beam 2T and at the cold neutron beam 4F1 of Laboratoire Léon Brillouin at Orphée (Saclay). Monochromator and analyzer were of pyrolytic graphite. The analyzer was focussed horizontally on 4F1, and horizontally and vertically on 2T to improve the intensity using collimations. Using the thermal beam, we worked in the constant-k_F mode with $k_F = 2.662$ $^{-1}$ leading to a typical energy resolution of 1 meV. A graphite filter was put on the scattered beam to remove neutrons with wave vectors $2k_F$. Cold neutron experiments were performed on the spectrometer 4F1 in the range 0.1-4.0 meV. We used both the constant-k_F and constant-k_I modes with $k_{I,F} = 1.55$ $^{-1}$ for the higher energy part and the constant-k_F mode at $k_F = 1.15$ $^{-1}$ for the lowest energies investigated. This gives a typical energy resolution (full width at half maximum) of 0.2 and 0.1 meV, respectively. Multiple order contamination in the scattered beam was removed by a beryllium filter and the results were corrected for efficiency and influence of multiple order on monitor as a function of k_I.

The strong in-plane superexchange coupling in cuprates ($J > 100$ meV) put severe constraints on the neutron scattering methods implying a very large in-plane spin-wave velocity, c_0. The key point is that the in-plane q-resolution is larger than the in-plane wave vector of the spin-waves $q_{SW} = \omega/c_0$. That is the case for the whole energy range investigated (0.1-12 meV) here. The separation of the counter propagating spin-waves (from q and $-q$) in the (q_a, q_b) plane may be done in the high-energy region only (see Ref.[1]). Accordingly, in our case spin-wave branches do not show two individual peaks in an in-plane q-scan at a constant energy. The measured intensity for such q-scans looks as a single peak of roughly the q-width of the resolution centered on the magnetic rod. Such sharp peaks around the AF wave vector point to the magnetic origin of the scattering. Another consequence is that the scattered intensity measured at the antiferromagnetic line, $\mathbf{Q} = (1/2, 1/2, q_c)$, is not proportional to the spin-wave cross section. Indeed, the ellipsoid of the triple axis spectrometer performs a $2D$ integration over in-plane (q_a, q_b) wave vectors. The measured intensity then corresponds to a $2D$ integrated spin susceptibility,

$$\bar{S}(\mathbf{Q}, \omega) = \int dq q \, R_{ab}(q) S(Q, \omega) , \qquad (17)$$

where R_{ab} is the resolution function in the ab plane.

Since the neutron cross sections always behaves in an antiferromagnet as $\epsilon_q^{-1} \propto 1/q$ for low energies, the $2D$ integration gives a result which is essentially energy independent apart from the temperature factor, $[1 - \exp(-\omega/k_B T)]^{-1}$. Therefore, the measured intensity in an energy scan at a constant wave vector would increase when the energy enters the spin-wave sheet and would remain constant multiplied by the temperature factor mentioned above. The existence of a spin-wave branch having a minimum energy is thus detected as a step (see Fig.2). This technique has been applied to investigate both the out-of-plane and in-plane branches.

We now deduce magnetic parameters: i) the pseudo-dipolar intraplane interaction P responsible for the in-plane gap, ii) the pseudo-dipolar interplane interaction Q responsible for the dispersion along c^* axis, iii) the planar XY anisotropy A related to the out-of-plane gap. Fig.3 resumes the whole dispersion scheme along the c^* axis within the first magnetic Brillouin zone. The two in-plane components and the twice degenerated out-of-plane components are represented. The out-of-plane branch does not depend on q_c and is located at the out-of-plane gap which is found to be $\Delta_{out} = 8 \pm 0.5$ meV at 10K. It has a larger value than the one previously reported [1]. However our data are got with much better statistical accuracy than in [1] and thus we believe that they are more reliable. The in-plane spin component exhibit a clear dispersion along c^*. Lines in Fig.3 correspond to the calculated dispersion on the basis of Eq.(15) with $I = 0$. It is clear, that the description in terms of pseudo-dipolar interactions alone reproduces very accurately the observed dispersion.

From the measurements at integer value of q_c we obtain $\Delta_0 = 0.36 \pm 0.03$ meV for the small gap of the in-plane excitations at $T = 18$K. Applying Eq.(12) for $S = 1/2$, we obtain $P_0 = 0.45 \pm 0.04$ meV for the intraplane pseudo-dipolar interaction. This value is by one order of magnitude larger than the value theoretically estimated in Ref.[14]. However, it is confirmed

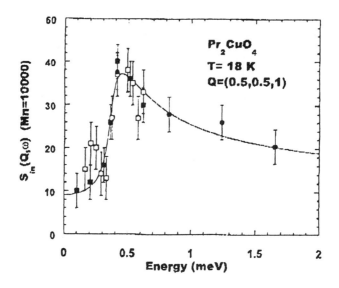

Figure 2. In-plane scattering functions at the AF wave vector $Q = (0.5, 0.5, 1)$. Measurements performed using different experimental conditions i) fixed $k_F = 1.15$ $^{-1}$ (squares) and ii) fixed $k_I = 1.55$ $^{-1}$ (circles) have been rescaled. Closed symbols have been obtained from q-scans. Open squares have been obtained from the energy scan at the AF wave vector where the background has been subtracted.

by an analysis of the spin-flops transition data of Ref.[1]. We discuss this problem in Sec.4.

According to Eqs. (15) and (16) all other spin-wave parameters depend on the strong in-plane nearest neighbor superexchange interaction J. We will put $J = 125$ meV.

From our experimental data and Eq.(15) the total amplitude for the dispersion is $\sqrt{16SJR_0} = 2.8 \pm 0.05$ meV, and for $S = 1/2$, one obtains $R_0 = 7.9$ meV. Then using the determination of R_0 given from Eq.(15) and values of lattice constants $a = 3.958$ and $c = 12.19$[1] we get the interplane pseudodipolar constant $Q = 0.023$ meV. This value is by an order of magnitude larger than corresponding dipolar magnetic interactions (see the last section). As we have shown above from the out-of-plane gap, $\Delta_{out} = 8 \pm 0.5$ meV, and Eq.(16) we get $A = 6.1 \cdot 10^{-2}(S/\langle S_t \rangle)^2$, and using experimental value of the ratio $\langle S_t \rangle/S = 0.8$ [1] we obtain $A = 9.5 \cdot 10^{-2}$ meV. This value of the easy-plane anisotropy is three times larger than that discussed in Refs. [13, 14].

We compare now these results for Pr_2CuO_4 with available data for Nd_2CuO_4 and point out the following striking differences. i) In $NdCuO_4$

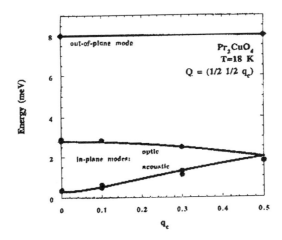

Figure 3. Spin-wave dispersion curves along the c^* direction in the non-collinear Pr_2CuO_2. Lines are fits from Eqs.(15) and (16) for the in-plane and the out-of-plane excitations, respectively.

there are two spin reorientation transitions at $T_1 \simeq 80K$ and $T_2 \simeq 30K$ [2, 3] (see Fig.1). It means that the interplane PD interaction Q changes sign twice. ii) Both in-plane and out-of-plane gaps are strongly T-dependent and the former is by an order of magnitude larger than in the case of Pr_2CuO_4 [10]. iii) The dispersion of the in-plane excitations along $(1/2, 1/2, q_c)$ direction has not been observed.

The first and the second peculiarities are related to the Kramers ground state of the Nd^{3+} ions and will be discussed below. The last may be explained by the large value of the in-plane gap. Indeed, in this case from Eq.(15) we obtain the following maximum value of the in-plane spin-wave energy

$$\epsilon_{in}^{(max)} = \Delta_0 + \frac{8SJ_0R_0}{\Delta_0} . \qquad (18)$$

According to Ref. [10] the minimum value of Δ_0 is approximately 3 meV and the experimental error is 0.25 meV. Hence, the dispersion cannot be visible if $(16SJR_0)^{1/2} \leq 1.2$ meV. This value is of the same order as determined above for Pr_2CuO_4.

4. Spin Configuration in Magnetic Field

In this Section we examine the spin configuration of the non-collinear AF R_2CuO_4 in the magnetic field applied in the basal plane at $T = 0$. We obtain the expression for the ground-state energy and analyze it in two practically important limiting cases studied in Refs. [1]–[5]: i) **H** along the [1,1,0] direction and ii) **H** at a small angle to the a (b) axis. We demonstrate that in the former case the spin-flop is a second order transition with the critical field given by Eq.(2).

The spin configuration in the magnetic field is shown in Fig.4. We obtain the ground-state energy of the system adding to Eq.(13) terms which describe the change of the energy connected with relative rotations of the sublattices at the angles ϑ_1 and ϑ_2 in the corresponding perpendicular magnetic fields. These terms are given by

$$\Delta E_{1,2} = SJ_0^2 N \left[\left(\vartheta_{1,2} + \frac{g\mu H_{\perp,1,2}}{2SJ_0} \right)^2 - \left(\frac{g\mu H_{\perp,1,2}}{2SJ_0} \right)^2 \right], \qquad (19)$$

where $H_{\perp,1,2} = H \sin(\psi - \varphi_{1,2})$. In equilibrium we have

$$\vartheta_{1,2} = -\vartheta_{01,2} = -\frac{g\mu H_{\perp,1,2}}{2SJ_0}$$

$$\Delta E_{1,2} = -2S^2 J_0 N \left(\frac{g\mu H_{\perp 1,2}}{2SJ_0} \right)^2. \qquad (20)$$

However, rotations of sublattices at the angles $\vartheta_{1,2}$ give rise to simultaneous rotations of corresponding y axis and an additional interaction appears between y spin components and $H_{\parallel 1,2} = H \cos(\psi - \varphi_{1,2})$. This interaction is linear in operators $(a_0 - a_0^+)/i$ and should be excluded.

As a result the ground state energy is represented by

$$\begin{aligned}
E &= S^2 N J_0 \left[\frac{\Delta_0^2}{16(SJ_0)^2} (\sin^2 2\varphi_1 + \sin^2 2\varphi_2) \right. \\
&\quad \left. - \left(\frac{g\mu H_{\perp,1}}{2SJ_0} \right)^2 - \left(\frac{g\mu H_{\perp,2}}{2SJ_0} \right)^2 - \frac{R_0}{SJ_0} \right] \\
&\quad - S^2 N J_0 \Big\{ (g\mu H_\parallel g\mu H_\perp)_1^2 \left[\Delta_{in,2}^2 + 2SJ_0 R_0 - (g\mu H_{\parallel,2})^2 \right] \\
&\quad + (g\mu H_\parallel g\mu H_\perp)_2^2 \left[\Delta_{in,1}^2 + 2SJ_0 R_0 - (g\mu H_{\parallel,1}^2 \right] \\
&\quad - 4SJ_0 R (g\mu H_\parallel g\mu H_\perp)_1 (g\mu H_\parallel g\mu H_\perp)_2 \Big\} \\
&\quad \times (2SJ_0)^{-2} \Big\{ \left[\Delta_{in,1}^2 + 2SJ_0 R_0 - (g\mu H_{\parallel,1})^2 \right]
\end{aligned} \qquad (21)$$

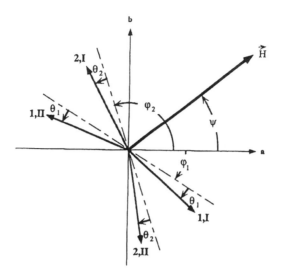

Figure 4. Spin configuration in the magnetic field. Indices (1I), (1II) and (2I), (2II) label neighboring spins I and II in the subsystems 1 and 2 respectively.

$$\times \ \left[\Delta_{in,2}^2 + 2SJ_0R_0 - (g\mu H_{\|,2})^2\right] - (2SJ_0R_0)^2\bigg\}^{-1},$$

where $R_0 = -8SQ_0 \sin(\varphi_1 + \varphi_2) > 0$ and

$$\Delta_{in,1,2}^2 = \Delta_0^2 \cos 4\varphi_{1,2} + (g\mu H_{\perp,1,2})^2. \tag{22}$$

Minimum of this energy as a function of φ_1 and φ_2 determines the ground-state spin configuration. If $g\mu H \ll \Delta_0$ the last term in Eq.(21) is small and may be neglected. However, it may be shown that the denominator of this term is a squared product of the in-plane spin-wave energies at $\mathbf{k} = 0$. As is well known the spin-flop transition takes place when one of the spin-wave branches becomes gapless. Hence, near the transition this denominator is small and the second term in Eq.(21) becomes essential.

If we put $R_0 = 0$ the energy E becomes a sum of two terms corresponding to independent subsystems 1 and 2. In particularly the ground state energy of conventional two-sublattice tetragonal antiferromagnets is determined by

$$E = S^2 N J_0 \left\{ \frac{\Delta_0^2}{16(SJ_0)^2} \sin^2 2\varphi - \left(\frac{g\mu H_\perp}{2SJ_0}\right)^2 \right.$$

$$-\frac{(g\mu H_\perp g\mu H_\|)^2}{(2SJ_0)^2[\Delta_{in}^2 - (g\mu H_\|)^2]}\right\}. \tag{23}$$

if $\hat{a}(\hat{b})$ is an easy axis. The same expression remains valid for $[1,1,0]$ easy direction if one changes sign before Δ_0^2 in Eqs.(22) and (23).

It should be noted also that Eq.(23) is applicable to the AF compound YBa$_2$Cu$_3$O$_{6+x}$ with antiferromagnetically coupled adjacent planes. Indeed in the magnetic field neighboring spins along the $\hat{c}$ axis rotate in opposite directions without violation their mutual antiparallel orientation. Hence, experimental studies similar to Refs.[1]–[5] provide a possibility to determine the magnetic anisotropy of the 1–2–3 compound as well as the value of the in-plane spin-wave gap using Eq.(2).

Let us consider now the ground-state spin configuration in two important limiting cases which have been studied experimentally [1]–[5].

1. The field is along the $[1,1,0]$ direction ($\psi = \pi/4$).
2. The field is applied at the angle $\psi \ll 1$ to the $[0,0,0]$ direction.

We begin with the case $\psi = \pi/4$. Due to the symmetry the spins of two subsystems rotate in opposite directions by the same angle $\varphi = \varphi_1$, the sum $\varphi_1 + \varphi_2$ remains unchanged and R_0 is a constant.

As a result, this energy differs by a factor two only from the energy of the collinear system given by Eq.(23). Hence, the subsequent analysis is applicable to both collinear and non-collinear systems. It is a result of a special symmetry of the problem at $\psi = \pi/4$.

For a small field one can neglect the second term in Eq.(23) and obtain

$$\sin 2\varphi = -\left(\frac{g\mu H}{\Delta_0}\right)^2. \tag{24}$$

This field dependence of $\sin 2\varphi$ has been used in [1] for the description of the low-temperature neutron scattering data in Pr$_2$CuO$_4$. From (24) we get also Eq.(2) for the critical field of the spin-flop transition. However, at $\sin 2\varphi = -1$ the denominator in the second term of Eq.(23) is equal to zero and the problem needs a more careful consideration. Near the spin-flop transition where $(\Delta_0 - g\mu H)/\Delta_0 \ll 1$ for the angle $\alpha = \pi/q + \varphi$ we get

$$\alpha = \left\{\frac{\Delta_0^2 - (g\mu H)^2}{2\Delta_0^2}\right\}^{1/4} \simeq \left(\frac{\Delta_0 - g\mu H}{\Delta_0}\right)^{1/4} \tag{25}$$

instead of $\alpha = \{[\Delta_0^2 - (g\mu H)^2]/(2\Delta_0)^2\}^{1/2} \simeq [(\Delta_0 - g\mu H)/\Delta_0]^{1/2}$ which follows from Eq.(24). This crossover from $\alpha \sim (\Delta_0 - g\mu H)^{1/2}$ to $\alpha \sim (\Delta_0 - g\mu H)^{1/4}$ behavior near the spin-flop transition has been observed recently in Pr$_2$CuO$_4$ [18] (see Fig.5).

We consider now the case of the field directed almost along the $\hat{a}$ axis, when $\psi \ll 1$. It is the case, studied experimentally in Ref.[3]. If $\psi = 0$ the second term in Eq.(21) disappears. However the spin-flop field may be determined from the condition of the positiveness of the denominator. As a result, for $\psi = 0$ we get

$$g\mu H_c = \left[\Delta_0^2(\Delta_0^2 + 4SJ_0R_0)\right]^{1/4} . \qquad (26)$$

Obviously the spin-flop transition is of the first order and if $H > H_c$ we have the state with all spins along the $\hat{b}$ direction.

If $\psi \ll 1$, for a small field we can neglect the second term in Eq.(21) and from the conditions $\partial E/\partial \varphi = \partial E/\partial \eta = 0$ we have

$$\varphi = -\frac{(g\mu H)^2(\Delta_0^2 + 3S^2 J_0 Q_0)\psi}{\Delta_0^2(\Delta_0^2 + 4S^2 J_0 Q_0)} \simeq -\frac{3(g\mu H)^2}{\Delta_0^2}\psi, \qquad (27)$$

$$\eta = \frac{(g\mu H)^2(\Delta_0^2 + 6S^2 J_0 Q_0)\psi}{2\Delta_0^2(\Delta_0^2 + 4S^2 J_0 Q_0)} \simeq \frac{3(g\mu H)^2}{\Delta_0^2}\psi , \qquad (28)$$

where $\varphi = \varphi_1$ and $\eta = \varphi_2 \pm \pi/2$. We see that φ and η are proportional to the angle ψ in agreement with the experimental data of Ref.[3], and $\varphi + \eta \simeq 0$ if $\Delta_0^2 \ll 2S^2 J_0 Q_0$.

With increasing ψ the spin-flop remains the first order transition and smoothly approaches the second order one at $\psi = \pi/4$.

Using Eq.(2) and the data of Ref.[1] one can determine the in-plane spin-wave gap for Pr_2CuO_4 independently from the inelastic scattering. Indeed from Fig.4 of Ref.[1] we obtain the critical field $H_c = 4.4T$ at 4.2K and $\Delta_0 = 0.51$ meV, if $g = 2$. This value is slightly larger than $\Delta_0 = 0.36 \pm 0.03$ meV given above. However from the same figure we get $H_c = 2.0T$ and $\Delta_0 = 0.23$ meV at $T = 25K$. In the case of linear T dependence of Δ_0 we obtain $\Delta_0 = 0.33$ meV in a satisfactory agreement with the inelastic scattering data.

However, this comparison is rather dubious due to the following reasons. i). The samples used in Ref.[1] and this study have different Neel temperatures: $T_N = 284K$ and $T_N = 252K$, respectively. ii) Preliminary inelastic scattering experiments demonstrate the T-independent in-plane gap in the wide range of T. iii) There are no clear theoretical reasons for the strong T-dependence of the in-plane gap in Pr_2CuO_4. Hence, we conclude that the data of Ref.[1] confirm the large value of the in-plane gap in comparison with the theoretical calculations of Ref..[14]. However, the observed T dependence of the critical field H_c remains unclear. Probably it is related to the temperature dependence of the g-factor. In this respect we wish to point out that in the sample with $T_N = 240K$ the reentrant behavior of H_c has been observed with a maximum at $T_m \simeq 5K$ [18].

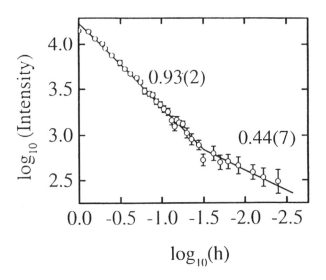

Figure 5. Logarithm of intensity of the antiferromagnetic Bragg scattering $(1/2, 1/2, 1)$ in Pr_2CuO_4 at $T = 4.5K$ as a function of the reduced magnetic field $h = (H - H_c)/H_c$. This intensity is proportional to $\sin^2 \varphi$. One can see a crossover to the low-h behaviour.

Concluding this section we present some results for the spin-wave energy in a magnetic field. As above, we consider two cases: $\psi = 0$ and $\psi = \pi/4$.

I. $\psi = 0$. Below the spin-flop transition where $g\mu H < \Delta_0$ the spin structure is determined by angles $\varphi_1 = 0$ and $\varphi_2 = \pm\pi/2$. In this case instead of (15) we have

$$\epsilon_\pm^2 = \Delta_0^2 + 2SJ_0R_0 + 4SJ_0I(1 - \cos k_c)$$
$$\pm \left\{ \left[2SJ_0R_0 \cos \frac{k_c}{2}\right]^2 + (g\mu H)^4 \right\}^{1/2}. \tag{29}$$

We see that now there is a gap at $k_c = \pi$ equal to $(g\mu H)^2(\Delta_0^2 + 2SJ_0R_0)^{-1/2}$ between acoustic and optical branches. At the same time the spin-flop field is determined by the condition $\epsilon_-^2(k_c = 0) = 0$ which coincides with Eq.(26). In the spin-flop phase we have $\varphi_1 = -\pi/2$, $\varphi_2 = \pm\pi/2$ and $R_0 = 0$, and we obtain the doubly-degenerate in-plane mode with the energy given by

$$\epsilon_{\mathbf{k}}^2 = \Delta_0^2 + (g\mu H)^2 + 4SJ_0I(1 - \cos k_c). \tag{30}$$

We see that now the spin-wave spectrum remains stable at all values of H, particularly at $H < H_c$. It means that a metastable state should exist which may be achieved if after the spin-flop transition one lowers the field below H_c.

II. $\psi = \pi/4$. In this case two subsystems rotate in opposite directions at the same angle $\varphi = -\pi/2 + \alpha$ and R_0 remains constant. Near the spin-flop transition where $g\mu H < \Delta_0$ we have

$$\begin{aligned}\epsilon_\pm^2 &= \left\{18\Delta_0^2[\Delta_0^2 - (g\mu H)^2]\right\}^{\frac{1}{2}} + 2SJ_0R_0\left[1 \pm \cos\frac{k_c}{2}\right] \\ &+ 4SJ_0I(1 - \cos k_c) \,.\end{aligned} \quad (31)$$

In the spin-flop phase $\varphi = -\pi/4$ and

$$\begin{aligned}\epsilon_\pm^2 &= (g\mu H)^2 - \Delta_0^2 + 2SJ_0R_0\left[1 \pm \cos\frac{k_c}{2}\right] \\ &+ 4SJ_0I(1 - \cos k_c) \,.\end{aligned} \quad (32)$$

Exactly in the spin-flop field, $g\mu H = \Delta_0$, the ϵ_- mode is a gapless one as it should be in the case of a second order transition. However due to critical fluctuations the field dependence of the spin-wave energies may deviate from the simple expressions given by Eqs.(30) and (31).

5. Results and Discussion

We investigate magnetic properties of non-collinear tetragonal antiferromagnets R_2CuO_4 with particular attention to Pr_2CuO_4 and Nd_2CuO_4. To conclude, we obtain the following principal results.

1. The non-collinear Cu^{2+} spin ordering below the Neel temperature is explained using the assumption of PD interaction between copper spins in CuO_2 planes as well as between the adjacent planes. The corresponding expression for the ground-state energy at $\mathbf{H} = 0$ is given by Eq.(13).

2. We demonstrate that the interplane PD interaction contribute to the out-of-plane spin-wave gap (16) and splits the in-plane spin-wave mode into acoustic and optical branches (15).

3. Interaction between spin-waves gives rise to the in-plane spin-wave gap which depends on the orientation of the sublattice magnetization in (ab) plane [see Eq.(14)]. This dependence is crucial for the determination of the spin orientation in the magnetic field. It may be shown that the φ dependence of the gap given by Eq.(14) is not a specific result connected with the PD interaction, but is a consequence of the square symmetry of CuO_2 planes.

4. Detailed inelastic neutron scattering data are presented for Pr_2CuO_4 in the case of the transferred momentum along $(1/2, 1/2, q_c)$ direction.

5. We determine the out-of-plane spin-wave gap $\Delta_{out} = (8.0 \pm 0.5)$meV. This value is larger than previously reported [1]. However, as shown in

Sec.3, our data are taken with much better statistical accuracy and thus we believe that our value of Δ_{out} is more reliable.

6. We observe acoustic and optical spin-wave branches and determine their dispersion (Fig.3). It is well described by Eq.(15) with $I = 0$.

7. From these data we determine the in-plane spin-wave gap $\Delta_0 = 0.36\pm 0.03$ meV and calculate the intra-plane PD interaction $P = 0.45\pm 0.04$ meV. This value is by an order of magnitude larger than theoretically predicted in Ref.[14]. However, it is confirmed by the analysis of the spin-flop transition data of Ref.[1] given in Sec.4. The discussion of this disagreement was given above.

8. From the observed acoustic and optical branches we find the value of the interplane PD interaction $Q = 0.023$ meV which is much larger than the corresponding magnetic interaction between neighboring Cu spins in the adjacent planes,

$$Q_m = \frac{24(g\mu_B)^2}{[2a^2 + c^2]^{3/2}} = 2.2 \cdot 10^{-3} \text{ meV} . \tag{33}$$

9. The spin configuration of the non-collinear spin structure is analyzed in the magnetic field parallel to the (ab) plane. For the field along [1,1,0] direction the spin-flop transition is of the second order and the critical field is given by Eq.(2). For other directions the transition is the first order one. It is also shown that for the [1,0,0] direction the metastable flop state should exist at $H < H_c$ where H_c is given by Eq.(26).

10. Analysis of the spin-flop data of Ref.[1] confirms the large value of the in-plane gap in Pr_2CuO_4 determined in the neutron scattering experiments. However, strong temperature dependence of the critical field H_c observed in [1] and confirmed in [18] remains unexplained. Presumably it is related to the T dependence of the g-factor.

11. The spin-wave spectrum in the magnetic field is considered.

The results listed above are discussed in the main part of the paper. Here we present some additional comments.

First of all we wish to point out that the intra-plane PD interaction maintains the long-range AF order at $T \neq 0$ in the isolated CuO_2 plane instead of the fact that in the linear theory the spin-wave spectrum is gapless. Indeed, we have seen in Sec.3 that the interaction between spin-waves gives rise to the in-plane gap. Similarly the gap appears in the out-of-plane branch too, if $A = 0$. Hence, we have a new example of the so-called ordering from the disorder (see Ref.[19] and references therein).

A comparison of the spin-wave gaps in Pr_2CuO_4 determined in this study and in Nd_2CuO_4 [10] reveals the following important features. i) The in-plane gap in Nd_2CuO_4 is by an order of magnitude larger than in Pr_2CuO_4. ii) In Nd_2CuO_4 both gaps are strongly T-dependent: $\Delta_{in,out}^2 \sim 1/T$. At the same time in Pr_2CuO_4 the gaps are apparently weakly dependent on T. As a result, we conclude that the values of both gaps are determined not only by the intrinsic properties of CuO_2 planes but also by the type of the RE ion. It explains the difference between the value of the intraplane PD interaction determined above and calculated in [14]. Along with the intraplane PD interaction the square anisotropy mediated by anisotropic interaction of the copper spins with RE ions may contribute to the values of both gaps. In this respect it would be very important to determine the in-plane spin-wave gap in compounds without RE ions such as $Sr_2CuO_2Cl_2$ and antiferromagnetic $YBa_2Cu_3O_{6+x}$. In particular, the spin-flop experiments are urgent with the magnetic field along [1,1,0] and [1,0,0] directions. This way one can find out if the magnetic structure of $Sr_2CuO_2Cl_2$ is non-collinear. In any case, Eq.(2) allows to determine the in-plane gap. Indeed, this equation holds at $\mathbf{H}\|[1,1,0]$ and $\mathbf{H}\|[1,0,0]$ for [1,0,0] and [1,1,0] easy axis, respectively.

As we have mentioned in the Introduction the Nd^{3+} crystal field ground state is a Kramers doublet and its local susceptibility behaves as $1/T$. It explains [10, 11] the spin reorientation transitions in Nd_2CuO_4 observed in [1, 3]. Indeed, there are two superexchange paths through one and two Nd ions respectively, which connect neighboring Cu^{2+} ions in the adjacent planes. Corresponding contributions to the PD interaction between these ions are proportional to $-Q_1/T$ and Q_2/T^2, where Q_1 and Q_2 are positive. As a result the strength of the interplane PD interaction has the form

$$Q = Q_0 - \frac{Q_1}{T} + \frac{Q_2}{T^2} = Q_0 \left(1 - \frac{T_1}{T}\right)\left(1 - \frac{T_2}{T}\right), \tag{34}$$

where $T_1 \simeq 80K$ and $T_2 \simeq 30K$ are the temperatures of the spin reorientation transitions.

It is tempting to describe in the same way the above mentioned T dependence of the spin-wave gaps in Nd_2CuO_4. Indeed, one can introduce the square anisotropy between copper spins as a result of the interaction with Nd^{3+} ions (cf.[7]). In the static limit this interaction should be proportional to $1/T$. However this behaviour holds if the measuring time t is larger than the inverse rate of the transition between levels of the Kramers doublet. In our case $t \sim \hbar/\Delta_{in(out)}$ is very short and this condition is hardly fulfilled. Hence, the T dependence of the spin-wave gaps in Nd_2CuO_4 remains unexplained.

In conclusion, we present a theoretical description of the spin configuration and spin-wave spectrum of the non-collinear tetragonal AF R_2CuO_4 assuming the PD interaction in CuO_2 planes as well as between them. The inelastic neutron scattering data for Pr_2CuO_4 are presented and allow us to determine the main parameters of the theory. A comparison of the data for Pr_2CuO_4 and Nd_2CuO_4 [10] with the theoretical results [14] reveals a strong dependence of the anisotropic interactions between Cu^{2+} spins on the type of RE ion. The spin configuration of the non-collinear R_2CuO_4 magnets in the magnetic field is analyzed too.

Acknowledgments

One of the authors (SM) is grateful for the financial support of the Russian Foundation for Basic Research (Grants 96.02 18037-a and 96-15-9675) and Russian State Program for Statistical Physics (Grant VIII-2). The authors (SM and AI) also thanks Russian Program "Neutron Studies of the Condensed Matter".

References

1. Sumarlin, I.W., Lynn, J.W., Chattopadhyay, T., Barilo, S.N., Zhigunov, D.I., and Peng, J.L. (1995) Magnetic structure and spin dynamics of the Pr and Cu in Pr_2CuO_4, *Phys.Rev.* **B51**, 5824.
2. Petitgrand, D., Moudden, A.H., Galez, P., and Boutrouille, P. (1990) Field-induced transformation of the spin-ordering in Nd_2CuO_4, *J.Less Common.Met.* **164–165**, 768.
3. Skanthakumar, S., Lynn, J.W., Peng, J.L., and Li, Z.Y. (1993) Observation of non-collinear magnetic structure for the Cu spins in Nd_2CuO_4- type systems, *Phys.Rev.* **B47**, 6173.
4. Skanthakumar, S., Lynn, J.W., Peng, J.L., and Li, Z.Y. (1993) Field dependence of the magnetic ordering of Cu in R_2CuO_4 (R-Nd,Sm), *J.Appl.Phys.* **73**, 6326.
5. Chattopadhyay, T., Lynn, J.W., Rosov, N., Grigereit, T.E., Barilo, S.N., and Zhigunov, D.I. (1994) Magnetic ordering in $EuCuO_4$, *Phys.Rev.* **B49**, 9944.
6. Vaknin, D., Sinha, S.K., Stassis, C., Miller, L.L., and Johnston, D.C. (1990) Antiferromagnetism in $Sr_2CuO_2Cl_2$, *Phys.Rev.* **B41**, 1926.
7. Sachidanandam, R., Yildirim, T.A., Harris, B., Aharony, A., and Entin-Wohlman, O. (1997) Single ion anisotropy, crystal field effects, spin reorientation transitions, and spin waves in R_2CuO_4, *Phys.Rev.* **B56**, 260.
8. Ivanov, A.S. (1995) Private Communication.
9. Plakhty, V.P., Stratilatov, A.B., and Beloglazov, S. (1997) Oxygen displacements in Eu_2CuO_4 by X-ray scattering, *Solid State Commun.* **103**, 685.
10. Ivanov, A.S., Bourges, Ph., Petitgrand, D., and Rossat-Mignod, J. (1995) Spin dynamics in Nd_2CuO_4 and Pr_2CuO_4, *Physica* **B213–214**, 60.
11. Bourges, P., Boudarene, L., and Petitgrand, D. (1992) Antiferromagnetic phase stability by Nd^{3+}–Cu^{2+} interplanar coupling in Nd_2CuO_4, *Physica* **B180–181**, 128.
12. Chou, F.C., Aharony, A., Birgeneau, B.J., Entin-Wohlman, O., Greven, M., Harris, A.B., Kastner, M.A., Kim, Y.J., Kleinberg, D.S., Lee, Y.S., and Zhu, Q. (1997) Ferromagnetic moment and spin rotational transition in tetragonal antiferromagnet $Sr_2Cu_3O_4Cl_2$, *Phys.Rev.Lett.* **78**, 535.
13. Yildirim, T., Harris, A.B., Entin-Wohlman, O., and Aharony, A. (1994) Spin structures of tetragonal lamellar copper oxides, *Phys.Rev.Lett.* **72**, 3710.
14. Yildirim, T., Harris, A.B., Aharony, A., and Entin-Wohlman, O. (1995) Anisotropic spin Hamiltonian due to spin-orbit and Coulomb exchange interactions, *Phys.Rev.* **B52**, 10239.
15. Maleyev, S.V. (1995) Pseudodipolar interaction in metals and alloys of rare

earths and actinides, *Pis'ma Zh.Eksp.Teor.Fiz.* **61**, 43; *JETP Lett.* **61**, 44.
16. Kondo, J. (1962) Anomalous Hall effect and magnetoresistance in ferromagnetic metals, *Prog.Theor.Phys.* **27**, 772.
17. Petitgrand, D., Maleyev, S.V., Bourges Ph., and Ivanov A.S. (1997) Pseudodipolar interaction and antiferromagnetism of R_2CuO_4 compounds (R=Pr, Nd, Sm and Cu), Submitted to *Phys.Rev.***B**.
18. Plakhty, V.P., Maleyev, S.V., Gavrilov, S., Smirnov, O.P., and Burlet, P. to be published.
19. Yildirim, T., Harris, A.B., and Shender, E.F. (1996) Three-dimensional ordering in *bct* antiferromagnets due to quantum disorder, *Phys.Rev.* **B53** 6455.

SOFT-MODE SPIN FLUCTUATIONS IN ITINERANT ELECTRON MAGNETS.

A. SOLONTSOV AND A. VASIL'EV
State Center for Condensed Matter Physics,
123060 Moscow, Rogova str.5, Russia
Theoretical Laboratory, A.A. Bochvar Institute for Inorganic Materials,
123060 Moscow, Russia

AND

D. WAGNER
Theoretische Physik III, Ruhr-Universität Bochum,
D-44780, Bochum, Germany

Abstract. This paper critically overviews recent developments of the theory of itinerant magnets with soft-mode spin fluctuations (SF) which are slowed down near some points of the Brillouin zone. With respect to the role of spatial dispersion and quantum effects a novel classification scheme for different regimes of SF with anisotropic spectrum is presented including generalized Fermi liquid (FL), soft-mode (SM), and localized moments (LM) ones. By introducing a dimensionless spin anharmonicity parameter it is shown that conventional SF theories are related to the weak coupling limit and cannot be applied to real magnets where mode-mode coupling of SF is strong. A variational SM theory is presented accounting for effects of strong spin anharmonicity induced by zero-point SF. Basing on the SM theory temperature dependencies of the spin anharmonicity parameter, local magnetic moments, magnetic susceptibility, thermal expansion, and specific heat are analyzed. It is shown that low temperature specific heat anomalies arising near critical points in the quantum SM regime are strongly affected by zero-point SF and caused by them strong anharmonicity, which may lead to a heavy fermion and non-Fermi liquid behavior. In the high temperature LM regimes the Curie-Weiss law for the susceptibility is shown to be accompanied by a saturation of local moments which may be affected by quantum dynamical effects reducing their values. This may lead to changes in the Rhodes-Wohlfarth plots and to a breakdown of the conventional criterion for itinerant magnetism. A novel criterion separating magnets with LM and soft-mode SF caused by itinerant electrons is suggested.

1. Introduction.

After the pioneering works of Frenkel [1] and Bloch [2] who first started to analyze ferromagnetism in itinerant electron systems and the classical paper of Heisenberg [3] forming the basis for understanding localized moments magnetism these two disciplines were rapidly developing quite separately (see, e.g., [4]), the first being attributed to transition metals with partially filled 3d electronic bands, the other was successfully used to describe magnetic isolators and rare-earth compounds where spins were regarded as localized.

The success of the Stoner model of ferromagnetic metals [5] focusing on the importance of individual fermionic degrees of freedom of an electron system instead of bosonic ones in localized magnets made the difference between itinerant electron and localized spin approaches principal.

In 70's Murata and Doniach [6] and Morya and Kawabata [7] suggested that in weak itinerant magnets close to an instability individual fermionic degrees of freedom may be in fact integrated out and one may use the Ginzburg-Landau or Hubbard models focusing on the overdamped bosonic spin fluctuations (SF). The formulated then theory of SF in itinerant magnets [8-11] (usually called a self-consistent renormalization (SCR) theory) has accounted for the mode-mode coupling of overdamped longwavelength fluctuations treating them in the lowest order approximation and surprisingly well described a large variety of experimentally observed properties of magnetic metals.

However, a number of important questions remains out of the scope of the SCR theory. First, the separation of the bosonic SF from the fermionic electron ones must be based on some energy scale. Near a phase transition SF are slowed down compared to individual Stoner excitations and may be considered as soft, like soft-mode phonons in anharmonic crystals [12]. On the other hand, far from a transition the energy of SF may be comparable to the energy of the Stoner electron-hole pairs. In this case bosonic and fermionic degrees of freedom must be treated on the equal footing, as, e.g., in the theory of electron-magnon scattering [13]. Nevertheless, the criterion for softening of SF was never worked out by the SCR theory.

Second, softening of SF implies an increasing role of zero-point motions neglected in the SCR theory. Both the experiment [14, 15] and theory [16,17] yield estimates of the amplitude of soft-mode magnetic fluctuations in a series of magnetic metals about one Bohr magneton, $1\mu_B$, which makes the weak coupling approximation used in the SCR theory inappropriate. Moreover, bearing in mind a similarity between soft-mode phonons in anharmonic crystals and soft-mode SF in itinerant magnets one may introduce a dimensionless spin anharmonicity parameter g_{SF} and analyze its behavior in different regimes of SF as it was recently done in [17,18]. The analysis showed that the spin anharmonicity parameter of soft-mode SF is of order unity, $g_{SF} \sim 1$, due to zero-point SF. On the other hand, the SCR theory based on the random phase approximation (RPA) perturbative arguments is valid only in the weakly coupling limit, $g_{SF} \ll 1$. Thus it was shown that unlike phonons in weakly anharmonic crystals [12] soft-mode SF in itinerant magnets should be treated as a Bose liquid of overdamped SF rather than a gas of weakly interacting paramagnons as was suggested by the SCR theory.

To account for large zero-point SF amplitudes and strong spin anharmonicity we later formulated a soft-mode (SM) theory of SF in itinerant magnets, which was based on both phenomenological [18] and microscopic [19] grounds and instead of a conventional perturbative approach used a variational procedure.

In the paramagnetic phase with the increase of temperature and the inverse susceptibility the SM features of SF may disappear and they behave as dispersionless weakly coupled excitations, localized in the real space. If a saturation of the localized magnetic moments takes place the susceptibility will show the Curie-Weiss behavior like in the Heisenberg magnets [20]. However, the saturated moments and Curie constants may be strongly affected by the quantum effects of SF dynamics in itinerant magnets.

Anyhow, concentrating on the collective overdamped spin variables, SF, and integrating out the individual electronic ones, which is provided by softening of the SF spectrum, enables one to work out a rather general physical picture of itinerant magnets with soft-mode SF, which at higher temperatures converges with the conventional Heisenberg model.

In the present paper we give a brief overview of recent findings in the theory of itinerant electron magnets with soft-mode SF. In Sec.2 we present a novel classification scheme for different SF regimes with respect to the role of anisotropic spatial dispersion and quantum effects. Then in Sec.3 we use the phenomenological Ginzburg-Landau and microscopic approaches to introduce a dimensionless spin anharmonicity parameter and estimate the validity of the conventional perturbative descriptions of SF. The temperature variation of spin anharmonicity is analyzed in Sec.4 basing of the fluctuation-dissipation theorem with account of both zero-point and thermal SF. In Sec. 5 we overview the recently worked out SM theory of SF and discuss its main results: the magnetic equation of state, temperature variation of the total squared local moments, magnetovolume effect, anomalies of the specific heat near quantum critical points, ground state properties, etc. In Sec. 6 we apply the fluctuation-dissipation theorem to analyze quantum effects of SF dynamics in the regime of localized magnetic moments. Finally, Sec. 7 presents a summary.

2. Regimes of SF behavior.

The spectrum of SF can be described by the imaginary part of the dynamical spin susceptibility, Im $\chi_\nu(\mathbf{Q},\omega)$, as a function of the wavevector $\mathbf{Q}$, frequency ω, and depends on a polarization ν ($\nu = t$ marks transverse and $\nu = l$ longitudinal SF). Here we assume that the SF spectrum is peaked near some wavevector $\mathbf{Q} = \mathbf{Q}_0$ ($\mathbf{Q}_0 = 0$ for a ferromagnetic and $\mathbf{Q}_0 \neq 0$ for antiferromagnetic instabilities) where SF soften (or slow down), as was reported for a variety of itinerant magnets close to magnetic instabilities [10,15,21-25]. Then we use the following phenomenological form for the inverse dynamical susceptibility,

$$\chi_\nu^{-1}(\mathbf{Q}_0 + \mathbf{q},\omega) = \chi_\nu^{-1}(T) + c(\mathbf{q}) - i\frac{\omega}{\Gamma(\mathbf{q})}, \qquad (1)$$

which accounts for an anisotropy of the SF spectrum. Here $\chi_\nu(\mathbf{Q}_0,\omega) = \chi_\nu(T)$ is the static susceptibility,

$$c(\mathbf{q}) = \sum_{i=x,y,z} a_i q_i^2 \leqslant c_0(\mathbf{q}_c) \equiv c_c \qquad (2)$$

and

$$\Gamma(\mathbf{q}) = \Gamma \left[c(\mathbf{q})/c_c\right]^{z/2-1} \qquad (3)$$

describe the spatial dispersion and relaxation rate which are assumed to be polarization independent, c_c is related to the volume of the Brillouin zone N_0,

$$N_0 = \frac{1}{6\pi^2} \left(\frac{c_c^3}{a_x a_y a_z}\right)^{1/2}, \qquad (4)$$

and z is the dynamical exponent ($z = 3$ for ferromagnetic and $z = 2$ for antiferromagnetic instabilities).

It should be emphasized that the form of the susceptibility (1) has essentially a phenomenological origin inferred mainly from neutron scattering experiments [10,15,21-25]. For isotropic ferromagnets ($a_x = a_y = a_z$) in the low frequency long wavelength limit $\omega \ll q v_F$, $q \ll p_F/\hbar$, where v_F and p_F

are the Fermi velocity and momentum the inverse susceptibility (1) may be justified in the RPA where the relaxation rate $\Gamma(\mathbf{q})$ is related to the Landau damping of SF [7,8,10].

The susceptibility given by (1) has a much more wider range of applicability than suggested by the RPA and may be used in a wider ω and $\mathbf{q}$ area as is supported by neutron scattering investigations [21,22]. However, (1) has a phenomenological origin and does not satisfy rigorous relations of the theory, e.g., the Kramers-Kronig relations [20], sum rules [26], etc.. Moreover, the moments of $\operatorname{Im}\chi_\nu(\mathbf{Q},\omega)$,

$$\int_{-\infty}^{+\infty} d\omega \operatorname{Im}\chi_\nu(\mathbf{Q},\omega)\omega^n$$

for odd $n > 1$ are divergent instead of being finite [27]. This means that the relaxation rate $\Gamma(\mathbf{q})$ in (1) must be ω dependent and at high frequencies vanish faster than any power of frequency. As we have shown recently [28] the relaxation rate $\Gamma(\mathbf{q})$ besides the Landau damping may be strongly affected by nonlinear damping mechanisms due to mode-mode scattering of SF which may give rise to a rather complicated ω dependence of $\Gamma(\mathbf{q})$. However, we shall not analyze this problem here and adopt a phenomenological relation (1). To avoid divergencies we introduce here an anisotropic cut-off frequency

$$\omega_c(\mathbf{q}) = \omega_c \left[c(\mathbf{q})/c_c\right]^{z/2-1}, \tag{5}$$

where ω_c is treated as a phenomenological parameter which is to be inferred from experiments. It should be also mentioned that as far as we use a phenomenological susceptibility given by (1), instead of a rigorous one, out-off frequencies may be different for different integrations with respect to ω. In the present paper we use the out-off frequency $\omega_c(\mathbf{q})$ only to evaluate the zero-point contributions to the fluctuation- dissipation theorem.

According to (1) the SF spectrum is characterized by the frequency

$$\omega_{SF}^{(\nu)}(\mathbf{q}) = \Gamma(\mathbf{q})\left[\chi_\nu^{-1} + c(\mathbf{q})\right], \tag{6}$$

which softens near $\mathbf{Q} = \mathbf{Q}_0$ where $c(\mathbf{q})$ vanishes. Comparing the SF energy $\hbar\omega_{SF}^{(\nu)}(\mathbf{q})$, to the thermal one, $k_B T$, one can introduce the characteristic SF temperature $T_{SF} \approx \hbar\Gamma c_c / k_B$ which separates the temperature scale into the quantum ($T < T_{SF}$) and classical ($T > T_{SF}$) regions and estimate the characteristic spatial dispersion of thermal SF,

$$c_T = c_c \begin{cases} \left(\frac{T}{T_{SF}}\right)^{2/z}, & T \leqslant T_{SF}, \\ 1, & T > T_{SF}. \end{cases} \tag{7}$$

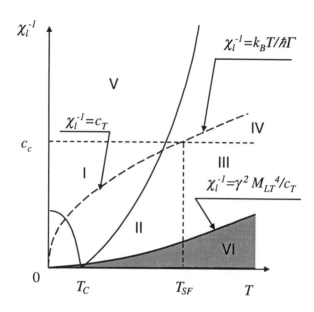

Figure 1. Phase diagram schematically showing SF regimes in itinerant electron magnets. Region I defined by $c_T \ll \chi_l^{-1} \ll c_c$ is the generalized Fermi liquid regime, FL arising in the low temperature limit $T < T_{FL}$. Region II and III specified by $\chi_l^{-1} \ll c_T$, $T \ll T_{SF}$ and $\chi_l^{-1} \ll c_c$, $T \gg T_{SF}$ are related to the quantum and classical soft-mode regimes, QSM and CSM. Regimes IV and V defined by $c_c \ll \chi_l^{-1} \ll k_B T/\hbar\Gamma$ and c_c, $k_B T/\hbar\Gamma \ll \chi_l^{-1}$ are the classical and quantum localized moments regimes, CLM and QLM. The solid curve illustrates a typical temperature dependence of the inverse longitudinal susceptibility, χ_l^{-1}, for a ferro- or antiferromagnet, the temperature T_C being the Curie or Néel temperature. The shaded area marks the critical region VI defined by $\chi_l^{-1} \ll \gamma^2 M_{LT}^2/c_T$, where the thermal contribution to the anharmonicity parameter diverges, $g_T \gg 1$, resulting in the Ginzburg-Levanyuk criterion (see Sec. 4.1)

Then it is straightforward to define a set of dimensionless parameters

$$\frac{\chi_l^{-1}}{c_T}, \quad \frac{\chi_l^{-1}}{c_c}, \quad \frac{T}{T_{SF}}, \quad \frac{\hbar\Gamma\chi_l^{-1}}{k_B T} \tag{8}$$

which allow to separate between different regimes of SF behavior. The first two of them describe the role of spatial dispersion, the other two define quantum effects. Different SF regimes are shown on a phase diagram in Fig.1.

Inequalities,

$$\chi_l^{-1}(T) \ll c_T, \quad T \ll T_{SF} \tag{9}$$

and

$$\chi_l^{-1}(T) \ll c_c, \quad T \gg T_{SF} \tag{10}$$

define the quantum (QSM) and classical (CSM) soft-mode regimes, respectively, which are dominated by strongly dispersive soft-mode(SM) fluctuations. The SM regimes take place in any ferro- or aniferromagnet relatively close to the critical (Curie or Néel) temperature T_C defined by the equality $\chi_l^{-1}(T_C) = 0$. For enhanced paramagnets it may exist if they are sufficiently close to low-temperature magnetic instabilities. Inside the SM regimes, in the vicinity of T_C, a critical region takes place defined by the Ginzburg-Levanyuk criterion where the thermal contribution to the spin anharmonicity parameter is divergent, $g_T \gg 1$ (see Sec. 4.1)

The low-temperature regime limited by

$$c_T \ll \chi_l^{-1} \ll c_c \tag{11}$$

is affected by high frequency quantum fluctuations, $\hbar\omega_{SF}^{(\nu)} \gg k_B T$, and may be treated within a generalized Fermi liquid (FL) picture below some FL temperature

$$T \ll (\chi_l c_c)^{-z/2} T_{SF} \equiv T_{FL} \tag{12}$$

(see, e.g., [29]). In the limit

$$\chi_l^{-1} \gg c_c \tag{13}$$

SF may be regarded as dispersionless and local in real space, forming localized moments (LM).

Depending on the parameter $z_L = \hbar\omega_{SF}^{(l)}/k_B T \approx \hbar\Gamma\chi_l^{-1}/k_B T$ this limit may be related to the quantum localized moments (QLM) regime, when $z_L \ll 1$ and classical localized moments (CLM) one arising when $z_L \gg 1$.

The temperature T_L defined by $\chi_l^{-1}(T_L) = c_c$ describes a crossover from the SM to LM regimes. A crossover between the quantum QSM and QLM regimes which takes place when $T < T_{SF}$ occupies a temperature interval near T_L defined by (11).

The presented classification of SF regimes with respect to the role of spatial dispersion and quantum effects exhaust all possible SF regimes. The abbreviated discussion of this classification scheme applied to isotropic itinerant magnets [30-32] and highly anisotropic antiferromagnetic systems with frustration [33] were presented earlier.

It should be mentioned that the previously presented classification of SF regimes by Zülicke and Millis [34] based on a scaling hypothesis is incomplete as we have pointed out recently [30].

The comparison of our phase diagram with the one presented in [34] shows that the CSM, QLM, and CLM regimes are missing in the latter. The low χ_l^{-1} and T parts of both diagrams are similar, though Zülicke and Millis use terms "quantum" and "classical" to define FL and QSM regimes, respectively. Moreover, they relate both regimes to the region of Gaussian fluctuations (aside the critical region which they call "non-Gaussian"). This terminology leads to a neglect of quantum effects in the QSM regime where they play an essential role.

Below we shall concentrate on the SM and LM regimes of SF which define the properties of itinerant magnets in a wide temperature range.

3. SF coupling: phenomenological and microscopic approaches.

The most straightforward way to describe SF coupling is to assume that individual fermionic variables are integrated out and to use a phenomenological approach focusing on fluctuations of the magnetic order parameter and accounting for SF dynamics as well. Following the approach of Murata and Doniach [6] we concentrate on the forth-order anharmonicity and use the effective Ginzburg-Landau (GL) Hamiltonian

$$\widehat{H}_{eff} = \frac{1}{2}\sum_{\mathbf{Q}} \chi_0^{-1}(\mathbf{Q})|\mathbf{M}(\mathbf{Q})|^2$$
$$+ \frac{\gamma_0}{4}\sum_{\mathbf{Q}_1+\mathbf{Q}_2+\mathbf{Q}_3+\mathbf{Q}_4=0} (\mathbf{M}(\mathbf{Q}_1)\mathbf{M}(\mathbf{Q}_2))(\mathbf{M}(\mathbf{Q}_3)\mathbf{M}(\mathbf{Q}_4)) \quad (14)$$

which should be accompanied by the time-dependent GL equations

$$\frac{1}{\Gamma(\mathbf{Q})}\frac{\partial \mathbf{M}(\mathbf{Q})}{\partial t} = -\frac{\delta \widehat{H}_{eff}}{\mathbf{M}(-\mathbf{Q})}, \quad (15)$$

where $\mathbf{M}(\mathbf{Q}) = M\delta_{\mathbf{Q},\mathbf{Q}_0} + \delta\mathbf{m}(\mathbf{q})$ is the magnetic order parameter, M is its average value defining the homogenous ($\mathbf{Q}_0 = 0$) or staggered ($\mathbf{Q}_0 \neq 0$) magnetization density, $\delta\mathbf{m}(\mathbf{q})$ account for SF, $\chi_0(\mathbf{Q})$ and γ_0 are the static inhomogeneous susceptibility and mode-mode coupling constant not affected by SF. The effective Hamiltonian was shown to arise from microscopic approaches after integrating out individual quasiparticles degrees of freedom, except for special cases, e.g., in models with nesting [34]. It should be mentioned that in (14) one may account for higher order anharmonicity as was suggested by Shimizu [9]. This however would not lead to principally new results, although may result in a better understanding of real magnets.

The approximation of the conventional theory [6,9,10,11] is to drop the coupling term in (14), $\sim \gamma_0(\delta\mathbf{m}^2)^2$, which leads to a Gaussian approximation for SF and to the following contribution to the free energy

$$\Delta F = 2\sum_{\nu}\sum_{\mathbf{q},\omega} F_0(\omega)\frac{\omega_{SF}^{(\nu)}(\mathbf{q})}{[\omega_{SF}^{(\nu)}(\mathbf{q})]^2+\omega^2} \equiv \sum_{\nu}\Delta F_{RPA}(\chi_\nu(\mathbf{Q},\omega)),\quad (16)$$

where $F_0(\omega)=k_BT\ln[1-\exp(-\hbar\omega/k_BT)]+\hbar\omega/2$ is the free energy of a harmonic oscillator and $\sum_{\mathbf{q},\omega}=\int_0^\infty(d\omega/2\pi)\sum_{\mathbf{q}}$. Here the dynamical susceptibility is given by (1) - (3), and the static susceptibility χ_ν at $\mathbf{Q}=\mathbf{Q}_0$ is assumed to be renormalized with respect to the RPA value $\chi_{0\nu}$ [10,11],

$$\chi_\nu^{-1} = \chi_{0\nu}^{-1} + \lambda_\nu,\quad (17)$$

self-consistently accounting for mode-mode coupling effects described by

$$\lambda_t \simeq \lambda_l = 2\frac{\partial(\Delta F)}{\partial(M^2)} = \frac{5}{3}\gamma_0 M_L^2.\quad (18)$$

Here $M_L^2=\langle\delta\mathbf{m}^2\rangle$ is the squared amplitude of SF, or squared magnetic moment, $\langle...\rangle$ means statistical average, and in the absence of the magnetic order $\chi_{0\nu}=\chi_0(\mathbf{Q}_0)=\chi_0$. Without mode-mode coupling (16) reduces to the familiar RPA expression for the free energy of uncoupled paramagnons (see, e.g., [11]).

Another, microscopic approach to analyze SF is traditionally formulated in terms of Fermi quasiparticles within the Hubbard [7,10,11] or Fermi liquid [8] models. Using this approach we treat the dynamical susceptibility of a Fermi liquid in the spirit of the Moriya-Kawabata theory [7] and take it in the conventional form

$$\chi_\nu^{-1}(\mathbf{Q},\omega) = [\chi_\nu^{(0)}(\mathbf{Q},\omega)]^{-1}+\Psi+\lambda_\nu.\quad (19)$$

Here Ψ describes the exchange interaction of quasiparticles, $\chi_\nu^{(0)}(\mathbf{Q},\omega)$ is the susceptibility at $\Psi=0$, $[\chi_\nu^{(0)}(\mathbf{Q}_0)]^{-1}+\Psi=\chi_{0\nu}^{-1}$, and λ_ν similarly to (18) describes effects of SF coupling. In the limit $\lambda_\nu\to 0$ (19) reduces to the usual RPA result. The approximation of the SCR theory is to neglect Ψ and M dependencies of λ, setting [11]

$$\left|\frac{\partial\lambda}{\partial\Psi}\right|\ll 1,\quad \left|\frac{\partial\lambda}{\partial M}\right|\ll\frac{\partial(\chi_\nu^{-1})}{\partial M},\quad (20)$$

which yields the RPA-like formula (16) for the free energy, (see also Sec. 5.2) and the same self-consistent expression (18) for λ_ν.

Formulae (18) and (19) for the susceptibility were widely used to calculate the Curie temperature and to explain the Curie-Weiss law in a series of metals. However, little was done to clarify the validity of approximations made in both phenomenological and microscopic approaches. One important constraint of the conventional SF theory straightforwardly comes from the neglect of the four-mode coupling terms in the GL model, which gives corrections $\sim \gamma_0 \partial(M_L^4)/\partial(M^2)$ to (18). Assuming this to be small compared with λ_ν, one gets the following condition of the validity of the conventional SF theory

$$\frac{\partial(M_L^2)}{\partial(M^2)} \sim \gamma_0 \sum_\nu \left|\frac{\partial(M_L^2)}{\partial(\chi_\nu^{-1})}\right| \ll 1 \quad . \tag{21}$$

The same constraint comes from inequalities (20) of the microscopic approach after substituting there the expression (18) for λ_ν. The l.h.s. of (21) is a natural measure of spin anharmonicity or mode-mode coupling expressed in a dimensionless form [16,17]. Thus inequality (21) relates the conventional SF or SCR theory, formulated within phenomenological model and microscopic approaches to the limit of weak spin anharmonicity.

As it follows from (21) it is possible to introduce a dimensionless spin anharmonicity parameter [17,18]

$$\begin{aligned} g_{SF} &= \frac{8}{3}\gamma_0 \hbar \sum_\nu \sum_{\omega,\mathbf{Q}} \mathrm{Re}\,\chi_\nu(\mathbf{Q},\omega)\,\mathrm{Im}\,\chi_\nu(\mathbf{Q},\omega)(N_\omega + \frac{1}{2}) \\ &= \frac{\gamma_0}{3} \sum_\nu \left|\frac{\partial(M_L^2)}{\partial(\chi_\nu^{-1})}\right| = g_{Z.P.} + g_T \quad , \end{aligned} \tag{22}$$

related to the stiffness of the squared local magnetic moment, which can be expressed in terms of the fluctuation-dissipation theorem,

$$M_L^2 = 4\hbar \sum_\nu \sum_{\omega,\mathbf{Q}} \mathrm{Im}\,\chi_\nu(\mathbf{Q},\omega)(N_\omega + \frac{1}{2}) = (M_L^2)_{Z.P.} + (M_L^2)_T \quad . \tag{23}$$

Here the factors $N_\omega = [\exp(\hbar\omega/k_B T) - 1]^{-1}$ and $1/2$ are related to thermal and zero-point SF, respectively, which provides a natural separation of g_{SF} and M_L^2 into the zero-point, $g_{Z.P.}$ and $(M_L^2)_{Z.P.}$, and thermal, g_T and $(M_L^2)_T$, contributions.

It should be emphasized that the spin anharmonicity parameter (22) is analogous to the anharmonicity parameter introduced in the theory of a crystal lattice [12,35]. However, unlike crystals where anharmonicity is usually small, mode-mode coupling of overdamped SF in itinerant electron magnets is essentially strong, and spin anharmonicity parameter (22) represents a novel important characteristic defining the SF behavior and providing a new insight into the nature of SF. In the proceeding Sections we will analyze the temperature dependencies of the anharmonicity parameter, squared local moments, and the SF behavior in different SF regimes.

4. Analysis of spin anharmonicity.

We start the analysis of the SF behavior with the SM regimes defined by inequalities (9) and (10), when it is possible to use expansions in terms of the small parameters χ_ν^{-1}/c_T and χ_ν^{-1}/c_c describing properties of thermal and zero-point SF, respectively. As will be shown in the next Section anharmonicity due to zero-point SF may be strong, $g_{Z.P.} \sim 1$, which makes the conventional SCR theory of SF inappropriate to describe this regime. On the other hand, anharmonicity due to thermal SF is rather weak aside the critical region, $g_T \ll 1$. This provides to formulate a well-established SM theory accounting for strong mode-mode coupling effects induced by zero-point SF.

4.1. SOFT-MODE REGIMES.

First, we discuss zero-point contributions to the magnetic moments and spin anharmonicity. Expanding expressions (22) and (23) for $g_{Z.P.}$ and $(M_L^2)_{Z.P.}$ in powers of χ_ν^{-1} we have

$$(M_L^2)_{Z.P.} = M_{L0}^2 - g_0 \sum_\nu (\gamma_0 \chi_\nu)^{-1} + \cdots , \qquad (24)$$

$$g_{Z.P.} = g_0 + O\left(\frac{\chi_\nu^{-1}}{ck_c^2}\right) + \cdots , \qquad (25)$$

where

$$M_{L0}^2 = \frac{3\hbar}{2\pi} \Gamma N_0 \alpha_1(f), \qquad (26)$$

$$g_0 = \frac{\hbar \gamma_0 \Gamma N_0}{\pi c_c} \alpha_2(f) \qquad (27)$$

Here $f = \omega_c/\Gamma c_c$ and α_1, and α_2 are dimensionless parameters of order unity:

$$\alpha_1 = 3 \int_0^1 dx x^z \ln(1 + f^2/x^4) , \qquad (28)$$

$$\alpha_2 = 3 \int_0^1 dx x^{z-2} \frac{f^2}{f^2 + x^4}. \qquad (29)$$

Using simple estimates based on the Landau damping mechanism of SF and the Stoner model, $\hbar\Gamma \sim \mu_B^2 N_e$, $\gamma_0 \sim \mu_B^2/\varepsilon_F^2 \chi_p^3$, $c_c \sim \chi_p^{-1}$ (where $N_e \sim N_0$ is the electron density, ε_F is the Fermi energy, and χ_p is the Pauli susceptibility) we can roughly evaluate the squared amplitude of zero-point SF,

$$M_{L0}^2 \sim \mu_B^2 N_e^2, \tag{30}$$

and spin anharmonicity due to zero-point effects,

$$g_{Z.P.} \approx g_0 \sim \frac{M_{L0}^2}{\mu_B^2 N_e^2} \sim 1 \quad . \tag{31}$$

These estimates hold also for the FL regime. Thus we may conclude that in the FL and SM regimes effects of zero-point SF give rise to large SF amplitudes and therefore to strong spin anharmonicity. On the other hand, according to (21) the conventional SCR theory is based on the weak coupling limit,

$$g_{SF} \ll 1, \tag{32}$$

and cannot be applicable to itinerant magnets exhibiting soft-mode features of SF, which results in strong spin anharmonicity.

It should be emphasized that the estimates for the local moments (30) and spin anharmonicity (31) induced by zero-point SF do not include any magnetic properties of materials and depend only on the electronic band structure of a metal. Therefore, we believe that these estimates and conclusions presented above are generally applicable to all itinerant electron magnets with soft-mode SF.

The analysis of spin anharmonicity in the SM regimes arising due to thermal SF is somewhat more conventional. Analogously to (24) and (25) we expand $(M_L^2)_T$ and g_T in terms of χ_ν^{-1}/c_T,

$$(M_L^2)_T = M_{LT}^2 \left[1 - \frac{\pi}{6} \sum_\nu \sqrt{\frac{\chi_\nu^{-1}}{c_T}} + \ldots \right] \quad , \tag{33}$$

$$g_T = \frac{\pi}{12} \frac{\gamma_0 (M_L^2)_T}{\sqrt{\chi_l^{-1} c_T}} \left[1 + O\left(\sqrt{\frac{\chi_l^{-1}}{c_T}}\right) \right] \quad , \tag{34}$$

where

$$M_{LT}^2 = 9\alpha_0 k_B T N_0 \frac{c_T^{1/2}}{c_c^{3/2}} \quad . \tag{35}$$

Here α_0 is a dimensionless parameter, $\alpha_0 = 4\pi\zeta(4/3)/27/\Gamma(2/3) = 1.2377$ for ferromagnetic instabilities (z=3) and $\alpha_0 = \sqrt{\pi/2}\zeta(3/2)/2 = 1.6371$ for antiferromagnetic ones (z=2).

It should also be emphasized that in the QSM regime the zero-point contribution to the squared moment, $M_{L0}^2 \sim \hbar\Gamma q_C^{z-2} N_0 \gg M_{LT}^2 \sim M_{L0}^2(T/T_{SF})^{1+1/z}$, dominates over the thermally excited one, which shows an important role of zero-point SF in the quantum region ($T \leq T_{SF}$).

Near the critical temperature T_C, where $\chi_l^{-1}(T) \sim (T - T_C)$, spin anharmonicity (34) is divergent,

$$g_T \sim \sqrt{\frac{\tau_G T_C}{T - T_C}} \quad . \tag{36}$$

Here

$$\tau_G = \frac{1}{32\pi^2}\left[\frac{k_B}{\Delta C \xi_x \xi_y \xi_z}\right]^2 \tag{37}$$

is the conventional Ginzburg parameter (see, e.g., [18]) expressed in terms of the specific-heat jump at T_C, $\Delta C = \alpha^2/2\gamma T_C$, and the magnetic correlation lengths $\xi_i = \sqrt{a_i/\alpha}$, where $\alpha = \partial(\chi^{-1})/\partial \ln T$, and $\chi^{-1}(T)$ and γ are coefficients in the Landau free energy (see Sec. 5.3). The constraint of weak coupling of thermal SF leads to the well-known Ginzburg-Levanyuk criterion

$$\left|\frac{T - T_C}{T_C}\right| \gg \tau_G, \tag{38}$$

which was previously obtained for itinerant ferromagnets by other means [6,8]. At temperatures higher than T_C, $T \gg T_C$, (34) yields the following value

$$g_T \approx \frac{\pi}{20}\sqrt{\frac{\chi_l^{-1}}{c_T}} \tag{39}$$

for the thermal contribution to g_{SF}, which according to (9) is small in the SM regime and is about unity when a crossover to the LM regimes takes place.

As we shall see below in Sec. 5, though thermal contributions to the SF amplitude and anharmonicity are small in the SM regimes, thermal SF strongly coupled due to zero-point effects essentially influence thermal properties of itinerant magnets in these regimes.

4.2. LOCALIZED MOMENTS REGIMES.

In the LM regime defined by (13) the spatial dispersion of SF is not important and they behave as localized magnetic moments. We start the analysis with the thermal SF contribution to M_L^2 using the fluctuation-dissipation theorem (23). Omitting the polarization index we get

$$(M_L^2)_T = 3k_B T N_0 \chi(T) \cdot \begin{cases} 1, & k_B T \gg \hbar\omega_c, \\ G(z_L), & k_B T \ll \hbar\omega_c, \end{cases} \qquad (40)$$

where the factor

$$G(z_L) = 3\int_0^1 g(z_L x^{z-2}) x^2 dx \qquad (41)$$

accounts for the quantum dynamical effects of SF and is a rapidly decreasing function of its argument $z_L = \hbar\Gamma\chi^{-1}(T)/2\pi k_B T$ (see Sec. 6), $g(x) = 2x[\ln x - 1/2x - \Psi(x)]$, $\Psi(x)$ is Euler's psi function.

As it follows from (40) a saturation of $(M_L^2)_T$,

$$(M_L^2)_T \approx M_{eff}^2 = const \quad , \qquad (42)$$

implies a Curie-Weiss type behavior for $\chi(T)$,

$$\chi(T) = \frac{C}{T + \Theta_L} \quad , \qquad (43)$$

where according to (40) the Curie constant C and the saturation moments account for the quantum effects below the temperature $\hbar\omega_c/k_B$. Vice versa, according to (40), the Curie-Weiss behavior for $\chi(T)$ gives rise to a saturation of local moments (42) above the paramagnetic Curie temperature Θ_L. It should be mentioned that reliable calculations of M_{eff}^2 and C in the LM regimes are still lacking (cf. [11]). Moreover, the hypothesis of separating the collective spin and individual fermionic degrees of freedom may not work here because the former have a rather high energy $\sim \hbar\Gamma\chi^{-1}(T)$ comparable with ε_F in the LM regimes and may couple with the latter. However, some useful qualitative conclusions may be driven for the system of coupled SF.

First, differentiating (40) with respect to $\chi^{-1}(T)$ one may get the thermal SF contribution to the spin anharmonicity parameter,

$$g_T \sim \frac{\gamma_0}{3}(M_L^2)_T \chi(T) \quad , \qquad (44)$$

which decreases $\sim 1/T$ if $\chi(T)$ shows a Curie-Weiss behavior.

Then, it is straightforward to analyze the effects of zero-point SF. From (23) it follows that in the LM regimes the zero-point contribution to the squared local moment is given by (26) with the dimensionless coefficient α_1 replaced by

$$\alpha_1 = \frac{3}{z+1} \ln\left[1 + f^2 c_c^2 \chi^2(T)\right] \quad . \qquad (45)$$

Obviously, in the high-temperature LM regimes the quantum contribution of zero-point SF to M_L^2 is decreasing as a squared susceptibility $\sim \chi^2(T)$, and goes as $\sim 1/T^2$ if $\chi(T)$ shows a Curie-Weiss law. Similarly, the zero-point contribution to the anharmonicity parameter in the LM regime is given by (27) with α_2 replaced by

$$\alpha_2 = \frac{3}{z+1} \frac{f^2 [\chi(T) c_c]^3}{1 + [f\chi(T) c_c]^2} \quad . \qquad (46)$$

According to (46) in the LM regimes $g_{Z.P.}$ is decreasing $\sim \chi^3(T)$. And if a Curie-Weiss law holds we have

$$g_{Z.P.} \sim \frac{1}{T^3} \quad , \qquad (47)$$

and the total spin anharmonicity parameter in the LM regimes goes as

$$g_{SF} \sim g_T \sim \frac{1}{T} \quad . \qquad (48)$$

So, it is possible to conclude, that in the relatively high temperature LM regime spin anharmonicity accounting for coupling of the overdamped SF is decreasing, and the ensemble of SF may be viewed as a system of weakly coupled localized magnetic moments. Finally, we schematically present the temperature variation of the spin anharmonicity parameter in different SF regimes on the diagram in Fig.2.

5. Soft-mode theory of spin fluctuations.

As we have already seen in Sec.3 the conventional SCR approach both in the phenomenological and microscopic formulations is based on perturbative arguments and is valid only in the weak spin anharmonicity limit, when inequality (32) holds. However, (32) does not hold in the FL and SM regimes of SF due to strong anharmonicity induced by zero-point fluctuations. Fortunately, using the concept of soft-mode fluctuations it is possible to work out a theory of SF in the SM regimes using a variational procedure.

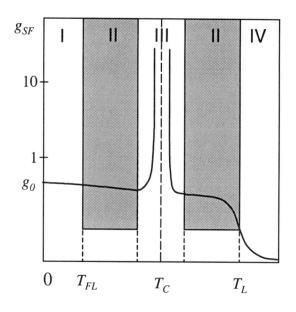

Figure 2. Temperature dependence of the spin anharmonicity parameter in itinerant ferro- and antiferromagnets. Regions I, II, III, and IV are related to the FL, SM, critical, and LM regimes of SF. The shaded regions define the area where the SM theory of SF is applicable. The unshaded regions in the SM regime where $g_{SF} \ll 1$ indicate the area where the SM theory reduces to the conventional SCR one. The solid curve illustrates the temperature dependence of g_{SF} for a typical itinerant ferro- or antiferromagnet.

5.1. PHENOMENOLOGICAL FORMULATION.

First, we formulate the main equations of the SM theory using a phenomenological GL approach [18]. The free energy related to the effective GL Hamiltonian (14) is given by

$$F(M,T) = F_0(T) + \frac{1}{2\chi_0}M^2 + \frac{\gamma_0}{4}M^4 + \Delta F. \qquad (49)$$

Here F_0 denotes the contribution independent on magnetism, terms with M^2 and M^4 are related to the Hartree-Fock approximation of the Stoner model, coefficients $\chi_0 \equiv \chi_0(\mathbf{Q}_0)$ and γ_0 incorporate static correlation effects and may be inferred from the band structure calculations. The term ΔF accounts for dynamical effects of SF and is the crucial point for the theory of fluctuations. In the SCR theory it was taken in the RPA like form (16) provided mode-mode coupling of SF is weak.

To go beyond this approximation we shall consider $\chi_0 \equiv \chi_0(\mathbf{Q}_0)$ in (14) as a parameter and integrate the equality

$$\frac{\partial F}{\partial(\chi_0^{-1})} = \frac{1}{2}(M^2 + M_L^2) \quad , \tag{50}$$

which results after averaging (14). Using the fluctuation-dissipation theorem, after integration (50) with respect to χ_0^{-1} we get

$$\Delta F = 2\hbar \sum_{\nu=t,l} \sum_{\omega,\mathbf{Q}} \left(N_\omega + \frac{1}{2}\right) \int d(\chi_0^{-1}) \operatorname{Im} \chi_\nu(\mathbf{Q},\omega) \quad . \tag{51}$$

The integration constant independent of magnetism is assumed to be incorporated into $F_0(T)$. In all the previous approaches to the SCR theory [6-11] static inverse susceptibilities χ_ν^{-1} were assumed just to be shifted from the Stoner value χ_0^{-1}. As far as they were based on the weak coupling approximation this shift was proportional to the coupling constant γ_0 and was independent on χ_0. This assumption applied to (51) immediately gives the RPA-like formula (16) for the SF contribution to the free energy and the conventional results of the SCR theory [11].

In the case of strong mode-mode coupling static susceptibilities χ_ν must be more complicated functions of the Stoner susceptibility χ_0. In the SM regimes of SF it is possible to find these functions self-consistently basing on the following arguments. The most important parameters defining the integration in the r.h.s. of (51) are

$$\zeta_\nu = \frac{\partial(\chi_\nu^{-1})}{\partial(\chi_0^{-1})} \quad . \tag{52}$$

As it was already mentioned, in the weak coupling limit of the SCR theory we have $\zeta_\nu = 1$. It is reasonable to assume that ζ_ν depend only on the spin anharmonicity parameter g_{SF} which in the SM regimes is defined by zero-point SF, $g_{SF} \approx g_0$, and is independent on χ_ν. In other words,

$$\zeta_\nu = \zeta_\nu(g_0) = const(\chi_\nu) \quad . \tag{53}$$

This assumption will be verified below within a self-consistent variational procedure.

Integrating (51) and using the assumption (53) we get the following explicit formula for the SF contribution to the free energy

$$\Delta F = \sum_\nu \frac{1}{\zeta_\nu} \Delta F_{RPA}\{\chi_\nu(\mathbf{Q},\omega)\} \tag{54}$$

where $\Delta F_{RPA}\{\chi_\nu\}$ is given by (16). In the weak anharmonicity limit when $\zeta_\nu = 1$ expression (54) reduces to (16). In principle, dynamical susceptibilities $\chi_\nu(\mathbf{Q},\omega)$ should be calculated on the basis of the nonlinear time-dependent equations (15) accounting for the effects of multi-mode scattering of SF and strong spin anharmonicity [28,36,37]. However, here we assume that these effects do not influence much the spatial and time dispersions of the dynamical susceptibilities and use a semi-phenomenological approach suggested by the SCR theory. Namely, we take the dynamical susceptibilities in the form (1) supported by neutron scattering experiments and the static susceptibilities define self-consistently by the conventional thermodynamic relations

$$\chi_t = \frac{M}{B} \quad , \quad \chi_l = \left(\frac{\partial M}{\partial B}\right)_V . \tag{55}$$

Here B is the magnetic (or staggered) field and V is the volume, which are defined by the equations of state

$$B = \left(\frac{\partial F}{\partial M}\right)_V , \tag{56}$$

$$P = -\left(\frac{\partial (FV)}{\partial V}\right)_{MV} , \tag{57}$$

where P is the pressure.

It should be emphasized that (55) are the thermodynamic susceptibilities at the constant volume and coincide with dynamical susceptibilities in the static limit provided the coupling between spin and electron density fluctuations is neglected, which is possible when one sets the screened Coulomb interaction of quasiparticles to infinity [19]. This situation cannot be accounted for by, e.g., the Hubbard Hamiltonian usually used in the SCR theory because both exchange and Coulomb interactions there are described by one and the same parameter Ψ which is fixed to satisfy the Stoner criterion.

Finally, we conclude that equalities (1), (49), and (53)-(57) form a complete set of differential equations to define the free energy.

5.2. MICROSCOPIC APPROACH.

To go beyond the SCR theory in its microscopic formulation we are to allow for the M and Ψ dependencies of λ_ν in the expression (19) for dynamical susceptibilities. Using the considerations similar to those discussed while getting (53) we introduce the dimensionless parameters

$$\xi_\nu = \frac{\partial \lambda_\nu}{\partial \Psi} \qquad (58)$$

which are assumed to dependent only on the spin anharmonicity parameter $g_{SF} \approx g_0$ and are independent on χ_ν and λ_ν. Comparing (52) and (58) we get

$$\zeta_\nu = 1 + \xi_\nu \quad . \qquad (59)$$

To calculate the SF contribution to the free energy we take the derivative (30) where χ_0^{-1} is replaced by Ψ and after the conventional Ψ- integration procedure, using the fluctuation-dissipation theorem and equalities (19) and (58), we get

$$\begin{aligned}\Delta F = & \; 2\hbar \sum_\nu \frac{1}{\zeta_\nu} \sum_{\omega,\mathbf{Q}} \mathrm{Im} \left\{ \ln\left[1 + (\Psi + \lambda_\nu)\chi_\nu^{(0)}(\mathbf{Q},\omega)\right] \right. \\ & \left. - \Psi\chi_\nu^{(0)}(\mathbf{Q},\omega) \right\} \left(N_\omega + \frac{1}{2}\right) \quad . \end{aligned} \qquad (60)$$

In the weak anharmonicity limit when $\zeta_\nu = 1$ expression (60) gives the SF free energy of the SCR theory [11]. The difference between the phenomenological (51) and microscopic (60) expressions for ΔF is due to the fermionic degrees of freedom of quasiparticles accounted for in (60). Here we assume them to be integrated out so, that expression (60) reduces to (51).

5.3. FREE ENERGY.

As we have already mentioned in Sec.5.2 a set of Eqs. (1), (49), and (52)-(57) is sufficient to calculate the free energy. However, the SF contribution to the free energy still remains an unknown functional of the static susceptibilities (55). However, the form of the functional can be easily found in the SM regimes where according to (9) and (10) it is possible to expand ΔF in terms of the inverse static susceptibilities χ_t^{-1} and χ_l^{-1},

$$\Delta F = \Delta F_0(T) + \frac{1}{2} \sum_\nu \frac{1}{\zeta_\nu} \left\{ \frac{1}{3} \left(M_{L0}^2 + M_{LT}^2\right) \chi_\nu^{-1} \right.$$
$$\left. - \frac{g_0}{2\gamma_0} \chi_\nu^{-2} - \frac{k_B T}{6\pi(a_x a_y a_z)^{1/2}} \chi_\nu^{-3/2} + \ldots \right\}, \qquad (61)$$

where the terms with M_{L0}^2 and g_0 describe effects of zero-point SF and the contribution containing M_{LT}^2 accounts for thermal SF. The GL term

$\sim \chi_\nu^{-3/2}$ is small in the SM regimes and gives divergent second thermodynamical derivatives in the critical region. Here $\Delta F_0(T)$ accounts for the SF contribution independent on the susceptibilities χ_ν and therefore does not affect magnetic properties. However, it is strongly temperature dependent and may influence thermal properties of magnets, e.g., the specific heat [30].

To illustrate this we present the explicit expressions for $\Delta F_0(T)$ in the low temperature ($T \ll T_{SF}$) limit of the QSM regime,

$$\Delta F_0(T) = -\frac{\pi}{6} N_0 k_B \sum_\nu \frac{T^2}{\zeta_\nu T_{SF}} \ln \frac{T_{SF}^*}{T} \quad (z=3), \qquad (62)$$

$$\Delta F_0(T) = -\frac{\pi}{2} N_0 k_B \sum_\nu \frac{T^2}{\zeta_\nu T_{SF}} (1 - \beta \sqrt{\frac{T}{T_{SF}}}) \quad (z=2), \qquad (63)$$

intending to comment further on the low temperature anomalies of the specific heat (see Sec. 5.5) in magnets with ferro ($z=3$) and antiferromagnetic ($z=2$) instabilities, where $T_{SF}^* \sim T_{SF}$ and $\beta = 0.80279$ is a dimensionless coefficient.

Neglecting the small GL contribution in (61) we may present the solution of Eqs. (49), (55), and (61) aside the critical region in the form of the Landau free energy [18]

$$F = F_0 + \Delta F_0(T) + \frac{1}{2\chi(T)} \left[M_{L0}^2 + M_{LT}^2 - \frac{3g}{2\gamma\chi(T)} \right]$$
$$+ \frac{1}{2\chi(T)} M^2 + \frac{\gamma}{4} M^4, \qquad (64)$$

where the coefficients $\chi(T)$ and γ are strongly renormalized with respect to the initial GL ones, χ_0 and γ_0,

$$\chi^{-1}(T) = \zeta \chi_0^{-1} + \frac{5}{3}\gamma \left(M_{L0}^2 + M_{LT}^2 \right) \qquad (65)$$

$$\gamma = \gamma_0 \frac{1 - 5g}{1 + 6g}, \qquad (66)$$

Here g is the renormalized spin anharmonicity parameter, defined by

$$g_0 = g \frac{1 + 6g}{1 - 5g} \qquad (67)$$

and

$$\zeta_t = \zeta_l = 1 - 5g \equiv \zeta > 0 \quad . \tag{68}$$

It should be emphasized that the free energy (64) is valid in the QSM and CSM regimes when inequalities (9) and (10) hold. Moreover, (64) describes also the low temperature ($T = 0$) ground state provided a magnet is close to a magnetic instability and the inequality (10) holds for $T = 0$ as we assume below.

In our treatment we have dropped the GL term $\sim \chi_\nu^{-3/2}$ in (61), which is small in the SM regimes. The account of it does not change the form of the free energy (64) and results in a small renormalization of ζ of order $5(\gamma/\gamma_0)g_T \leqslant 5g_T \ll 1$, where g_T is the thermal contribution to the spin anharmonicity parameter given by (34). This verifies the self-consistent variational procedure used above.

In the limit M_{L0}^2, $g_0 \to 0$ when all zero-point SF effects are neglected, (64) reduces to the free energy of the conventional SCR theory of SF [11]. Below we shall analyze the novel effects arising due to zero-point SF and induced by them strong spin anharmonicity.

5.4. CRITERION FOR MAGNETIC INSTABILITY.

As it follows from (64) zero-point fluctuations essentially modify the initial GL coefficients χ_0 and γ_0 defined by the electronic band structure and therefore essentially affect the ground state of itinerant magnets. The effects of spin anharmonicity reduce the coupling constant γ which according to (66) and (67) is approximately given by

$$\gamma \approx \begin{cases} \gamma_0, & g_0 \ll 1, \\ \gamma_0/5g_0, & g_0 \gg 1, \end{cases} \tag{69}$$

and decreases in the strongly anharmonic limit. Similarly, for the coefficient ζ which enhances the SF free energy (54) with respect to the RPA value, we have

$$\zeta \approx \begin{cases} 1 - 5g_0, & g_0 \ll 1, \\ \sim 1/g_0, & g_0 \gg 1. \end{cases} \tag{70}$$

In the strong anharmonicity limit ($g_0 \gg 1$) it vanishes, which results in the increase of the SF contribution to the free energy $\sim g_0$.

The criterion for a magnetic instability in the ground state is also strongly modified comparing with Stoner or Hartree-Fock one, $\chi_0^{-1} < 1$. According to (65) now it reads

$$\chi^{-1}(0) = \zeta \chi_0^{-1} + \frac{5}{3}\gamma M_{L0}^2 < 0 \quad . \tag{71}$$

The role of zero-point SF here is twofold. First, they tend to suppress an instability by adding a positive contribution $\sim \gamma M_{L0}^2$. Second, zero-point SF effects reduce the negative Stoner term χ_0^{-1} by a factor $\zeta < 1$.

Summarizing these results we would like to emphasize the necessity of incorporating dynamical SF effects into electronic band structure calculations [38] for itinerant magnets with soft-mode fluctuations. Anyway, Eqs. (65) and (66) establish a link between the initial GL coefficients χ_0^{-1} and γ_0 which may be calculated within the band structure calculations employing fixed-spin-moment method [39,40] and the values $\chi^{-1}(0)$ and γ renormalized by zero-point effects, which can be inferred, e.g., from magnetic measurements. The coupling constants γ, γ_0, spin anharmonicity parameters g, g_0, and the value ζ illustrating the importance of zero-point effects and spin anharmonicity are presented for a series of itinerant magnets in Table I taken from [18].

TABLE 1. Spin anharmonicity effects induced by zero-point SF in itinerant electron magnets MnSi, Ni$_3$Al, and ZrZn$_2$.

	g	g_0	ζ	γ (G^{-2})	γ_0 (G^{-2})
MnSi	0.18	3.7	0.1	0.15	3.1
Ni$_3$Al	0.15	1.1	0.25	0.53	4.0
ZrZn$_2$	0.10	0.32	0.5	2.0	6.4

5.5. LOW TEMPERATURE SPECIFIC HEAT.

Zero-point SF and caused by them strong spin anharmonicity may also essentially affect finite temperature properties of itinerant magnets in the SM regimes. Here we discuss anomalies of the specific heat in the low temperature limit of the QSM regime which according to (9) and (12) takes place in the following temperature range

$$T_{FL} \ll T \ll T_{SF} \quad . \tag{72}$$

Usually, the low temperature specific heat is interpreted in terms of interacting quasiparticles of a Fermi liquid,

$$C = \gamma_e T \left[1 + AT^2 + BT^2 \ln T + \ldots \right] \quad , \tag{73}$$

where the coefficients A and B are expressed in terms of Fermi liquid interactions [29]. In metals with magnetic instabilities they were also related to the SF spectrum [41,42] within the RPA or Gaussian approximation where mode-mode coupling is neglected. Although these approximations are not valid at low temperatures due to strong anharmonicity induced by zero-point SF the result (73) seems to be generally applicable below T_{FL} in the FL regime of SF [29, 42]. Above T_{FL} in the QSM regime (73) does not hold, which may be interpreted in terms of the non-Fermi liquid behavior [34] exhibited by many strongly correlated electron systems, heavy fermion metals in particular [29].

From the expressions (62)- (64) for the free energy one easily gets the SF contribution to the specific heat in the low temperature ($T \ll T_{SF}$) QSM regime:

$$C_{SF} = \frac{\pi}{\zeta} N_0 k_B \frac{T}{T_{SF}} \ln \frac{T^*_{SF}}{T}$$
$$- \frac{10}{27\zeta} \frac{\gamma M^2_{L0} M^2_{LT}}{T} \left\{ 1 + O\left(\frac{\chi^{-1}}{\gamma M^2_{L0}}\right) + O\left(\frac{M^2_{LT}}{M^2_{L0}}\right) \right\} \quad (74)$$

for ferromagnetic instabilities ($z = 3$) and

$$C_{SF} = \frac{3\pi}{\zeta} N_0 k_B \frac{T}{T_{SF}} \left\{ 1 - \frac{15\beta}{8} \sqrt{\frac{T}{T_{SF}}} \right\}$$
$$- \frac{5}{8\zeta} \frac{\gamma M^2_{L0} M^2_{LT}}{T} \left\{ 1 + O\left(\frac{\chi^{-1}}{\gamma M^2_{L0}}\right) + O\left(\frac{M^2_{LT}}{M^2_{L0}}\right) \right\} \quad (75)$$

for antiferromagnetic ones ($z = 2$). The first terms in the r.h.s. of (74) and (75), proportional to $T \ln T_{SF}/T$ and T, may be regarded in the spirit of [34] as a "non-Fermi liquid" and "Fermi liquid" ones. However, they are by a factor

$$\sim \frac{\gamma M^4_{L0}}{N_0 k_B T_{SF}} \left(\frac{T_{SF}}{T}\right)^{\frac{z-1}{z}} \ln^{2-z} \frac{T_{SF}}{T} \sim \left(\frac{T_{SF}}{T}\right)^{\frac{z-1}{z}} \ln^{2-z} \frac{T_{SF}}{T} \gg 1 \quad (76)$$

smaller than the leading contributions $\sim \gamma M^2_{L0} M^2_{LT}/T \sim T^{1/z}$ defined by the large amplitude of zero-point SF. In the ordered state there arise corrections to (74) and (75) of order $\chi^{-1}/\gamma M^2_{L0}$, $M^2_{LT}/M^2_{L0} \ll 1$, which may be neglected in the QSM regime. It should be emphasized that the specific heat in the QSM regime is essentially affected by the non-Gaussian behavior of SF: the leading terms in (74) and (75) are proportional to the mode-mode coupling constant and the whole specific heat is enhanced by a factor $\zeta^{-1} = (1 - 5g)^{-1} > 1$ due to spin anharmonicity induced by zero-point SF.

Recently the SF contribution to the specific heat of itinerant magnets close to a zero-point phase transition (quantum critical point), which in our classification scheme is related to the QSM regime, was analyzed in [34] using the scaling procedure and treating mode-mode coupling of SF within the GL model as a perturbation. The main result of [34] was to show that the leading temperature dependencies result from Gaussian non-interacting SF, and mode-mode coupling effects yield negligible corrections to them. Similar results are given by the SCR theory [11]. However, though the temperature dependencies of the leading terms in the specific heat presented in [11,34] are correct, the coefficients are not due to quantum effects of zero-point SF and spin anharmonicity neglected in the scaling procedure [34] and SCR approach [11].

Comparing the results (74) and (75) with those got, e.g., in [34] we conclude that the first terms in (74) and (75) contain the enhancement factor ζ^{-1} which is absent in the Gaussian approximation of [34]. The leading temperature behavior of (74) and (75) given by $M_{LT}^2/T \sim T^{1/z}$ also agrees with the presented in [34]. However, the coefficients in the leading terms are essentially different: in [34] they are proportional to χ^{-1} which according to (74) and (75) gives negligible corrections $\sim \chi^{-1}/\gamma M_{L0}^2 \ll 1$ to the calculated here coefficients $\sim \gamma M_{L0}^2/\zeta$ defined by quantum effects of zero-point SF. The term $\sim T^{3/2}$ in (75) which in [34] is thought to play an important role in metals with antiferromagnetic instabilities is also smaller by a factor $T_{FL}/T \gg 1$ than our leading term $\sim \gamma M_{L0}^2 M_{LT}^2/T \sim T^{1/2}$.

It should be emphasized that the FL temperature $T_{FL} \sim T_{SF}(\chi_l c_c)^{-z/2}$ is scaled with the characteristic SF energy [33] $k_B T_{SF} \sim \varepsilon_F$ proportionally to the Fermi energy ε_F, and in the parabolic electron band model - to the inverse mass of quasiparticles. Thus, the low values of T_{SF} may imply a heavy fermion behavior in the QSM regime above the temperature T_{FL} which may be rather low giving rise to interpretations of the low-temperature specific heat anomalies in terms of quantum critical effects. The description of the heavy fermion behavior based on our formula (74) and (75) essentially differs from the previous attempts [11,34] where zero-point effects were ignored.

It should be mentioned that below T_{FL} in the generalized FL regime a meaningful description of SF is lacking though it is obvious that zero-point effects and caused by them strong spin anharmonicity must play an important role there.

5.6. MAGNETIC EQUATION OF STATE AND TOTAL LOCAL MAGNETIC MOMENT.

Now we turn to analyze magnetic properties of itinerant magnets in the SM regimes. Substituting the free energy (64) into (56) we get the magnetic equation of state

$$\frac{B}{M} = \chi^{-1}(0) + \gamma \left(M^2 + \frac{5}{3} M_{LT}^2 \right), \tag{77}$$

which accounts for both zero-point and thermal SF and for strong spin anharmonicity effects. Resulting from the Landau free energy (64) equation of state (77) has essentially the same form as in the rotationally invariant form of the SCR theory [8,10,11]. The effects of zero-point SF and spin anharmonicity are incorporated into the renormalized GL coefficients $\chi^{-1}(0)$ and γ, which essentially differ from the initial ones, χ_0^{-1} and γ_0 (see (66) and (71)) related to conventional band structure calculations. We emphasize that $\chi^{-1}(0)$ and γ, not χ_0^{-1} and γ_0 are measurable quantities.

According to (77) the temperature dependence of the paramagnetic susceptibility is described by (65), and similar to the SCR theory [11], arises due to mode-mode coupling of SF in the SM regimes, $\chi^{-1}(T) \sim (5/3)\gamma M_{LT}^2$. However, unlike [11] based on the weak coupling constraint (32), the result (65) of the SM theory holds for arbitrary spin anharmonicity and accounts for effects of zero-point SF.

Basing on the magnetic equation of state (77) it is possible to analyze the temperature dependence of the total squared magnetic moment

$$(M_L^2)_{tot} = M^2 + (M_L^2)_{Z.P.} + (M_L^2)_T . \tag{78}$$

For the SM regimes, using the expansions (24) and (33) and the susceptibilities χ_t and χ_l resulting from (77) we get

$$(M_L^2)_{tot} = \zeta \left(M^2 + M_{LT}^2 \right) + M_{L0}^2 - \frac{3g}{\chi(0)\gamma} . \tag{79}$$

As it follows from (79) zero-point contribution to M_L^2 partially compensates the temperature dependence of M_{LT}, which results in the factor $0 < \zeta < 1$ in the first term of the r.h.s., defined by (68). However, this compensation may be complete,

$$(M_L^2)_{tot} \simeq M_{L0}^2 = const., \tag{80}$$

only in the limit of strong spin anharmonicity, $g_0 \to \infty$, when $\zeta \to 0$.

Recently this effect of compensation called later "a sum rule" [43] was used by Takahashi [44] as a basis to describe thermal properties of itinerant magnets with account of zero-point SF. The magnetic equation of state in his model may be written in the form (77) with a coefficient $\gamma = \gamma_0/5g_0$ which according to (66) arises in our approach in the limit of strong spin anharmonicity. However, the "sum rule" approach [43,44] is not borne out of thermodynamics of itinerant magnets with soft-mode SF and probably may be related to magnets with localized atomic spins fixed by the Hund's rule [3,4,20].

According to (40) and (79) in the paramagnetic state $(M_L^2)_{tot}$ increases with temperature in the SM regimes. In the LM regimes the zero-point contribution to $(M_L^2)_{tot}$ vanishes $(M_L^2)_{Z.P.} \sim T^{-2}$ and the thermal one, $(M_L^2)_T$, saturates (see (42)) provided the susceptibility follows the Curie-Weiss law, which results in a saturation of the total squared moment $(M_L^2)_{tot} \to M_{eff}^2$. We would like to point out to a maximum in the temperature dependence of $(M_L^2)_{tot}$ which is likely to arise near the temperature T_L of a crossover from the SM to LM regimes due to the relatively slow decrease of $(M_L^2)_{Z.P.}$.

As it follows from the magnetic equation of state (77) in the ordered state of ferro- and antiferromagnets the temperature dependencies of the squared magnetization

$$M^2(T) = M^2(0) - \frac{5}{3}M_{LT}^2 \qquad (81)$$

and the combination

$$M^2(T) + M_{LT}^2 = M^2(0) - \frac{2}{3}M_{LT}^2 \qquad (82)$$

are similar to those predicted by the SCR theory [11]. Near the critical temperature T_C, $(M_L^2)_{tot}$ has a minimum where $\zeta(M^2 + M_{LT}^2)$ equals $3\zeta/5$ of its value at $T = 0$. Temperature dependencies of local magnetic moments in the ordered and paramagnetic states of itinerant magnets are illustrated by Figs.3,4 and 5. It is worthy to note that the temperature dependence of $(M_L^2)_{tot}$ measured by inelastic neutron scattering for a paramagnetic system Y(Sc)Mn$_2$ with an antiferromagnetic instability [14] agrees with the presented in Fig.3. The inferred from the measurements zero-point contribution to $(M_L^2)_{tot}$ decreases rapidly above the temperature $100K$ [14], which is in good agreement with the theoretical estimate of T_L for this system [33].

5.7. MAGNETOVOLUME EFFECT.

In the previous Sec. 5.6 we analyzed the temperature dependence of the total square magnetic moment $(M_L^2)_{tot}$ which also defines the magnetovolume effect in itinerant electron magnets [18]. Here we assume that in itinerant magnets with soft-mode SF the parameter most sensitive to the volume

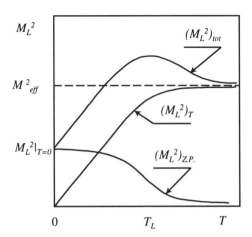

Figure 3. Temperature dependencies of the total squared local magnetic moments $(M_L^2)_{tot.}$ and zero-point and thermal contributions to it, $(M_L^2)_{Z.P.}$, and $(M_L^2)_T$, for paramagnets with $M_{eff}^2/(M_L^2)_{tot}|_{T=0} > 1$. M_{eff}^2 and T_L indicate the saturated moment and the temperature of a crossover to the LM regimes of SF.

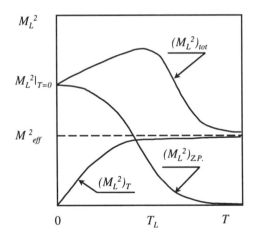

Figure 4. Temperature dependencies of the squared local magnetic moments $(M_L^2)_{tot}$, $(M_L^2)_{Z.P.}$, and $(M_L^2)_T$ for paramagnets with $M_{eff}^2/(M_L^2)_{tot}|_{T=0} < 1$.

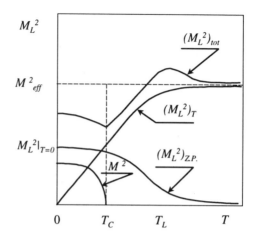

Figure 5. Temperature dependencies of the squared local magnetic moments $(M_L^2)_{tot.}$, $(M_L^2)_{Z.P.}$, and $(M_L^2)_T$ for ferro- or antiferromagnets with $M_{eff}^2/(M_L^2)_{tot}|_{T=0} > 1$. Here M^2 and T_C are the squared magnetization and the critical (Curie or Néel) temperature.

change is $\chi_0 = \chi_0(V)$, and the volume dependence of the free energy comes mainly from it provided

$$\left|\frac{\partial \ln \chi_0}{\partial \ln V}\right| \gg 1 \ . \tag{83}$$

Therefore, here we neglect weak volume dependencies of $c(\mathbf{q})$, $\Gamma(\mathbf{q})$, $\omega_c(\mathbf{q})$, and γ_0. Substituting then the free energy density (64) into (57) we get the equation of state in the following explicit form

$$P = P_0(V,T) + C_0(M_L^2)_{tot} \ , \tag{84}$$

where $P_0 = -\partial[(F_0+\Delta F)V]/\partial V$ is the nonmagnetic contribution and $C_0 = (1/2)V^2\partial(\chi V)^{-1}/\partial V$ is the conventional magnetoelastic coupling constant related to χ_0. Then for the magnetic contribution to the volume strain from (84) we get

$$\omega_m = \frac{C_0}{K}(M_L^2)_{tot} \tag{85}$$

(here K is the bulk modulus) which differs from the Moriya-Usami result based on the SCR theory [11] where the zero-point contribution to $(M_L^2)_{tot}$ was neglected. Using the explicit expression (79) for the total squared moment we have [18]

$$\omega_m = \frac{C}{K}(M^2 + M_{LT}^2) + \frac{C_0}{K}\left[M_{L0}^2 - \frac{3g}{\chi(0)\gamma}\right], \qquad (86)$$

where $C = \zeta C_0$ is the renormalized magnetoelastic constant. Comparing (86) with the result of the SCR theory [11] we may conclude that zero-point SF give rise to a huge temperature independent term in the r.h.s. containing M_{L0}^2 and $\chi^{-1}(0)$. Another effect of zero-point SF is to reduce the temperature dependence of the magnetovolume effect by a factor $\zeta^{-1} = (1-5g)^{-1} > 1$ which can be incorporated into the renormalized magnetoelastic constant C - a similar effect we discussed with respect to the coefficients $\chi^{-1}(0)$ and γ in the magnetic equation of state (77).

6. Quantum effects in the localized moments regimes of SF.

As we have already mentioned in Sec. 4.2 in the LM regimes existing at relatively high temperatures $T > T_L$ quantum zero-point contributions to M_L^2 and g_{SF} are vanishing. However, quantum effects may be still important for the thermal contribution to the squared local moment (40) where they are described by the factor $G(z) \leq 1$ given by (41), which accounts for the quantum reduction of the SF phase space and reduces $(M_L^2)_T$.

Provided local moments are saturated and the susceptibility has the Curie-Weiss form, $\chi^{-1} \sim T$, (40) gives a link between the squared saturated moments M_{eff}^2 and the Curie constant C,

$$M_{eff}^2 = 3k_B N_0 C G(z_c), \qquad (87)$$

where

$$z_c = \frac{\hbar \Gamma}{2\pi k_B C} \qquad (88)$$

is a constant describing the role of quantum effects. We would like to mention that according to the phase diagram (see Fig.1) the equality $z_c = 1$ divides the localized moments regimes in the temperature range we discuss into the quantum, QLM, $z_c > 1$, and classical, $z_c < 1$, ones. For high temperatures $T \gg \hbar \omega_c / k_B$, where quantum effects are negligible one should set $G(z_c) \to 1$ in (87).

Thus the values of z_c and $G(z_c)$ are a measure of quantum effects in the LM regimes: for $z_c \ll 1$ we have $G(z_c) \approx 1$, and quantum effects may be neglected, for $z_c \gg 1$ when $G(z_c) \sim 1/z_c$ they may essentially reduce the value of M_{eff}^2. The parameter z_c also determines the ratio of the squared local moments in the low and high temperature limits,

$$\frac{M_{eff}^2}{M_{L0}^2} = \frac{G(z_c)}{z_c \alpha_1} , \qquad (89)$$

which according to Figs.3,4 and 5 determines the temperature dependence of the total squared local moment through the whole temperature range. Estimating roughly $\alpha_1 \sim 1$ from (89) we get

$$\frac{M_{eff}^2}{M_{L0}^2} \sim \begin{cases} 1/z_c, & z_c \ll 1, \\ 1/z_c^2, & z_c \gg 1. \end{cases} \qquad (90)$$

Thus, e.g., in the CLM regime, when $z_c \ll 1$, quantum zero-point effects may be neglected and SF may be treated on a classical basis. When $z_c \gg 1$ zero-point contribution to $(M_L^2)_{tot}$ may be still important in the QLM regime at relatively high temperatures $T \gtrsim T_L$.

Finally, we would like to comment on the effective moments p_{eff} and p_c defined usually by [4,11]

$$M_{eff}^2 = \mu_B^2 N_0^2 \, p_{eff}^2 = \mu_B^2 N_0^2 \, p_c(p_c + 2) \qquad (91)$$

and compare them with the spontaneous atomic moment p_0 defined by

$$M^2 = \mu_B^2 N_0^2 \, p_0^2 . \qquad (92)$$

In Table II we present the calculated values z_c, $G(z_c)$, p_c/p_0, and p_{eff}/p_0 comparing the latter with $(p_{eff}/p_0)'$ taken from the previous estimates [10,11].

TABLE 2. Quantum effects in the LM regimes of SF.

	z_c	$G(z_c)$	$\frac{p_{eff}}{p_0}$	$\frac{p_c}{p_0}$	$\left(\frac{p_{eff}}{p_0}\right)'$
MnSi	0.97	0.21	2.5	1.0	5.5[a]
Ni$_3$Al	0.029	0.89	15	6.6	16[a]
ZrZn$_2$	4.41	0.053	2.1	0.34	8.9[b]

[a] taken from Ref. [10] ; [b] taken from Refs. [11,18]

As it follows from Table II quantum effects in the LM regimes may play an essential role as is the case with ZrZn$_2$. In this case the parameter p_c/p_0 shown as a function of T_C on the well-known Rhodes-Wohlfarth plot (see, e.g., Ref.[11]), where it is assumed to be always larger than 1, looses its role as a criterion for separating between localized ($p_c/p_0 \approx 1$) and itinerant ($p_c/p_0 \gg 1$) magnets, Relating that, it is possible to state that ZrZn$_2$ with $p_c/p_0 = 0.32 < 1$ is a good illustration for a break down of a previous physical picture for the LM behavior where quantum effects of SF dynamics were neglected.

7. Summary

We have presented a physical picture for the SF phenomena in itinerant electron magnets exhibiting strongly coupled soft-mode fluctuations, which essentially differs from the one previously worked out by the SCR theory [11], based on the weak coupling constraint. With respect to the role of spatial dispersion the SF behavior may be separated into three main regimes: a generalized FL, SM and LM ones. The low temperature FL regime is affected by relatively high frequency quantum SF leading to a FL behavior of the specific heat (73). The SM regime is dominated by strongly coupled low-frequency SF giving rise to an increase (decrease) of unsaturated local magnetic moments with temperature in the disordered (ordered) state. Strong mode-mode coupling induced by zero-point SF defines the mechanism of the temperature variation of the magnetic susceptibility which is often interpreted in terms of an effective Curie-Weiss law [11,33]. The true Curie-Weiss behavior associated with saturated and localized magnetic moments arises at higher temperatures, in the LM regimes of SF, where it is possible to neglect their spatial dispersion and treat them as excitations localized in real space. Quantum zero-point SF effects and caused by them spin anharmonicity are vanishing here. However, quantum dynamical effects may be still important for thermal SF, reducing saturated moments. Finally, we have outlined the main results of the recently worked out SM theory of SF [18,19] which unlike the conventional SCR approach [11] accounts for zero-point SF and induced by them strong spin anharmonicity and founds the basis for understanding the SF phenomena in itinerant magnets with soft-mode fluctuations.

Acknowledgments.

This work was supported by MINATOM of Russia and Bundesministerium für Forschung und Technologie of Germany (grant No X 221.32), and NATO Sience Committee (grant No HTECH. ARW 871251). One of us (A.S.) is grateful to C. Lacroix for helpful discussions and acknowledges the support of the Russian foundation for Basic Research (grant No 96-02-17693) and the State Support of the Leading Scientific Schools of Russia (grant No 96-15-96750)

References

1. Frenkel, Ya. I. (1928) Elementare Theorie Magnetischer und Electrischer Eigenschaften der Metalle beim absoluten Nullpunkt der Temperature, *Z. Physik* **49**, 31-45.
2. Bloch, F. (1929) Remarks on the electron theory of ferromagnetism and electrical conductivity, *Z. Physik* **57**, 545-557.
3. Heisenberg, W. (1928) Zur Theorie des Ferromagnetismus, *Z. Physik* **49**, 619-636.
4. Herring, C. (1964) Exchange interactions among itinerant electrons, in G. T. Rado and H. Suhl (eds.), *Magnetism*, **vol.4**, Academic Press, New York.
5. Stoner, E. C. (1938) Collective electron ferromagnetism, *Proc. Roy. Soc* **165A**, 372-414.
6. Murata, K.K. and Doniach, S. (1972) Theory of magnetic fluctuations in itinerant ferromagnets, *Phys. Rev. Lett.* 29, 285-288.
7. Moriya, T. and Kawabata, A. (1973) Effects of spin fluctuations on itinerant electron ferromagnetism, *J. Phys. Soc. Jap.* **34**, 639-651.
8. Dzyaloshinski, I. E. and Kondratenko, P. S. (1976) On the theory of weak ferromagnetism of a Fermi liquid, *Sov. Phys. - JETP* **43**, 1036-1043.
9. Shimizu, M. (1981) Itinerant electron magnetism,. *Rep. Prog. Phys.* **44**, 329-409.
10. Lonzarich, G. G. and Taillefer, L. (1985) Effect of spin fluctuations on the magnetic equation of state of ferromagnetic and nearly ferromagnetic metals, *J. Phys.* **180**, 4339-4371.
11. Moryia, T. (1985) *Spin Fluctuations in Itinerant Electron Magnetism*, Springer, Berlin.
12. Bruce, A. D. and Cowley, R. A. (1981) *Structural Phase Transitions*, Taylor and Francis, London.
13. Silin,V. P. and Solontsov, A. Z. (1981) On the relaxation of magnons in itinerant electron ferromagnets, *Fiz. Met. Metalloved.* **52**, 231-242.
14. Shiga, M., Fujisawa, K., and Wada, H. (1993) Spin liquid behaviour of highly frustrated Y(Sc)Mn$_2$ and effects of nonmagnetic impurity, *J. Phys. Soc. Jap.* **62**, 1329-1336.
15. Shiga, M. (1998) Spin fluctuations in itinerant frustrated systems in D. Wagner, W. Brauneck, and A. Z.Solontsov (eds.), *Itinerant Electron Magnetism: Fluctuation Effects*, Kluwer Academic Publisher, Dordrecht, this volume.
16. Solontsov, A. Z. (1993) Effects of zero-point spin fluctuations and Landau theory of weak itinerant magnets, *Sov. Phys. Met. Metallogr.* **75**, 1-5.
17. Solontsov, A. Z. (1993) Zero-point spin fluctuation effects in anharmonic weak itinerant electron magnets, *Int. J. Mod. Phys.* **7** , 585-589.
18. Solontsov, A. Z. and Wagner, D. (1995) Zero-point spin fluctuations and the magnetovolume effect in itinerant electron magnetism, *Phys. Rev.* **B51**, 12410-12417.
19. Solontsov, A. Z. and Wagner, D. (1994) Spin anharmonicity and zero-point fluctuations in weak itinerant electron magnets, *J. Phys.: Condens. Matter* **6**, 7395-7402.
20. Wagner, D. (1985) *Introduction to the Theory of Magnetism*, Spinger, Berlin.
21. Ishikawa, Y., Noda, Y., Uemura, Y. J., Majkrzak, C. F., and Shirane, G. (1985) Paramagnetic spin fluctuations in weak itinerant electron ferromagnet MnSi, *Phys. Rev.* **B31**, 5884-5893.
22. Lonzarich, G. G., Bernhoeft, N. R., and Paul, D. McK. (1989) Spin density fluctuations in magnetic metals, *Physica* **155B-157B**, 699-705.
23. Rainford, B. D., Dakin, S., and Cywinski,R. (1992) Spin fluctuations in YMn2 and related alloys, *J. Magn. Magn. Mat.* **104-107**, 1257-1261.
24. Ballou, R., Lelievre-Berna, E., and Fak, B. (1996) Spin fluctuaions in $(Y_{0.97}Sc_{0.03})Mn_2$: A geometrically frustrated, hearly antiferromagnetic, itinerant electron system, *Phys. Rev. Lett.* **67**, 2125-2128.
25. Bernhoeft, N. R, Roessli, B., Sato, N., Aso, N., Hiess, A., Lander, G. H., Endoh, Y., Komatsubara, T. (1998) UPd$_2$Al$_3$: An analysis of the Inelastic Neutron Scat-

tering Spektra in D. Wagner, W. Brauneck, and A. Z. Solontsov (eds.), *Itinerant Electron Magnetism: Fluctuation Effects*, Kluwer Academic Publisher, Dordrecht, this volume.

26. Kalashnikov, O. K. and Fradkin, E. S. (1973) The spectral density method applied to systems showing phase transitions, *Phys. Stat. Sol.* **59b**, 7-46.
27. Martin, P. (1967) Measurements and Correlaton Functions, in C. de Witt and R. Ballian (eds.), *Many Body Physics*, Gordon and Brench Sci. Publishers, New York, pp. 39-136.
28. Solontsov, A. Z., Vasil'ev, A. N., and Wagner, D. (1995) Longitudinal spin fluctuations in itinerant electron ferromagnets, *J. Phys.: Condens. Matter* **7**, 1855-1862.
29. Pethick, C. J. and Pines, D. (1989) Thoughs on heavy fermion systems, in L.V. Keldysh and V. Ya. Fainberg (eds.), *Problems of Theoretical Physics and Astrophysics*, Nauka, Moscow, 304-324.
30. Solontsov, A. Z. and Lacroix, C. (1997) Specific heat of soft-mode spin fluctuations in itinerant electron magnets, *Phys. Lett.* **224A**, 298-302.
31. Solontsov, A. Z.., Lacroix,C., and Wagner, D. (1997) Soft-mode spin fluctuations in itinerant electron ferro- and antiferromagnetism, *Physica* **237-238B**, 480-481.
32. Solontsov,A. Z., Lacroix, C., and Wagner,D. (1997) Soft-mode vs. localized moments regimes of spin flucuations in itinerant electron magnetism, *Acta Physica Polonica* **92**, 359-362.
33. Lacroix, C., Solontsov, A. Z., and Ballou, R. (1996) Spin fluctuations in itinerant electron magnetism and anomalous properties of $Y(Sc)Mn_2$, *Phys. Rev.* **B54**, 15178-15184.
34. Zülicke, U. and Millis, A. J. (1995) Specific heat of a three-dimentional metal near a zero-temperature magnetic phase transition with dynamic exponent z = 2, 3, or 4, *Phys. Rev.* **B51**, 8996-9004.
35. Silin, V. P. and Solontsov, A. Z. (1990) Theory of structural-volume effect and the invar anomaly systems with structural phase transitions, *Sov. Phys. - JETP* **71**, 610-618.
36. Solontsov, A. Z. and Vasil'ev, A. N. (1993) Spin fluctuation damping of magnons in itinerant electron magnets, *Phys. Lett.* **177A**, 362-366.
37. Solontsov, A. Z. and Vasil'ev, A. N. (1995) Magnon-paramagnon scattering and damping of magnons in itinerant ferromagnets, *J. Magn. Magn. Mat.* **140-144**, 213-214.
38. Steiner, M. M., Albers, R. C., and Sham, L. J. (1992) Quasiparticle properties of Fe, Co, and Ni, *Phys. Rev.* **45**, 13272-13284.
39. Mohn, P., Schwarz, K., and Wagner, D. (1991) Magnetoelastic anomalies in Fe-Ni Invar alloys, *Phys. Rev.* **43**, 3318-3324.
40. Entel, P., Hoffmann, E., Mohn, P., Schwarz, K., and Moruzzi, V.L. (1993) First-principles calculations of the instability leading to the Invar effect, *Phys. Rev.* **47**, 8706-8720.
41. Coffey, D. and Pethick, C. J. (1988) Quasiparticle spectra and specific heat of a normal Fermi liquid in a spin-fluctuation model, *Phys. Rev.* **B37**, 1647-1665.
42. Edwards, D. M. and Lonzarich, G. G. (1992) The entropy of fluctuating moments at low temperatures, *Phil. Mag.* **65**, 1185-1189.
43. Moryia, T. and Takimoto, T. (1995) Anomalous properties around magnetic instability in heavy electron systems, *J. Phys. Soc. Jap.* **64**, 960-969.
44. Takahashi, Y. (1994) Spin fluctuation theory of nearly ferromagnetic metals, *J. Phys.: Condens. Matter* **6**, 7063-7073.

EFFECTS OF SPIN FLUCTUATIONS IN TRANSITION METALS AND ALLOYS

M. SHIMIZU
2-10, Shinike-cho, Chikusa-ku,
Nagoya 464-0027, Japan

Abstract. The paramagnetic susceptibility for transition metals, alloys and compounds is calculated from a realistic itinerant electron model taking into account the effect of spin fluctuations (SF) in the static Gaussian approximation (SGA). The magnetic free energy based on the Stoner model is averaged over longitudinal and transverse SF. On the basis of values which were calculated for the electronic density of states, molecular field coefficient, exchange stiffness coefficient, and the orbital susceptibility, the temperature variation of the paramagnetic susceptibility is calculated for several transition metals, alloys, and compounds and is compared with experiment. It is concluded that SGA applied to a realistic Stoner model is sufficient to explain the temperature variation of the paramagnetic susceptibility in itinerant electron magnets not only qualitatively, but very often even quantitatively. The present theory is compared with other theories of SF in itinerant electron magnets.

1. Introduction.

After the magnetism of metals at absolute zero temperature was first studied by Frenkel [1] in Russia, Bloch, Mott, Stoner, Slater and others, (cf. references [2, 3]), adopted the itinerant electron model to explain the ferromagnetism of iron, cobalt and nickel. Among many theories on itinerant electron magnetism, the Stoner model, as a molecular field theory, is very simple and very useful to understand various magnetic and static properties of transition metals (TM), as for example temperature variation of electronic specific heat C_E, paramagnetic susceptibility χ and spontaneous magnetization M, forced magnetostriction h, high-field (or para-process) susceptibility χ_{hf} and various magnetovolume effects including the Invar effects [3].

On the other hand, the electronic density of states (DOS), cohesive energy E, lattice constant a, bulk modulus B, molecular field coefficient α, χ_{hf} and M at 0 K have been calculated for many TM in the local density approximation (LDA) or local spin density approximation (LSDA) for the exchange and correlation energy without introducing any adjustable parameters [4-6]. The calculated values of E, a, B, M, and h are close to the experimental ones and the general calculated trend of their dependence on atomic number is very similar to experiment, but the calculated values of α [4] predict very high Curie temperatures T_c in Ni and Fe if inserted in a simple Stoner model.

The temperature variation of χ for TM so far has been calculated from the Stoner model by making use of the calculated DOS [4], the contribution of the orbital paramagnetic susceptibility χ_{orb} to χ, which is fairly large in most TM, being calculated within a simple method [7]. The results are summarized in [3]. It has been found that the calculated temperature dependence of $1/\chi$ and C_E is qualitatively similar to the experimental results, provided that α is taken as an adjustable parameter. It then turns out that the values of α for the minus or plus group TM, (cf. [3]), are smaller or larger, respectively, than calculated in [4]. In the case of C_E there are no adjustable parameters. It is shown below that the difference between the calculated and empirical values of α can be attributed to the effect of spin fluctuations (SF) both for the minus and plus group TM. At finite temperature the effect of SF is very important in determining the temperature dependence of various magnetic and elastic properties and the thermal expansion in TM, in addition to the temperature variation of the Fermi-Dirac distribution of the electrons over DOS of the d-band.

The effect of SF has been extensively studied by many researchers, especially, for weak itinerant electron ferromagnets (WIEF) [8-20]. Murata and Doniach [8] first applied the Landau theory of fluctuations [21, 22] in the static Gaussian approximation (SGA) to WIEF and explained the observed Curie-Weiss type behaviour of χ. We have extended the Murata-Doniach theory to general itinerant electron magnets like Ni and Fe and have calculated the temperature variation of χ for various paramagnetic and ferromagnetic TM and also for WIEF, alloys and compounds of TM at high temperatures [15-18, 23-33]. Our main previous results as well as new results are discussed below. In our calculations we use, whenever available, the calculated results of DOS and α, for TM, their alloys and compounds, and determine the temperature variation of the coefficients of the magnetic free energy expansion with respect to M^2.

In the Murata-Doniach theory, the transition from the ferromagnetic state to the paramagnetic state is of first order. Therefore, in order to discuss the ferromagnetic properties of itinerant electron ferrromagnets, like Ni and Fe, it is important first to get rid of the defect of this first-order transition. It is shown below that if small terms, which were neglected, as far as the author knows, in all conventional theories of SGA [34,35], are included in the calculation the first-order transition can be avoided and a second-order transition is obtained, but the Kubo formula [36] is not satisfied. If the small terms, which are of order $1/N$, where N is the number of atoms, are neglected, as in the conventional theories of SGA, it is shown that the results for the longitudinal and transverse spin susceptibilities, χ_l and χ_t, satisfy the Kubo formula, but we cannot avoid the first-order transition. In this paper we will discuss only the paramagnetic state for various kinds of itinerant electron magnets.

In section 2, our SGA for SF with the inclusion of the small terms mentioned above both for the paramagnetic and the ferromagnetic state is explained. In Section 3, new numerical results for the temperature variation of $1/\chi$ for Ni and bcc Fe are shown and discussed. In Section 4, the ferromagnetic properties are discussed and the values of the magnetization and the high-field susceptibility at 0 K are estimated for Ni and Fe. In Section 5 our previous results for the temperature variation of the paramagnetic susceptibility for various transition metals, their alloys and compounds [15-18, 23-33] are summarized.

In Section 6, the case of negative (so-called mode-mode) coupling (a_3 in our calculations) of SF is discussed. According to our theory of SF in itinerant electron systems in this case it is possible that SF enhances the value of α and in some cases paramagnets can become ferromagnetic because of SF, as perhaps disordered Fe-Ni Invar alloys, Y(Co-Al)$_2$ compounds, etc. In the plus group TM, e.g., Cr, the sign of a_3 becomes negative and, due to the effect of SF, the empirical value of α becomes larger than the value of α at 0 K calculated in LSDA, as already noticed in the empirical analysis of χ for Cr [3]. Moreover, in the case of a negative value of the exchange stiffness coefficient A, a superposition of sinusoidal spin density waves may appear at low T, as observed in fcc Fe. Finally, the present theory is compared with other theories and main results and conclusions are summarized.

2. Static Gaussian Statistics.

In order to study the effect of SF in the ferromagnetic state, the longitudinal SF along the spontaneous magnetization M_z in the z direction and the transverse SF perpendicular to M_z are distinguished, as shown before [15]. The magnetic free energy f of an itinerant electron magnet with the magnetic density $\boldsymbol{M(r)}$ and volume V is given in the Stoner model as

$$f = \int d^3r \left[\frac{1}{2}(a_1 - \alpha)\boldsymbol{M}(\boldsymbol{r})^2 + \frac{1}{2}A|\nabla \boldsymbol{M}(\boldsymbol{r})|^2 \right.$$
$$\left. + \frac{1}{4}a_3\boldsymbol{M}(\boldsymbol{r})^4 + \frac{1}{6}a_5\boldsymbol{M}(\boldsymbol{r})^6 + \frac{1}{8}a_7\boldsymbol{M}(\boldsymbol{r})^8\right], \quad (1)$$

where a_1 is the inverse of the unenhanced spin susceptibility $\chi_0(T)$. The temperature variation of the coefficients a_1, a_3, a_5 and a_7 is determined from the numerical results for the exchange splitting $\Delta\zeta$, which is the difference between the chemical potentials of plus and minus spin bands and is calculated as a function of the temperature T from the calculated DOS curves [4]. The above coefficients then result from the best fitting as $\Delta\zeta/(2\mu_B) = a_1 M + a_3 M^3 + a_5 M^5 + a_7 M^7$. By the same method as that of Murata and Doniach [8] the average of f, $\bar{f}$, can be calculated as a function of T, M_z and the averages over the Gaussian distribution of the squares of the longitudinal and transverse fluctuating magnetizations $\xi_{lq}^2 =< m_{zq}m_{z-q}>$ and $\xi_{tq}^2 =< m_{xq}m_{x-q}>=< m_{yq}m_{y-q}>$, where m_{iq} ($i = z$, x and y) are the Fourier components of fluctuating parts of $\boldsymbol{M}(\boldsymbol{r})$, $m_z(\boldsymbol{r}) = M_z(\boldsymbol{r}) - M_z$, $m_x(\boldsymbol{r})$ and $m_y(\boldsymbol{r})$. If terms of order higher than the term $\sim a_7$ are neglected, $\bar{f}$ is then given by

$$\begin{aligned}\bar{f} = & -\frac{1}{2}k_BT\sum_q\{\ln(2\pi\xi_{lq}^2) + 2\ln(2\pi\xi_{tq}^2)\} - \frac{3}{2}k_BT\sum_q 1 \\ & + V[\frac{1}{2}(a_1-\alpha)(\xi_l^2 + 2\xi_t^2) + \frac{1}{2}A\sum_q q^2(\xi_{lq}^2 + 2\xi_{tq}^2) \\ & + \frac{1}{4}a_3(3\xi_l^4 + 4\xi_l^2\xi_t^2 + 8\xi_t^4) \\ & + \frac{1}{2}a_5(5\xi_l^6 + 6\xi_l^4\xi_t^2 + 8\xi_l^2\xi_t^4 + 16\xi_t^6) \\ & + \frac{1}{8}a_7(105\xi_l^8 + 120\xi_l^6\xi_t^2 + 144\xi_l^4\xi_t^4 + 192\xi_l^2\xi_t^6 + 384\xi_t^8) \\ & + \frac{1}{2}\{a_1 - \alpha + a_3(3\xi_l^2 + 2\xi_t^2) + a_5(15\xi_l^4 + 12\xi_l^2\xi_t^2 + 8\xi_t^4) \\ & + a_7(105\xi_l^6 + 90\xi_l^4\xi_t^2 + 72\xi_l^2\xi_t^4 + 48\xi_t^6)\}M_z^2 \\ & + \frac{1}{4}\{a_3 + 2a_5(5\xi_l^2 + 2\xi_t^2) + 3a_7(35\xi_l^4 + 20\xi_l^2\xi_t^2 + 8\xi_t^4)\}M_z^4 \\ & + \frac{1}{6}\{a_5 + 3a_7(7\xi_l^2 + 2\xi_t^2)\}M_z^6 + \frac{1}{8}a_7M_z^8 + \frac{3}{4}a_3\sum_q(\xi_{lq}^4 + 2\xi_{tq}^4) \\ & + \frac{1}{2}a_5\{3(\xi_l^2 + M_z^2)\sum_q(5\xi_{lq}^4 + 2\xi_{tq}^4) + 6\xi_t^2\sum_q(\xi_{lq}^4 + 6\xi_{tq}^4)\}\end{aligned}$$

$$+ 10\sum_q(\xi_{lq}^6 + 2\xi_{tq}^6)\}], \qquad (2)$$

with $\xi_l^2 = \sum_q \xi_{lq}^2$ and $\xi_t^2 = \sum_q \xi_{tq}^2$. The last four terms including ξ_{lq}^n and ξ_{tq}^n with $n = 4$ and 6 in (2) where the contributions, which are similar to them and include the coefficient a_7, are neglected for simplicity, are new, they were neglected in most conventional calculations in SGA [8,34,35] as well as in our previous calculations [15]. In the previous theories using SGA the Gaussian averages $< (m_{iq}m_{i-q})^n >$ with $n \geq 2$ are treated as ξ_{iq}^{2n}, but it should be $< (m_{iq}m_{i-q})^n >= n!\xi_{iq}^{2n}$ for $n \geq 2$. By including these terms, we can get a second-order transition at T_c as shown below, although these terms are of order of magnitude of $1/N$ as compared with other main terms, where N is the number of atoms in volume *newtermsandincludingthecoefficientsuptoa*5 or for the paramagnetic case $M_z = 0$ and $\xi_l = \xi_t$ is the same as given before in [15].

The uniform external field H_z is given by $H_z = \partial(\bar{f}/V)/\partial M_z$ as follows,

$$\begin{aligned}H_z &= M_z\{(a_1 - \alpha + a_3(3\xi_l^2 + 2\xi_t^2) + a_5(15\xi_l^4 + 12\xi_l^2\xi_t^2 + 8\xi_t^4) \\ &+ a_7(105\xi_l^6 + 90\xi_l^4\xi_t^2 + 72\xi_l^2\xi_t^4 + 48\xi_t^6) \\ &+ M_z^3\{a_3 + 2a_5(5\xi_l^2 + 2\xi_t^2) + 3a_7(35\xi_l^4 + 20\xi_l^2\xi_t^2 + 8\xi_t^4)\} \\ &+ M_z^5\{a_5 + 3a_7(7\xi_l^2 + 2\xi_t^2)\} + M_z^7 a_7 + 3a_5\sum_q(5\xi_{lq}^4 + 2\xi_{tq}^4)\}. \quad (3)\end{aligned}$$

The longitudinal and transverse spin susceptibilities, $1/\chi_l = \partial H_z/\partial M_z$ and $1/\chi_t$, are given, respectively, as

$$\begin{aligned}\frac{1}{\chi_l} &= a_1 - \alpha + a_3(3\xi_l^2 + 2\xi_t^2) + a_5(15\xi_l^4 + 12\xi_l^2\xi_t^2 + 8\xi_t^4) \\ &+ a_7(105\xi_l^6 + 90\xi_l^4\xi_t^2 + 72\xi_l^2\xi_t^4 + 48\xi_t^6) \\ &+ 3M_z^2\{a_3 + 2a_5(5\xi_l^2 + 2\xi_t^2) + 3a_7(35\xi_l^4 + 20\xi_l^2\xi_t^2 + 8\xi_t^4)\} \\ &+ 5M_z^4\{a_5 + 3a_7(7\xi_l^2 + 2\xi_t^2)\} + 7M_z^6 a_7 + 3a_5\sum_q(5\xi_{lq}^4 + 2\xi_{tq}^4) \quad (4)\end{aligned}$$

and

$$\begin{aligned}\frac{1}{\chi_t} &= a_1 - \alpha + a_3(\xi_l^2 + 4\xi_t^2) + a_5(3\xi_l^4 + 8\xi_l^2\xi_t^2 + 24\xi_t^4) \\ &+ a_7(15\xi_l^6 + 36\xi_l^4\xi_t^2 + 72\xi_l^2\xi_t^4 + 192\xi_t^6) \\ &+ M_z^2\{a_3 + 2a_5(3\xi_l^2 + 4\xi_t^2) + 3a_7(15\xi_l^4 + 24\xi_l^2\xi_t^2 + 24\xi_t^4)\} \\ &+ M_z^4\{a_5 + 3a_7(5\xi_l^2 + 4\xi_t^2)\} + M_z^6 a_7 + 3a_5\sum_q(\xi_{lq}^4 + 6\xi_{tq}^4). \quad (5)\end{aligned}$$

The expression (5) for $1/\chi_t$ is different from that shown in [19], where χ_t was defined by M_z/H_z, but this cannot be used in the presence of longitudinal and transverse SF. The equation (5) is derived from (2) by $\partial(\bar{f}/V)/\partial M_x = H_x$ and $1/\chi_t = \partial H_x/\partial M_x$ with including a small external magnetic field H_x and with the induced magnetization M_x in the transverse x direction. From $\partial \bar{f}/\partial \xi_{lq}^2 = 0$ and $\partial \bar{f}/\partial \xi_{tq}^2 = 0$, we have

$$\xi_{l,tq}^2 = \frac{k_B T}{V}(\chi_{l,t}^{-1} + Aq^2 + 3a_3\xi_{l,tq}^2)^{-1}. \tag{6}$$

In (6) the new additional small terms $\sum_q \xi_{lq}^n$ and $\sum_q \xi_{tq}^n$ ($n \geq 4$) are neglected. The equations (6) are different from the Kubo formula [36].[1] This fact shows that SGA is not appropriate to calculate the higher-order couplings of SF.

If the small correction terms $3a_3\xi_{l,tq}^2$ in (6) are neglected, the equation (6) becomes consistent with the Kubo formula and it should be noted that the expression of the Kubo formula for χ_t^{-1} is given by (5) which is different from the expression for χ_t^{-1} defined by H_z/M_z. And then, the integration of (6) over the wave vector q gives the usual equations [8,15],

$$\xi_{l,t}^2 = \frac{q_m k_B T}{2\pi^2 A}(1 - y_{l,t}^{-1}\tan^{-1} y_{l,t}), \tag{7}$$

with $y_{l,t} = q_m(A\chi_{l,t})^{1/2}$, where q_m is the cut-off wave vector, which is the only adjustable parameter, its introduction being inevitable as the present model is too simple to describe the real behavior of the generalized spin susceptibility $\chi(q)$ at large q.

If the small correction terms in (6) are remained, the integration of $\xi_{l,tq}^2$ over the wave vector can only be carried out approximately. In order to examine the effect of the small correction terms in (6) at T_c, the solution (6) is expanded in a power series of $1/\chi_l$ or $1/\chi_t$ in the neighborhood of T_c, where $1/\chi_l = 1/\chi_t = 0$, and then integrated over q up to q_m and for the small factor $6a_3 k_B T/(VA^2) \cong (\pi/a)(20/N)$ is used at T_c. Then we can get the following equations for ξ_l^2 and ξ_t^2

$$\xi_{l,t}^2 = \frac{q_m k_B T}{2\pi^2 A}\left(1 + y_{l,t}^{-2} - \frac{1}{3}y_{l,t}^{-4}\right). \tag{8}$$

It should be noted that (8) can be used only near T_c and we can confirm from (3) with $H_z = 0$, (4), (5) and (8) that the transition at T_c is of the second-order type. In the paramagnetic state and also in the ferromagnetic state at low temperatures (8) will not be useful as $1/\chi_l$ is finite and (7) can be used. However, as the equations (6) do not satisfy the Kubo formula, the small terms $3a_5\sum_q(5\xi_{lq}^4 + 2\xi_{tq}^4)$ in (3) and (4), $3a_5\sum_q(\xi_{lq}^4 + 6\xi_{tq}^4)$ in (5), and $3a_3\xi_{l,tq}^2$ in (6) are all neglected in the subsequent sections.

---

[1] This fact was pointed out to the author by Prof. D. Wagner.

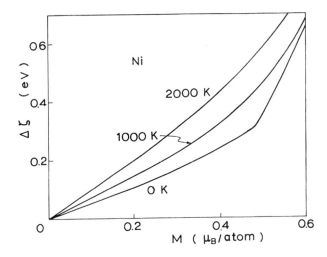

Figure 1. Exchange splitting $\Delta\zeta$ calculated from the calculated density of states [4] as a function of magnetization M and temperature T for Ni.

It is clear from the expressions (4) for $1/\chi_l$ and (5) for $1/\chi_t$ that the effect of α or the exchange interaction, is reduced by SF when the sign of a_3 is positive and is enhanced by the effect of SF when a_3 is negative. The former case gives the effect of lowering of T_c and the latter gives the effect of rising of T_c or appearance of ferromagnetism. This fact was already known in the theory of critical phenomena [34,35], but the researchers in this field were mainly interested in the critical indexes for various systems rather than in a lowering of T_c by the effect of SF in the usual case of $a_3 > 0$.

3. Paramagnetic Susceptibility for Ni and Fe.

The temperature variation of χ for TM, their alloys and compounds is calculated from (4) without the small term $3a_5 \sum_q (5\xi_{lq}^4 + 2\xi_{tq}^4)$ using (7). With $M_z = 0$, $\xi_{lq}^2 = \xi_{tq}^2 = \frac{1}{3}\xi_q^2$, $\xi_l^2 = \xi_t^2 = \frac{1}{3}\xi^2$, $\chi_l = \chi_t = \chi_s$ and $y_l = y_t = y$ then follows

$$\frac{1}{\chi_s} = a_1 - \alpha + \frac{5}{3}a_3\xi^2 + \frac{35}{9}a_5\xi^4 + \frac{35}{3}a_7\xi^6, \qquad (9)$$

$$\xi_q^2 = \frac{3k_BT}{V}(\chi_s^{-1} + Aq^2)^{-1}, \qquad (10)$$

and

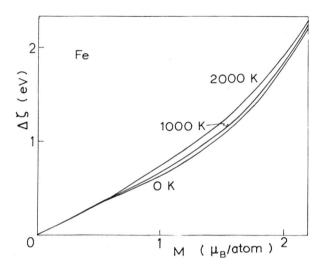

Figure 2. Exchange splitting $\Delta\zeta$ calculated from the calculated density of states [4] as a function of magnetization M and temperature T for bcc Fe.

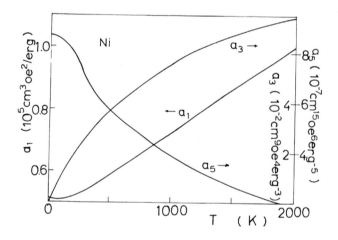

Figure 3. Temperature variation of a_1, a_3 and a_5 estimated from the fit of $\Delta\zeta/(2\mu_B) = a_1 M + a_3 M^3 + a_5 M^5$ to the curve $\Delta\zeta$ in Figure 1 for Ni.

$$\xi^2 = \frac{3q_m k_B T}{2\pi^2 A}(1 - y^{-1}\tan^{-1} y), \qquad (11)$$

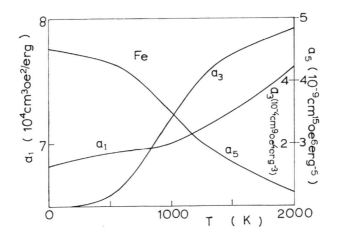

Figure 4. Temperature variation of a_1, a_3 and a_5 estimated from the fit of $\Delta\zeta/(2\mu_B) = a_1 M + a_3 M^3 + a_5 M^5$ to the curve $\Delta\zeta$ in Figure 2 for bcc Fe.

with $y = q_m(A\chi_s)^{1/2}$. The equations of (9)-(11) are the same as given in [15]. In this section, first new calculated results for Ni and bcc Fe are shown and in section 5 some of our previous results [15-18, 23-33] are reviewed and discussed.

In the numerical calculation of χ_s, the values of the coefficients introduced in the expansion of f, a_1, a_3, a_5, a_7, α and A, are as far as possible determined from the theoretical values obtained from calculations of the electronic structure in LSDA etc., as shown in the respective papers. The temperature variation of the exchange splitting $\Delta\zeta$ for Ni and Fe, which is calculated from the DOS calculated in [4], is shown in Figures 1 and 2, respectively [37]. The values of a_1, a_3 and a_5 for Ni and Fe are determined by fitting the $\Delta\zeta$ in Figures 1 and 2 to the formula $\Delta\zeta/(2\mu_B) = a_1 M + a_3 M^3 + a_5 M^5$. The results for a_1, a_3 and a_5 for Ni and Fe are shown in Figures 3 and 4, respectively. The values of α, A and χ_{orb} are presented in Table 1 and are the same as given in [15, 33]. The value of $q_m = \pi/\lambda a$, where a is the lattice constant and λ is a parameter, is determined in this paper in such a way that the value of T_c determined from the relation, $a_1 - \alpha + 5a_3\xi_c^2 + 35a_5\xi_c^4 = 0$ at T_c, where $\xi_c^2 = q_m k_B T_c/(2\pi^2 A)$, fits to the observed values, $T_c = 630$ K for Ni or 1043 K for Fe. We get $\lambda = 1.72$ for Ni and 0.99 for Fe, respectively.

It can be easily shown that at T_c, M_z, $1/\chi_l$ and $1/\chi_t$ become zero. Setting $\xi_l^2 = \xi_t^2 = \frac{1}{3}\xi^2 = \xi_c^2$ at T_c, the values of ξ_c are estimated as 0.239

TABLE 1. Numerical values of a_1 (in 10^5 cm^{3}Oe^2/erg) at 0 K, a_3 (in 10^{-1}cm^{9}Oe^4/erg^3), a_5 (in 10^{-7}cm^{15}Oe^6/erg^5), a_7 (in 10^{-12}cm^{21}Oe^8/erg^7), α (in 10^5cm^{3}Oe^2/erg), A (in 10^{-12}cm^{5}Oe^2/erg), $\lambda=\pi/(aq_m)$ and χ_{orb} (in $10^{-6}/(\text{cm}^3\text{Oe}^2)$) and the calculated results for ξ (in μ_B/atom or formula unit) at 1000 K for bcc Fe, Co, Ni, fcc Fe, V, Cr, Nb, Mo, Rh, Pd, ZrZn$_2$ and TiBe$_2$. The lower figures of λ and ξ for bcc Fe and Ni are the results obtained in this paper.

	a_1 (0 K)	a_3	a_5	a_7	α	A	λ	χ_{orb}	ξ (10^3K)
bcc Fe	0.68	0.0174	0.0378	–	1.02	1.8	0.90	9.93	1.47
							0.99		1.46
Co	1.02	-0.235	0.152	–	1.01	5.0	0.65	7.51	1.06
Ni	0.51	0.836	4.15	–	1.02	5.5	1.34	6.72	0.38
							1.72		0.45
fcc Fe	1.07	-0.0853	0.0549	–	1.06	-0.7	1.29	7.91	1.23
V	1.59	1.18	3.33	–	0.95	2.2	1.90	20.0	0.18
Cr	3.15	-2.76	4.56	–	1.87	2.2	1.24	14.8	0.27
Nb	2.37	1.24	30.3	–	1.48	2.2	0.79	11.1	0.53
Mo	4.42	-12.1	74.7	–	0.86	2.2	1.20	6.07	0.19
Rh	1.87	-2.77	3.54	–	0.84	-2.4	0.93	1.63	0.46
Pd	1.18	0	124	–	1.05	110	0.57	3.52	0.23
ZrZn$_2$	1.95	18.4	-3830	21100	1.95	5.8	0.85	0	0.55
TiBe$_2$	1.13	19.0	-3090	251	1.10	3.4	1.75	0	0.05

and 0.845 μ_B/atom for Ni and Fe, respectively. The calculated results for $1/\chi = 1/(\chi_s + \chi_{orb})$ and ξ for Ni and Fe are shown in Figures 5 and 6, respectively, and in Figure 7 these results are compared with the previous ones obtained in [15, 33], where the values of a_3 and a_5 were assumed to be constant and the values of q_m were determined such that an agreement between the calculated and observed values of χ at the highest available temperatures is obtained. The obtained values of λ and ξ at 1000 K are very similar to the ones reported above, as seen from Table 1. The general trends of the temperature variations of $1/\chi$ in both present and previous calculated results for Ni and Fe are similar, but both results for $1/\chi$ show a concaveness downward just above T_c and this trend is different from the observed results. To get a better agreement between the calculated and observed results for $1/\chi$, we should include the effect of critical fluctuations. It is interesting, as seen from Fig. 6, that the temperature dependence of ξ in Fe is very small, but is a bit larger in Ni than in Fe. This is due to the difference in the shape of the DOS curves at the Fermi level, i.e., a picky structure of DOS near the position of the Fermi level in Ni, but not in Fe [4].

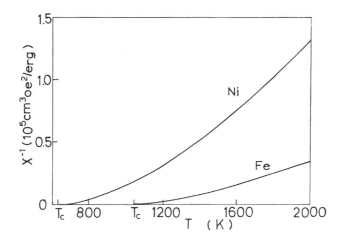

Figure 5. Calculated results for the inverse total paramagnetic susceptibility $1/\chi = 1/(\chi_s + \chi_{orb})$ for Ni and bcc Fe.

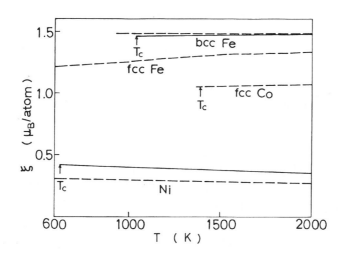

Figure 6. Calculated results for the root mean square of the fluctuating magnetic moment ξ in the paramagnetic state for Ni, fcc Co, bcc Fe and fcc Fe. The solid curves are results from this paper and the broken curves are previous results [15,18].

4. Ferromagnetic Properties of Ni and Fe.

The temperature variations of M_z in the itinerant electron model for Fe-Pt alloys have been calculated in the SGA of SF, including the effect of

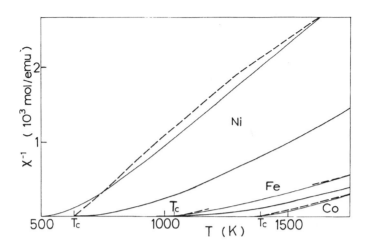

Figure 7. Comparison between the present results (solid curves) in Figure 5 and the previous results (thin solid curves) in [15] and the observed results (broken curves) for the inverse total paramagnetic susceptibility $1/\chi$ of Ni, bcc Fe and fcc Co.

volume fluctuations, by Podgórny et al. [38,39]. However, they have shown that as expected the transition is of first-order and the discontinuity of M_z at the transition temperature becomes smaller due to the effect of volume fluctuations.

As the effects of SF at 0 K become zero in the SGA, we can estimate the values of M_z and χ_l at 0 K by using the same values of coefficients as used in the previous section. Our calculated values of M_z and χ_l at 0 K are 0.576 μ_B/atom and 3.23×10^{-5} emu/mol for Ni and 2.08 μ_B/atom and 5.28×10^{-5} emu/mol for Fe, respectively. These calculated values of M_z for Ni and Fe are very near to the observed values of M_z, 0.56 and 2.13 μ_B/atom for Ni and Fe, respectively [3]. The good agreement for M_z at 0 K is due to the results of the band structure calculations [4].

The value of the high-field susceptibility χ_{hf} at 0 K is estimated as a sum of χ_l and the calculated values of the orbital susceptibility in the ferromagnetic state [3,40]. The obtained values of χ_{hf} are 5.93×10^{-5} and 1.54×10^{-4} emu/mol for Ni and Fe, respectively. The observed values of χ_{hf} at 0 K for Ni and Fe [41] are about two times larger than the calculated values.

The χ_t given by (5) is expected to be finite below T_c, although the anisotropy energy is not taken into account in the present model, because χ_t in (5) is different from M_x/H_z in (3). If we include the anisotropy energy in (1) we can easily show that χ_t becomes finite at 0 K.

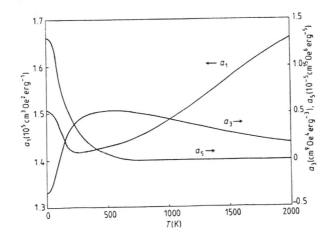

Figure 8. Temperature variation of a_1, a_3 and a_5 in YCo_2 [23].

In the present model the phase transitions are of first-order as in other theories [8,38,39], and from a preliminary numerical calculation the discontinuity of M_z at T_c for Fe and Ni seems to be rather large. Anyhow, before we calculate the temperature dependence of M_z for Ni and Fe in our realistic Stoner model including the effect of SF we must find a reasonable way to get rid of the first-order transition.

5. Summary of our Previous Results on Paramagnetic Susceptibility of Transition Metals, Alloys and Compounds.

In this section, our previous results for the temperature variation of χ calculated by using the same approximation as in Section 3 and the comparison between these results and the observed ones are summarized. The calculations have been carried out for various transition metals, bcc Fe, fcc Fe, fcc Co, Ni, V, Cr, Nb, Mo, Rh and Pd [15,18,33], $ZrZn_2$, [16], $TiBe_2$ [15], MnSi and ordered and disordered Fe_3Ni alloys [17], various intermetallic compounds with the cubic Laves phase structure, YMn_2, YFe_2, YCo_2, YNi_2, $ScCo_2$, $TiCo_2$, $ZrCo_2$, $LuCo_2$, $HfCo_2$, $ScNi_2$, $NbCo_2$, $TaCo_2$, $ZrFe_2$, $LuFe_2$, $HfFe_2$, $Y(Fe-Co)_2$ and $Zr(Fe-Co)_2$ and $Zr(Fe-Co)_2$ [23], Y compounds with Ni, Co, Fe and Mn [24,25,27], $Y_2Fe_{14}B$ in [26], FeSi [28], (Fe-Co)Si [29], $NbFe_2$ with the C14 Laves phase structure [30], UPt_3 [31] and $(Cr-Mn)Pt_3$ [32].

The numerical values of the coefficients or quantities a_1 at $T=0$ K, a_3,

TABLE 2. Numerical values of a_1 (in $10^5 \mathrm{cm}^3 \mathrm{Oe}^2/\mathrm{erg}$) at 0 K, a_3 (in $10^{-1} \mathrm{cm}^9 \mathrm{Oe}^4/\mathrm{erg}^3$), a_5 (in $10^{-7} \mathrm{cm}^{15} \mathrm{Oe}^6/\mathrm{erg}^5$), a_7 (in $10^{-12} \mathrm{cm}^{21} \mathrm{Oe}^8/\mathrm{erg}^7$), α (in $10^5 \mathrm{cm}^3 \mathrm{Oe}^2 \mathrm{erg}$), A (in $10^{-12} \mathrm{cm}^5 \mathrm{Oe}^2/\mathrm{erg}$), $\lambda = \pi/(aq_m)$, χ_{orb} (in $10^{-6} \mathrm{erg}/(\mathrm{cm}^3 \mathrm{Oe}^2)$) and the calculated results for ξ (in μ_B/formula unit) at 1000K for alloys and intermetallic compounds of transition metals. (o) and (d) denote the ordered and disordered states of Fe_3Ni, respectively.

	a_1	a_3	a_5	a_7	α	A	λ	χ_{orb}	ξ
YMn_2	1.9	9.8	-1570	3750	1.40	4.0	0.98	27.6	0.7
YFe_2	0.60	3.5	–	–	1.51	3.8	0.76	26.7	1.5
YCo_2	1.51	-4~5	0~1005	–	1.35	5.5	2.0	28.3	0.3
YNi_2	4.3	0	200000	–	0.89	4.0	0.9.	10.0	0.3
$ScCo_2$	1.6	-6.0	165	–	1.35	5.5	1.5	32.1	0.5
$LuCo_2$	1.55	-6.5	232	–	1.36	5.5	1.5	32.1	0.5
$TiCo_2$	1.19	18.8	-938	1000	1.08	5.5	1.2	47.6	0.4
$ZrCo_2$	1.46	10.5	-250	333	1.22	5.5	1.2	60.4	0.4
$HfCo_2$	1.54	5.2	0	–	1.37	5.5	1.3	55.9	0.4
$ScNi_2$	2.9	40	12700	–	1.00	4.0	1.3	8.8	0.2
$NbCo_2$	1.6	36	-1600	1900	1.19	5.5	1.3	800	0.4
$TaCo_2$	1.7	24	-1260	2650	1.56	5.5	1.5	800	0.4
$ZrFe_2$	1.05	2.0	-80	0.08	1.41	3.8	0.56	30.0	2.3
$LuFe_2$	0.60	3.5	0	0	1.66	3.8	0.76	28.3	1.7
$HfFe_2$	1.07	2.7	-10.8	0.01	1.39	3.8	0.58	30.6	1.7
$NbFe_2$	1.3	3.13	-10.7	1.20	1.11	3.8	1.3	28.5	0.8
Y_2Ni_{17}	0.38	97.8	-4410	37900	0.72	56	3.1	–	0.5
Y_2Fe_{17}	0.83	0.297	-0.23	0.0062	0.98	0.49	0.55	–	1.0
Y_2Fe_{23}	1.66	-0.381	0.307	–	1.23	0.98	0.27	–	1.4
Y_2Fe_3	0.92	0.206	0.141	–	1.34	0.98	0.78	–	1.2
$FeSi$	∞	-50.2	448000	–	1.55	0.99	0.48	–	1.5
$MnSi$	1.62	-0.534	10.6	–	1.63	2.9	1.2	–	0.7
UPt_3	0.79	0.30	94	–	0.70	8.9	0.38	3.77	2.4
$CrPt_3$	0.90	4.4	-20.6	3.26	1.16	8.9	1.5	160	1.2
Fe_3Ni(o)	1.29	-0.228	0.2	–	1.52	0.54	1.1	–	1.2
Fe_3Ni(d)	0.95	-0.122	0.054	–	0.91	0.54	1.0	–	1.2

a_5, a_7 (if necessary), α, A and $\lambda = \pi/(aq_m)$, which are introduced in our calculation of χ_s in SGA, as well as χ_{orb} and ξ at 1000 K are summarized in Tables 1 and 2, including the new results given in Section 3 for Ni and Fe. The values of a_1, a_3, a_5 and a_7 are determined from the calculated results of DOS in various approximations, e.g., LSDA, tight-binding approximation, coherent potential method, or recursion method, etc. Some values of α, A and χ_{orb} are determined by theoretical calculations, others empirically. The values for λ or q_m, which is the only adjustable parameter, are deter-

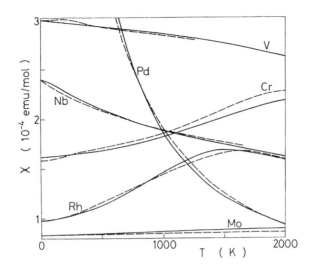

Figure 9. Calculated (solid curves) and observed (broken curves) results for the total paramagnetic susceptibility χ of V, Cr, Nb, Mo, Rh and Pd [15,18,33].

mined such that an agreement between the calculated and observed values of χ at the highest available temperature is obtained. In the case of YCo_2 the coefficients a_1, a_3 and a_5 are dependent on T [23], their temperature dependences are shown in Figure 8.

The agreement between the calculated and observed results of χ are at least qualitatively and sometimes even quantitatively satisfactory in almost all magnetic materials treated in our previous papers [15-18, 23-33], and all the obtained values of ξ and λ seem to be reasonable as seen from Tables 1 and 2.

In the following, several interesting results are shown and discussed and new effects of SF, which we have discovered, are emphasized. Examples treated below are Ni, fcc Co, bcc Fe, fcc Fe, paramagnetic transition metals, V, Cr, Nb, Mo, Rh and Pd, ordered and disordered Fe_3Ni alloys, intermetallic compounds $ZrZn_2$, MnSi, FeSi, and YCo_2.

Previous results of χ and ξ calculated using the same approximation as in Section 3 and the comparison with the observed results of χ for V, Cr, Nb, Mo, Rh and Pd and for Ni, fcc Co, bcc Fe, fcc Fe, $ZrZn_2$ and $TiBe_2$ are shown in Figures 6, 9-11, respectively. The agreement between the calculated and observed results of χ in Figures 9 and 10 is quite satisfactory.

For the minus-group transition metals, e.g., V, bcc Fe, Ni, Nb, Pd etc, where a_1 increases with increasing temperature and the sign of a_3 is positive (see Table 1), we can see from Figures 9 and 10 that χ decreases with

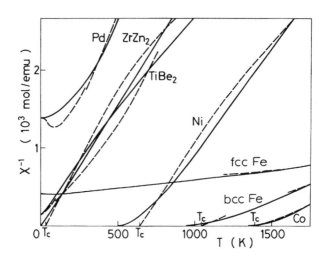

Figure 10. Calculated (solid curves) and observed (broken curves) results for the inverse total paramagnetic susceptibility $1/\chi$ of Pd, Ni, fcc Co, bcc Fe, fcc Fe, $ZrZn_2$ and $TiBe_2$ [15,16,18,33].

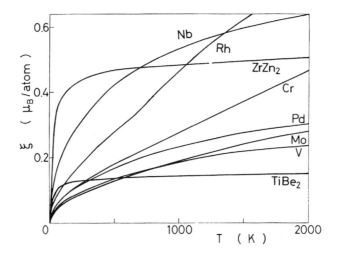

Figure 11. Calculated results for the root mean square of the fluctuating moment ξ of V, Cr, Nb, Mo, Rh, Pd, $ZrZn_2$ and $TiBe_2$ [15,16,18,33].

increasing temperature and it is expected from (10) that the temperature variation of ξ at higher temperature becomes very small as shown in Figures 6 and 11. The effect of α in the minus-group is reduced by taking into

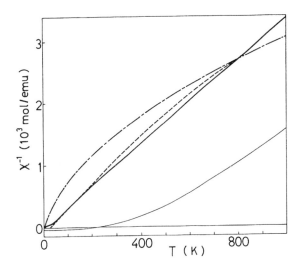

Figure 12. Calculated temperature variation of $1/\chi$ (solid curve), $a_1 - \alpha$ (thin solid curve), $1/\chi$ obtained in the approximations $a_1 - \alpha =$const., $a_3 \neq 0$ and $a_5=a_7=0$ (short-long dashed curve), and the observed result of $1/\chi$ (dashed curve) for ZrZn$_2$ [16].

account the effect of SF, as $a_3 > 0$.

On the other hand, in the plus-group transition metals, e.g., Cr, Mo and Rh, where a_1 decreases with increasing temperature at low temperatures and the sign of a_3 is negative (see Table 1), we can see from Figures 9 and 10 that χ increases with increasing temperature at low temperatures and shows a maximum at a certain temperature. It is expected from (10) that ξ increases with increasing temperature, as is shown in Figure 11. The effect of α in the plus-group is enhanced by taking into account the effect of SF, as $a_3 > 0$. Especially, in Rh, as discussed below, the role of short-wavelength SF is important, because $A < 0$ as well as $a_3 < 0$.

In Figures 12 and 13, the previously calculated results for χ and ξ for ZrZn$_2$ [16] are shown, using the values of coefficients in Table 1. Although the estimated value of T_c was too low as compared with the observed value of T_c, the calculated result of χ agrees fairly well with the observed result in a wide range of temperatures. It was shown in Figure 12 that if the temperature variation of a_1 is neglected, as assumed in other theories of SF [8,9,11,19], or if the effects of SF are neglected and the value of α is reduced empirically as in the theory of WIEF [42], the calculated results of χ do not agree with experiment. In Figure 13 it is shown that the obtained value of ξ is about 0.5 μ_B per f.u. in ZrZn$_2$ at higher temperatures and is larger than the spontaneous magnetization 0.15 μ_B for ZrZn$_2$ at 0 K. So it was concluded that the effects of SF are very important, because they lead

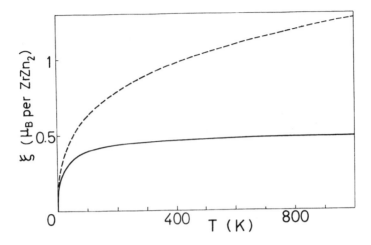

Figure 13. Calculated temperature variation of ξ (solid curve) and ξ obtained from the approximation $a_1-\alpha=$const., $a_3\neq 0$ and $a_5=a_7=0$ (broken curve) for ZrZn$_2$ [16].

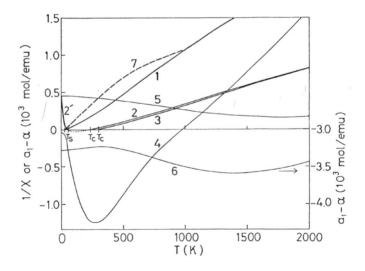

Figure 14. Calculated results for $1/\chi$ of MnSi (curve 1) and disordered (curves 2 and 2') and ordered (curve 3) Fe$_3$Ni alloys, and for $a_1-\alpha$ of MnSi (curve 4) and disordered (curve 5) and ordered (curve 6, with the right-hand side abscissa) Fe$_3$Ni alloys. The broken curve 7 is the observed $1/\chi$ for MnSi [17].

to the reduction of the value of T_c resulting from the simple Stoner model. The temperature variation of a_1 is important, as well.

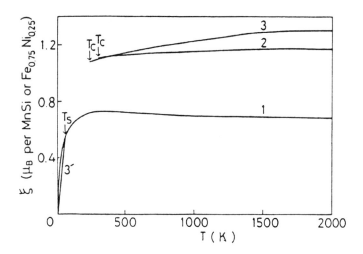

Figure 15. Calculated results for ξ of MnSi or $Fe_{0.75}Ni_{0.25}$ (curve 1) and ordered (curve 2) and disordered (curves 3 and 3') Fe_3Ni alloys [17].

Previous results for χ and ξ, respectively, for ordered and disordered fcc Fe_3Ni alloys and MnSi [17], are shown in Figures 14 and 15. They have been calculated using the values of coefficients from Table 2. In all these alloys the values of a_3 are not so largely negative, but for the ordered Fe_3Ni alloy and MnSi the calculated values of T_c become fairly lower than the values of T_c expected from the Stoner condition for ferromagnetism, $a_1 - \alpha = 0$, due to the effects of higher-order SF coupling terms with a_5 and rather large values of ξ as seen from Figure 15. For the disordered Fe_3Ni alloy, the Stoner condition for ferromagnetism, $a_1 - \alpha = 0$, is not satisfied for any temperature, but because the value of a_3 is negative and the value of a_5 is positive and not so large, due to the effects of SF the alloy becomes ferromagnetic between $T_s = 55$ K and $T_c = 235$ K, i.e., shows the so-called thermal spontaneous magnetism [3,43]. It is interesting that there is a possibility of the appearance of ferromagnetism by the effect of SF and that the effect of SF, as well as the magnetic mixing with the volume change [3], may be very important in Fe-Ni Invar alloys. The calculated values of ξ at higher temperatures in Figure 15 are consistent with the values obtained from the measurements of the paramagnetic diffuse scattering of polarized neutrons [44].

For fcc Fe and Rh the values of A are estimated to be negative from the calculated generalized spin susceptibility $\chi(q)$ [45], therefore the above theory was extended to the case where $A < 0$ in [18]. In this case the

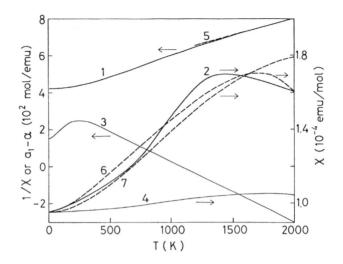

Figure 16. Calculated results for $1/\chi$ of fcc Fe (curve 1) and χ of Rh (curve 2), $a_1 - \alpha$ of fcc Fe (curve 3) and $(a_1-\alpha)^{-1}+\chi_{orb}$ of Rh (curve 4) and the observed results (broken curves) for $1/\chi$ of fcc Fe (curve 5) and for χ of Rh (curves 6 and 7). The abscissa for χ and $(a_1-\alpha)^{-1}+\chi_{orb}$ of Rh is the right-hand side.[18]

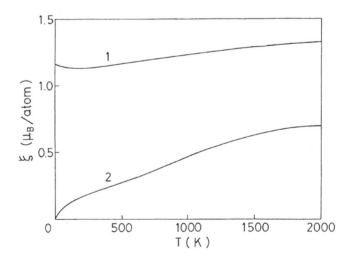

Figure 17. Calculated results for ξ of fcc Fe (curve 1) and Rh (curve 2)[18].

equation (11) is replaced by

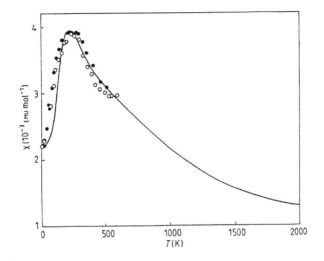

Figure 18. Calculted results for χ of YCo$_2$. Full and open small circles denote the observed results [23].

$$\xi^2 = \frac{3q_m k_B T}{2\pi^2 |A|} \{\frac{1}{2} y^{-1} ln(\frac{|1+y|}{|1-y|}) - 1\} \tag{12}$$

with $y = q_m(|A|\chi_s)^{1/2}$.

Using equations (9) and (12) and the values of the coefficients in Table 1, χ and ξ for fcc Fe and Rh were calculated, and the results are shown and compared with the experiment in Figures 16 and 17 [18]. In this case the cut-off wave vector is sufficient to get a finite contribution of SF. As seen from Figure 16 the agreement between the calculated and observed results of χ or $1/\chi$ is satisfactory. In the case of Rh the largely negative value of a_3 enhances the value of χ.

In the case of fcc Fe, even if the Stoner condition for ferromagnetism is satisfied, as A is negative ferromagnetism is not stabilized, ξ remains finite at 0 K and there exists a sinusoidal spin density with a certain value of q_m similar to the spinodal decomposition in alloys [46]. This result will be related to the existence of antiferromagnetic ordering in fcc Fe, which was observed in neutron measurements [47] and in Mössbauer spectroscopy [48].

The calculated result for χ in YCo$_2$ is shown in Figure 18 using the temperature variation of a_1, a_3 and a_5 given in Figure 8 and the values of other coefficients from Table 2. The value of ξ calculated in the same way is about 0.3 μ_B per f.u. at high temperatures and remains nearly

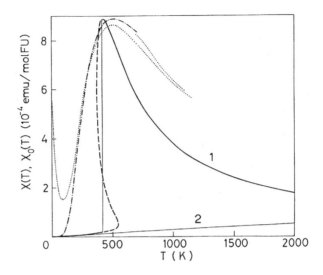

Figure 19. Calculated results for χ (curve 1) and $1/a_1$ (curve 2) and observed results for χ (chained curve and dotted curve) of FeSi. Broken curve shows unstable solutions [28].

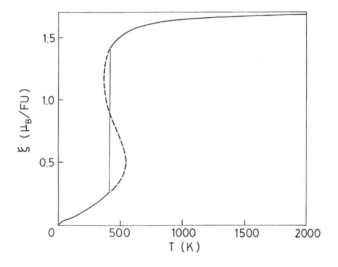

Figure 20. Calculated results for ξ of FeSi [28].

constant there [23]. a_3 is negative at low temperature and becomes positive at higher temperatures, χ shows a maximum and the agreement between the calculated and observed temperature variation of χ is very good.

Finally, it is shown that the calculated result for χ of FeSi shows anomalous temperature behavior, as is illustrated in Figure 19, and it was suggested that the rapid increase of χ with increasing temperature may be related to a discontinuous change in χ_s from a low fluctuating spin state to a high fluctuating spin state, as is shown in Figures 19 and 20. This anomalous behavior of χ_s may be due to a largely negative value of a_3, which was estimated from the special shape of the electronic density of states of FeSi which is intrinsic semiconductor like. Then, the values of a_3 and a_5 are expected to be strongly dependent on T, as well as a_1, as is seen from Table 2. The calculation of χ_s by taking into account the temperature variation of a_3 and a_5 is desirable and this is a future problem. Futhermore, as the value of a_3 is largely negative, there is a possibility that FeSi may have a strong tendency to ferromagnetism due to the effects of SF similar to the case of the disordered Fe_3Ni alloy discussed above.

6. Discussions and Conclusions.

In our previous calculations of the effect of SF in disordered Fe_3Ni alloys [17], the spontaneous magnetization becomes nonzero at temperatures between 55 K and 235 K due to the effect of SF, as the value of a_3 is negative, even though the Stoner condition for ferromagnetism is not satisfied at any temperature. Unfortunately, fcc Fe_3Ni is not stable in bulk samples. Then, if a sample of a fine particle or thin film of stable Fe_3Ni alloy can be made, as for fcc Fe_3Pt alloys [49], it would be interesting to examine the possibility of thermal spontaneous ferromagnetism [3, 43] in fcc Fe_3Ni alloys experimentally. The appearance of ferromagnetism in Y(Co-Al) with a small amount of Al [50] may be explained by the increase of effective α because of SF and the negative sign of a_3 in YCo_2 at low temperature.

In some TM and their alloys the value of A is expected to be negative, e.g., from the calculated results for the wave-vector dependence of the generalized spin susceptibility.In this case ξ remains finite at 0 K and there exists a sinusoidal spin density with a certain value of q_m similar to the spinodal decomposition in alloys. This result will be related to the appearance of antiferromagnetic ordering in fcc Fe, observed in neutron diffraction measurements [47] and in Mössbauer spectroscopy [48]. In order to determine the ordering of spin densities in fcc Fe of short- or long-range, besides the experimental determinations, it will be necessary to compare the free energies of fcc Fe calculated in an antiferromagnetic state with a long-range order and in the present model with spontaneous spin fluctuations.

There are many theoretical works on the effect of SF in itinerant electron magnetism. The results are mostly qualitative and there are very few attempts to calculate or estimate the effect of SF based on a realistic model of respective itinerant electron magnets. In most theories, all coefficients appearing in the magnetic free energy have been assumed to be constant including $a_1 = 1/\chi_0(T)$, which temperature variation is very important in any itinerant electron magnet. In ZrZn$_2$, it was shown in [16] that if the temperature variation of a_1 is neglected we cannot obtain satisfactory agreement between the calculated and observed results of $1/\chi$, as shown in Figure 12. Most WIFF show very picky structures of the electronic density of states, then the degeneracy temperature of electrons becomes low and $a_1 = 1/\chi_0(T)$ increases rapidly with increasing temperature similar to the Curie law. The effective value of the molecular field coefficient α, which is reduced by the effect of electron correlation and further by the effect of thermal spin fluctuations in the Landau theory of fluctuations [22] at finite temperatures, will give a low T_c, as shown in this paper. In the neighborhood of the transition temperature of any ferromagnet thermal spin fluctuations should be treated as critical fluctuations.

Our SGA of SF and the Murata-Doniach theory [8] based on the Landau theory of fluctuations [21,22] are quite different from the Moriya-Kawabata (M-K) theory [9] based on the paramagnon theory. For example, if there is no exchange interaction, as an extreme case, the effect of SF remains in our theory of SF for itinerant electron systems, but in the M-K theory of SF perhaps it does not remain and the theory will only yield Pauli paramagnetism for non-interacting electron systems. In the M-K theory, as the Hubbard interaction is given in a quadratic form of spin density or electron density, the magnetic free energy can be given rigorously in terms of the dynamical susceptibility by making use of the fluctuation dissipation theorem and the effect of SF for second powers of spin densities would be included, but powers higher than the second, e.g., ξ_l^n and ξ_t^n of SF for $n \geq 4$, are not included. As can be seen from our results, the magnitude of the fluctuating moment is not necessarily small even in WIEF, although M_z is very small. Moreover, in the M-K theory the longitudinal and transverse SF are not distinguished in the ferromagnetic state as is the case in the present theory and in [19], although it is clear that χ_t diverges and χ_l becomes finite at 0 K. Recently, it has been pointed out by Solontsov and Wagner [51] that the effect of zero-point SF, which was neglected in most theories based on the paramagnon theory, is expected to give an important temperature variation of χ through a_1.

The form of the dynamical susceptibility in the M-K theory is assumed to be consistent with the form of the static thermodynamical susceptibility and to give a reduction of the effective interation, which is expected to be proportional to, at least, square of the interaction parameter as a first term in the theory, but this was neglected, the theory becomes very complicated and the degree of approximation used in their theory is not clear. Because of the complexity of the theory in most cases the values of various coefficients introduced in the theory are estimated based on the free electron model, which is very far from the real itinerant electrons in metallic magnetic materials. Roughly speaking, we may understand that in the M-K theory the temperature dependent correlation energy would reduce the effect of exchange interaction and give lowering of T_c and then an apparent Curie-Weiss law of χ, even if the effect of SF is not taken into account as in the Landau theory of SF and the temperature variation of $a_1 = 1/\chi_0(T)$ is neglected.

On the other hand, the Stoner model, which is based on the calculated results of DOS, α, $\chi(q)$ and χ_{orb} in the respective magnetic materials and includes the temperature variations due to the Fermi-Dirac distributions of electrons over DOS, can be easily amended so as to include the effect of SF in SGA, as shown in the present theory. And it is very simple, although the calculations of higher order terms of SF coupling in SGA are very tedious, and transparent, as compared with other theories based on the paramagnon theory and on the Onsager reaction field concept, etc. In conclusion, the static Gaussian approximation applied to a realistic Stoner model can satisfactorily explain the temperature variation of the paramagnetic susceptibility of transition metals, their alloys and compounds, except just above T_c, as shown above and before [15-18, 23-33], but in the ferromagnetic state the agreement between theory and experiment is qualitative and the present approximation of SF for the ferromagnetic state must be improved to get a quantitative agreement.

Finally, the author would like to express his thanks to Professor D. Wagner in Bochum and Professor E. F. Wassermann in Duisburg for kind hospitality and valuable discussions on the effect of spin fluctuations and financial support through the project of Sonderforschungsbereich 166 Duisburg/Bochum in Germany.

References

1. Frenkel, J. (1928) Elementare Theorie magnetischer und elektrischer Eigenschaften der Metalle beim absoluten Nullpunkt der Temperature, *Zeits. f. Phys.* **49**, 31-45.
2. Wohlfarth, E.P. (1953) The theoretical and experimental status of the collective electron theory of ferromagnetism, *Rev. Mod. Phys.* **25** 211-219, (1980) Iron, Cobalt and Nickel, in E. P. Wohlfarth (ed.), *Ferromagnetic Materials* Vol.1, North-Holland Publishing, Amsterdam, pp. 1-70.
3. Shimizu, M. (1981) Itinerant electron magnetism, *Rep. Prog. Phys.* **44**, 329-409.
4. Moruzzi,V. L., Janak, J. F. and Williams, A. R. (1978) Calculated Electronic Properties of Metals, Pergamon Press, New York; Janak, J. F. (1977) Uniform susceptibilities of metallic elements, *Phys. Rev.* **B16**, 255-262.
5. Andersen, O. K., Madsen, J., Poulsen, U. K., Jepsen, O. and Kollar, J. (1977) Magnetic ground state properties of transition metals, *Physica* **B86-88**, 249-256.
6. Callaway, J. and Wang, C. S. (1977) Energy bands in ferromagnetic iron, *Phys. Rev.* **B16**, 2095-2105.
7. Yasui, M. and Shimizu, M. (1979) A numerical calculation of the orbital mag-

netic susceptibility for BCC 3d series transition metals, *J. Phys. F: Met. Phys.* **9**, 1653-1664, (1985) Relativistic formulae for the wavevector-dependent magnetic susceptibilities and the numerical calculation of the orbital and spin-orbit magnetic susceptibilities in vanadium, *J. Phys. F: Met. Phys.* **15**, 2365-2373, (1986) The orbital and spin-orbit magnetic susceptibilities in vanadium, *J. Magn. Magn. Mat.* **54-57**, 989-990; Yasui, M. (1988) Orbital and spin-orbit magnetic susceptibilities in bcc transition metals: V, Nb and W, *Physica B* **149**, 139-142.

8. Murata, K. K. and Doniach, S. (1972) Theory of magnetic fluctuations in itinerant ferromagnets, *Phys. Rev. Lett.* **29**, 285-288.
9. Moriya, T. and Kawabata, A. (1973) Effect of spin fluctuations on itinerant electron ferromagnetism, *J. Phys. Soc. Japan* **34**, 639-651, (1973) Effect of spin fluctuations on itinerant electron ferromagnetism. II, *J. Phys. Soc. Japan* **35**, 669-676.
10. Hertz, J. A. and Klenin, M. A. (1974) Fluctuations in itinerant-electron paramagnets, *Phys. Rev.* **B10**, 1084-1096.
11. Gumbs, G. and Griffin, A. (1976) Effect of spin fluctuations on the Stoner transition temperature, *Phys. Rev.* **B13**, 5054-5064.
12. Dzyaloshinskii, I. E. and Kondratenko, P. S. (1976) Theory of weak ferromagnetism of a Fermi liquid, *Sov. Phys. JETP* **43**, 1036-1045.
13. Prange, R. E. and Korenman, V. (1979) Local-band theory of itinerant ferromagnetism. IV. Equivalent Heisenberg model, *Phys. Rev.* **B19**, 4691-4697.
14. Cyrot, M. (1981) Comments on weakly ferromagnetic metals in T. Moriya (ed.) *Electron Correlation and Magnetism in Narrow Band Systems*, Springer Verlag, Berlin, 51-54.
15. Shimizu, M. (1981) Effects of spin fluctuations in itinerant electron ferromagnets, *Phys. Lett.* **A81**, 87-90, (1984) Effects of spin fluctuations in transition metals and compounds, *J. Magn. Magn. Mat.* **31-34**, 299-300, (1984) Effect of spin fluctuations in the photoemission spectrum, *Phys. Lett.* **A106**, 143-145.
16. Shimizu, M. and Okada, H. (1981) Effects of spin fluctuations on the paramagnetic susceptibility in $ZrZn_2$, *Phys. Lett.* **A85**, 474-476.
17. Shimizu, M., Kunihara, A., and Tamaoki, A. (1983) Effects of spin fluctuations on the paramagnetic susceptibility in MnSi and fcc Fe_3Ni alloys, *Phys. Lett.* **A99**, 107-110.
18. Shimizu, M. and Kunihara, A. (1984) Effects of spin fluctuations on the paramagnetic susceptibility in fcc Fe and Rh, *Phys. Lett.* **A100**, 218-220.
19. Lonzarich, G. G. and Taillefer, L. (1985) Effect of spin fluctuations on the magnetic equation of state of ferromagnetic or nearly ferromagnetic metals, *J. Phys. C: Solid State Phys.* **18**, 4339-4371.
20. Wagner, D. (1989) The fixed-spin-moment method and fluctuations, *J. Phys. Condens. Matter* **1**, 4635-4642.
21. Levanyuk, A. P. (1964) Theory of second-order phase transitions, *Sov. Phys. Solid State* **5**, 1294-1298.
22. Lifshitz, E. M. and Pitaevskii, L. P. (1980) *Statistical Physics*, 3rd edition, Part 1, Chaps. XII and XIII, Pergamon Press, Oxford.
23. Yamada, H., Inoue, J., Terao, K., Kanda, S., and Shimizu, M. (1984) Electronic structure and magnetic properties of YM_2 compounds (M=Mn, Fe, Co and Ni), *J. Phys. F: Metal Phys.* **14**, 1943-1960; Yamada, H., Inoue, J., and Shimizu, M. (1985) Electronic structure and magnetic properties of the cubic Laves phase compounds ACo_2 (A=Sc, Ti, Zr, Lu and H_f) and $ScNi_2$, *J. Phys. F: Met. Phys.* **15**, 169-180, (1986) Electronic structure and magnetic properties of the cubic Laves-phase intermetallic compounds, *J. Magn. Magn. Mat.* **54-57**, 961-962; Yamada, H. and Shimizu, M. (1986) Electronic structure and magnetic properteis of the cubic Laves phase compounds AFe_2, (A=Zr, Lu and H_f), *J. Phys. F: Metal Phys.* **16**, 1039-1050; Yamada, H. Tohyama, T., and Shimizu, M. (1987) Magnetic properties of $Y(Fe-Co)_2$, *J. Magn. Magn. Mat.* **66**, 409-412, Magnetic properties of $Zr(Fe-Co)_2$, *J. Magn. Magn. Mat.* **66**, 413-416.

24. Shimizu, M., Inoue, J., and Nagasawa, S. (1984) Electronic structure and magnetic properties of Y-Ni intermetallic compounds, *J. Phys. F: Met. Phys.* **14**, 2673-2687.
25. Inoue, J. and Shimizu, M. (1985) Electronic structure and magnetic properties of Y-Co, Y-Fe and Y-Mn intermetallic compounds, *J. Phys. F: Met. Phys.* **15**, 1511-1524.
26. Inoue, J. and Shimizu, M. (1986) Electronic structure and magnetic properties of $Y_2Fe_{14}B$, *J. Phys. F: Met. Phys.* **16**, 1051-1058.
27. Shimizu, M. and Inoue, J. (1987) Thermal spontaneous ferromagnetism in Y_2Ni_7 intermetallic compound, *J. Phys. F: Met. Phys.* **17**, 1221-1229.
28. Inoue, J., Tamura, T. and Shimizu, M. (1987) On the anomaly in the magnetic susceptibility of iron monosilicide, *Phys. Lett.* **A 121**, 39-42.
29. Inoue, J., Hattori, T., and Shimizu, M. (1987) Effects of spin fluctuations on the magnetism of $(Fe_{1-x}Co_x)Si$, *phys. stat. sol. (b)* **143**, 243-248.
30. Takayama, N. and Shimizu, M.(1988) Electronic structure and paramagnetic susceptibility of the C14 Laves phase compound $NbFe_2$, *J. Phys. F: Met. Phys.* **18**, L83-L86.
31. Shimizu, M. and Mukaigawa, T. (1988) Effects of spin fluctuations on the paramagentic susceptibility and metamagnetic transition in UPt_3, *Phys. Lett.* **A 131**, 61-63.
32. Tohyama, T. and Shimizu, M. (1989) Magnetic properties of the pseudobinary alloy $(Cr-Mn)Pt_3$, *J. Magn. Magn. Mat.* **78**, 412-414.
33. Shimizu, M. (1989) Relation between electronic structure and magnetic properties in transition metals, *Physica B* **159**, 26-34.
34. Ma, S. (1976) *Modern Theory of Critical Phenomena*, Benjamin, Reading. pp.77-100 and 163-218.
35. Wallace, D. J. (1976) The ε-expansion for exponents and the equations of state in isotropic systems, in (eds.) C. Domb and M. S. Green, *Phase Transition and Critical Phenomena* Vol. 6, Academic Press, London.
36. Kubo, R. (1957) Statistical-mechanical theory of irreversible processes. I. General theory and simple applications to magentic and conduction problems, *J. Phys. Soc. Japan* **12**, 570-586.
37. Sakoh, M. unpublished work.
38. Podgórny, M., Tohn, M. and Wagner, D. (1992) Electronic structure and thermodynamic properties of Fe-Pt alloys, *J. Magn. Magn. Mat.* **104-107**, 703-704.
39. Thon, M. (1996) Allegemeine Fluktuationstheorie zur Beschreibung magnetoelastischer Systeme, Thesis, Ruhr-Universität Bochum.
40. Yamada, H., Yasui, M., and Shimizu M. (1981) High-field susceptibility for Fe and Ni, *Inst. Phys. Conf. Ser. No.* **55**: Chap. 4, 177-180.
41. Stoelinga, J. H. M., Gersdorf, R., and de Vries, G. (1965) Forced magnetostriction and its temperature-dependence of binary alloys between iron, cobalt and nickel, *Physica* **31**, 349-361; Foner, S., Freeman, A. J., Blum, A. J., Frankel, N. A., McNiff, R. B., and Praddaude, E. J. (1969) High-field studies of band ferromagnetism in Fe and Ni by Mössbauer and magnetic moment measurements, *Phys. Rev.* **181**, 863-882; Rebouillat, P. (1972) High resolution automatic magnetometer using a superconducting magnet: Application to high field susceptibility measurements, *IEEE Trans. Magn.* MAG-8, 630-633; Acker, F. and Huguenin, R. (1972) Magnetic properties of ferromagnetic NiCu alloys, *Phys. Lett.* **A38**, 343-344.
42. Edwards, D. M. and Wohlfarth, E. P. (1968) Magnetic isotherms in the band model of ferromagnetism, *Proc. Roy. Soc.* **A303**, 127-137.
43. Shimizu, M. (1965) On the conditions of ferromagnetism by the band model:II, *Proc. Phys. Soc.* **86**, 147-157, (1982) Magnetic phase transitions in itinerant electron system, *J. Physique* **43**, 681-683.
44. Ziebeck, K. R. A., Brown, P. J., Booth, J. G., and Bland, J. A. C. (1981) Observation of spatial magnetic correlations in the paramagnetic phase of weak ferromagnets; short-range magnetic order in MnSi at T_N, *J. Phys. F: Met. Phys.* **11**, L127-L30,

Ziebeck, K. R. A., Webstar, P. J., Brown, P. J., and Cappellmann, H. (1983) The invar effect and spin fluctuations in disordered Fe$_3$Pt, *J. Magn. Magn. Mat.* **36**, 151-159.

45. Lipton, D. (1971) The generalized spin susceptibility function for the transition metal series, *J. Phys. F: Metal Phys.* **1**, 469-479.
46. Cahn, J. W. (1968) Spinodal decomposition, *Trans. Met. Soc. A. I. M. E.* **242**, 167-180.
47. Abrahams, S. C., Guttman, L., and Kasper, S. S. (1962) Newtron diffraction determination of antiferromagnetism in face-centered cubic (γ) iron, *Phys. Rev.* **127**, 2052-2055.
48. Gonser, U., Meechan, C. J., Muir, A. H,. and Wiedersich, H. (1963) Determination of Néel temperatures in fcc iron, *J. Appl. Phys.* **34**, 2373-2378.
49. Tang, W., Gerhards, Ch., Heise, J., and Zabel, H. (1966) Structure and magnetic properties of Fe$_{1-x}$Ni$_x$/Cu Invar superlattices, *J. App. Phys.* **80**, 2327-2333.
50. Aleksandryan, V. V. Lagutin, A. S., Levitin, R. Z., Markosyan, A. S., and Snegirev, V. V. (1985) Metamagnetism of itinerant d-electron in YCO$_2$: Investigation of metamagnetic transitions in Y(Co, Al)$_2$, *Sov. Phys. JETP* **62**, 153-155.
51. Solontsov, A. and Wagner, D. (1994) Spin anharmonicity and zero-point fluctuations in weak itinerant electron magnets, *J. Phys. Condens. Matter* **6**, 7395-7402.

THE TEMPERATURE DEPENDENCE OF THE ENHANCED PARAMAGNETIC SUSCEPTIBILITY AT FINITE MAGNETIC FIELD

E. PAMYATNYKH, A. POLTAVETS AND M. SHABALIN
Ural State University,
Ekaterinburg, 620083, Russia

1. Introduction

The problem of the influence of spin fluctuations (SF) on the susceptibility of enhanced paramagnets is discussed for many years, and a lot of results exists now which give a good understanding of some features of its observable temperature and field behavior in wide temperature and magnetic field ranges. Nevertheless, a number of discrepancies between theories and experimental data as well as numerous contradictions between the results of different theoretical approaches have been the subject of much controversy.

The existence of detailed reviews of these problems (see, for example, [1-3]) permits us to restrict ourselves here to a brief overview of the main difficulties.

The theoretical efforts were directed basically towards the explanation of: (i) abnormal strong temperature dependence of $\chi(T)$ of nearly ferromagnetic Fermi systems at $T \ll T_F$ (T_F is the Fermi temperature), and, especially, (ii) a nonmonotonous temperature behavior of $\chi(T)$ (in particular, the existence of a maximum of the susceptibility for Pd, U_2C_3, $CeMn_3$, etc.). The confirmation of the key role of the incoherent SF in formation of low-temperature thermodynamics of such materials is the commonly recognized result of these efforts.

There are two groups of contradicting results concerning the temperature behavior of $\chi(T)$. In a number of papers [5-8] it was shown that the SF give rise to "doubly enhanced" T^2-terms in the temperature dependence of $\chi(T)$ (as compared with the "singly enhanced" T^2-term in Stoner's theory):

$$\chi(T) = \chi(0)\left(1 - \alpha S^2 T^2\right) \qquad (1)$$

(here and below the system of units $k_B = \mu_B = \hbar = 1$ is used, all values are normalized to T_F), S is the Stoner enhancement factor, $\chi(0) = S\chi_p$, and χ_p is the Pauli susceptibility. According to (1), the characteristic temperature scale of changing $\chi(T)$ decreases from T_F to $T_{SF} = T_F/S \ll T_F$, and the observable susceptibility maximum was tried to be explained either through the effects of complicated band structure or through the presence of impurities and defects in samples [9]. The explanations were based either on the Fermi-liquid approach [8], or on particular microscopic models [5-7].

In a series of other works based either on microscopic approaches [1,2], or on the general Fermi-liquid theory [4], it was claimed that in the temperature dependence of $\chi(T)$ along with the usual T^2-term a logarithmic term has to be present :

$$\chi(T) = \chi(0)\left(1 - \gamma S^n T^2 \ln \frac{T}{T^*}\right) \qquad (2)$$

where $n = 1$ [4], or $n = 3$ [1,2].

This kind of temperature dependence can explain the existence of a maximum in $\chi(T)$, however, the problem is still far from being well understood. In a number of works the conclusion about appearance of logarithmic terms was shown to be erroneous because when these terms are carefully collected the total elimination of various ones (to quasiparticle density of states, effective mass, the vertex part etc.) takes place [8,10]. It was also indicated that the presence of such logarithmic terms must cause the divergency of $\gamma(H)$ (the coefficient of linear in T term in the specific heat) at $T \to 0$ [11]. At last, the relation of these results to the conclusions of the first group of works [5-8] about the absence of logarithmic contributions is not obvious, and the ranges of applicability of these results are not clear.

In the present work, the Fermi-liquid approach is used to show that at finite magnetic fields the SF contribution to the free energy F is a complicated function of the parameter $\ell = 2B/T$ (where $2B$ is an energy of spin splitting), and, therefore, the behavior of the thermodynamic characteristics for different temperature regions ($\ell \ll 1$ and $\ell \geq 1$) differs essentially. Namely, in weak but finite magnetic fields (when $\ell \geq 1$ and $T \gg H$) the new temperature contributions arise, and, for $\ell \gg 1$, these ones can be evaluated analytically. In the previous works devoted to SF contributions to the thermodynamic characteristics of enhanced paramagnets, however, only the case of extremely weak magnetic fields ($\ell \ll 1$) was considered. We have found the "singly enhanced" logarithmic term $T^2 \ln T$ at finite H, which at $H \to 0$ corresponds to the zero-field $ST^2 \ln S$-term of Beal-Monod et al [5]. The width of the temperature range of the existence of this new contribution is proportional to S, i.e. this range may be large enough for the strongly enhanced paramagnets.

The technique of extracting the temperature contributions to the free energy described in the present paper can be used for a calculation of the whole spectrum of the thermodynamic parameters of the enhanced paramagnets in finite magnetic fields.

2. Main equations.

In the model of a Fermi-liquid, when only the spin part of the interelectronic interaction approximated by a single constant Ψ_0 is taken into account, the SF correction to the free energy $\delta F(M,T)$ (M is a magnetization) has the form [12]:

$$\delta F = \frac{1}{2} V \sum_k \int_0^1 \frac{d\lambda}{\lambda} \lambda \Psi_0 \langle \delta m \delta m \rangle_{\lambda k}, \qquad (3)$$

Here is an averaged square of spectral density of the SF (in which Ψ_0 is replaced by $\lambda \Psi_0$), $k = (\mathbf{k}, \omega)$.

By making use of the Fluctuation-Dissipation Theorem, the expression for $\langle \delta m \delta m \rangle_{\lambda k}$ in the random phase approximation, and after integrating over the coupling constant in (3), one can find:

$$\frac{1}{V} \delta F = \frac{1}{2} \sum_{k,\omega} cth \frac{\omega}{2T} \Im \Big(\ln(1 + \Psi_0 \chi_0^{zz}) -$$

$$\Psi_0 \chi_0^{zz} + \sum_{\sigma} [\ln(1 + \Psi_0 \chi_0^{\sigma,-\sigma}) - \Psi_0 \chi_0^{\sigma,-\sigma}] \Big). \qquad (4)$$

Here χ_0^{zz} and $\chi_0^{\sigma,-\sigma}$ are the longitudinal and transverse dynamical spin susceptibilities of quasiparticles in the absence of interaction given by:

$$\chi_0^{zz}(k,\omega) = \sum_{p,\sigma} \frac{n_{p+k/2}^\sigma - n_{p-k/2}^\sigma}{\omega - \varepsilon_{p+k/2} + \varepsilon_{p-k/2}}$$

$$\chi_0^{\sigma,-\sigma}(k,\omega) = 2 \sum_p \frac{n_{p+k/2}^\sigma - n_{p-k/2}^{-\sigma}}{\omega^\sigma - \varepsilon_{p+k/2} + \varepsilon_{p-k/2}}.$$

In the latter expressions, the following notations are used: $\omega^\sigma = \omega - 2\sigma B$, $\sigma = \pm 1$ is a spin quantum number, $n_p^\sigma = n(\epsilon_{p\sigma})$ is the Fermi function for quasiparticles with the energy $\epsilon_{p\sigma} = \epsilon_p + \sigma B$.

The transverse dynamical susceptibilities χ^{+-} and χ^{-+} contain explicitly the magnetic field through ω^σ in the denominator (in contrast to the longitudinal function χ^{zz}). As will be shown below only this explicit field dependence is important for the appearance of the new contributions to

$\chi(T)$ in finite magnetic fields, and therefore we restrict ourselves to a further consideration of the transverse SF only.

The following expression for evaluating the magnetic susceptibility will be used:

$$\frac{1}{\chi} = \left(\frac{\partial H}{\partial M}\right) = \left(\frac{\partial^2 F}{\partial m^2}\right). \tag{5}$$

Assuming δF to be a small correction, we get the contribution of the transverse SF to the susceptibility,

$$\chi(T) = \chi(0)\left(1 - \chi(0)\frac{1}{V}\frac{\partial^2 \delta F^{tr}}{\partial m^2}\right) = \chi(0)\left(1 - \chi(0)\delta(T)\right), \tag{6}$$

where the second term is defined by the following expression:

$$\delta(T) = \frac{1}{V}\frac{\partial^2 \delta F^{tr}}{\partial \mathbf{m}^2} = -\frac{8}{\pi\chi_p}\frac{\partial^2}{\partial B^2}\sum_\sigma \int_0^\infty d\omega N_\omega \int_0^\infty dk\, k^2 \arctan\left(\frac{\Im\chi^{\sigma,-\sigma}}{-\frac{1}{F_0^a} - \Re\chi^{\sigma,-\sigma}}\right) \tag{7}$$

Here $F_0^a = \eta(0)\Psi_0$ ($\eta(0)$ is a density of states at the Fermi level), N_w is the Planck function, summation over σ takes into account only transverse SF. The term, which is proportional to $\sum \chi^{\sigma,-\sigma}$, does not give a valuable contribution to the temperature behavior of the susceptibility and thus is excluded from the further consideration.

The expressions (5)–(7) allow us to calculate the SF contribution to the susceptibility (see [5,9,12]).

3. Temperature dependence of $\chi(T)$ due to SF

For the strongly enhanced paramagnets ($S \gg 1$) the main contribution to the integrals over k and ω is given by the long-wavelength and low-frequency SF, and this permits us to expand the dynamical susceptibilities in powers of $s = \omega/4k \ll 1$, $s^\sigma = \omega^\sigma/4k \ll 1$, $k \ll 1$, and to keep only a finite number of terms in this expansion. For $\chi^{\sigma,-\sigma}(k,\omega,B)$ one can obtain [12]:

$$\chi^{\sigma,-\sigma}(k,\omega,B) = 1 - ss^\sigma - \frac{1}{3}s(s^\sigma)^3 - \frac{1}{3}k^2 + \ldots + i\frac{\pi}{2}s\,\Theta(k - \frac{|\omega^\sigma|}{4})\Theta(1-k), \tag{8}$$

where the expansion coefficients correspond to a parabolic form of the quasiparticle spectrum.

Substitution of (8) into (7) gives:

$$\delta(T) = \frac{8}{\pi \chi_p} \frac{\partial^2}{\partial B^2} \sum_\sigma \int_0^\infty d\omega N_\omega \int_{|\omega^\sigma|/4}^1 dk\, k^2 \arctan \Phi_\sigma(k,\omega), \qquad (9)$$

Here we use the notations

$$\Phi^\sigma(k,\omega) = \frac{k}{\beta\omega} + \frac{1}{2\pi}\frac{\omega^\sigma}{k}\left[1 + \frac{1}{3}\left(\frac{\omega^\sigma}{4k}\right)^2\right], \quad \beta = -\left(\frac{\pi}{8}\right)\left(\frac{F_o^a}{1+F_o^a}\right),$$

and the relation: $\arctan x = (\pi/2)\text{sign}(x) - \arctan(1/x)$. It should be noted here that in the previous author's work [14] the additional term proportional to $\text{sign}\Phi_\sigma(k,\omega)$, which gives a $T^{5/2}/H^{1/2}$-term in the susceptibility, was erroneously written out.

The integration over k in (9) can be performed explicitly. However, in the general case the result will have a complicated form and so we do not write it here. Let us remark only that the character of the dependence of this expression on ω and B will be determined by the parameter $\omega^\sigma/(\beta\omega)$. The new contribution, which we are interested in, arises only at small values of this parameter, $\omega^\sigma/(\beta\omega) \ll 1$, or, if one notes that the main contribution to the integral over ω arises from the region $\omega \sim T$, at $T \gg H$. This new contribution is proportional to $(\omega^\sigma)^3 \ln |\omega^\sigma|/\omega$.

For an illustration of these statements, we describe a simplified calculation of these terms, based on the expansion of arctangent in a series of its argument. Here one must take into account the possibility of a change of the character of the arctangent's expansion at the point k^*, where the argument is equal to 1 (in our case k^* approximately equals $\beta\omega$). Thus, the condition $\omega^\sigma/(\beta\omega) < 1$ for the appearance of the new contributions corresponds to the case that k^* falls within the range of integration over k.

We can therefore split the region of the integration over k into $|\omega^\sigma|/4 < k < \beta\omega$ and $\beta\omega < k < 1$, and consider further only the first interval (where arctangent's argument is less than 1). Thus, the fluctuation factor (9) may be written in the form:

$$\delta(T) = \frac{8}{\pi \chi_p} \frac{\partial^2}{\partial B^2} \sum_\sigma \int_0^\infty d\omega N_\omega \int_{|\omega^\sigma|/4}^{\beta\omega} dk\, k^2 \left\{\frac{k}{\beta\omega} + \frac{2}{3\pi}\left(1 - \frac{4}{\pi^2}\right)(s^\sigma)^3\right\}, (10)$$

where only the terms responsible for the essential contributions to the temperature dependence of the susceptibility are left.

Let's now consider the integrals over k for each term in the bracket in (10).

(i) The result of the integration over k from the first term in (10) is:

$$I_1 = -\frac{1}{\beta\omega}\int_{|\omega^\sigma|/4} dk\, k^3 = \frac{(\omega^\sigma)^4}{4^5\beta\omega} \tag{11}$$

Evaluating the contribution from I_1 to $\chi(T)$ one can conclude that no significant temperature corrections occur.

(ii) The second integral contains the terms which are non-analytical when $H \to 0$:

$$I_2^\sigma = -\frac{2}{3\pi}\left(1-\frac{4}{\pi^2}\right)\int_{|\omega^\sigma|/4}^{\beta\omega} dk\, k^2 (s^\sigma)^3 =$$
$$\frac{2}{3\pi}\left(1-\frac{4}{\pi^2}\right)\left(\frac{\omega^\sigma}{4}\right)^3 \ln\frac{|\omega^\sigma|}{4\beta\omega}. \tag{12}$$

After integrating over ω this expression yields the logarithmic term in the susceptibility.

If the condition $H \ll T$ is satisfied, the value of the integral I_1 is much smaller than I_2, and therefore for $\delta(T)$ we can write:

$$\delta(T) = -\frac{8}{\pi\chi_p}\frac{\partial^2}{\partial B^2}\int_0^\infty d\omega N_\omega \sum_\sigma I_2^\sigma \tag{13}$$

Differentiating (13) twice with respect to B and carrying out simple transformations taking (12) into account, we get:

$$\delta(T) = \frac{1}{\chi_p}T^2\left\{-\frac{2}{\pi^4}\left(1-\frac{4}{\pi^2}\right)(J_1(\ell) - \ell J_2(\ell))\right\}, \tag{14}$$

where

$$J_1(\ell) = \int_0^\infty \frac{dz}{e^z-1}z\ln\left|\frac{1-(\ell/z)^2}{\beta^2}\right|, \tag{15}$$

$$J_2(\ell) = \int_0^\infty \frac{dz}{e^z-1}\ln\left|\frac{z+\ell}{z-\ell}\right|. \tag{16}$$

Thus, one can see that $\delta(T)$ really contains non-analytical (with respect to ℓ) terms. We note once again that these results (which lead to the new contributions to the susceptibility) could be obtained without making use of the approximate arctangent expansion just by performing the explicit integration over k in the initial expression (9) (using the expansion (8) for $\chi^{\sigma,-\sigma}(k,\omega,B)$ and the condition $|\omega^\sigma|/4 \ll k^* \ll 1$).

Now the behavior of $\delta(T)$ and $\chi(T)$ must be analyzed in the extreme cases where the explicit form for the new SF terms in the susceptibility may be derived.

(i) $\ell \ll 1$ (the extremely weak magnetic fields)

The approximate expression for the integral $J_1(\ell)$ has the form:

$$J_1(\ell) = \int_0^\infty dz \frac{z}{e^z - 1} \ln \frac{1}{\beta^2} = -\frac{\pi^2}{3} \ln \beta. \tag{17}$$

The numerical analysis of the integral $J_2(\ell)$ vs ℓ shows that at small ℓ ($\ell \sim 10^{-3} \div 10^{-4}$) $J_2(\ell)$ behaves as a smooth function of ℓ being approximately equal to a constant: $J_2(\ell) \approx const = A(\sim 4.7)$.

Then for $\delta(T)$ we get:

$$\delta(T) = \frac{1}{\chi_p} \left[\left(\frac{4 \cdot A}{\pi^4}\left(1 - \frac{4}{\pi^2}\right)\right) BT + \frac{2}{3\pi^2}\left(1 - \frac{4}{\pi^2}\right) T^2 \ln \beta \right] \tag{18}$$

Substituting (18) into the expression (10) for the susceptibility, we get the result which partially reproduces the contribution obtained in the previous work [5] (a total coincidence will take place if the longitudinal SF would be taken into account and the quantum approach rather than a quasiclassical one would be used). Thus, it is not the case we are interested here.

(ii) $\ell \gg 1$ (the finite fields)

For the integrals J_1–J_2 the following approximate expressions are valid:

$$J_1(\ell) = 2\int_0^\infty dz \frac{z}{e^z - 1} \ln \frac{\ell}{\beta z} = -\frac{\pi^2}{3} \ln \frac{T}{T^*}, \tag{19}$$

$$J_2(\ell) = \frac{2}{\ell}\int_0^\infty dz \frac{z}{e^z - 1} = \frac{\pi^2}{6}\frac{T}{B}, \tag{20}$$

where $\zeta(x)$ is the Riemann zeta-function, T^* is a characteristic temperature ($T^* \sim B$). For $\delta(T)$ we have:

$$\delta(T) = \frac{1}{\chi_p} \frac{2}{3\pi^2}\left(1 - \frac{4}{\pi^2}\right) ST^2 \ln \frac{T}{T^*}. \tag{21}$$

Therefore, the temperature expansion of the susceptibility for $\ell \gg 1$ has the form:

$$\chi(T) = \chi(0)\left[1 - \frac{2}{3\pi^2}\left(1 - \frac{4}{\pi^2}\right)ST^2 \ln \frac{T}{T^*}\right]. \tag{22}$$

Here the T^2-term from $J_2(\ell)$ was introduced under the logarithm renormalizing the temperature T^*. It should be noted that the S^2T^2 contribution to $\chi(T)$ obtained by Beal-Monod et al [5] is really present in the temperature expansion both for $\ell \ll 1$ and $\ell \gg 1$, but we shall not discuss it here (as already mentioned, to obtain it we must take into account the longitudinal SF and finite k). This T^2-term will be included in the final expressions only. We emphasize also that our $ST^2 \ln T$-term ($\ell \gg 1$) originates from the same free energy terms which produce $ST^2 \ln S$-term obtained by Beal-Monod et al [5] for $B = 0$.

At the end of this section we additionally emphasize that this new contribution was not revealed in the previous works since only the case of extreme weak magnetic field ($\ell \ll 1$) was considered there.

4. Discussion

Thus, the temperature dependence of the susceptibility of exchange enhanced paramagnets, $\chi(T)$, has the form:

$$\frac{\chi(T)}{\chi(0)} = \begin{cases} 1 - \alpha S^2 T^2, & \text{when } T \ll H \text{ or } T \gg 2SH \\ 1 - \alpha S^2 T^2 - \gamma ST^2 \ln \frac{T}{T^*}, & \text{when } H \ll T \ll 2SH, \end{cases} \quad (23)$$

where $\alpha = 7.7\pi^2/24$ [5,6], $\gamma = (2/3\pi^2)(1 - 4/\pi^2)$.

So, the consideration of the case of a finite magnetic field has resulted in the appearance of the "singly enhanced" logarithmic contribution to $\chi(T)$, which is in agreement with a qualitative result of Misawa [4] obtained with the help of a phenomenological Fermi-liquid approach. Moreover, this logarithmic contribution exists only at finite temperatures ($T \gg H$) which makes it possible to avoid the problem indicated by Beal-Monod [11] (singularity of $\gamma(H) = C_v(H)/T$ at $T \to 0$).

References

1. Barnea, G. (1971) Microscopic Fermi-liquid Theory of the Temperature Dependence of the Susceptibility of the Normal Paramagnetic Metals, *J.Phys.C: Solid State Phys.*, **7**, pp. 315–337
2. Grempel, D. R. (1983) Temperature Dependence of the Magnetic Susceptibility of Almost Ferromagnetic Materials, *Phys. Rev* **B27**, pp. 4281–4287
3. Misawa, S. (1988) Susceptibility Maximum as a Fermi-liquid Effect in Paramagnetic Fermi-liquid, *Physica* **B 149**, pp. 162–168
4. Misawa, S. (1970) Anomalous Properties of a Normal Fermi-liquid at Low Temperatures, *Phys. Letters* **32A**, pp. 153–154
5. Beal-Monod, M. T., Ma, S. K. and Fredkin, D. R. (1968) Temperature Dependence of the Spin Susceptibility of a Nearly Ferromagnetic Fermi-liquids, *Phys. Rev. Letters* **20**, pp. 929–932
6. Kawabata, A. (1974) Interaction of Spin Fluctuations in Ferromagnetic and Nearly Ferromagnetic Itinerant Systems, *J. Phys. F: Metal Phys.* **4**, pp. 1477–1488

7. Mishra, S. G. and Ramakrishnan, T. V. (1978) Temperature Dependence of the Spin Susceptibility of a Nearly Ferromagnetic Fermi System, *Phys. Rev* **B18**, pp. 2308–2317
8. Carneiro, G. M. and Pethick, C. J. (1977) Finite Temperature Contributions to the Magnetic Susceptibility Nearly Ferromagnetic Fermi-systems, *Phys. Rev* **B16**, pp. 1933–1943
9. Beal-Monod, M. T. and Lawrence, J. M. (1980) Paramagnon Picture of the Low-Temperature Susceptibility of Some Intermediate-Valence Compounds, *Phys. Rev* **B21**, pp. 5400–5409
10. Mishra, S. G. and Ramakrishnan, T. V. (1977) Absence of $T^2 \ln T$ term in Spin Susceptibility of Nearly Ferromagnetic Fermi Systems, *J.Phys.C: Solid State Phys.* **10**, pp. L667–670
11. Beal-Monod, M. T. (1982) Field Effects in Strongly Enhanced Paramagnets, *Physica* **109&110B**, pp. 1837-1841
12. Moriya, T. (1988) *Spin Fluctuations in Itinerant Electron Magnetism* . Mir, Moscow.
13. Pethick, C. J. and Carneiro, G. M. (1973) Specific Heat of a Normal Fermi-liquid. Landau-Theory Approach, *Phys.Rev.* **A7**, pp. 304–318
14. Pamyatnykh, E., Poltavets, A. and Shabalin, M. (1997) The Temperature and Field Dependence of the Enhanced Paramagnetic Susceptibility due to Spin Fluctuations, *J.Phys.C: Solid State Phys.* **9**, pp. 715–722

FIRST-PRINCIPLES STUDY OF ITINERANT-ELECTRON MAGNETS: GROUND STATE AND THERMAL PROPERTIES

L.M. SANDRATSKII, M. UHL AND J. KÜBLER
*Institut für Festkörperphysik, Technische Universität
D-64289 Darmstadt, Germany*

Abstract. We briefly review modern applications of density-functional theory for studies of both ground state magnetic structures and finite-temperature properties of itinerant-electron magnets. In the first part we deal with first-principles calculations of complex non-collinear magnetic structures taking as examples spiral structures like those found experimentally in fcc-Fe, the non-collinear magnetic states in U compounds and weak ferromagnetism in α-Fe$_2$O$_3$ and Mn$_3$Sn. We show that these different phenomena are explained by means of a single theoretical approach. In the second part we describe a statistical mechanics scheme which is based essentially on using total-energy parameters obtained from first-principles calculations for particular non-collinear magnetic structures. This information is used to model spin fluctuations that determine the thermal properties of itinerant-electron magnets. As an example we determine the magnetic phase diagram of cobalt and show how the hcp–fcc phase transition can be described in terms of spin fluctuations. Furthermore, we explain the antiferromagnetic–to–ferromagnetic phase transition of FeRh by spin fluctuations obtaining the transition temperature as well as the Curie temperature in good agreement with experiment.

1. Introduction

It was more than forty years ago that non-collinear magnetic structures were first discovered experimentally [2, 3], yet this physical phenomenon was investigated theoretically almost entirely in the framework of model Hamiltonians assuming localized atomic moments [3] in contrast to collinear magnets which were intensively studied using first-principles calculational schemes provided by the density functional theory [4, 5, 6, 7]. This is perhaps due to the fact that non-collinear magnetic order seemed rather exotic and rare. But it attracted renewed interest in the late seventies and beginning eighties when it became clear that Stoner theory, which explains ground-state properties quite well, fails to describe the temperature behavior of itinerant magnets. The reason for this failure was found in the neglect of transverse fluctuations of the magnetization density that, if viewed as originating from well-formed atomic moments, is equivalent to neglecting fluctuations of the directions of atomic moments (see e.g. [7]).

Detailed investigations of particular non-collinear magnetic configurations were started by Heine and his group using the cluster recursion method applied to a simplified tight-binding Hamiltonian [8]. The line of this work was continued with the use of the KKR method for a periodic solid (see e.g. [9]). A thorough discussion of density functional theory for non-collinear magnets was subsequently given in [10] and applied to investigate the ground-state electronic and magnetic structure of Mn_3Sn. It was the experimental discovery of a spiral magnetic structure in fcc-Fe [11] that made Fe an interesting object for theoretical studies [12, 13, 14]. The last years showed a boom in investigations of non-collinear magnetism. Thus, for instance, successful studies of disordered systems [15, 16, 17] as well as multilayers [18] were reported and a noteworthy and interesting step was made in describing the spin dynamics in itinerant-electron systems [19]. In spite of its rather short history the theory of non-collinear magnetism involves quite different methods and is applied to a multitude of different physical problems. In this article we are not trying to review all aspects of the presently used first-principles studies of non-collinear magnets, but rather select aspects which possess interesting symmetry properties. In the last Section of this paper we show how our technique of determining total-energy properties of spiral magnetic structures incommensurate with the crystal lattice can be used to model spin fluctuations and thus develop a first-principles thermodynamics of itinerant magnets.

2. Kohn-Sham Hamiltonian of a non-collinear magnet

The derivation of the Kohn-Sham Hamiltonian for non-collinear magnets follows the standard procedure of density functional theory. However, in this case the total energy is considered as a functional of the two-by-two density matrix and all single-particle wave functions are consequently treated as two-component spinor functions [10]. After variation of the functional with respect to the components of the spinor function one obtains equations that describe a magnetization continuously varying in space. For practical calculations it is common to use the atomic sphere approximation for the magnetization direction, i.e. the direction of the magnetization is supposed to be constant within the atomic sphere of every atom and different for different atoms. The scalar-relativistic Kohn-Sham Hamiltonian of a non-collinear magnet then takes the form

$$\hat{H}_{sc}(a_\nu, e_\nu) = \sum_\nu U^+(\theta_\nu, \phi_\nu) \begin{pmatrix} H_{sc}^{\nu\uparrow}(r_\nu) & 0 \\ 0 & H_{sc}^{\nu\downarrow}(r_\nu) \end{pmatrix} U(\theta_\nu, \phi_\nu) \ . \quad (1)$$

Here $U(\theta_\nu, \phi_\nu)$ is the standard spin-$\frac{1}{2}$-rotation matrix which describes the transformation between a global and a local coordinate system of the νth atom whose spin orientation is given by the polar angles θ_ν and ϕ_ν with respect to the z-axis of the global system. The quantities $H_{sc}^{\nu\uparrow}(r_\nu)$ and $H_{sc}^{\nu\downarrow}(r_\nu)$ are the standard atomic scalar-relativistic Hamiltonians (spin up, spin down) [20] in the local frame of reference for the atom at site ν. They contain the mass-velocity, the Darwin term and the effective one-particle potential which, as usual, is given by functional derivatives of the total energy and is spin-diagonal in the local frame of this atom.

Note that the scalar-relativistic Hamiltonian possesses the same symmetry properties as the non-relativistic Hamiltonian which is obtained from the former by neglecting the mass-velocity and Darwin terms. Therefore the symmetry aspects discussed in this paper will be equally valid for both cases. Our calculations were carried out with the scalar-relativistic Hamiltonian.

A substantial part of the discussion will be devoted to the role of spin-orbit coupling which we will write in the form

$$\hat{H}_{so} = \sum_\nu U^+(\theta_\nu, \phi_\nu) \left\{ \sum_\alpha M_\alpha \sigma_\alpha \hat{l}_\alpha \right\} U(\theta_\nu, \phi_\nu). \quad (2)$$

Here σ_α and $\hat{l}_\alpha$ are the Cartesian components of the Pauli spin matrices and the angular momentum, respectively, in the local frame of reference and the coefficients M can be found in Ref. [21].

One of the purposes of our work is the first-principles determination of the magnetic structure of a crystal. The basic steps to achieve this can be summarized as follows [10]: in each iteration step the eigenstates $\begin{pmatrix} \psi_1(r) \\ \psi_2(r) \end{pmatrix}_i$ of the Hamiltonian $\hat{H}_{sc}$ in the scalar-relativistic or $\hat{H}_{sc}+\hat{H}_{so}$ in the relativistic case are calculated which enables us to determine the two-dimensional density matrix of the system:

$$\rho(r) = \sum_{i \ occ} \begin{pmatrix} \psi_1(r)^*\psi_1(r) & \psi_1(r)^*\psi_2(r) \\ \psi_2(r)^*\psi_1(r) & \psi_2(r)^*\psi_2(r) \end{pmatrix}_i \quad (3)$$

where the sum extends over all occupied states. The density matrix contains information on the charge density, the directions of the atomic magnetic moments and the magnetization density all of which are necessary to redetermine the Hamiltonian for the next iteration step. The iterations are repeated until self-consistency is achieved. The new degrees of freedom connected with the variation of the direction of the magnetic moments make the problem numerically more involved than in the case of a collinear magnetic state.

3. Generalization of the symmetry basis

Traditionally the symmetry properties of non-relativistic Hamiltonians are described in terms of ordinary irreducible representations of the relevant space group whereas for the relativistic problem the so-called double-valued irreducible representations are used. However, this difference in the type of representation merely reflects the difference in the choice of the functions subjected to the transformation. In the relativistic case the transformed function is always a spinor function whereas it is a scalar wave function in the non-relativistic case for which, correspondingly, only the transformation properties in real space need be considered.

It is the spinor form of the wave function that is of prime importance for the case of non-collinear magnetism. Hence, as in the relativistic case, one must use the double-valued irreducible representation. Still there is a subtle difference in the symmetry properties of the problem depending on whether or not the spin-orbit coupling is retained in the Hamiltonian. To describe this properly one introduces a generalized set of operators which allows an independent transformation of the spin and space variables [22]. For the group of such operators we will use the term spin-space group (SSG).

We define the action of an operator of the SSG on a two-component spinor function as follows

$$\{\alpha_S|\alpha_R|\mathbf{t}\}\,\psi(\mathbf{r}) = U(\alpha_S)\,\psi(\{\alpha_R|\mathbf{t}\}^{-1}\mathbf{r}) \qquad (4)$$

where ψ is a two-component spinor function, $\mathbf{U}$ is the spin-$\frac{1}{2}$-rotation matrix, α_S and α_R are, respectively, spin and space rotations, and $\mathbf{t}$ is a space translation. Operators of the usual space group are those with $\alpha_S = \alpha_R$.

One easily proves that a transformation of the scalar-relativistic Hamiltonian (i.e. not including spin-orbit coupling) of a non-collinear magnet with the operations (4) leaves the form of the Hamiltonian invariant, i.e.

$$\hat{H}_{sc}(\mathbf{a}'_\nu, \mathbf{e}'_\nu) = \{\alpha_S|\alpha_R|\mathbf{t}\}\hat{H}_{sc}(\mathbf{a}_\nu, \mathbf{e}_\nu)\{\alpha_S|\alpha_R|\mathbf{t}\}^{-1} \qquad (5)$$

where $\mathbf{a}_\nu$ are atomic positions and $\mathbf{e}_\nu$ directions of atomic moments corresponding to the untransformed Hamiltonian. ¿From (5) it follows that two magnetic crystals with atomic positions connected by the relation

$\mathbf{a}'_\nu = \alpha_R \mathbf{a}_\nu + \mathbf{t}$ and directions of magnetic moment connected by $\mathbf{e}'_\nu = \alpha_S \mathbf{e}_\nu$ are equivalent in the scalar-relativistic case.

However, a transformation of the spin-orbit coupling term with an SSG operator does not reproduce the form of this term. Only when $\alpha_S = \alpha_R$ do we restore the form-invariance of the Hamiltonian and find equivalent magnetic crystals:

$$\{\alpha_R|\alpha_R|\mathbf{t}\}\,\hat{\mathbf{H}}_{so}(\mathbf{a}_\nu, \mathbf{e}_\nu)\,\{\alpha_R|\alpha_R|\mathbf{t}\}^{-1} = \hat{\mathbf{H}}_{so}(\mathbf{a}'_\nu, \mathbf{e}'_\nu) \qquad (6)$$

In many cases we are interested in the symmetry properties of one particular magnetic configuration and not in establishing the equivalence of different magnetic configurations. In this case the atomic positions and corresponding atomic moments must be the same before and after the transformation and the equations (5,6) reduce to commuting Hamiltonian and symmetry operators.

In what follows we will show how symmetry arguments help us to make calculations and facilitate to analyze the calculational results for a number of different physical problems. In the discussion of the calculated ground state magnetic structures we focus our attention on the direction of the magnetic moments. For the information concerning calculations of the length of the magnetic moments we refer reader to the original papers: γ-Fe [12, 13, 23], U_3P_4 [24, 25], U_3As_4 [25], U_3Sb_4 [25], UPdSn [26], Fe_2O_3 [27, 28], Mn_3Sn [10, 29].

Note, however, that for all systems discussed in this review the agreement between theoretical and experimental values of the length of magnetic moments was found to be very good.

4. Incommensurate spiral structure

A spiral magnetic structure is defined by

$$\mathbf{M}_n = M\,(\cos(\mathbf{q} \cdot \mathbf{R}_n)\sin\theta, \sin(\mathbf{q} \cdot \mathbf{R}_n)\sin\theta, \cos\theta) \qquad (7)$$

where $\mathbf{M}_n$ is the magnetic moment of magnitude M of the n-th atom, and $(\mathbf{q} \cdot \mathbf{R}_n)$, θ are polar coordinates.

A picture of a spiral structure with vector $\mathbf{q}$ parallel to the z axis is shown in Fig. 1.

An apparent difficulty facing first-principles calculations for a spiral structure is the loss of periodicity with respect to lattice translations non-orthogonal to $\mathbf{q}$. One should notice, however, that in formula (7) all atoms of the spiral structure are equivalent, in particular because of the equal length of all atomic moments. But atoms can be equivalent only if they are connected by a symmetry transformation. A solution of the problem is suggested by using the operators of the SSG.

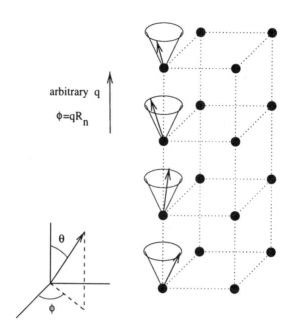

Figure 1. A spiral structure with vector **q** parallel to the z axis.

Indeed, transformations combining a lattice translation $\mathbf{R}_n$ and a spin rotation about the z axis by an angle $\mathbf{q} \cdot \mathbf{R}_n$ leave the spiral structure invariant. The corresponding operators $\{\mathbf{q} \cdot \mathbf{R}_n | \varepsilon | \mathbf{R}_n\}$ commute with the Hamiltonian of the spiral structure and therefore supply a symmetry transformation of the Hamiltonian. Here ε denotes the identity operation. These generalized translations form an Abelian group isomorphic to the group of ordinary space translations by vectors $\mathbf{R}_n$. Therefore the irreducible representations of both groups coincide and for the eigenfunctions of the Hamiltonian (1) there exists a generalized Bloch theorem [30, 31]

$$\{\mathbf{q} \cdot \mathbf{R}_n | \varepsilon | \mathbf{R}_n\}\psi_k(\mathbf{r}) = exp(-i\mathbf{k} \cdot \mathbf{R}_n)\psi_k(\mathbf{r}) \qquad (8)$$

where the vectors **k** lie in the first Brillouin zone which is defined in the usual way by the vectors $\mathbf{R}_n$. These properties permit to restrict our considerations of real space to a chemical unit cell, not a supercell.

Generalizations of modern methods employing density functional theory to the case of spiral structures were done in Ref. [31] using KKR, APW and tight binding methods, in Ref. [12] using the LMTO method, and in Ref. [13] using the ASW method. This approach was successfully applied to the description of the ground state of fcc-Fe [12, 13, 14] which was experimentally observed to have a spiral magnetic structure [11].

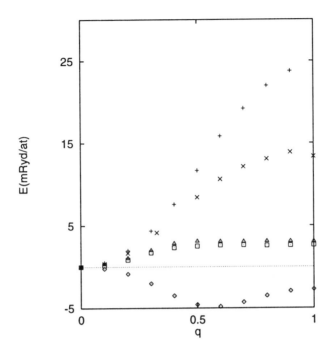

Figure 2. Total energy difference as a function of q for the (001) direction. ◇ - fcc-Fe; + - bcc-Fe; □ - fcc-Co; × - hcp-Co; △ - Ni.

In Fig. 2 we show the results of calculations for the q-dependence of the total energy for a number of transition metals [23]. In full agreement with experiment for all metals, excluding the case of fcc-Fe, the ground state was found to be ferromagnetic which is the case $q=0$. In fcc-Fe the minimum of the total energy occurs at a finite value of q, i.e. the ground state consists of a spiral spin arrangement.

We complete this section with a discussion of the role of spin-orbit coupling in a spiral structure. The symmetry analysis of the previous section shows that spin-orbit coupling does not allow a separate transformation of the spin and space variables. This means the generalized translations do not apply here and we conclude that spin-orbit coupling is destructive for the spiral structure. Indeed, experimental evidence seems in favor of our point of view since spiral structures are not observed in cases where the spin-orbit coupling is strong in the valence states. In particular, no spiral structures are observed in the U compounds where the U 5f electrons are itinerant and where the spin-orbit coupling is of the same order of magnitude as the exchange splitting. In contrast to this is the strong spin-orbit coupling for the core states which does not lead to an increase of the magnetic anisotropy and therefore seems not to be an obstacle for the use of the generalized symmetry. Thus, in our opinion, the observation of spiral structures in 4f elements [2] is an argument in favor of treating the 4f states as core states.

5. Non-collinear magnetic structures in Uranium compounds

5.1. NON-COLLINEAR MAGNETIC STRUCTURES IN U_3X_4

Up to date no spiral structure was observed in Uranium compounds but one finds a great variety of non-collinear magnetic configurations. It is in particular in U_3P_4 and U_3As_4 that complex non-collinear structures were observed [32] and we will start with a discussion of U_3P_4 subsequently turning to results of calculations for magnetic properties of U_3As_4 and U_3Sb_4.

The magnetic moments of the individual atoms of U_3P_4 do not compensate but rather possess a large ferromagnetic component along the (111)-axis and the angles between the magnetic moments and the (111) axis seem to assume some accidental value. When calculations were started aligning all magnetic moments along the easy (111) axis then the scalar-relativistic and the relativistic Hamiltonians lead to drastically different results. Free to rotate, the magnetic moments deviate from the initial parallel directions in the relativistic case but stay parallel in the scalar-relativistic case. To appreciate the difference between these two cases we formulate the following nearly self-evident statement: "*The symmetry of the initial Kohn-Sham Hamiltonian must be preserved during calculations*". This means on the one hand that if the combined symmetry of the crystal and magnetic structure is so high that a deviation of magnetic moments from the initial directions leads to perturbing the invariance of the Hamiltonian with respect to at least one symmetry operator this deviation cannot take place. On the other hand, if a deviation of the magnetic moments from the initial directions is allowed by all symmetry operations present, then there are no symmetry reasons for keeping the initial magnetic configuration and the magnetic moments will start to rotate tending to assume the state of lowest total energy. This simulated annealing of magnetic moments will continue until the "accidental" - from a symmetry point of view - ground state magnetic structure will be found by the system.

These general statements will help us to explain the behavior of magnetic moments in U_3P_4. We start with the scalar-relativistic case. As was shown in Sect. 3 the symmetry basis of a scalar-relativistic problem is formed by the spin-space group, i.e. separate transformations of the spin and space variables are allowed. We can formulate the result in a very general way: starting the scalar-relativistic calculation with a collinear configuration we will never obtain a deviation of magnetic moments from the initial direction. Indeed, independent of the crystal structure any spin rotation by an arbitrary angle ϕ about the direction of the magnetic moments $\{C_\phi|\varepsilon|0\}$ is a symmetry operation. This group of symmetry operations gives the symmetry basis for treating the spin projection of an electron state as a good quantum number [22]. Deviations of any magnetic moment from this direction would destroy the symmetry with respect to operations $\{C_\phi|\varepsilon|0\}$ and are, therefore, forbidden.

The situation changes drastically in the presence of spin orbit coupling (SOC) because of the reduction of the symmetry basis from the SSG to the usual space groups which transforms spin and space variables in an identical way i.e. $\alpha_S=\alpha_R$ in Sect. 3. This means that the question of stability of a magnetic structure cannot be answered without an analysis of the particular crystal structure. Therefore we note that U_3P_4 has a bcc lattice with a basis formed by two formula units, i.e. the unit cell consists of 6 U and 8 P atoms. The crystal structure is rather complicated but need not be discussed here in detail. Instead, to illustrate the important symmetry properties it is sufficient to consider the simple picture of Fig. 3 where the projections onto the (111) plane of the positions of the six U atoms are shown. Let us assume that initially all magnetic moments are parallel to the (111) axis which is perpendicular to the plane of the paper. Then the following operations leave the magnetic and crystal structures invariant: the rotations by 120° and 240° about the (111) axis and the reflections in the planes containing the (111) axis accompanied by time reversal. Of importance is the observation that none of these operations leaves the position of any particular atom unchanged. Because of this, symmetry imposes no restrictions on the direction of the magnetic moment of a particular atom but only on the orientation of the atomic moments relative to each other and to the crystal lattice. The deviation of the moments from the (111) axis resulting in a non-collinear magnetic configuration does not change the symmetry of the crystal. Therefore, the ferromagnetic ($\theta=0$) structure, from a point of view of symmetry, is not isolated from structures possessing a non-zero θ. In Fig. 3 we show the projections of the magnetic moments on the (111) plane which are obtained in the self-consistent calculation; these calculated deviations from the (111) axis evidently do not destroy the symmetry of the atomic configuration. To emphasize that the magnetic moments possess components parallel to the (111) axis, we also show the cone formed by the six U-moments with the axis of the cone parallel to (111)-direction.

More understanding of the system's behavior can be gained with the help of Fig. 4 which shows the result for total energies when the directions

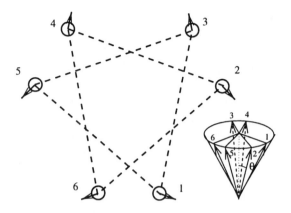

Figure 3. Projection of the atomic positions and magnetic moments of the U atoms in U_3X_4 onto the (111) plane. At the right the cone formed by the magnetic moments of the U atoms is shown, the cone's axis being parallel to the (111) axis.

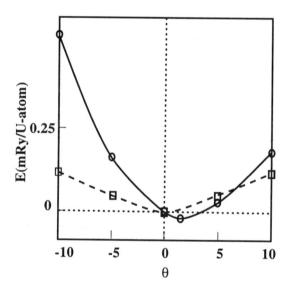

Figure 4. Total energy of U_3P_4 as a function of angle θ. Solid line was calculated with and dashed line without SOC.

of the magnetic moments are constrained to some values of θ near equilibrium (see Fig. 3 for a definition of θ). Fig. 4 shows that scalar-relativistic calculations give a total-energy curve symmetric with respect to a change of the sign of θ. This means an extremum of the total energy for the ferromagnetic configuration $\theta=0$ is predetermined by symmetry. In the relativistic case, however, the total energy as a function of θ is not symmetric about

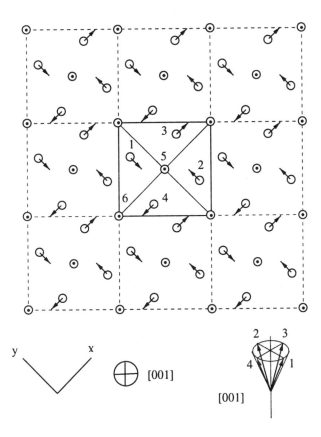

Figure 5. Projection of the atomic positions and magnetic moments of the U atoms in U_3X_4 onto the (001) plane. At the right the cone formed by the magnetic moments of the U atoms 1-4 is shown, the cone's axis being parallel to the (001) axis.

$\theta=0$. In fact, for reasons of symmetry no extremum of the total energy at $\theta=0$ is expected. As a result, the position of the extremum of the the total energy curve as a function of θ is "accidental" i.e. not determined by the symmetry of the problem.

Results of calculations for U_3As_4 [25] are very similar to those for U_3P_4 and again, in agreement with experiment, supply a non-collinear cone structure with axis of the cone parallel to the (111) direction.

In contrast to U_3P_4 and U_3As_4, the experimental properties of U_3Sb_4 (see e.g. [33, 34]) are discussed in terms of a collinear magnetic structure with all magnetic moments parallel to the (001) direction. Our calculation does not support this point of view. Starting calculations with all moments parallel to the (001) direction one sees that the magnetic moments of the first group of the U atoms (atoms 1-4), if free to rotate, deviate from their initial direction and form a kind of cone structure shown schematically in Fig. 5 where the projection of the U atoms on the (001) plane is represented. Atoms 5-6 keep their directions along the (001) axis. Our symmetry principle allows easy understanding of this observation. Indeed, as is seen in Fig. 5, the system has a fourfold symmetry axis which passes through the position of atom 5. (A detailed symmetry analysis shows that, to keep the crystal structure invariant, the 90°-rotation about the z axis needs to be accompanied by the inversion.) To preserve this symmetry the direction of the moment of atom 5 (and equivalent to it atom 6) must be kept along the (001) axis. On the other hand, atoms 1-4 transform into each other under rotation about this axis. Therefore, only relative directions of the moments of these atoms and not the directions of the particular moments are restricted by this symmetry axis. In particular, the non-collinear structure shown in Fig. 5 has the entire symmetry of the initial structure.

Although the possibility of the magnetic structure of U_3Sb_4 being non-collinear had already been discussed in Ref. [35] on the basis of a symmetry analysis of a phenomenological spin-Hamiltonian, in later treatments of experimental results this structure was considered as one of the possible alternatives not realized in reality. Our calculation and our symmetry analysis show unequivocally that in the case of U_3Sb_4 we deal again with the non-collinearity predetermined by symmetry, hence the magnetic structure of U_3Sb_4 cannot be collinear. Note that the result of a very recent neutron diffraction measurements [36] of U_3Sb_4 gave an estimate of the deviation angle somewhat bigger than the error bar of the experiment which can be treated as a qualitative confirmation of our conclusion.

5.2. NON-COLLINEAR MAGNETIC STRUCTURE IN UPDSN

In the last years UPdSn was extensively studied experimentally revealing a number of interesting physical properties [37, 38, 39, 40, 41, 42, 43]. The main attention focussed on two magnetic phase transitions which were found to be accompanied by lattice distortions. Thus, the paramagnetic state has the hexagonal GaGeLi crystal lattice. Below 45 K UPdSn becomes magnetic with a non-collinear antiferromagnetic structure. In this phase (which we will refer to as structure I) all magnetic moments of the Uranium atoms lie parallel to a plane and compensate one another completely. Simultaneously orthorhombic lattice distortions are detected. At 20 K a second phase transition is observed. Here the magnetic structure (structure II) is still non-collinear and compensated, however, the magnetic moments deviate from the plane developing components perpendicular to it.

Thus, contrary to the U_3X_4 compounds, both non-collinear magnetic

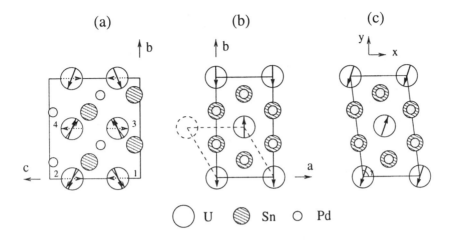

Figure 6. Projections of the crystal and magnetic structure onto the xy and yz planes. (a) Projection of the orthorhombic unit cell onto the yz plane. Dotted arrows show the initial magnetic structure used to start the calculation, thick arrows show the resulting self-consistent directions of the magnetic moments. Thin arrows show the experimental magnetic structure [37]. (b) Projection of the orthorhombic unit cell onto the xy plane. Both experimental and theoretical projections of the magnetic moments are parallel to the y axis. (c) Projection of the monoclinic unit cell onto the xy plane. Arrows show schematically the deviations of the magnetic moments from the yz plane.

structures of UPdSn possess no ferromagnetic component.

We start our calculations with the undistorted hexagonal lattice having initially all uranium magnetic moments directed along the z axis and forming a collinear antiferromagnetic structure (see Figs. 6 a) with an orthorhombic magnetic unit cell. Subsequently, allowing the moments to rotate, they deviated immediately from the z axis keeping however their equivalence and the compensated character of the magnetic structure. The resulting self-consistent directions of the magnetic moments are shown in Fig. 6 a; they form a magnetic structure which is very similar to the experimental structure I. The symmetry principle, formulated above helps us to expose the physical reasons for the instability of the initial collinear structure.

To apply the symmetry principle to this problem we begin with a symmetry analysis of the initial collinear antiferromagnetic structure. We will consider two types of symmetry transformations: unitary transformations which do not contain time reversal and anti-unitary transformations which are products of a unitary transformation with the time-reversal operation. The system at hand possesses four unitary symmetry transformations and four anti-unitary transformations collected in Table 1. Note that the vector $\vec{\tau}_2 = \frac{1}{2}(a, b, 0)$ is a lattice translation of the hexagonal chemical lattice which in the magnetic case must be accompanied by time-reversal to be a symmetry operator (operator 5 in Table 1).

The following symmetry property of the initial collinear structure is important in our case: all non-trivial unitary transformations, 2-4, and three of the four non-unitary transformations, 5,6 and 8, do not leave the atomic positions unchanged and transfer every atom into the position of another atom. Simultaneously, the moment of the atom is transformed to assume a direction corresponding to the new atomic position. Therefore these operations do not impose any restrictions on the directions of particular atomic moments but only on their relative directions. These restricting relations are collected in the last column of Table 1.

The only non-trivial symmetry operation which keeps atoms in their initial positions is the anti-unitary transformation 7. This symmetry operation requires that $m_x^i=0$ for each atom i. Because of this the restrictions on the relative directions of the atomic moments resulting from the other symmetry operations take the following form: $m_y^1 = m_y^2 = -m_y^3 = -m_y^4$ and $m_z^1 = -m_z^2 = m_z^3 = -m_z^4$. This means that the initial collinear structure has the same symmetry as the non-collinear structures satisfying these relations. The non-collinear magnetic structure detected experimentally belongs to this type of structures. Thus, according to the symmetry principle, the realization of an extremum of the total energy by the collinear structure is improbable and hence the moments will deviate from collinearity. The details of this deviation can only be ascertained by self-consistent calculations which establish the minimum of the total energy.

Next we study the influence of lattice distortions on the magnetic structure. We start with the orthorhombic distortions which were observed to accompany the magnetic structure I. Following the experiment [41] we introduce a small variation of the lattice parameters a and b such that the relation $b = \sqrt{3}a$ valid for the ideal hexagonal lattice is no longer satisfied. This distortion does not affect the symmetry of the system because the magnetic structure has already lowered the symmetry of the crystal from hexagonal to orthorhombic. As a result no qualitative changes of the magnetic structure were observed due to the orthorhombic lattice distortion and quantitative changes appeared to be very small,too. The magnetic structure is very close to the structure shown in Fig. 6 a and is still in good agreement with the experimental structure I.

A basically different response was obtained to the monoclinic distortions. In agreement with experiment the b-side of the basal rectangle (Fig. 6) was rotated by 0.4° about the c axis. We start the calculations with the magnetic structure I. Already after the first iteration all Uranium magnetic moments deviated from the yz plane staying however mutually equivalent and compensating the magnetic structure.

TABLE 1. Symmetry properties of the orthorhombic UPdSn.

	Operation	Transposition of U atoms	Restriction on U moments	
1	$\{\varepsilon\|0\}$	no	no	
2	$\{C_{2z}\|\vec{\tau}_1\}$	1 ↔ 4 ; 2 ↔ 3	$\begin{pmatrix} m_x \\ m_y \\ m_z \end{pmatrix}_i = \begin{pmatrix} -m_x \\ -m_y \\ m_z \end{pmatrix}_j$	; $i \leftrightarrow j$
3	$\{\sigma_x\|\vec{\tau}_2\}$	1 ↔ 3 ; 2 ↔ 4	$\begin{pmatrix} m_x \\ m_y \\ m_z \end{pmatrix}_i = \begin{pmatrix} m_x \\ -m_y \\ -m_z \end{pmatrix}_j$	; $i \leftrightarrow j$
4	$\{\sigma_y\|\vec{\tau}_3\}$	1 ↔ 4 ; 2 ↔ 3	$\begin{pmatrix} m_x \\ m_y \\ m_z \end{pmatrix}_i = \begin{pmatrix} -m_x \\ m_y \\ -m_z \end{pmatrix}_j$	; $i \leftrightarrow j$
5	$\{\varepsilon\|\vec{\tau}_2\}R$	1 ↔ 3 ; 2 ↔ 4	$\vec{m}_i = -\vec{m}_j$;	$i \leftrightarrow j$
6	$\{C_{2z}\|\vec{\tau}_3\}R$	1 ↔ 2 ; 3 ↔ 4	$\begin{pmatrix} m_x \\ m_y \\ m_z \end{pmatrix}_i = \begin{pmatrix} m_x \\ m_y \\ -m_z \end{pmatrix}_j$	; $i \leftrightarrow j$
7	$\{\sigma_x\|0\}R$	no	$m_x = 0$;	all i
8	$\{\sigma_y\|\vec{\tau}_1\}R$	1 ↔ 4 ; 2 ↔ 3	$\begin{pmatrix} m_x \\ m_y \\ m_z \end{pmatrix}_i = \begin{pmatrix} m_x \\ -m_y \\ m_z \end{pmatrix}_j$	; $i \leftrightarrow j$

$$\varepsilon = \begin{pmatrix} 1 & & \\ & 1 & \\ & & 1 \end{pmatrix}; \quad C_{2z} = \begin{pmatrix} -1 & & \\ & -1 & \\ & & 1 \end{pmatrix}; \quad \sigma_x = \begin{pmatrix} -1 & & \\ & 1 & \\ & & 1 \end{pmatrix}; \quad \sigma_y = \begin{pmatrix} 1 & & \\ & -1 & \\ & & 1 \end{pmatrix};$$

$\vec{\tau}_1 = \frac{1}{2}(a,b,c); \quad \vec{\tau}_2 = \frac{1}{2}(a,b,0); \quad \vec{\tau}_3 = \frac{1}{2}(0,0,c);$

R: time reversal operation.

Again, a symmetry analysis helps us to understand this process. The monoclinic distortion decreases the symmetry of the system such that from the eight operations of the orthorhombic structure only four are left over in this case. These are the operations numbered in Table 1 as 1, 2, 5 and 6. Operation 5 demands equivalence of atom 1 to atom 3 and atom 2 to atom 4. Simultaneously, the moments of the equivalent atoms must be antiparallel: $\vec{m}_1 = -\vec{m}_3$ and $\vec{m}_2 = -\vec{m}_4$. Operation 2 is responsible for the equivalence of atoms 1 and 4 and the following relation between the components of the magnetic moments: $m_x^1 = -m_x^4$, $m_y^1 = -m_y^4$, $m_z^1 = m_z^4$. Further symmetry operations do not lead to additional restrictions. Thus we see the important difference between the orthorhombic and the monoclinic structures of UPdSn: in the monoclinic structure there is no symmetry operation demanding the x component of the magnetic moments to be zero. This means that a deviation of the magnetic moments from the yz plane does not change the symmetry of the system and therefore will take place according to our symmetry principle. Thus the result of the calculation for the monoclinically distorted lattice and the corresponding symmetry analysis are in agreement with the experimental data.

The deviations of the magnetic moments from parallel directions in the case of U_3X_4 and UPdSn which are caused by symmetry properties of the crystal and magnetic structures reminds us of the effect of weak ferromagnetism in Fe_2O_3, connected with the names Dzialoshinski and Moriya. We therefore show in the next Sect. that our symmetry analysis together with the method of calculation also explains this interesting case.

6. First-principles study of the effect of weak ferromagnetism

The phenomenon of weak ferromagnetism has been known for more than forty years. It is characterized by a small net magnetic moment resulting from a collection of atomic magnetic moments that nearly cancel each other. Weak ferromagnetism was first observed in hematite, α-Fe_2O_3 [44]. Dzialoshinski [45] showed that it was an intrinsic effect due to particular symmetry properties of the crystal structure and the magnetic moment arrangements. Dzialoshinski suggested a model Hamiltonian

$$H = I_{ij}\mathbf{S}_i\mathbf{S}_j + \mathbf{d}_{ij} \cdot [\mathbf{S}_i \times \mathbf{S}_j] + \mathbf{S}_i \cdot \mathbf{K}_i \cdot \mathbf{S}_i \qquad (9)$$

that gave a basis for most of further work on weak ferromagnetism. In (9) the indices i and j (implying appropriate summation) number the atoms in the lattice, I_{ij} and $\mathbf{d}_{ij}$ are the symmetric and antisymmetric exchange constants, respectively, and the tensor $\mathbf{K}_i$ contains information about the single-ion magneto-crystalline anisotropy. The first term of the Hamiltonian (9), the symmetric exchange, is supposed to lead to a compensated magnetic configuration. The next two terms, the anisotropic exchange and the magneto-crystalline anisotropy terms, respectively, can lead to a small ferromagnetic moment in an otherwise antiferromagnetic crystal.

Moriya [46] showed that Dzialoshinski's explanation can be given a microscopic footing by means of Anderson's perturbation approach to magnetic superexchange. Developments in calculational methods in the framework of the local approximation to density functional theory discussed in Sect. 2 provide all components necessary for a first-principles study of weak ferromagnetism. Considering spin-orbit coupling and allowing for arbitrary non-collinear configurations of the atomic magnetic moments we succeed in explaining the phenomenon without any phenomenological parameters or a perturbation treatment; a canted magnetic structure appears within the usual DFT self-consistency cycle as that magnetic configuration which supplies the minimum of the total energy.

Moriya [46] showed that depending on the type of crystal structure either of two mechanisms, antisymmetric exchange or magneto-crystalline anisotropy, can be the origin for the canting of magnetic moments. Thus, in the case of α-Fe_2O_3 it is the antisymmetric exchange that plays the dominant role, whereas in the case of NiF_2 antisymmetric exchange is ruled out in favor of the magneto-crystalline anisotropy which here is responsible for the appearance of the ferromagnetic component.

In the present section the first-principle study of both types of weak ferromagnetism is discussed. α-Fe_2O_3 is chosen as a classical example of the weak ferromagnetism caused by the antisymmetric exchange. Triangular antiferromagnet Mn_3Sn is considered as an example where the antisymmetric exchange contributions from different atoms cancel perfectly and cannot be a reason of the observed weak ferromagnetism. In this case the magneto-crystalline anisotropy term is seen to be responsible for weak ferromagnetism.

6.1. WEAK FERROMAGNETISM OF α-FE_2O_3

The crystal structure of Fe_2O_3 is shown in Fig. 7. This is a rhombohedral lattice with a basis of two formula units per unit cell. The following 12 point group operations enter the space group characterizing the symmetry of the atomic positions: the identity operation, rotations by 120 and 240° about the z axis, rotations by 180° about three axes in the xy plane (y=0 and y=$\pm\sqrt{3}x$) and these 6 operations multiplied by the inversion.

We start the discussion with the results of a scalar relativistic calculation [27] for four collinear magnetic structures where the moments of Fe along the symmetry axis in Fig. 7 can be arranged in the following sequences: $(+ + + +)$, $(+ - + -)$, $(+ - - +)$, and $(+ + - -)$. Here + and − designate

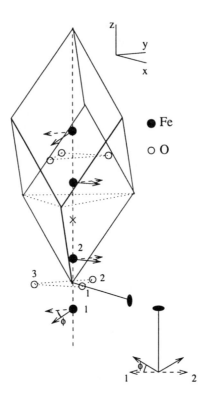

Figure 7. The unit cell of Fe_2O_3. The cross on the diagonal of the rhombohedron shows the point of inversion. The solid line passing through the 1st Oxygen atom indicates a twofold symmetry axis. The collinear (dashed arrow) and canted (solid arrow) directions of the Fe atoms are shown. The canting of the Fe moments in the xy plane is demonstrated differently in the lower right corner of the Figure.

up and down directions with respect to some chosen axis, e.g. the z-axis in Fig. 7. In agreement with experiment it was found that the structure $(+--+)$ possesses the lowest total energy. Therefore only calculations for the $(+--+)$ configuration will be discussed further. As was shown in Sect. 5.1 any collinear structure is stable within a scalar-relativistic calculation. So, no deviation of the magnetic moments were obtained in this case.

If we next choose the z-axis as the direction of the magnetic moments and switch on SOC, the following operations remain in the symmetry group of the Kohn-Sham Hamiltonian: the identity, rotations about the z axis and these operations multiplied by the inversion. The symmetry operations which are of special importance for us are the rotations about the z-axis: none of them change the position of any of the four Fe atoms lying on the axis of rotation. The directions of the magnetic moments are parallel to this axis and are not changed either. It is clear that any deviation of the magnetic moments from the z-axis will destroy the invariance of the crystal with respect to this operation. As the symmetry of the Hamiltonian cannot become lower during iterations this change is forbidden by symmetry. This result agrees with the experimental observation of a collinear antiferromagnetic structure with moments oriented parallel to the z axis below 260 K [47].

The situation is changed completely when the moments are parallel to the y-axis arranged again in the sequence $(+ - -+)$. Now the following symmetry operations are left in the group of the KSH: the identity, the 180°-rotation about the x-axis and these operations multiplied by the inversion. Inversion transforms the atoms of the upper Fe_2O_3 molecule into the atoms of the lower molecule, see Fig. 7. Since the magnetic moments are axial vectors, they do not change under this transformation. Hence, corresponding atoms of two molecules must keep parallel moments and one may restrict the consideration to the lower molecule in Fig. 7. The only condition imposed on the moments of the Fe atoms by symmetry is the transformation of the moment of atom 1 into the moment of atom 2 by the rotation through 180° about the x-axis (see Fig. 7). However to fulfill this condition it is not necessary for the atomic moments to be parallel to the y-axis nor to remain collinear. Indeed, calculations show that a collinear starting configuration does not lead to zero off-diagonal elements of the atomic density matrices, ρ_ν. Correspondingly, in this simulated annealing process, the magnetic moments move and deviate from their collinear starting directions toward the direction of the x-axis (see Fig. 7) until an "accidental" magnetic structure with the lowest energy will be achieved.

6.2. MAGNETIC STRUCTURE OF MN_3SN

Mn_3Sn belongs to the compounds where the experimentally observed weak ferromagnetism cannot be attributed to the anisotropic exchange interaction because of complete compensation of the contributions of different atoms [48].

In an earlier paper on Mn_3Sn Ref. [10], a number of different magnetic configurations allowed by Landau's theory of phase transitions were studied, but SOC was neglected. These calculations showed that, compared to other magnetic states, the triangular configurations have a distinctly lower total energy. This simplifies our task and allows to restrict the relativistic calculations to triangular structures only, of which four configurations are depicted in Fig. 8 together with the crystal structure of Mn_3Sn. Calculations without SOC showed that all four configurations are equivalent, i.e. they supply equal values of the total energy and the local magnetic moments. Another property of all four configurations is their stability during

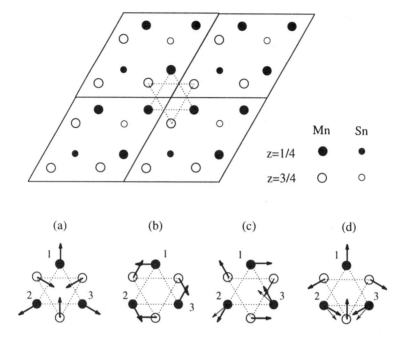

Figure 8. Crystal and magnetic structure of Mn_3Sn. Rotations of the magnetic moments leading to weak ferromagnetism in structures (c) and (d) are shown only for atoms in the z=0.25 plane (thin arrows). Moments of the atoms in the z=0.75 plane are parallel to the moments of the corresponding atoms of the z=0.25 plane.

iterations.

Again, as in the case of U_3X_4 or Fe_2O_3 the situation changes drastically when SOC is taken into account: First, the degeneracy is lifted, i.e. all four magnetic configurations become inequivalent and, second, for two of them (c and d) the magnetic moments deviate slightly from the initial directions as is depicted in Fig. 8.

Again the stability of the antiferromagnetic triangular structure and their equivalence within scalar-relativistic calculations can be explained on the basis of the theory of the SSG (Sect. 3) and the principle of preserving of the Hamiltonian symmetry during the iteration process as formulated in Sect. 5.1. Indeed, structure (a) transforms into structure (b) and (c) into (d) by a pure spin rotation through 90° about the hexagonal axis and (a) transforms into (d) by a 180° spin rotation about the direction of the spin of atom 1. Thus, in the scalar relativistic approximation, all four structures are equivalent.

Concerning the property of the stability of a given magnetic structure, it is easily seen that for all four structures depicted in Fig. 3 a deviation of any magnetic moment from the directions shown destroys at least one symmetry operation, in particular, for all atoms of structures (a) and (b) and for atom 1 of structures (c) and (d) these operations are the combined spin-space rotations through 180° about the spins of these atoms. For atoms 2 and 3 of structures (c) and (d) the symmetry operation is more involved and consists of a 180° spin-rotation about the direction of the spin of the atom accompanied by a 180 ° space rotation about another axis, *i.e.* the spin and space parts of the transformation are different.

In the fully-relativistic case the symmetry analysis is based on the same general principles as in the scalar-relativistic approach. However, the operations of the spin-space group do not leave invariant the form of the relativistic Kohn-Sham Hamiltonian, the operations that remain being those of the usual space groups with $\alpha_S=\alpha_R$. No space-group operation transforms any of the four magnetic structures into another one whence all four magnetic structures become inequivalent when SOC is switched on.

In the relativistic case only the operations of the usual space groups with $\alpha_S=\alpha_R$ leave the Kohn-Sham Hamiltonian invariant. To decide about the stability of a given magnetic configuration in the relativistic case one again must check whether one of the symmetry operations of the configuration will be destroyed by a deviation of the moment. Now we see that the two types of structures (a), (b) ("direct structures") and (c), (d) ("inverse structures") are distinguished by the "handedness" of the spin rotations: in the case of (a), (b) a $+120°$ rotations turns the spin of atom 1 into that of atom 2 *etc.*, whereas it is a $-120°$ rotation in the case of (c), (d). Indeed, transformations guaranteeing the stability of the directions of all moments of the direct structures in the scalar-relativistic case have coinciding spin and space transformations and continue to be symmetry operations also in the fully relativistic case. The same holds true for atom 1 of the inverse structures. However, the operations responsible for the stability of the other two moments in the scalar-relativistic case do not apply here and the moments can move without decreasing the symmetry of the Hamiltonian although the deviations of the moments of two atoms are not independent.

7. Quantitative model of spin fluctuations

Leaving now the realm of ground-state considerations and their symmetry properties we turn to a brief discussion of finite-temperature properties of itinerant-electron magnets. The basic assumption is that total-energy differences obtainable from constrained spin-spiral configurations can be employed to model the energetics of spin-fluctuations. Thus it is the spiral-spin configuration that methodologically links the previous sections with the present one.

Concerning the physics of spin fluctuations it is of importance to stress a central assumption. This is the adiabatic approximation for magnetic moments which we assume to be valid; it allows us to separate slow and fast electron motion, the time scale for spin fluctuations being much longer than typical electronic hopping times. This approximation receives support by the difference in energy scales, which fixes the scale of spin fluctuations.

7.1. MATHEMATICAL DETAILS

We begin by writing the local magnetization of the atom at site **R** in the form

$$\mathbf{M}(\mathbf{R}) = M\mathbf{e}_z + \sum_{j,\mathbf{k}\in 1.BZ} m_{j\mathbf{k}} \exp(i\mathbf{k}\cdot\mathbf{R})\mathbf{e}_j \quad , \tag{10}$$

where the first term on the right-hand side is the magnetic order parameter assumed to be in the z-direction and the second term on the right-hand side comprises time-averaged fluctuations which are connected with the susceptibility in linear response theory by the formally exact relation

$$\langle |m_{j\mathbf{k}}|^2 \rangle = \frac{\hbar}{2\pi} \int_0^\infty d\omega \, \coth(\frac{\hbar\omega}{2k_BT}) \, \mathrm{Im}\chi_j(\mathbf{k},\omega) \quad . \tag{11}$$

Here $j = 1, 2, 3$ labels Cartesian coordinates and $\chi_j(\mathbf{k},\omega)$ is the temperature dependent magnetic susceptibility. This is the dissipation-fluctuation theorem. For sufficiently high temperatures this expression becomes classical and the integral in (11) can be carried out to give after summing over all fluctuations

$$\tilde{m}_j^2 = \sum_{\mathbf{k}} \langle |m_{j\mathbf{k}}|^2 \rangle = k_BT \sum_{\mathbf{k}} \chi_j(\mathbf{k}) \quad . \tag{12}$$

Because of the Kramers-Kronig relation the real part of the non-uniform, static susceptibility, $\chi_j(\mathbf{k})$ appears and $\tilde{m}_j^2$ is the average squared fluctuation in the j-direction. So in what follows we must remember that we treat spin fluctuations as classical variables.

To proceed further we obviously must determine the susceptibility $\chi_j(\mathbf{k})$. Lonzarich and Taillefer [49] chose the simple form

$$\chi_j^{-1}(\mathbf{k}) = \chi_j^{-1} + c\,k^2 + \ldots \quad , \tag{13}$$

where the first term on the right-hand side is the inverse homogeneous, static susceptibility and the coefficient c is taken from experiment. Lonzarich and Taillefer then made the theory self-consistent using a Ginzburg-Landau type free energy, see also the work of Moriya [50].

Our approach [51] consists essentially of replacing the term $c\,k^2$ by a computed function, $J(\mathbf{k})$, that is obtained *ab initio* from total-energy differences of spin-spiral configurations. The theory is made self-consistent by constructing an appropriate free energy similar to previous work [52, 53], but the gradient term in the Ginzburg-Landau expression is – as in the susceptibility – replaced by an expression originating from the computed function $J(\mathbf{k})$.

The salient steps of our theory are as follows. The magnetization of a spin spiral is given by (7). The total energy is then obtained as described in Sect. 4 and is written as

$$E(M, \mathbf{q}, \theta) = AM^2 + BM^4 + \sin^2\theta\, J(\mathbf{q})\, M^2 \quad , \tag{14}$$

where we count the energy from its $M = 0$-value. The coefficients A, B and the function $J(\mathbf{q})$ are extracted numerically. We call $J(\mathbf{q})$ the exchange function and identify it with the left-hand side of

$$J(\mathbf{k}) = \frac{1}{N} \sum_{\mathbf{R},\mathbf{R}'} J(\mathbf{R}-\mathbf{R}')\, \exp(-i\mathbf{k}\cdot(\mathbf{R}-\mathbf{R}')) \quad . \tag{15}$$

This specifies all terms in a classical Hamiltonian which we write as

$$\mathcal{H} = \frac{1}{N} \sum_{\mathbf{R}} \left(A M^2(\mathbf{R}) + B M^4(\mathbf{R}) \right) + \frac{1}{N} \sum_{\mathbf{R},\mathbf{R}'} \left(J(\mathbf{R}-\mathbf{R}')(\mathbf{M}(\mathbf{R})\cdot\mathbf{M}(\mathbf{R}')) \right) \quad . \tag{16}$$

Note that the gradient term in this Ginzburg-Landau type of Hamiltonian is now replaced by the computed exchange function $J(\mathbf{R}-\mathbf{R}')$. We remark that for sake of simplicity we have written out only the lowest-order expansion in (14) and (16). A general formulation of arbitrarily high order can be found in Ref. [54] where also coupling to volume fluctuations is taken into account.

The statistical mechanics calculations are now carried out following the work of Murata, Doniach and Wagner [52, 53]. In particular, the model Hamiltonian needed to set up the Bogoliubov-Peierls inequality is written as

$$\mathcal{H}_0 = \sum_{j,\mathbf{k}} \chi_{j\mathbf{k}} |m_{j\mathbf{k}}|^2 \quad , \tag{17}$$

where the quantity $\chi_{j\mathbf{k}}$ is determined by varying the free energy that is calculated to be

$$\mathcal{F}_{SF} = -\frac{k_B T}{2} \sum_{j,\mathbf{k}} \left(1 + \ln\left(\pi \langle |m_{j\mathbf{k}}|^2\rangle_0\right) \right) + \langle \mathcal{H} \rangle_0 \quad , \tag{18}$$

where $\mathcal{F}_{SF}$ is an upper bound to the correct free energy and the thermal average implied in the second term on the right-hand side is obtained using the model Hamiltonian (17). One finds

$$\begin{aligned}\langle \mathcal{H}\rangle_0 &= A\Big(M^2 + \sum_j \tilde{m}_j^2\Big)\\ &+ B\Big(M^4 + 2M^2\sum_j(1+2\delta_{jz})\tilde{m}_j^2 + \sum_{j,j'}(1+2\delta_{jj'})\tilde{m}_j^2\tilde{m}_{j'}^2\Big)\\ &+ \sum_{j,\mathbf{k}} J(\mathbf{k})\langle|m_{j\mathbf{k}}|^2\rangle_0 \;,\end{aligned} \quad (19)$$

where

$$\tilde{m}_j^2 \equiv \sum_{\mathbf{k}} \langle|m_{j\mathbf{k}}|^2\rangle_0 \;. \quad (20)$$

The self-consistency equations now give

$$\chi_j^{-1}(\mathbf{k}) = \frac{k_B T}{\langle|m_{j\mathbf{k}}|^2\rangle_0} = \frac{2\partial\langle\mathcal{H}\rangle_0}{\partial\langle|m_{j\mathbf{k}}|^2\rangle_0} \quad (21)$$

which yields the inverse susceptibility

$$\chi_j^{-1}(\mathbf{k}) = 2A + 4B((1+2\delta_{jz})M^2 + \sum_{j'}(1+2\delta_{jj'})\tilde{m}_{j'}^2) + 2J(\mathbf{k}) \;. \quad (22)$$

The magnetization follows as usual from $\partial \mathcal{F}_{SF}/\partial M = 0$ giving

$$2A + 4B(M^2 + \sum_j(1+2\delta_{jz})\tilde{m}_j^2) = 0 \qquad (T < T_C) \;, \quad (23)$$

$$M = 0 \qquad (T > T_C) \;. \quad (24)$$

Thus we obtain below T_C: $\chi_x^{-1}(\mathbf{k}) = \chi_y^{-1}(\mathbf{k}) \equiv \chi_t^{-1}(\mathbf{k})$ and $\chi_z^{-1}(\mathbf{k}) \equiv \chi_l^{-1}(\mathbf{k})$, whereas above T_C (where $M = 0$): $\chi_x^{-1}(\mathbf{k}) = \chi_y^{-1}(\mathbf{k}) = \chi_z^{-1}(\mathbf{k}) \equiv \chi_p^{-1}(\mathbf{k})$.

We collect our results for the inverse susceptibility writing:

$$\chi_j^{-1}(\mathbf{k}) = \chi_j^{-1}(0) + 2J(\mathbf{k}) \quad (25)$$

with

$$\chi_t^{-1}(0) = 8B(\tilde{m}_t^2 - \tilde{m}_l^2) \;, \quad (26)$$
$$\chi_l^{-1}(0) = 8BM^2 \;, \quad (27)$$
$$\chi_p^{-1}(0) = 2A + 20B\tilde{m}_p^2 \;. \quad (28)$$

To compare the previous expression for the inverse susceptibility, (13), with our new results we show in Fig. 9 the exchange function, $J(\mathbf{k})$ calculated for bcc-Fe, fcc-Co, and fcc-Ni. Note that in contrast to the previous

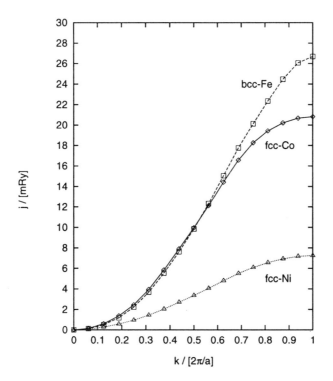

Figure 9. Exchange functions $j = M_0^2 J(\mathbf{k})$ for bcc-Fe, fcc-Co, and fcc-Ni along the (001) axis

theories, [52, 49, 53], no cut-off radius exists in our theory, the function $J(\mathbf{k})$ being defined in the Brillouin zone of the crystal.

7.2. RESULTS

We now discuss some results here that were reported in detail in previous work [51, 54] adding results not yet published before.

To begin with we collect in Table 2 relevant computed data for the transition metals bcc-Fe, fcc-Ni, hcp-Co, and fcc-Co.

The calculated Curie temperatures are given and compared with experimental values. We see that the agreement between theory and experiment is fair: our value for bcc-Fe is slightly too large whereas the calculated Curie temperatures of fcc-Co and fcc-Ni are smaller than the experimental values. In Table 2 we also list computed characteristic $T = 0$ values that serve as input for our description of spin fluctuations. These are the magnetic moments at $T = 0$, M_0, and the ground-state energy, $E_G = AM_0^2 + BM_0^4$, which describes the energy difference between the ferromagnetic ground state and the non-magnetic state. Furthermore, we list the average mag-

TABLE 2. Calculated magnetic properties of bcc-Fe, fcc-Ni, hcp-Co and fcc-Co, compared with experimental data.

		bcc-Fe	fcc-Ni	hcp-Co	fcc-Co
T_C / K	calc.	1095	412	995	1012
	exp.	1044	627	—	1388
M_0 / μ_B	calc.	2.20	0.61	1.52	1.61
	exp.	2.216	0.616	1.715	—
E_G / mRy		25.4	2.9	10.8	6.5
J_0 / mRy		10.3	5.6	12.8	13.8

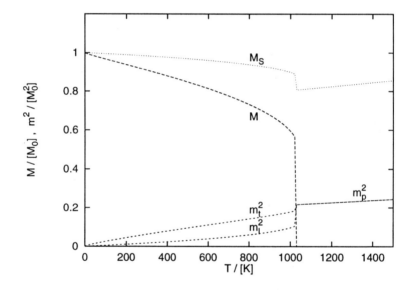

Figure 10. The transverse, longitudinal and paramagnetic fluctuations $\tilde{m}_t^2$, $\tilde{m}_l^2$, $\tilde{m}_p^2$ (in units of M_0^2), the magnetization M and the local magnetic moment M_s (both in units of M_0) of fcc-Co

netic exchange energy, defined by $J_0 = M_0^2 \sum_{\mathbf{k}} J(\mathbf{k})$, from which in first order the Curie temperature can be estimated by means of $k_b T_C = \frac{2}{5} J_0$, in agreement with Moriya's formula.

In Fig. 10 the magnetization M, the local magnetic moment $M_s = (M^2 + \sum_j \tilde{m}_j^2)^{1/2}$, and the fluctuations $\tilde{m}_t^2$, $\tilde{m}_l^2$, $\tilde{m}_p^2$, are shown as a function of temperature for fcc cobalt.

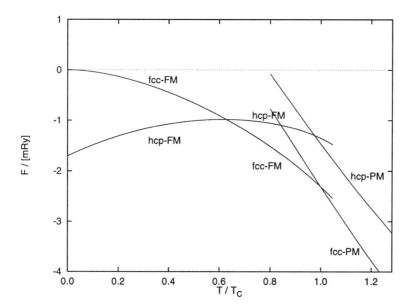

Figure 11. Calculated free energies for ferromagnetic and paramagnetic hcp- and fcc-Co. The energy origin ($\mathcal{F} = 0$) is chosen to be the value of the free energy of the constrained fcc phase with $M = M_0$ for all temperatures.

The decreasing magnetization vanishes abruptly at $T_C = 1012$K where the free-energy of paramagnetic Co becomes lowest. We stress and clearly see in the Figure that the local magnetic moments remain large (but disordered) even above the Curie temperature. Thus itinerant electrons support the paramagnetic phase just as they constitute the magnetic moment at lower temperatures. Of course, the abrupt decrease of the magentization is an artifact of our theory. It is interesting to note that this type of incorrect, first-order transition was commented on before, see for instance the references [49, 50, 55]. In the phenomenological approach [53] this behavior can be suppressed almost completely by using a large cut-off radius k_c; other remedies are described by Lonzarich and Taillefer[49]. In the *ab-initio* method presented here there is no simple way to correct this weakness since the moments reside on the crystal lattice whence the number of modes are restricted to the first Brillouin zone.

Next we may ask if our ability to compute the free energy *ab initio* can be used to throw light on the hcp–fcc phase transition of Co. By comparing with the fcc structure of Rh and Ir it is common to argue that the hcp structure of Co is stabilized by its being magnetic; indeed, the calculated ground-state energy of ferromagnetic hcp-Co is lower by about 1.7mRy per atom than that of ferromagnetic fcc-Co.

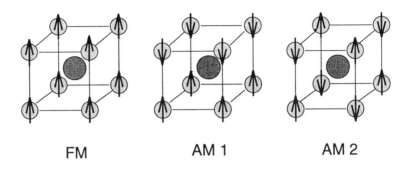

Figure 12. The three possible spin configurations (FM, AM1, AM2) of FeRh

Considering nothing but spin fluctuations we find the calculated free energy for ferromagnetic hcp-Co to be lower than that of ferromagnetic fcc-Co at low temperatures crossing at 590 K such that fcc-Co becomes the stable structure at higher temperatures, see Fig. 11. The calculated transition temperature of 590 K should be compared with the experimental temperature of 703 K [56].

Although we cannot definitely rule out other mechanisms like those due to phonons, our calculations lend weight to magnetism as constituting the dominant mechanism. We may thus conclude that the hexagonal phase of Co at low temperatures is stabilized dominantly by its being magnetic; spin fluctuations and the decreasing magnetization at higher temperatures restore the normal tendency of Co to be face-centered cubic.

Finally we investigate the first-order phase transition in FeRh from the antiferromagnetic (AM) phase at lower temperatures to the ferromagnetic (FM) phase at higher temperatures. Our band-structure calculations give three local minima in the total energy $E(\mathbf{q})$, these are: *a)* the FM phase, where $\mathbf{q} = (0,0,0)\frac{\pi}{a}$; *b)* the AM1 phase, where $\mathbf{q} = (0,0,1)\frac{\pi}{a}$; and *c)* the AM2 phase, where $\mathbf{q} = (1,1,1)\frac{\pi}{a}$; for the notation of the various magnetic phases see Fig. 12

Contrary to the FM case, in the AM case the fluctuations couple to the magnetization in a different way, leading to different expressions for the susceptibility. Thus, one can show that the Eqs. 26,27 must be replaced by $\chi_t^{-1}(0) = 4BM^2$ and $\chi_l^{-1}(0) = 8B(\tilde{m}_l^2 - \tilde{m}_t^2)$. We show in Fig. 13 the free energies of the magnetic phases of FeRh in a manner analogous to the hcp–fcc transition in Co.

In agreement with experiment, we observe at $T_{AM/FM} = 435K$ (exp. value: 328K) a first-order phase transition from the AM2 to the FM phase, the AM2 phase being stable at $T = 0$. The Curie temperature is seen to be $T_C = 885K$ which should be compared with the experimental value of 670K.

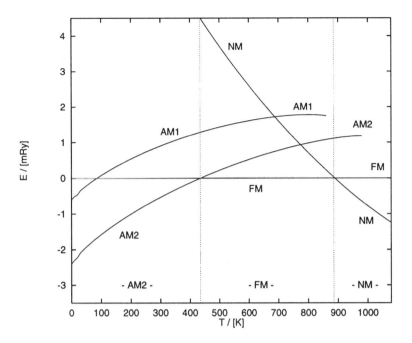

Figure 13. Calculated free energies for the FM, AM1, AM2, and the paramagnetic (PM) phase of FeRh. The energy origin is chosen to be the free energy of the FM phase ($\mathcal{F}_{FM} \equiv 0$).

Analyzing the computed data we find that the AM–FM phase transition in FeRh is driven by the increasing occupation of fluctuations in the FM phase, thus leading to an entropy stabilized FM phase at $T_{AM/FM}$.

8. Conclusion

In this short review we have shown that the local density approximation to density functional theory supplies a reliable basis for first-principles studies of non-collinear magnetic configurations in crystals. We restricted our discussion of the ground-state magnetic structures to ordered non-collinear structures which allowed us to use symmetry arguments to both make accurate calculations possible and to render the physics transparent.

We have shown that a consequent development of density functional theory allows to explain in the itinerant-electron picture the physical phenomena which were traditionally discussed in terms of a model Hamiltonian on the basis of localized moments. We have demonstrated how the results of first-principles calculation for particular non-collinear magnetic structures can be used to develop a parameter-free statistical mechanics scheme. We discussed thermodynamic applications for a number of itinerant-electron magnets concentrating in this review on *ab initio* calculations of the free energy for understanding in terms of spin fluctuations different types of phase transitions which we chose to be the hcp–fcc transition of Co and the AM–FM transition of FeRh.

Acknowledgment

This work was supported by the SFB 252 Darmstadt, Frankfurt, Mainz. It also benefited from collaboration within the European Union's Human Capital and Mobility Network on "Ab-initio (from electronic structure) calculation of complex processes in materials" (contract: ERBCHRXCT930369)

References

1. Coey, J.M.D. (1987) Noncollinear spin structures, *Can. J. Phys.* **65**, 1210.
2. Keffer, F. (1966) Ferromagnetism, in H.P.J. Wijn (ed.), *Encyclopedia of Physics*, vol. 18/2, Springer, Berlin.
3. Hohenberg, P. and Kohn, W. (1964) Inhomgenous electron gas, *Phys. Rev.* **B 136**, 864.
4. Kohn, W. and Sham, L.J. (1965) Self-consistent equations including exchange and correlation effects, *Phys. Rev.* **A 140**, 1133.
5. Kübler, J. and Eyert, V. (1992) Electronic structure calculations, in K.H.J. Buschow (ed.), *Electronic and magnetic properties of metals and ceramics*, VCH-Verlag, Weinheim.
6. Staunton, J.B. (1994) The electronic structure of magnetic transition metallic materials, *Rep. Prog. Phys.* **57**, 1289.
7. Gyorffy, B.L., Pindor, A.J., Staunton, J., Stocks, G.M., and Winter, H. (1985) A first-principles theory of ferromagnetic phase transitions in metals, *J. Phys.* **F 15**, 1337.
8. You, M.V. and Heine, V. (1982) Magnetism in transition metals at finite temperatures: I, *J. Phys.* **F 12**, 177.
9. Sandratskii, L.M. and Guletskii, P.G. (1989) Energy band structure of BCC iron at finite temperatures, *J. Magn. Magn. Mater.* **79**, 306.
10. Sticht, J., Höck, K.H., and Kübler, J. (1989) Non-collinear magnetism, *J. Phys. Condens. Matter* **1**, 8155.
11. Tsunoda, Y. (1989) Spin-density wave in cubic Fe and FeCo precipitates in Cu, *J. Phys. Condens. Matter* **1**, 10427.
12. Mryasov, O.N., Liechtenstein, A.I., Sandratskii, L.M., and Gubanov, V.A. (1991) Magnetic structure of FCC iron, *J. Phys. Condens. Matter* **3**, 7683.
13. Uhl, M., Sandratskii, L.M., and Kübler, J. (1992) Electronic and magnetic states of gamma-Fe, *J. Magn. Magn. Mater.* **103**, 314.
14. Uhl, M., Sandratskii, L.M., and Kübler, J. (1994) Spin fluctuations in gamma-Fe and in Fe3Pt Invar from local-density-functional calculations, *Phys. Rev.* **B 50**, 291.
15. Lorenz, R., Hafner, J., Jaswal, S.S., and Sellmyer, D.J. (1995) Disorder and noncollinear magnetism in permanent-magnet materials with the ThMn12 structure, *Phys. Rev. Lett.* **74**, 3688.
16. Liebs, Hummer, M., and Fähnle, M. (1995) Influence of structural disorder on the magnetic order (Fe, Co, Ni), *Phys. Rev.* **B 51**, 8664.
17. Smirnov, A.V. and Bratkovsky, A.M. (1996) Non-collinear magnetism in Al-Mn topologically disordered systems, *Europhys. Lett.* **33**, 527.
18. Vega, A., Stoeffler, D., Dreysse, H., and Demangeat, C. (1995) Magnetic-order tran-

sition in thin Fe overlayers on Cr: role of the interfacial roughness, *Europhys. Lett.* **31**, 561.
19. Antropov, V.P., Katsnelson, M.I., van Schilfgaarde, M., and Harmon, B.N. (1995) Ab-initio spin dynamics in magnets, *Phys. Rev. Lett.* **75**, 729; Antropov, V.P., Katsnelson, M.I., van Schilfgaarde, M., Harmon, B.N., and Kuznezov, D. (1995) Ab-initio description of spin dynamics in magnets, *Phys. Rev.* **B 54**, 1019.
20. Koelling, D.D. and Harmon, B.N. (1977) A technique for relativistic spin-polarised calculations, *J. Phys.* **C 10**, 3107.
21. Sandratskii, L.M., Kübler, J., Zahn, P., and Mertig, I. (1994) Electronic structure, magnetic, and Fermi-surface properties of UPd2Al3, *Phys. Rev.* **B 50**, 15834.
22. Sandratskii, L.M. (1991) Symmetry analysis of electronic states for crystals with spiral magnetic order. I. General properties, *J. Phys. Condens. Matter* **3**, 8565; Sandratskii, L.M. (1991) Symmetry analysis of electronic states for crystals with spiral magnetic order. II. Connection with limiting cases, *J. Phys. Condens. Matter* **3**, 8587.
23. Sandratskii, L.M. and Kübler, J. (1992) Static nonuniform magnetic susceptibility of selected transition metals, *J. Phys. Condens. Matter* **4**, 6927.
24. Sandratskii, L.M. and Kübler, J. (1995) Magnetic structures of uranium compounds: effects of relativity and symmetry, *Phys. Rev. Lett.* **75**, 946; Sandratskii, L.M. and Kübler, J. (1996) Noncollinear magnetic order and electronic properties of U_2Pd_2Sn and U_3P_4, *Physica* **B 217**, 167.
25. Sandratskii, L.M. and Kübler, J. (1997) Electronic Properties and Magnetic Structure of U_3X_4 (X=P,As,Sb), *Phys. Rev.* **B 55**, 11395.
26. Sandratskii, L.M. and Kübler, J. (1997) Electronic and magnetic properties of UPdSn: itinerant 5f electrons approach, *J. Phys. Condens. Matter* **9**, 4897.
27. Sandratskii, L.M., Uhl, M., and Kübler, J. (1996) Band theory for electronic and magnetic properties of α-Fe_2O_3, *J. J. Phys. Condens. Matter* **8**, 983.
28. Sandratskii, L.M. and Kübler, J. (1996) First-principles DF study of weak ferromagnetism in Fe_2O_3, *Europhys. Lett.* **33**, 447 (1996).
29. Sandratskii, L.M. and Kübler, J. (1996) The role of orbital polarization in weak ferromagnetism, *Phys. Rev. Lett.* **76**, 4963 (1996).
30. Herring, C. (1966) Exchange interaction among itinerant electrons, in G. Rado and H. Suhl (eds.), *Magnetism IV, Chapt. V and XIII*, Academic Press, New York, London.
31. Sandratskii, L.M. (1986) Energy band structure calculations for crystals with spiral magnetic structure, *Phys. Status Solidi b* **135**, 167.
32. Burlet, P., Rossat-Mignod, J., Troc, R., and Henkie, Z. (1981) Non-collinear magnetic structure of U_3P_4 and U_3As_4, *Solid State Commun.* **39**, 745.
33. Henkie, Z., Maslanka, R., Oleksy, Cz., Przystawa, J., de Boer, F.R., and Franse, J.J.M. (1987) High field magnetisation and hall effect of U_3Sb_4 single crystals, *J. Magn. Magn. Mater.* **68**, 54.
34. Maslanka, R., Henkie, Z., Franse, J.J.M., Verhoff, R., Oleksy, Cz., and Przystawa, J. (1989) Magnetic field induced spin reorientation transition in U_3Sb_4, *Physica* **B 159**, 181.
35. Oleksy, Cz. (1984) Critical phenomena in systems with bilinearly coupled order parameters. Application: magnetic structure of uranium compounds U_3X_4, *Acta Phys. Polonica* **A 66**, 665.
36. Gukasov, A., Wisniewski, P., and Henkie, Z. (1996) Neutron diffraction study of magnetic structure of U_3Bi_4 and U_3Sb_4, *J. Phys. Condens. Matter* **8**, 10589.
37. Robinson, R.A., Lawson, A.C., Buschow, K.H.J., de Boer, F.R., Sechovsky, V., and Von Dreele, R.B. (1991) Low temperature magnetic structures of UPdSn, *J. Magn. Magn. Mater.* **98**, 147.
38. de Boer, F.R., Brück, E., Nakotte, H., Andreev, A.V., Sechovsky, V., Havela, L., Nozar, P., Denissen, C.J.M., Buschow, K.H.J., Vaziri, B., Meissner, M., Malette H.,

and Rogl, P. (1992) Magnetic, electrical, and specific-heat properties of UPdSn and UAuSn, *Physica* B **176**, 275.
39. Robinson, R.A., Lawson, A.C., Lynn, J.W., and Buschow, K.H.J. (1992) Temperature dependence of magnetic order in UPdSn, *Phys. Rev.* B **45**, 2939.
40. Havela, L., Almeida, T., Naegele, J.R., Sechovsky, V., and Brück, E. (1992) Hihg resolution photoemission on UPdSn and UNiAl, *J. Alloys Compounds* **181**, 205.
41. Robinson, R.A., Lawson, A.C., Goldstone, J.A., and Buschow, K.H.J. (1993) Magneto-structural distortions in the noncollinear hexagonal antiferromagnet UPdSn, *J. Magn. Magn. Mater.* **128**, 143.
42. Robinson, R.A., Lynn, J.W., Lawson, A.C., and Nakotte, H. (1994) Themperature dependence of magnetic order in single-crystalline UPdSn, *J. Appl. Phys.* **75**, 6589.
43. Troc, R., Tran, V.H., Kolenda, M., Kruk, R., Latka, K., Szytula, A., Rossat-Mignod, J., Bonnet, M., and Büchner, B. (1995) X-ray and neutron diffraction studies of UPdSn, *J. Magn. Magn. Mater.* **151**, 102.
44. Néel, L. (1953) Some new results on antiferromagnetism and ferromagnetism, *Rev. Modern. Phys.* **25**, 58.
45. Dzialoshinski, I.J. (1958) A thermodynamic theory of weak ferromagnetism of antiferromagnetics, *Phys. Chem. Solids* **4**, 241.
46. Moriya, T. (1960) Anisotropic superexchange interaction and weak ferromagnetism, *Phys. Rev.* **120**, 91.
47. Morin, F.J. (1950) Magnetic susceptibility of α-Fe$_2$O$_3$ and α-Fe$_2$O$_3$ with added Titanium, *Phys. Rev.* **78**, 819.
48. Tomiyoshi, S. and Yamaguchi, Y. (1982) Magnetic structure and weak ferromagnetism of Mn3Sn studied by polarized neutron diffraction, *J. Phys. Soc. Japan* **51**, 2478.
49. Lonzarich, G.G. and Taillefer, L. (1985) Effect of spin fluctuations on the magnetic equation of state of ferromagnetic or nearly ferromagnetic metals, *J. Phys.* C **18**, 4339.
50. Moriya, T. (1985) *Spin fluctuations in itinerant electron magnetism* Springer, Berlin.
51. Uhl, M. and Kübler, J. (1996) Exchange-coupled spin-fluctuation theory: Application to Fe, Co and Ni, *Phys. Rev. Lett.* **77**, 334.
52. Murata, K.K. and Doniach, S., (1972) Theory of magnetic fluctuations in itinerant ferromagnets, *Phys. Rev. Lett.* **29**, 285.
53. Wagner, D. (1989) The fixed-spin-moment method and fluctuations, *J. Phys. Condens. Matter* **1**, 4635.
54. Uhl, M. and Kübler, J. (1997) Exchange-coupled spin-fluctuation theory: Calculation of magneto-elastic properties, *J. Phys. Condens. Matter* **9**, 7885.
55. Entel, P. and Schröter, M. (1989) Volume instabilities in itinerant magnetic systems, *Physica B* **161**, 160.
56. Stearns, M.B. (1986) Fe, Co, Ni, in K.H. Hellwege (ed.), *Landolt-Börnstein: New Series III/19a*, Springer, Berlin.

MOLECULAR DYNAMICS APPROACH TO COMPLEX MAGNETIC STRUCTURES IN ITINERANT-ELECTRON SYSTEMS

Y. KAKEHASHI, S. AKBAR AND N. KIMURA
Hokkaido Institute of Technology
Maeda, Teine-ku, Sapporo 006, Japan

Abstract. A molecular dynamics (MD) theory of itinerant electron magnetism which has recently been proposed to describe the complex magnetic structures is reviewed. The isothermal MD approach based on the functional integral method is shown to predict automatically the complex magnetic structure with a few hundred atoms in a unit cell at finite temperatures. It is demonstrated by the numerical calculations for bcc Fe that the MD approach describes the second order phase transition as a function of temperature because of a selfconsistent effective medium in the theory. The numerical results of the MD calculations for the fcc transition metals with use of 108 atoms and 256 atoms show the existence of various complex magnetic structures for the d electron numbers between 6.0 and 7.0, and the strong spin frustrations for γ-Fe. The theory is extended to the magnetic alloys and is found to explain the basic feature of γ-FeMn alloys.

1. Introduction

In the itinerant electron systems such as 3d transition metals and alloys, the electrons move from site to site via transfer integrals and form the energy bands. The Coulomb interactions comparable to the band width and the Pauli principle are considered to build up the magnetic order in the itinerant electron systems. Fe, Co, and Ni among the 3d transition metals are well-known to show the simple ferromagnetism, which is stabilized by a large kinetic energy gain associated with the high densities of states (DOS) at the Fermi levels in the nonmagnetic states. The bcc Cr with a few percent Mn shows the CsCl-type antiferromagnetic (AF) structure with the wave vector $Q = (1, 0, 0)2\pi/a$ (a : lattice constant) which is stabilized by the energy gain due to the nesting Fermi surfaces around $\Gamma(0,0,0)$ and H(1,0,0) points being connected via the $\boldsymbol{Q}$ vector.

The other transition metals, *i.e.* pure Cr, α-Mn, γ-Fe, and their alloys show the complex magnetic structures such as the spin-density waves, complex AF structures, and noncollinear magnetic structures, which prevented us from understanding their magnetic properties. α-Mn, for example, has 29 atoms per unit cell with 4 different crystallographic sites I~IV, and shows a noncollinear complex magnetic structure in which all the magnetic moments have different directions and magnitudes from 1.9 μ_B on site I to 0.25 μ_B on site IV, according to the neutron analysis [1]. In the case of γ-Fe, the magnetic structure was determined to be the first-kind AF (AFI) structure in which the ferromagnetic (0,0,1) planes vary alternatively along a [0,0,1] axis [2, 3]. This structure however is found in the tetragonal phase. Recent neutron experiments [4] suggest that the cubic γ-Fe shows more complex magnetic structure, *i.e.* the helical structure with the wave vector $\boldsymbol{Q} = (0,0.1,1)2\pi/a$, in which LM's on the antiferromagnetic plane rotate along the axis perpendicular to the plane. In Fe-rich amorphous alloys, the spin glasses (SG) are found beyond 90 at.% Fe irrespective of the second elements [5, 6, 7], suggesting that the amorphous Fe shows the itinerant-electron SG which has a random spin configuration and has many local minima of free energy in configuration space due to competing interactions.

The stability of these complex magnetic structures and their magnetic properties associated with the peculiar magnetic structures have not been understood from the microscopic point of view. Traditional approach to this problem has been based on a simple energy comparison among possible magnetic structures [8, 9], the energy minimization with respect to the order parameter found in experiments [10, 11, 12], or the susceptibility analysis for the magnetic instability to the fluctuations around the nonmagnetic state [13]. These methods are useful for finding the electronic origin of the stability of magnetic structures, and made considerable contribution to understanding the formation of the ferromagnetism, the antiferromagnetism, and the helical magnetic structure. But, they do not guarantee that the obtained structure yields the lowest energy of the systems. This problem becomes significant when there are many local minima in free energy, and it is the reason why one could not established the phase diagram even for the simplified model Hamiltonian in a region between the ferro- and antiferro-magnetic structures or between the AF and paramagnetic states. We need more advanced theory of the itinerant electron magnetism with complex magnetic structures towards predicting new magnetic structures and developing new magnetic materials.

We have recently proposed the molecular dynamics (MD) approach to complex magnetic structures of the itinerant electron systems with competing magnetic interactions [14], which allows us to treat the magnetic structure with a large number of atoms in a unit cell at finite temperatures, and automatically find the stable magnetic structure. In this article, we review our MD approach to the itinerant electron systems and discuss the present status of the theory showing some numerical results for transition metals and alloys.

In the next section, we derive the expressions of the free energy and related magnetic quantities on the basis of the functional integral method [15, 16, 17, 18, 19, 20] describing the spin fluctuations at finite temperatures. The locally rotated coordinates will be introduced there so that the free energy reduces to that of the generalized Hartree-Fock approximation at the ground state, being suitable for the description of noncollinear magnetic structures. In section 3, we present the MD approach to the itinerant electron systems and describe the electronic structure calculations for the magnetic force in equations of motion. Numerical results for bcc Fe, the fcc transition metals, and γ-FeMn alloys are presented in section 4 to demonstrate the validity and limitation of the present approach. Various complex magnetic structures appear for the d electron numbers between 6.0 and 7.0. The strong spin frustrations in γ-Fe, and the effects of disorder on the complex magnetic structures will be discussed. The last section is devoted to summary and discussion on the future problems.

2. Functional integral technique

The itinerant electron systems are well-known to be described by the tight binding Hamiltonian as follows.

$$H = H_0 + H_1 . \quad (1)$$

Here H_0 is the noninteracting Hamiltonian with the atomic level ϵ_i^0 on site i and the transfer integral $t_{i\nu j\nu'}$ between the orbitals ν on site i and ν' on site j:

$$H_0 = \sum_{i\nu\sigma} \epsilon_i^0 n_{i\nu\sigma} + \sum_{i\nu j\nu'\sigma} t_{i\nu j\nu'} a_{i\nu\sigma}^\dagger a_{j\nu'\sigma} , \quad (2)$$

and H_1 is the interacting Hamiltonian with the intraatomic Coulomb integral U_i and exchange integral J_i on site i:

$$H_1 = \frac{1}{4}\sum_i U_i n_i^2 - \sum_i J_i \boldsymbol{S}_i^2 . \quad (3)$$

Here $a_{i\nu\sigma}^\dagger(a_{i\nu\sigma})$ denotes the creation (annihilation) operator for an electron with spin σ and orbital ν on site i. $n_{i\nu\sigma} = a_{i\nu\sigma}^\dagger a_{i\nu\sigma}$ is the number operator for the electrons on site i, orbital ν, and spin σ. n_i and $\boldsymbol{S}_i$ denote the charge and spin densities on site i, and are defined by $n_i = \sum_{\nu\sigma} n_{i\nu\sigma}$ and $\boldsymbol{S}_i = \sum_{\nu\sigma\sigma'} a^\dagger_{i\nu\sigma}(\boldsymbol{\sigma})_{\sigma\sigma'} a_{i\nu\sigma'}/2$, respectively.

To derive the free energy which reduces to the generalized Hartree-Fock approximation at the ground state, we introduce first the locally rotated coordinates at each site, and rewrite the interacting Hamiltonian in terms of the operators on the rotated coordinates as

$$H_1 = \frac{1}{8}\sum_{i\nu}(U_i + 3J_i)\hat{n}_{i\nu}^2 + \frac{1}{4}\sum_i U_i \sum_{\nu\nu'}{}' \hat{n}_{i\nu}\hat{n}_{i\nu'}$$
$$- \frac{1}{8}\sum_{i\nu}(U_i + 3J_i)\hat{m}_{i\nu z}^2 - \frac{1}{4}\sum_i J_i \sum_{\nu\nu'}{}' \sum_\alpha \hat{m}_{i\nu\alpha}\hat{m}_{i\nu'\alpha} \,. \quad (4)$$

Here the hats on the operators mean the operators defined on the rotated coordinates; $\hat{n}_{i\nu} = \sum_\sigma \hat{a}^\dagger_{i\nu\sigma}\hat{a}_{i\nu\sigma}$ and $\hat{m}_{i\nu\alpha} = \sum_{\sigma\sigma'} \hat{a}^\dagger_{i\nu\sigma}(\sigma_\alpha)_{\sigma\sigma'}\hat{a}_{i\nu\sigma'}$. $\hat{a}^\dagger_{i\nu\sigma}$ and $\hat{a}_{i\nu\sigma}$ are the creation and annihilation operators for an electron whose spin σ is quantized along the z axis of the rotated coordinates on site i.

The itinerant electron systems generally show the localized as well as itinerant features [21]; the ground state properties of Fe, Co, and Ni are well explained by the band model, while the finite temperature properties such as the transition temperatures and the Curie constant are rather close to those expected from the localized model. One has to take into account the thermal spin fluctuations to describe the latter feature as well as the itinerant one. The functional integral method is suitable for the description of the spin fluctuations at finite temperatures. We adopted the method, and transformed the interacting Hamiltonian (4) into the one electron Hamiltonian with time-dependent random charge and exchange fields $\{\zeta_{i\nu}(\tau), \boldsymbol{\xi}_{i\nu}(\tau)\}$ by means of the Hubbard-Stratonovich transformation [15, 16]:

$$e^{-\beta H_1(\tau)\Delta\tau}$$
$$= \left[\prod_{i=1}^N \left(\frac{(\Delta\tau)^D \det A_i}{(4\pi)^D}\prod_\alpha \frac{(\Delta\tau)^D \det B_i^\alpha}{(4\pi)^D}\right)\right]\int\left[\prod_{i=1}^N \prod_{\nu=1}^D d\boldsymbol{\xi}_{i\nu}(\tau)d\zeta_{i\nu}(\tau)\right]$$
$$\times \exp\left[-\frac{\Delta\tau}{4}\sum_{i\nu\nu'}\left(\zeta_{i\nu}(\tau)A_{i\nu\nu'}\zeta_{i\nu'}(\tau) + \sum_\alpha \xi_{i\nu\alpha}(\tau)B^\alpha_{i\nu\nu'}\xi_{i\nu'\alpha}(\tau)\right)\right.$$
$$\left.+\frac{\Delta\tau}{2}\sum_{i\nu\nu'}\left(i\zeta_{i\nu}(\tau)A_{i\nu\nu'}\hat{n}_{i\nu'}(\tau) + \sum_\alpha \xi_{i\nu\alpha}(\tau)B^\alpha_{i\nu\nu'}\hat{m}_{i\nu'\alpha}(\tau)\right)\right], \quad (5)$$

$$A_{i\nu\nu'} = \frac{1}{2}(U_i + 3J_i)\delta_{\nu\nu'} + U_i(1-\delta_{\nu\nu'}) \,, \quad (6)$$
$$B^\alpha_{i\nu\nu'} = J_i(1-\delta_{\nu\nu'}) \quad (\alpha = x, y) \,, \quad (7)$$
$$B^z_{i\nu\nu'} = \frac{1}{2}(U_i + 3J_i)\delta_{\nu\nu'} + J_i(1-\delta_{\nu\nu'}) \,. \quad (8)$$

Here β is the inverse temperature defined by T^{-1}. $H_1(\tau)$ denotes the Hamiltonian H_1 in the interaction representation, $\Delta\tau$ is an infinitesimal time interval. N (D) denotes the number of site (orbital degeneracy), and $\det A_i$ denotes the determinant of matrix $A_{i\nu\nu'}$ for orbital indices.

The partition function is then written as

$$Z = \int \left[\prod_{i=1}^{N}\prod_{\nu=1}^{D} \delta\xi_{i\nu}(\tau)\delta\zeta_{i\nu}(\tau)\right] Z^0\left(\boldsymbol{\xi}(\tau),\zeta(\tau)\right)$$

$$\times \exp\left[-\frac{1}{4}\sum_{i\nu\nu'}\int_0^\beta d\tau \left(\zeta_{i\nu}(\tau)A_{i\nu\nu'}\zeta_{i\nu'}(\tau) + \sum_\alpha \xi_{i\nu\alpha}(\tau)B^\alpha_{i\nu\nu'}\xi_{i\nu'\alpha}(\tau)\right)\right], \quad (9)$$

$$Z^0\left(\boldsymbol{\xi}(\tau),\zeta(\tau)\right) = \text{Tr}\left[\mathcal{T}\exp\left(-\int_0^\beta H\left(\tau,\boldsymbol{\xi}(\tau),\zeta(\tau)\right)d\tau\right)\right], \quad (10)$$

$$H(\tau,\boldsymbol{\xi}(\tau),\zeta(\tau)) = K_0(\tau)$$
$$-\frac{1}{2}\sum_{i\nu\nu'}\left(i\zeta_{i\nu}(\tau)A_{i\nu\nu'}\hat{n}_{i\nu'}(\tau) - \sum_\alpha \xi_{i\nu\alpha}(\tau)B^\alpha_{i\nu\nu'}\hat{m}_{i\nu'\alpha}(\tau)\right). \quad (11)$$

Here $\mathcal{T}$ in Eq. (10) denotes the time order product. The operator K_0 in Eq. (11) is defined by $H_0 - \mu N_e$, μ and N_e being the chemical potential and the total electron number respectively. The functional integral is defined as

$$\int \prod_{n=1}^{N'} \left[\left(\frac{(\Delta\tau)^D \det A_i}{(4\pi)^D}\right)^{1/2} \prod_{\nu=1}^{D} d\zeta_{i\nu}(\tau_n)\right] \to \int \left[\prod_{\nu=1}^{D} \delta\zeta_{i\nu}(\tau)\right], \quad (12)$$

where the imaginary time τ is divided into N' points in the range $[0,\beta]$, and τ_n denotes the n-th point.

We adopt the two-field static approximation [19, 20] among various approximation schemes in the functional integral technique; we approximate the time-dependent field variables on the rotated coordinates $\{\xi_{i\nu x}(\tau), \xi_{i\nu y}(\tau), \xi_{i\nu z}(\tau), \zeta_{i\nu}(\tau)\}$ with the time-independent variables $\{0, 0, \xi_i/D, \zeta_i/D\}$ in the one-electron Hamiltonian (11).

The partition function is then given by

$$Z(\{e_i\}) = \int \prod_{i=1}^{N}\left[\left(\frac{\beta\tilde{J}_i}{4\pi}\right)^{1/2} d\xi_i de_i\right] \text{Tr}\left(e^{-\beta H_{\text{st}}(\xi e)}\right)$$

$$\times \exp\left[-\beta\sum_i\left(-n_i w_i(\boldsymbol{\xi}) + \frac{1}{4}\tilde{J}_i\xi_i^2\right)\right], \quad (13)$$

$$H_{\text{st}}(\xi e) = \sum_{i\nu\sigma} \left(\epsilon_i^0 - \mu + w_i(\boldsymbol{\xi})\right) n_{i\nu\sigma}$$
$$-\frac{1}{2}\sum_{i\nu}\tilde{J}_i\xi_i\boldsymbol{e}_i\cdot\boldsymbol{m}_i + \sum_{i\nu j\nu'\sigma} t_{i\nu j\nu'} a^\dagger_{i\nu\sigma} a_{j\nu'\sigma} \,. \qquad (14)$$

Here $\boldsymbol{e}_i$ is the unit vector showing the direction of the rotated z axis on site i and $\boldsymbol{\xi}_i = \xi_i \boldsymbol{e}_i$. We adopted the saddle point approximation in the strong Coulomb interaction limit for the charge fields for brevity, so that $w_i(\boldsymbol{\xi})$ is determined from the charge neutrality condition:

$$n_i = \langle n_i\rangle_0 = \frac{\text{Tr}\left(n_i\, e^{-\beta H_{\text{st}}(\boldsymbol{\xi})}\right)}{\text{Tr}\left(e^{-\beta H_{\text{st}}(\boldsymbol{\xi})}\right)}, \qquad (15)$$

where n_i at the left hand side denotes the average charge of pure metal for the atom on site i.

The partition function $Z(\{\boldsymbol{e}_i\})$ does not satisfy the rotational invariance since we neglected the transverse spin fluctuations on the rotated coordinates. We therefore obtain the free energy F_{st} averaging $Z(\{\boldsymbol{e}_i\})$ over all the directions $\{\boldsymbol{e}_i\}$.

$$F_{\text{st}} = -\beta^{-1}\int\left[\prod_i^N\left(\frac{\beta\tilde{J}_i}{4\pi}\right)^{1/2} d\xi_i de_i\right] e^{-\beta E(\boldsymbol{\xi})}, \qquad (16)$$

$$E(\boldsymbol{\xi}) = -\beta^{-1}\ln\text{Tr}\left(e^{-\beta H_{\text{st}}(\boldsymbol{\xi})}\right) - \sum_i\left(n_i w_i(\boldsymbol{\xi}) - \frac{1}{4}\tilde{J}_i\xi_i^2\right). \qquad (17)$$

Here $de_i = (4\pi)^{-1}\sin\theta_i d\theta_i d\phi_i$.

The local magnetic moment (LM) is obtained by taking the derivatives of F_{st} with respect to the local magnetic field $\boldsymbol{h}_i$ as follows.

$$\langle\boldsymbol{m}_i\rangle = \left\langle\left(1+\frac{4}{\beta\tilde{J}_i\xi_i^2}\right)\boldsymbol{\xi}_i\right\rangle. \qquad (18)$$

Here the average $\langle\sim\rangle$ at the right hand side of the above equation is defined by

$$\langle\sim\rangle = \frac{\int\left[\prod_i d\boldsymbol{\xi}_i\right](\sim)\, e^{-\beta\Psi(\boldsymbol{\xi})}}{\int\left[\prod_i d\boldsymbol{\xi}_i\right] e^{-\beta\Psi(\boldsymbol{\xi})}}, \qquad (19)$$

$$\Psi(\boldsymbol{\xi}) = E(\boldsymbol{\xi}) + 2T \sum_i \ln \xi_i . \qquad (20)$$

Note that we adopted the spherical coordinates in the above average $\langle \sim \rangle$.

The simplest method to evaluate the thermal average of LM (18) is the single-site approximation [19, 20], in which the LM on site i is treated as an impurity in an effective medium $\mathcal{L}_\sigma^{-1}$.

$$\langle m \rangle = \frac{\int d\boldsymbol{\xi} \xi^{-2} \left(1 + \frac{4}{\beta \tilde{J}\xi^2}\right) \boldsymbol{\xi} \, e^{-\beta E_0(\boldsymbol{\xi})}}{\int d\boldsymbol{\xi} \xi^{-2} e^{-\beta E_0(\boldsymbol{\xi})}}, \qquad (21)$$

$$E_0(\boldsymbol{\xi}) = \int d\omega f(\omega) \frac{1}{\pi} \mathrm{Im} \sum_\Gamma d_\Gamma \ln \det{}_\Gamma(\omega + i\delta, \xi_z, \xi_\perp^2)$$
$$- nw(\boldsymbol{\xi}) + \frac{1}{4}\tilde{J}\xi^2 , \qquad (22)$$

$$\det{}_\Gamma(\omega + i\delta, \xi_z, \xi_\perp^2) = \left(L_\uparrow^{-1} - \mathcal{L}_\uparrow^{-1} + F_{\Gamma\uparrow}^{-1}\right)\left(L_\downarrow^{-1} - \mathcal{L}_\downarrow^{-1} + F_{\Gamma\downarrow}^{-1}\right)$$
$$- \frac{1}{4}\tilde{J}^2 \xi_\perp^2 , \qquad (23)$$

$$L_\sigma^{-1} = \omega + i\delta - \epsilon_0 + \mu - w(\boldsymbol{\xi}) + \frac{1}{2}\tilde{J}\xi_z \sigma , \qquad (24)$$

$$F_{\Gamma\sigma} = \left[(\mathcal{L}^{-1} - t)^{-1}\right]_{i\Gamma\nu\sigma, i\Gamma\nu\sigma} = \int \frac{\rho_\Gamma(\epsilon)d\epsilon}{\mathcal{L}_\sigma^{-1} - \epsilon} . \qquad (25)$$

Here we assumed that all the sites are equivalent for brevity, and omitted the site index i. δ is an infinitesimal positive number. $\xi_\perp$ denotes the transverse component of field variable defined by $\xi_\perp = (\xi_x^2 + \xi_y^2)^{1/2}$. d_Γ denotes the number of orbitals ν belonging to the irreducible representation Γ of a point symmetry in crystal, $F_{\Gamma\sigma}$ is the diagonal coherent Green function for the orbital ($i\Gamma\nu\sigma$). $\rho_\Gamma(\epsilon)$ is the density of states for the transfer matrix $(t)_{i\nu j\nu'} = t_{i\nu j\nu'}$ which is projected onto the orbital ($i\Gamma\nu$).

The charge potential $w(\boldsymbol{\xi})$ is determined from the charge neutrality condition (15):

$$n = \int d\omega f(\omega) \frac{(-)}{\pi} \mathrm{Im} \sum_{\Gamma\sigma} d_\Gamma G_{\Gamma\sigma}(\omega + i\delta, \xi_z, \xi_\perp^2, \mathcal{L}^{-1}) . \qquad (26)$$

Here $G_{\Gamma\sigma}(z, \xi_z, \xi_\perp^2, \mathcal{L}^{-1})$ is the impurity Green function defined by

$$G_{\Gamma\sigma}(\omega + i\delta, \xi_z, \xi_\perp^2, \mathcal{L}^{-1})$$
$$= \left(L_\sigma^{-1} - \mathcal{L}_\sigma^{-1} + F_{\Gamma\sigma}^{-1} - \frac{\frac{1}{4}\tilde{J}^2 \xi_\perp^2}{\left(L_{-\sigma}^{-1} - \mathcal{L}_{-\sigma}^{-1} + F_{\Gamma-\sigma}^{-1}\right)}\right)^{-1} . \qquad (27)$$

The effective medium $\mathcal{L}_\sigma^{-1}(z)$ is determined selfconsistently by the CPA (coherent-potential-approximation) equation [22] as

$$\sum_\Gamma \frac{d_\Gamma}{D} \sum_{q=\pm 1} \frac{1}{2}\left(1 + q\frac{\langle\xi_z\rangle}{\langle\xi_z^2\rangle^{1/2}}\right) G_{\Gamma\sigma}(\omega + i\delta, q\langle\xi_z^2\rangle^{1/2}, \langle\xi_\perp^2\rangle, \mathcal{L}^{-1})$$

$$= \sum_\Gamma \frac{d_\Gamma}{D} F_{\Gamma\sigma} \ . \tag{28}$$

Here $\langle\xi_z\rangle$, $\langle\xi_z^2\rangle$, and $\langle\xi_\perp^2\rangle$ are calculated from $E_0(\boldsymbol{\xi})$ selfconsistently.

The single-site theory is useful for the finite-temperature problems with rather simple magnetic structure. However, we have to introduce site-dependent effective medium for the systems with complex magnetic structures. The determination of the medium is then difficult since we do not know the direction of polarization on each site in advance. Thus, we have to rely on more direct methods such as the molecular dynamics (MD) method and the Monte-Carlo method for the systems with complex magnetic structures. A remarkable difference between the two methods is that one needs in the former the memory proportional to the number of magnetic atoms N in the system, while one needs the memory proportional to N^2 in the latter. Thus, the MD method is more suitable for the description of complex magnetic structures with a large number of atoms in a unit cell. We describe the MD approach to the itinerant electron systems with complex magnetic structures in the following section.

3. Molecular dynamics approach

The thermal average of LM given by Eq. (18) has a semi-classical form described by a potential energy $\Psi(\boldsymbol{\xi})$. We adopt the isothermal MD method [23, 24] to the calculation of average LM's. The method assumes the ergodicity of the system, and calculate the ensemble average of LM (18) by means of the time average of fictitious LM's $\boldsymbol{\xi}(t)$.

$$\langle \boldsymbol{m}_i\rangle = \lim_{t_0\to\infty} \frac{1}{t_0}\int_0^{t_0}\left(1 + \frac{4}{\beta\tilde{J}_i\xi_i^2(t)}\right)\boldsymbol{\xi}_i(t)dt \ . \tag{29}$$

The time development of the fictitious LM's is determined by the Nosé-Hoover dynamics whose equations of motion are given by

$$\dot{\xi}_{i\alpha} = \frac{p_{i\alpha}}{\mu_{LM}} \ , \tag{30}$$

$$\dot{p}_{i\alpha} = -\frac{\partial\Psi(\boldsymbol{\xi})}{\partial\xi_{i\alpha}} - \eta_\alpha \cdot p_{i\alpha} \ , \tag{31}$$

$$\dot{\eta}_\alpha = \frac{1}{Q}\left(\sum_i \frac{p_{i\alpha}^2}{\mu_{LM}} - NT\right) . \tag{32}$$

Here μ_{LM} is the effective mass for the LM on site i and Q in Eq. (32) denotes a constant parameter. $p_{i\alpha}$ is the fictitious momentum conjugate to the exchange field $\xi_{i\alpha}$. The first term at the right-hand-side of Eq. (31) is a magnetic force, and the second term is the friction force which describes the heat bath by changing the coefficient η_α according to Eq. (32).

It should be noted that we introduced three independent variables η_α instead of a uniform friction coefficient η in the above equations since the dynamics with the uniform friction coefficient violates the ergodicity when the MD is applied to an impurity problem with the initial conditions of the zero angular momentum of the system.

The magnetic force in Eq. (31) is obtained from Eqs. (17) and (20) as

$$-\frac{\partial \Psi(\xi)}{\partial \xi_{i\alpha}} = \frac{1}{2}\tilde{J}_i(\langle m_{i\alpha}\rangle_0 - \xi_{i\alpha}) - \frac{2T\xi_{i\alpha}}{\xi_i^2}. \tag{33}$$

Here $\langle m_{i\alpha}\rangle_0$ is the average magnetic moment with respect to one-electron Hamiltonian (14) in the random exchange fields. It is clear from Eq. (33) that the ground state given by the zero magnetic force obeys the Hartree-Fock equation. Therefore the dynamics of fictitious spins given by Eqs. (30) - (32) is regarded as a nonlinear fluctuations around the Hartree-Fock value in the present approach.

All the electronic structures appear in the magnetic moment $\langle m_{i\alpha}\rangle_0$. It is written by the Green function G as

$$\langle m_{i\alpha}\rangle_0 = \int d\omega f(\omega)\frac{(-)}{\pi}\text{Im}\sum_{\nu\sigma}(\sigma_\alpha G)_{i\nu\sigma i\nu\sigma}, \tag{34}$$

$$G = \left(L^{-1} - t\right)^{-1}, \tag{35}$$

$$\left(L^{-1}\right)_{i\nu\sigma j\nu'\sigma'} = \left(\omega + i\delta - \epsilon_i^0 + \mu - w_i(\xi)\right)\delta_{ij}\delta_{\nu\nu'}\delta_{\sigma\sigma'}$$
$$+ \frac{1}{2}\tilde{J}_i\xi_i\cdot(\sigma)_{\sigma\sigma'}\delta_{ij}\delta_{\nu\nu'}. \tag{36}$$

Here the products $(\sigma_\alpha G)_{i\nu\sigma i\nu\sigma}$ in Eq. (35) is expressed by means of the diagonal Green functions in the representation which diagonalizes Pauli's spin matrix $\sigma_\alpha(\alpha = x,y,z)$.

It should be noted that the locator L in the Green function randomly changes in space and time in the MD method. Therefore, we adopt the recursion method [25, 26] for electronic structure calculations, which does not need any translational or point symmetry. The diagonal Green function $G_{i\nu\alpha i\nu\alpha}$ is then given by

$$G_{i\nu\alpha i\nu\alpha} = \tag{37}$$

$$\cfrac{1}{\omega + i\delta - a_{1i\nu\alpha}(\boldsymbol{\xi}) - \cfrac{|b_{1i\nu\alpha}(\boldsymbol{\xi})|^2}{\omega + i\delta - a_{2i\nu\alpha}(\boldsymbol{\xi}) - \cfrac{|b_{2i\nu\alpha}(\boldsymbol{\xi})|^2}{\cdots - \cfrac{|b_{l-1i\nu\alpha}(\boldsymbol{\xi})|^2}{\omega + i\delta - a_{li\nu\alpha}(\boldsymbol{\xi}) - T_{li\nu\alpha}}}}}.$$

Here $a_{li\nu\alpha}(\boldsymbol{\xi})$ and $b_{li\nu\alpha}(\boldsymbol{\xi})$ are the recursion coefficients of the l-th order. $T_{li\nu\alpha}$ is the terminator at the l-th level.

Obviously, the number of LM's which are treated in the MD method is limited to a few hundred or several hundred in maximum at the present stage because of the limited memory and computing time. The choice of boundary condition therefore becomes significant in the MD method. The direct extension of the single-site theory mentioned in the last section would lead to the free boundary condition of a cluster in the effective medium. The free boundary condition, however, does not lead to the well defined magnetic structure for crystals. We adopt here the periodic boundary condition. In this case, we consider a crystal with N atoms in a unit cell, and solve Eqs. (30)~(32) for the N LM's calculating the magnetic force at each step. The periodicity of the LM's introduces artificial magnetic forces due to self-interactions and double-counting interactions via the recursion coefficients in Eq. (37) because of the existence of equivalent LM's in other unit cells. We therefore stop the calculation of the recursion coefficients at the largest level l with no artificial interaction due to the boundary condition, and replace the terminator $T_{li\nu\alpha}$ with the effective one $\tilde{T}_{l\nu\alpha}$ i.e. the terminator of the coherent Green function $F_{i\nu\alpha i\nu\alpha} = [(\mathcal{L}^{-1} - t)^{-1}]_{i\nu\alpha i\nu\alpha}$.

$$T_{li\nu\alpha} \approx \tilde{T}_{l\nu\alpha} . \tag{38}$$

The effective medium $\mathcal{L}_\sigma^{-1}$ is determined by solving the CPA equation (28).

The selfconsistent MD calculations described above are now performed by taking the following steps. First, we assume the average LM's $[\langle\xi_z\rangle]_{\rm av}$, $[\langle\xi_z^2\rangle]_{\rm av}$, $[\langle\xi_\perp^2\rangle]_{\rm av}$, and the charge potentials $w(\boldsymbol{\xi})$ for the configuration $\xi_z = \pm[\langle\xi_z^2\rangle]_{\rm av}^{1/2}$, $\xi_\perp^2 = [\langle\xi_\perp^2\rangle]_{\rm av}$ within the single-site approximation. Here $[\ ]_{\rm av}$ denotes the average over sites. Next, we solve the CPA equation (28) in which $\langle\xi_z\rangle$, $\langle\xi_z^2\rangle$, $\langle\xi_\perp^2\rangle$ have been replaced by $[\langle\xi_z\rangle]_{\rm av}$, $[\langle\xi_z^2\rangle]_{\rm av}$, $[\langle\xi_\perp^2\rangle]_{\rm av}$. We then calculate the effective terminators $\tilde{T}_{l\nu\alpha}$ from the effective medium and the coherent Green function $F_{\Gamma\sigma}$. Under the given effective medium and terminator, we solve the equations of motion (30)~(32) in which we calculate the magnetic force (33) by means of the recursion method at each time step. With use of the MD data obtained in the stationary state, we calculate again $[\langle \xi_z \rangle]_{av}$, $[\langle \xi_z^2 \rangle]_{av}$, $[\langle \xi_\perp^2 \rangle]_{av}$, and $w(\boldsymbol{\xi})$ for $\xi_z = \pm[\langle \xi_z^2 \rangle]_{av}^{1/2}$ and $\xi_\perp^2 = [\langle \xi_\perp^2 \rangle]_{av}$. If these output values agree with the input, we calculate the magnetic moments at finite temperatures according to Eq. (29).

4. Numerical examples

4.1. BCC Fe

As a test of the MD approach, we first performed the calculations for bcc Fe with use of two schemes : the MD in the single-site approximation and the MD with periodic boundary condition. In the former, we calculated the thermal averages in the single-site approximation (Eqs. (21) and (28)) using the MD method solving Eqs. (30)~(32) for $N = 1$ and Eq. (26).

We adopted the Slater-Koster tight-binding parameters used by Pettifor [27], the band width 0.45 Ry [19], the d electron number $n = 7.0$, and the effective exchange energy parameter $\tilde{J} = 0.065$ Ry [28] for the electronic structure calculations of bcc Fe. Note that the present choice of parameters leads to the ground-state magnetization 2.15 μ_B, which is consistent with the experimental value 2.21 μ_B [29].

The time development of fictitious LM $\xi_z(t)$ for bcc Fe reveals that the LM $\xi_z(t)$ rapidly reaches the equilibrium state in a few hundred steps, and fluctuates around the equilibrium value. We dropped the first 1000 steps from the data, and obtained the time average from the remaining data of 10000 steps selfconsistently at each temperature. The calculated temperature dependences of the magnetization (M) as well as amplitude of LM's are presented in Fig. 1, together with those obtained by the direct numerical integration of Eq. (21). The MD approach quantitatively reproduces $\langle m_z \rangle$ and $\langle \boldsymbol{m}_i^2 \rangle^{1/2}$ within a few percent error over all temperatures. The results verify the quantitative aspect of the MD approach based on the ergodicity. Calculated Curie temperature $T_C = 800$ K is somewhat smaller than the experimental value 1040 K [29]. The discrepancy originates in the overestimate of the magnetic entropy due to a semi-classical treatment in obtaining the free energy (16).

In the MD calculations with periodic boundary condition, we adopted $N = 128$ atoms in a unit cell, which corresponds to a $4 \times 4 \times 4$ bcc lattice. Solving the equations of motion (30)~(32) together with Eqs. (26) and (28) selfconsistently, we obtained the time average of LM at each site. Calculated distributions of 128 Fe LM's are presented in Fig. 2 for various temperatures. At low temperatures, the distribution due to statistical error

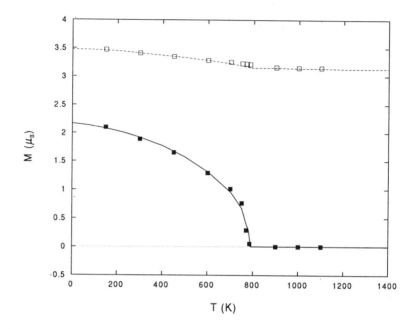

Figure 1. Temperature dependence of magnetization ($M = \langle m_z \rangle$: ■) and amplitude of local moment ($\langle \mathbf{m}^2 \rangle^{1/2}$: □) for α-Fe calculated by the MD approach in the single-site approximation. Solid and dashed curves show the results obtained by the direct numerical integration.

is limited to $0.1\mu_B$ around the average magnetization even when the number of MD steps are several hundred (see the distribution at 150 K in Fig. 2). With increasing temperature we have more thermal spin fluctuations, therefore need more number of time steps. In particular, we need more than a few thousand steps near the Curie temperature to obtain the statistical error smaller than 0.15 μ_B.

Calculated magnetization vs. temperature curves as well as temperature dependence of amplitude of LM are shown in Fig. 3. Amplitudes of LM ($\langle \mathbf{m}_i^2 \rangle^{1/2}$) show a good agreement with those in the single-site approximation. The present approach leads to the second order phase transition with increasing temperature. This is because the effective medium is self-consistently determined to describe the long range magnetic interactions. Calculated Curie temperature is 950 K, which is 1.2 times as large as that of the single-site approximation. This is probably caused by the underestimate of entropy due to the periodic boundary condition with rather small number of atoms N in a unit cell. In fact, we obtained $T_C = 1020$ K for smaller number of atoms $N = 54$, which is 1.3 times as large as that of the single-site approximation.

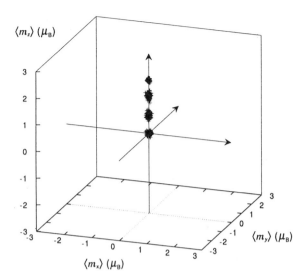

Figure 2. The distribution of local magnetic moment (LM) for α-Fe obtained by the MD approach with 128 atoms in a unit cell at various temperatures. The distributions from the top to the bottom in the figure are obtained at 150 K, 600 K, 900 K, and 1050 K, and after 800, 1000, 2000, 2800, 2000 MD steps, respectively.

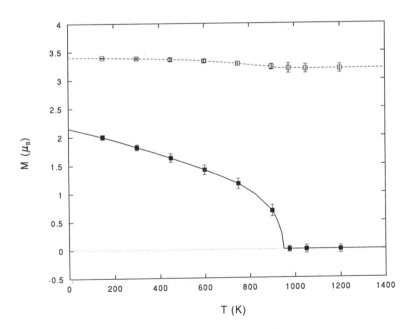

Figure 3. Temperature dependence of magnetization (■) and the amplitudes of LM (□) for α-Fe calculated by the MD with $N = 128$ atoms in a unit cell.

4.2. MAGNETIC STRUCTURE IN FCC TRANSITION METALS

The fcc transition metals are expected to form complex magnetic structures around the d electron numbers n between 6.0 and 7.0 due to competing interactions. We adopted $N = 108$ atoms per unit cell ($3 \times 3 \times 3$ fcc lattice) and the Slater-Koster tight binding parameters used by Pettifor [27]. The band width and effective exchange energy parameter were fixed to be the values of γ-Mn : $W = 0.443$Ry [30] and $\tilde{J} = 0.060$Ry [31]. The effective medium was not determined selfconsistently for brevity. It was constructed from the values $\langle \xi_z \rangle = 0$, $\langle \xi_z^2 \rangle$ and $\langle \xi_\perp^2 \rangle$ obtained in the single-site approximation. For the other parts of calculations, we adopted the same procedure as in the MD calculations for bcc Fe.

The calculated magnetic structure for $n = 6.0$ shows a complex structure in which the AFI structure is modulated by $\pi/2$ rotations of the LM's on a (1,0,0) antiferromagnetic plane as ABA$_{90}$BA$_{90}$B$_{90}$, where the subscript 90 denotes the $\pi/2$ rotation. In the case of $n = 6.2$, we obtain the AFI structure characterized by a stacking ABABAB as shown in Fig. 4(a). It explains the magnetic structure of γ-Mn. The calculated magnetic moment $|\langle m_i \rangle| = 2.5$ μ_B is also consistent with the experimental value 2.3 μ_B [3, 32] and the theoretical one 2.32 μ_B [13, 33].

When the d electron number n is increased, the AFI structure changes to the helical structures due to competing magnetic interactions. For $n = 6.4$, the LM's in an antiferromagnetic plane rotate by 120° with a translation of the lattice constant a along the axis perpendicular to the plane as shown in Figs. 4(b) and 5(a). We obtained the helical structure with the wave vector $Q = (0, 1/3, 1)2\pi/a$ and $|\langle m_i \rangle| = 2.0$ μ_B. Further increase of d electron number reduces the average magnetic moments and leads to the helical structure with amplitude modulations. Figure 4(c) shows the modulated helical structure for $n = 6.6$; the wave vector Q is the same as in $n = 6.4$, but the magnitudes of LM $[\langle m_i \rangle^2]_{av}^{1/2}$ are spatially modulated from 0.75 μ_B to 1.30 μ_B. This feature is more clearly seen from Fig. 5(b) as compared to 5(a). The modulated structure is characteristic of the itinerant electron systems because such a modulation should be suppressed by a large energy loss of Coulomb interactions in the insulator systems. It is stabilized by the energy gain caused by the break of a frustrated magnetic structure with equal amplitudes of LM's [9].

The DOS for $n = 6.0 \sim 6.4$ are characterized by a dip near the Fermi energy and the bonding-antibonding two peak structures as shown in Figs. 6(a) and (b). As has been found in the previous band calculations for γMn [13], these band structures cause the band-energy gain as compared to those

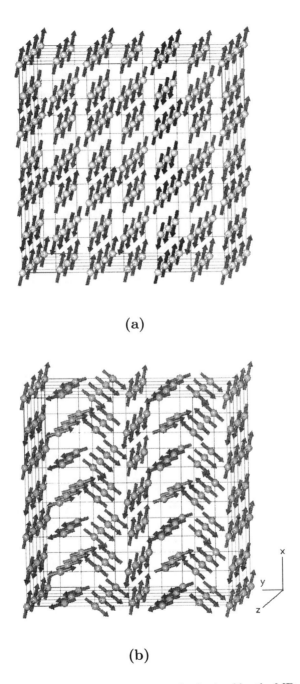

Figure 4. Magnetic structures of fcc transition metals obtained by the MD with $N = 108$ atoms at 50 K. (a) $n = 6.2$: the first-kind antiferromagnetic structure (AFI). (b) $n = 6.4$: the helical structure with the wave vector $(0, 1/3, 1)2\pi/a$. Here and in the following, the magnitudes of arrows are drawn in arbitrary unit.

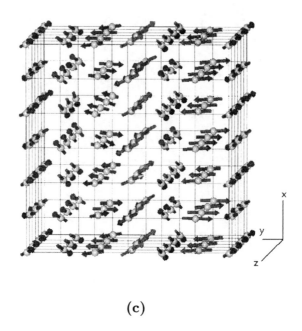

(c)

Figure 4. (c) $n = 6.6$: the helical structure with the wave vector $(0, 1/3, 1)2\pi/a$ and with the spatial modulation of the amplitudes of LM's.

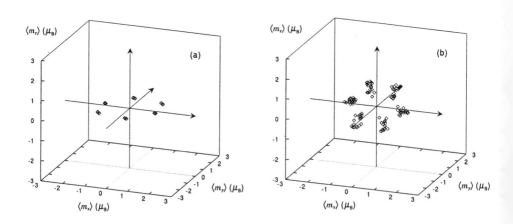

Figure 5. Distributions of LM for (a) $n = 6.4$, and (b) $n = 6.6$.

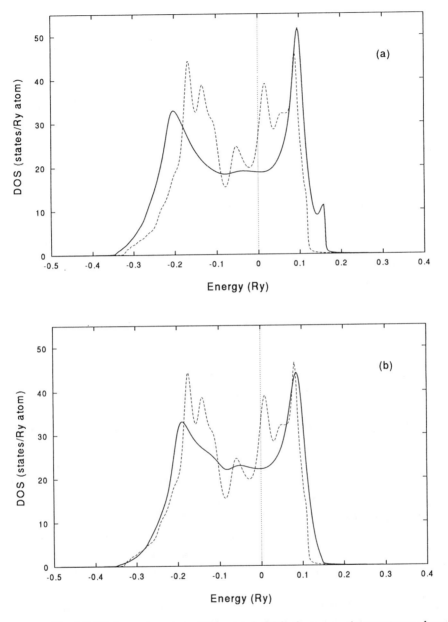

Figure 6. Total DOS for various transition metals with the magnetic structures shown in figures 4(a)–(c). (a) $n = 6.2$, (b) $n = 6.4$. The dashed lines show the DOS in the nonmagnetic state. The vertical dotted lines indicate the Fermi level.

in the nonmagnetic and ferromagnetic states, therefore they are considered to be a reason for the stability of the magnetic structures. For $n = 6.6$, a valley appears below the Fermi level (see Fig. 6(c)), so that a large band-

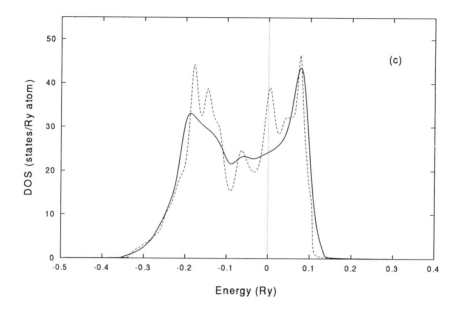

Figure 6. Total DOS for transition metals. (c) $n = 6.6$.

energy gain may not be expected for the present magnetic structure. On the other hand, the Fermi level is just on the peak in the nonmagnetic DOS, therefore the nonmagnetic state is not stable. The modulated structure for $n = 6.6$ may be realized by a small energy gain as compared with other magnetic structures.

It is not easy, except for γ-Mn, to compare the calculated magnetic structures with the experimental data within the rigid band model. The average magnetic moments, however, may be less sensitive to the details of the structure. We compared the results with the experimental data of γ-FeMn alloys in Fig. 7. The calculated average magnetic moments $[\langle \boldsymbol{m}_i \rangle^2]_{\mathrm{av}}^{1/2}$ qualitatively explain the concentration dependence in γ-FeMn alloys with less than 50 at.% Fe, but fail to explain a peculiar minimum of the averaged LM vs concentration curve at 50 at.% Fe and the behaviors above 50 at.% Fe. This feature was also found in the ground state band calculations with the AFI structure by Asano and Yamashita [13] as shown in Fig. 7.

γ-Fe is reported to have a helical magnetic structure with $\boldsymbol{Q} = (0, 0.1, 1)$ $2\pi/a$ [4]. The stability of the structure, however, has been a controversial question in the recent band theory. Fujii, Ishida, and Asano [8] performed the ground-state band calculation for γ-Fe assuming the three possible magnetic structures : AFI, 2-$\boldsymbol{Q}$ state with $\boldsymbol{Q} = (1, 0, 0)2\pi/a$ and $(0, 1, 0)2\pi/a$, and 3-$\boldsymbol{Q}$ state with $\boldsymbol{Q} = (1, 0, 0)2\pi/a$, $(0, 1, 0)2\pi/a$, and $(0, 0, 1)2\pi/a$. They

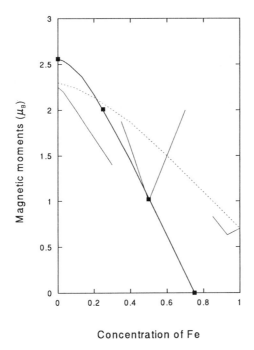

Figure 7. Concentration dependence of the average magnetic moment $[\langle \mathbf{m}_i \rangle^2]_{av}^{1/2}$ in γ-FeMn alloys calculated by the rigid band approximation (solid line with ■) The dotted line shows the result obtained by Asano and Yamashita [13] with use of the rigid band calculation and the AFI structure. The thin lines represent the experimental values [3].

found that the 3-Q structure with LM 1.2 μ_B is the most stable one among the possible states. Mryasov et al. [10] and Uhl et al. [11] minimized the energy with respect to the wave vector along Γ-X-W line of the fcc Brillouin zone, assuming the helical magnetic structure. They obtained $Q = (0, 0, 0.6)2\pi/a$, while Körling and Ergon [12] obtained $Q = (0, 0.5, 1)2\pi/a$ using gradient-corrected exchange correlation potential.

Antropov et al. [34] recently obtained the 3-Q state with LM 1.6 μ_B at $a = 3.59$ Å and the ↑↑↓↓ [001] structure with LM 2.3 μ_B at $a = 3.73$ Å, using the spin dynamics method with 32 atoms ($2 \times 2 \times 2$ fcc lattice). Here ↑↑↓↓ [001] means the structure in which the ferromagnetic xz planes are stacked along y axis as ↑↑↓↓. They also calculated the structure with use of the gradient-corrected potential, and obtained the ↑↑↓↓ structure at $a = 3.61$ and 3.65 Å, and obtained a noncollinear AF at $a = 3.57$ Å, in which the ferromagnetic xz plane with low LM 0.8 μ_B and the xz plane with rotation of LM's on the same plane are alternatively stacked along y axis. Finally, they concluded that the 3-Q plus helical structure with $Q = (0, 0.1, 1/6)2\pi/a$ at $a = 3.44$ Å may be the most stable one for

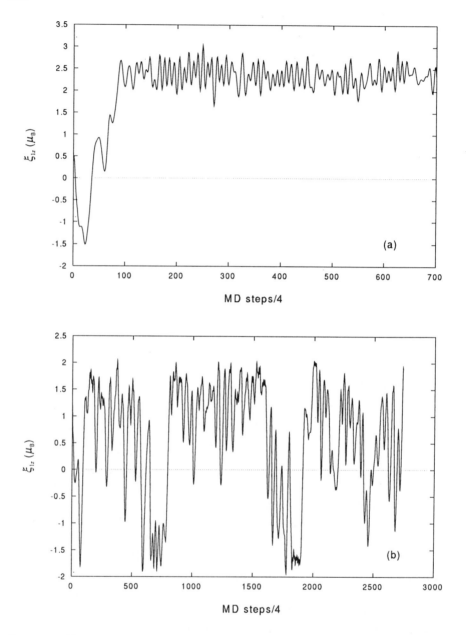

Figure 8. The time development of fictitious magnetic moment $\xi_{1z}(t)$ on site 1 obtained by the MD with 108 atoms in a unit cell. (a) γ-Mn at 50 K, (b) γ-Fe at 25 K.

local-spin density potential, and the ↑↑↓↓ state for the gradient-corrected potential, by comparing the energies of various possible structures, although both of them disagree with the helical structure suggested from the neutron

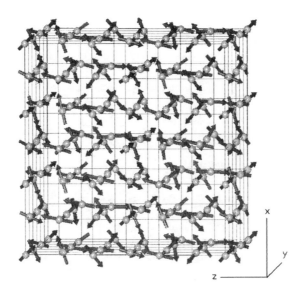

Figure 9. Helical magnetic structure of γ-Fe obtained by the MD with $N = 108$ ($3\times3\times3$ fcc lattice) at $T = 25$K. The wave vector is specified by $(0, 1/3, 2/3)2\pi/a$.

experiments [4].

We performed the MD calculations with use of $n = 7.0$, $W = 0.390$ Ry [30], $\tilde{J} = 0.065$ Ry and $N = 108$ atoms ($3 \times 3 \times 3$ fcc lattice), which is larger than 32 atoms ($2 \times 2 \times 2$ fcc lattice) used by Antropov et al. We found a strong spin frustration for γ-Fe. Figures 8(a) and 8(b) show the time development of a fictitious LM for γ-Mn and γ-Fe to clarify the difference in spin fluctuations between γ-Mn and γ-Fe. In the case of γ-Mn, the LM rapidly reaches an equilibrium value 2.5 μ_B and shows a small fluctuations around it as shown in Fig. 8(a). The LM for γ-Fe, however, shows a large spin fluctuations including spin reversal with long period as shown in Fig. 8(b). This means that γ-Fe shows a strong spin fluctuations, therefore we need a large number of MD steps to obtain the time average. The magnetic structure calculated at 25 K shows a helical structure with $\bm{Q} = (0, 1/3, 2/3)2\pi/a$ and $|\langle \bm{m}_i \rangle| = 0.75$ μ_B (see Fig. 9), which are compared with 0.7 μ_B for tetragonal γ-Fe [2] and $\bm{Q} = (0, 0.1, 1)2\pi/a$ for the cubic γ-Fe$_{100-x}$Co$_x$ ($x < 4$) alloy precipitated in Cu [4].

Although the result seems to explain qualitatively the magnetic structure of γ-Fe, we extended the size N in our MD calculation from 108 atoms ($3 \times 3 \times 3$ fcc lattice) to 256 atoms ($4 \times 4 \times 4$ fcc lattice) to examine the N

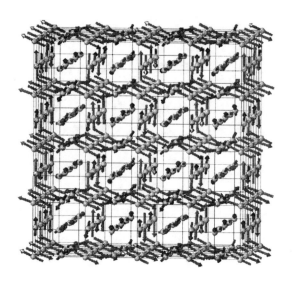

Figure 10. Noncollinear antiferromagnetic structure of γ-Fe obtained by the MD with $N = 256$ ($4 \times 4 \times 4$ fcc lattice) at $T = 25$K.

dependence. We found that the magnetic structure for $N = 256$ atoms are quite different from the helical structure, and shows a noncollinear antiferromagnetic (AF) structure with a unit cell consisting of 32 atoms ($2 \times 2 \times 2$ fcc lattice) as shown in Fig. 10. The LM's have 16 nonequivalent directions. The distribution of the magnitudes of LM is extended from 0.7 μ_B to 1.7 μ_B, and shows two peaks at 0.85 and 1.5μ_B, which might correspond to the low-spin and high-spin states, respectively. Calculated DOS again show a dip at the Fermi level for both $N = 108$ and $N = 256$.

The structure presented in Fig. 10 was not obtained in the spin dynamics calculations [34] with use of 32 atoms ($2 \times 2 \times 2$ fcc lattice). Apart from the detailed difference in electronic structure, there are essential differences between the two approaches. First, we replaced the long range magnetic force by the effective one to correct the artificial magnetic force caused by the periodic boundary condition. Second, the present calculations are performed at finite temperature (25 K) so that the system can reach global minimum jumping the energy barriers of order of 25 K per LM. We conclude at the present stage that γ-Fe is a strongly frustrated system with the long-range competing magnetic interactions, and the helical or the noncollinear AF structure seems to be stable. More detailed MD calculations are left for future work to resolve the problem of the stability for γ-Fe.

4.3. NONCOLLINEAR MAGNETISM OF γ-FeMn ALLOYS

The MD approach is more suitable for the disordered alloys as compared to the pure metals, since the random potentials cause the damping of long-range magnetic interactions and the random atomic configuration generally suppresses the spin frustrations. In this subsection we present the results of calculation for disordered γ-FeMn alloys as an example, and discuss the validity of the MD approach.

The γ-FeMn alloys are well-known to show the complex magnetic structures in the concentrated region due to competing magnetic interactions and random atomic configuration. The neutron experiments [3] reported that there are three different types of spin structure for the alloys : (1) the γ-Mn type for the concentration up to 30 at.% Fe, (2) the noncollinear spin arrangement in the range between 35 to 75 at.% Fe and (3) γ-Fe type beyond 80 at.% Fe. In particular, the noncollinear structure which is called a multiple spin density wave with the 3-Q wave vectors $(1,0,0)2\pi/a$, $(0,1,0)2\pi/a$, and $(0,0,1)2\pi/a$, was proposed around 50 at.% Fe [3, 35], while the collinear magnetic structure was reported for a single crystal of γ-Fe$_{66}$Mn$_{34}$ alloy in the inelastic neutron scattering experiment [36, 37].

The theoretical investigation for γ-FeMn alloys has been done first by Asano and Yamashita [13]. They explained the monotonical decrease of average magnetic moment up to 50 at.% Fe within the rigid band calculation assuming the AFI structure, and obtained the DOS of γ-Mn showing the bonding-antibonding due to local exchange splitting as well as the DOS having the gap-like dip near the Fermi level for γ-Fe$_{50}$Mn$_{50}$ alloy. The rigid band calculation, however, did not explain the minimum of the average magnetic moment at 50 at.% Fe and the rapid increase of the moment beyond 50 at.% Fe. Kübler et al. [38] and Fujii et al. [8] performed the total-energy band calculations for γ-Fe$_{50}$Mn$_{50}$ alloys assuming a few possible noncollinear magnetic structures with the CuAu-type crystal structure, but their calculations did not yield the 3-Q spin density wave proposed in the experiments. F. Süss and U. Krey [39] calculated the ordered γ-FeMn alloys with the collinear spin structure. They failed to obtain the convergence in their selfconsistent equations, therefore the magnetic structure. No theory to explain the magnetic structures as well as the magnetic moments over all concentrations has been proposed.

It is noted that the disorder on the LM originates in the random configuration of the atomic levels ϵ_i^0 and the transfer integrals $t_{i\nu j\nu'}$ in the substitutional binary alloys. The off-diagonal disorder associated with the transfer integral is important when the relative bandwidths of the constituent atoms A and B are considerably different. We considered a geometrical-mean model for the transfer integrals. If the constituent metals A and B have the similar energy band, one can write the transfer integral $t_{i\nu j\nu'}$ in this model as

$$t_{i\nu j\nu'} = r_\lambda^* t_{i\nu j\nu'}^0 r_{\lambda'} , \qquad (39)$$

where $t_{i\nu j\nu'}^0$ represents the transfer integral of pure metal B, which is expressed by a linear combination of the Slater-Koster parameters for B atom $dd\sigma^{BB}$, $dd\pi^{BB}$ and $dd\delta^{BB}$. The parameter $|r_\lambda|$ is defined by

$$|r_\lambda| = \sqrt{\frac{(ddm)^{\lambda\lambda}}{(ddm)^{BB}}} , \qquad (40)$$

which depends on the type λ of atom on site i. If site i is occupied by A atom, $|r_\lambda|^2$ would be the ratio of the bandwidth of pure metal A to that of B metal, otherwise it would be 1.

For random alloys, we determined the random atomic configuration in the unit cell to realize the alloy with a given concentration c_λ and a given atomic short range order τ_{SRO} parameter, which is defined by

$$p^{AA} = c_A + c_B \tau_{SRO} . \qquad (41)$$

Here p^{AA} denotes the probability of finding A atom at the nearest neighbor site of A atom.

We adopted again the bandwidths 0.390Ry for γ-Fe and 0.443Ry for γ-Mn [30], the exchange energy parameters 0.065Ry [28] and 0.060Ry [31], and the d electron numbers 7.0 and 6.25 for γ-Fe and γ-Mn, respectively. The effective medium was not determined selfconsistently in the present calculations; we used the medium obtained by the single-site approximation over all concentrations for brevity. Moreover, we adopted $N = 108$ atoms per MD unit cell ($3 \times 3 \times 3$ fcc lattice).

Figures 11(a) - 11(g) and 12(a) - 12(f) show the distributions of magnetic moments and the magnetic structures calculated at 25 K for various concentrations, respectively. γ-Mn shows the AFI structure in agreement with the experiments [3, 40]. When Fe atoms are added to γ-Mn, the Fe magnetic moments are arranged without violation of the AFI structure. The magnitude of Fe LM is calculated as 1.6 μ_B in the impurity limit, which is smaller than the calculated Mn LM 2.4 μ_B. This feature does not change up to 10 at.% Fe as shown in Fig. 11(a).

The collinear AF structure starts to collapse beyond 10 at.% Fe and changes to a helix-like structure as shown in Figs. 11(b) and 12(a), which are obtained at 20 at.% Fe. The obtained structure at 20 at.% Fe may

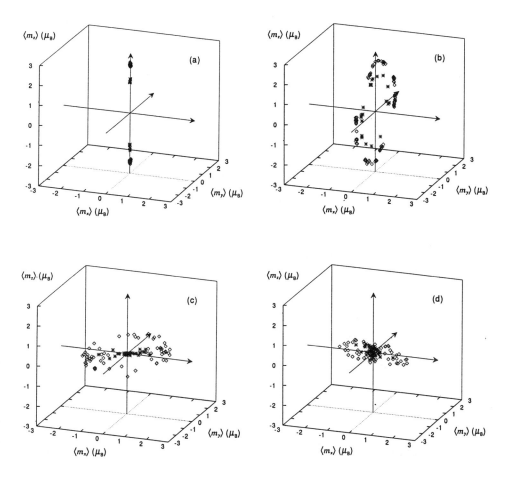

Figure 11. The distribution of local magnetic moments (LM) in γ-FeMn alloys at various concentrations obtained by the MD approach at 25 K : (a) 10 at.% Fe, (b) 20 at.% Fe, (c) 40 at.% Fe, (d) 50 at.% Fe. The symbols ◊ and * represent the Mn LM and Fe LM respectively.

be regarded as a mixture of two kinds of helical structures with the wave vectors $Q = (0, 1/3, 1)2\pi/a$ and $Q' = (0, 1/2, 1)2\pi/a$ so that LM's on the three AF planes remain collinear and those on the other AF planes rotate on the planes. Note that the helical structure with $Q = (0, 1/3, 1)2\pi/a$ was obtained in the previous rigid band calculation for the d electron number $n = 6.4$ [14], which corresponds to the 20 at.% Fe alloy. The alloying effect is therefore considered to be partly to remain the collinear AF planes and partly to cause the distribution of the direction of LM's.

The AF planes are almost destroyed around 30 at.% Fe, and a noncollinear structure which can not be expressed by simple Q vectors is obtained. Although the average Mn LM have still a large value 2.5 μ_B there,

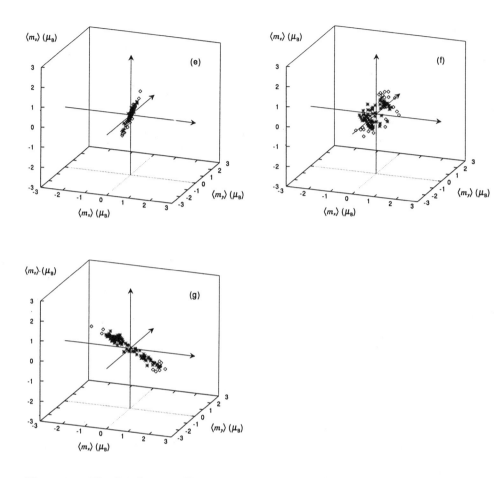

Figure 11. The distribution of local magnetic moments (LM) in γ-FeMn alloys at 25 K : (e) 65 at.% Fe, (f) 80 at.% Fe, and (g) 90 at.% Fe. The symbols ◇ and ∗ represent the Mn LM and Fe LM respectively.

the average Fe LM show a small value 0.5 μ_B. The structure at 40 at.% Fe is also noncollinear as shown in Figs. 11(c) and 12(b), but the average Mn LM rapidly decrease there and take a small value 1.5 μ_B. At 50 at.% Fe, the magnetic structure shows a noncollinear structure with large amplitude fluctuations as well as directional fluctuations, as shown in Figs. 11(d) and 12(c). The concentration 50 at.% Fe corresponds to the average d electron number 6.625. The magnetic structure obtained by the rigid band calculations for d electron number $n = 6.6$ was found to be the helical structure with amplitude fluctuations from 0.75 μ_B to 1.3 μ_B [9]. The effect of configurational disorder with competing interactions is therefore considered to remove the characteristic **Q** vector and to enhance the amplitude fluctuations.

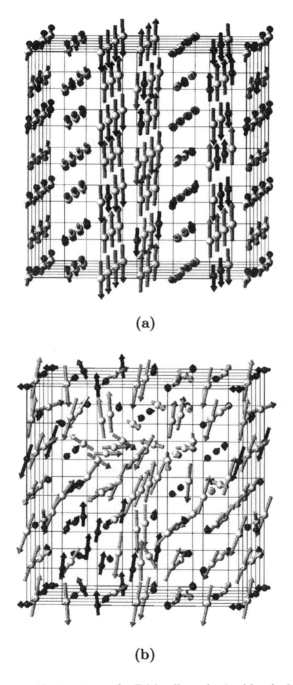

Figure 12. The magnetic structures of γ-FeMn alloys obtained by the MD at 25 K : (a) 20 at.% Fe, (b) 40 at.% Fe. The dark (light) gray spheres and arrows show the Fe (Mn) atoms and LM. The magnitudes of LM's are drawn in arbitrary unit.

With further increase of Fe concentration, the magnetic structure becomes collinear-like around 65 at.% Fe as shown in Figs. 11(e) and 12(d), and changes to the noncollinear ones as shown in Figs. 11(f) and 12(e) for 80 at.% Fe. Between 85 at.% Fe and 95 at.% Fe, the collinear AF structures are stabilized as shown in Figs. 11(g) and 12(f) for 90 at.% Fe. The locally ferromagnetic planes are found there, while the LM's show a large distribution in magnitude (see Fig. 11(g)). These structures are changed to the helical structure with $\boldsymbol{Q} = (0, 1/3, 2/3)2\pi/a$ beyond 95 at.% Fe [14].

Calculated magnetic structures mentioned above qualitatively agree with the experimental results. In fact, they explain three different regions for magnetic structure obtained from the neutron diffraction and Mössbauer techniques: the AFI collinear region between 0 and 30 at.% Fe (corresponding to our result Fig. 11(a)), the noncollinear region between 35 and 75 at.% Fe (corresponding to our result Figs. 11(b) - 11(d) and 12(a) - 2(c)), and the collinear region between 80 and 100 at.% Fe (corresponding to our result Figs. 11(g) and 12(f)). Using inelastic neutron scattering, Bisanti et al. [36, 37] reported that the fcc $Fe_{66}Mn_{34}$ alloy has a collinear magnetic structure, which may be explained by our result Fig. 11(e) and 12(d). Tsunoda [4] investigated the cubic γ-$Fe_{100-x}Co_x$ ($x < 4$) alloys precipitated in Cu and suggested that γ-Fe should have the helical structure with $\boldsymbol{Q} = (0, 0.1, 1)2\pi/a$, which is qualitatively in agreement with our result [14].

There are some disagreements in detailed magnetic structures between theory and experiment. For example, calculated magnetic structure at 50 at.% Fe shows the turbulent noncollinear magnetic structure due to configurational disorder as shown in Fig. 12(c), while the 3-$\boldsymbol{Q}$ spin density wave with the wave vectors $(1, 0, 0)2\pi/a$, $(0, 1, 0)2\pi/a$, and $(0, 0, 1)2\pi/a$ was suggested from the neutron experiment by Endoh and Ishikawa [3]. Moreover, $\boldsymbol{Q}$ vector in γ-Fe is considerably different between our result and experiment. The effect of more detailed electronic structure, the size effect of the MD unit cell, and the effect of spin-orbit interaction have to be examined in the future investigations to clarify the disagreements from the theoretical point of view.

The DOS for γ-Mn with AFI structure is characterized by the bonding and antibonding peaks due to the local exchange splitting, which generally stabilizes the structure in the systems with nearly half-filled bands [13]. This feature remains in the DOS at 10 and 20 at.% Fe. The small dip near the Fermi level as well as the bonding-antibonding structure seems to stabilize the magnetic structures in Mn-rich alloys. The DOS above 50 at.% Fe are characterized by the deep valley at the Fermi level in the magnetic state and a sharp peak at the Fermi level in the nonmagnetic state. This feature enhances the kinetic energy gain, and therefore stabilizes the structures

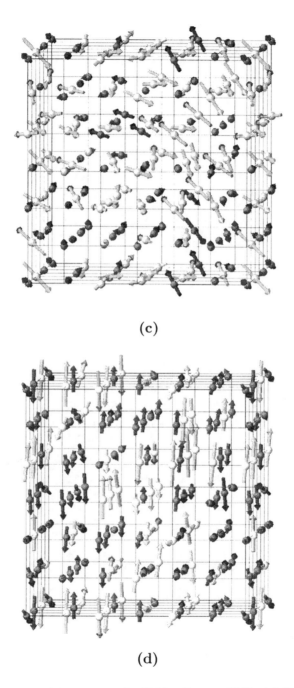

Figure 12. The magnetic structures of γ-FeMn alloys at 25 K : (c) 50 at.% Fe, (d) 65 at.% Fe. The dark (light) gray spheres and arrows show the Fe (Mn) atoms and LM.

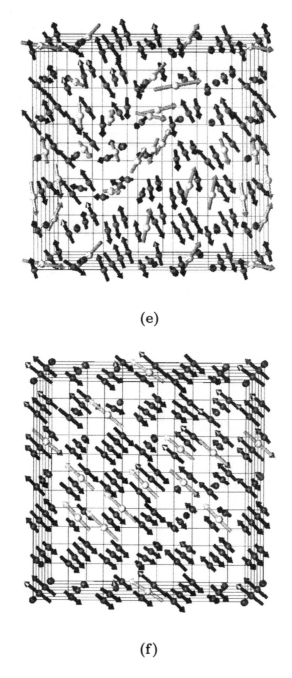

Figure 12. The magnetic structures of γ-FeMn alloys at 25 K : (e) 80 at.% Fe, and (f) 90 at.% Fe. The dark (light) gray spheres and arrows show the Fe (Mn) atoms and LM.

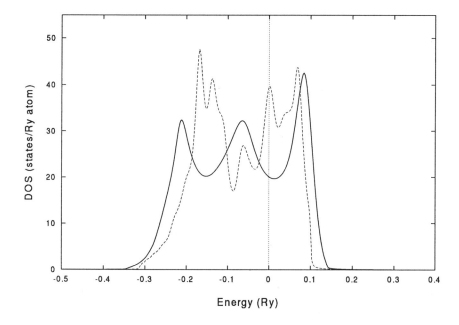

Figure 13. Total DOS at 60 at.% Fe and the ground state. The dashed curves show the DOS in the nonmagnetic state. The vertical dotted lines indicate the Fermi level.

obtained in our calculations. In particular, the Fermi level lies just on the sharp nonmagnetic peak at 60 at.% Fe and the DOS in the magnetic state creates a deeper valley (see Fig. 13). It may explain the enhancement of average magnetic moments at 60 at.% Fe.

Calculated magnetic moments are presented in Fig. 14 together with the experimental data [3]. When Fe atoms are added to γ-Mn, the average magnetic moment decreases linearly from 2.4 μ_B according to the simple dilution line up to 30 at.% Fe. The calculated Fe LM is 1.6 μ_B in the impurity limit. It remains constant until 20 at.% Fe and starts to decrease rapidly beyond 20 at.% Fe, while that of Mn is almost constant up to 30 at.% Fe.

The calculated average magnetic moment vs concentration curve explains the experimental curve up to 50 at.% Fe. The rigid band calculations led to the similar concentration dependence up to 50 at.% Fe, but it failed to explain the minimum at 50 at.% Fe and the behavior beyond 50 at.% Fe [13, 14]. The present result shows a minimum at 50 at.% Fe in agreement with the experimental data, and once increases at 60 at.% Fe. The Fermi level lies there just on the peak of the nonmagnetic DOS. The enhancement of average magnetic moment can be attributed to a large kinetic energy gain due to the change of the peak to the dip at the Fermi

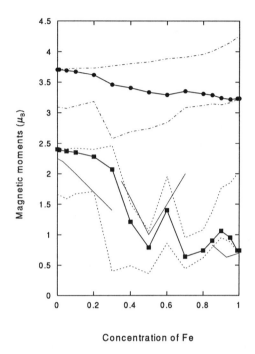

Figure 14. Concentration dependence of the magnetic moments in γ-FeMn alloys calculated by the MD approach at 25 K. The solid line with ■ represents the average magnetic moments, the upper (lower) dashed line represents the average Mn (Fe) magnetic moments. The thin lines represent the experimental average magnetic moments obtained by the neutron experiments [3]. The average and partial amplitudes of LM $[\langle \mathbf{m}_i^2 \rangle]_c^{1/2}$ are also shown by the solid line with ● and the dot-dashed lines (the upper line: Mn, the lower one: Fe).

level with polarization, as has been mentioned before (see Fig. 13). The average magnetic moment does not show the strong concentration dependence beyond 70 at.% Fe, although the Mn LM start to increase beyond 80 at.% Fe with the localization of electrons on Mn sites due to decreasing transfer integrals.

Calculated average LM in the concentration regions 0 - 30 at.% Fe and 90 - 100 at.% Fe are overestimated as compared with the experimental data. In both regions, the γ-FeMn alloys show the tetragonal distortion. Recent ground state calculations [41] show that the Fe LM changes the sign in the CsCl-type antiferromagnetic Mn, which is obtained by 10 % contraction of c axis in γ-Mn. Thus, there is a possibility that the Fe LM reduce the magnitudes with tetragonal distortion. Our results of calculations do not cause large magnetic moment ($\sim 2.0\mu_B$) found in neutron experiment around 70 at.% Fe. This might be attributed to the atomic-short range order or some local lattice distortion. The effects should be investigated in the future to clarify the disagreement between the theory and experiments.

5. Summary and discussion

We have proposed a MD approach to the complex magnetic structures in the itinerant electron systems on the basis of the isothermal MD method and the functional integral technique. The theory reduces to the generalized Hartree-Fock approximation at the ground state, and takes into account spin fluctuations at finite temperatures. The MD approach assumes the ergodicity for the motion of fictitious LM's in phase space, therefore the thermal average of LM is obtained by a time average, solving the $6N + 3$ equations of motion. The electronic structure in the magnetic force was calculated with use of the recursion method at each MD step. The method allows us to treat the magnetic structures with a few hundred atoms in a unit cell at finite temperatures, and automatically find the stable magnetic structure in principle.

We have verified that the MD approach quantitatively reproduces the $M-T$ curve in the single-site approximation, and describes the second order phase transition with increasing temperature, with a help of a selfconsistent effective medium introduced into the magnetic force. The calculated Curie temperatures for bcc Fe are comparable to that of the single-site approximation ; $T_C(N = 54) = 1.3 T_C$ (SSA) and $T_C(N = 128) = 1.2 T_C$ (SSA), although they are somewhat overestimated because of the periodic boundary condition for a unit cell with rather small number of atoms.

The MD approach yields the magnetic structure within a few thousand steps at low temperatures in most cases. The MD calculations with use of $N = 108$ atoms revealed that the fcc transition metals with the d electron number between 6.0 and 7.0 show various complex magnetic structures due to competing magnetic interactions : a complex AF with 90° spin rotation on the AF planes for $N = 6.0$, the AFI for $n = 6.2$, the helical structure with $\boldsymbol{Q} = (0, 1/3, 1) 2\pi/a$ for $n = 6.4$, and the helical structure with the amplitude modulation for $n = 6.6$.

In the case of γ-Fe, we found a strong spin frustration and the nature of long-range competing magnetic interactions, so that the structure shows the size dependence: a helical structure with $\boldsymbol{Q} = (0, 1/3, 2/3) 2\pi/a$ for $N = 108$ ($3 \times 3 \times 3$) and the noncollinear AF structure with 32 atoms in a unit cell for $N = 256$ ($4 \times 4 \times 4$). One needs more detailed MD calculations with more detailed electronic structure and larger N to resolve the structure of γ-Fe.

The application to the disordered γ-FeMn alloys yields various complex magnetic structures due to competing interactions: the AFI structure up to 10 at.% Fe, the helix-like structure around 20 at.% Fe, the non-collinear structures between 30 and 60 at.% Fe, a collinear-like structure at 65 at.% Fe, the noncollinear structures between 70 and 80 at.% Fe, the collinear structures between 85 and 95 at.% Fe, and the helical structures with $\boldsymbol{Q} = (0, 1/3, 2/3)2\pi/a$ beyond 95 at.% Fe, which qualitatively explain the results obtained by the neutron diffraction experiments. The calculated magnetic moments also explained the concentration dependence obtained by the neutron experiments, in particular, the existence of the peculiar minimum at 50 at.% Fe, which was not obtained by the rigid band calculations. The enhancement of magnetic moment at 60 at.% Fe was explained by a large kinetic energy gain associated with a peak in the nonmagnetic DOS and a valley in the magnetic DOS on the Fermi level.

The numerical examples mentioned above show that the MD approach is useful for the theoretical investigations of itinerant electron magnetism with complex magnetic structures. It is desired to improve further the present approach towards more quantitative description of the magnetism. One of the important problems is to increase the number of atoms N in a unit cell to describe more complex magnetic structures and to relax the periodic boundary condition. In the present calculations, most of the computing time is spent for the calculation of magnetic force with use of the recursion method, which needs the computing time proportional to N^2. We could treat the systems with larger N using the parallel computers more effectively because the diagonal Green functions can be calculated at each site, each orbital, and each spin, independently. It is also desired to use the first principles tight-binding LMTO Hamiltonian at the ground state for more detailed electronic structure calculations, which is possible in the present framework of the theory at the ground state.

Another problem for the improvement of theory is to take into account the dynamical effects towards more quantitative description of electron correlations and thermodynamics. This could be done within the theoretical framework presented here by improving the effective energy potential $E(\boldsymbol{\xi}, T)$ for the thermal average of LM. After these improvements, the MD approach will become a powerful tool for the investigations of the itinerant electron magnetism with complex magnetic structures.

Acknowledgments

A part of this paper has been presented in the International Symposium in honor of Martin C. Gutzwiller "From Correlated Electrons to the Quantum Mechanics of Complex Systems" held in Dresden, June 1997. One of the authors (Y. K) expresses his sincere thanks to Professor Martin C. Gutzwiller for encouragement of the present work. The work has been done partly with use of the facilities of the Supercomputer Center, Institute for Solid State Physics, University of Tokyo.

References

1. Yamada, T. (1970) Magnetism and Cristal Symmetry of α-Mn, *J. Phys. Soc. Jpn.* **28**, 596 ; Yamada, T., Kunitomi, N., and Nakai, Y. (1971) Magnetic Structure of α-Mn, *J. Phys. Soc. Jpn.* **30**, 1614.
2. Abrahams, S.C., Gutman, L., and Kaksper, J.S. (1962) Neutron diffraction determination of antiferromagnetism in face-centered cubic (γ)Iron, *Phys. Rev.* **127**, 2052.
3. Endoh, Y. and Ishikawa, Y. (1971) Antiferromagnetism of γ Iron Manganes Alloys, *J. Phys. Soc. Jpn.* **30**, 1614.
4. Tsunoda, Y. (1989) Spin-density wave in cubic γ-Fe and γ-Fe$_{100-x}$Co$_x$ precipitates in Cu, *J. Phys.: Condens. Matter* **1**, 10427.
5. Hiroyoshi, H. and Fukamichi, K. (1981) Spin-glass like behavior in Fe-Zr amorphous alloys, *Phys. Lett.* **85A**, 242 ; (1982) Ferromagnetic-spin glass transition in Fe-Zr amorphous alloy system, *J. Appl. Phys.* **53**, 2226.
6. Saito, N., Hiroyoshi, H., Fukamichi, K., and Nakagawa Y. (1986) Micromagnetism of Fe-rich Fe-Zr amorphous alloys studied by AC susceptibility in a superposed DC field, *J. Phys.* **F 16**, 911.
7. Fukamichi, K., Goto, T., Komatsu, H. and Wakabayashi, H. (1988) Spin Glass and Invar Properties of Iron-rich Amorphous Alloys, *Proc. 4th Int. Conf. on Phys. Magn. Mater. (Poland)*, ed. Gorkowski, W., Lachowics, H.K., and Szymczak, H., World Scientific Pub., Singapore, 1989, p. 354.
8. Fujii, S., Ishida, S., and Asano, S. (1991) Band Calculations for Non-Collinear Spin Arrangements in Gamma-Phase Manganese-Iron Alloys, *J. Phys. Soc. Jpn.* **60**, 4300.
9. Lacroix, C. and Pinettes, C. (1992) Itinerant antiferromagnetism in a frustrated lattice, *J. Magn. Magn. Mater.* **104-107**, 751.
10. Mryasov, O.N., Lichtenstein, A.I., Sandratskii, L.M., and Gubanov, V.A. (1991) Magnetic structure of FCC iron, *J. Phys.: Condens. Matter* **3**, 7683 ; Mryasov, O.N., Gubanov, V.A., and Lichtenstein, A.I. (1992) Spiral-spin-density-wave states in fcc iron: Linear-muffin-tin-orbitals band-structure approach, *J. Appl. Phys.* **45**, 12330.
11. Uhl, M., Sandratskii, L.M., and Kübler, J. (1992) Electronic and magnetic states of γ-Fe, *J. Magn. Magn. Mater.* **103**, 314.
12. Körling, M. and Ergon, J. (1997) Gradient-Corrected Ab-initio Calculations of Spin-Spiral States in FCC-Fe, *Physica* **B 237-238**, 353.
13. Asano, S. and Yamashita, J. (1971) Band Theory of Antiferromagnetism in 3d f.c.c. Transition Metals, *J. Phys. Soc. Jpn.* **31**, 1000.
14. Kakehashi, Y., Akbar, S., and Kimura, N. (1998) Molecular dynamics approach to itinerant magnetism with complex magnetic structures, *Phys. Rev.* **B 57**,1 ; Akbar, S., Kakehashi, Y., and Kimura, N. (1998) A molecular dynamics approach to the magnetic alloys with turbulent complex magnetic structures: γ-FeMn alloys, *J. Phys.: Condens. Matter* **10**, 2081.
15. Hubbard, J. (1959) Calculation of Partition Functions, *Phys. Rev. Lett.* **3**, 77.
16. Stratonovich, R.L. (1958) On a method of calculating quantum distribution functions, *Dokl. Akad. Nauk SSSR* **115**, 1097 [*Sov. Phys. - Dokl.* **2** (1958) 416].
17. Hubbard, J. (1979) Magnetism of Iron, *Phys. Rev.* **B 19**, 2626 ; (1979) Magnetism of Iron II, *ibid* **20**, 4584 ; (1981) Magnetism of Nickel, *ibid* **23**, 5974.
18. Hasegawa, H. (1979) Single-Site Functional-Integral Approach to Itinerant-Electron Ferromagnetisms, *J. Phys. Soc. Japan.* **46**, 1504 ; (1980) Single-Site Spin Fluctuation Theory of Itinerant-Electron Systems with Narrow Bands, *ibid.* **49**, 178.

19. Hasegawa, H. (1983) A spin fluctuation theory of degenerate narrow bands – finite-temperature magnetism of iron, *J. Phys.* **F 13**, 1915.
20. Kakehashi, Y. (1986) Degeneracy and quantum effects in the Hubbard model, *Phys. Rev.* **B 34**, 3243.
21. See, for example, Fulde, P. (1995), Electron Correlations in Molecules and Solids, *Solid State Sciences* **Vol. 100**, Springer-Verlag, Berlin, Chap. 11.
22. Soven, P. (1967) Coherent-Potential Model of Substitutional Disordered Alloys, *Phys. Rev.* **156**, 809 ; Velický, B., Kirkpatrick, S., and Ehrenreich, H. (1968) Single-Site Approximations in the Electronic Theory of Sinple Binary Alloys, *ibid.* **175**, 747.
23. Nosé, S. (1984) A molecular dynamics method for simulations in the canonical ensembre, *J. Chem. Phys.* **81**, 511.
24. Hoover, W.G. (1991), *Computational Statistical Mechanics*, Elsevier, Amsterdam.
25. Haydock, R., Heine, V., and Kelly, M.J. (1975) Electronic structure based on the local atomic environment for tight-binding bands, *J. Phys.* **C 8**, 591.
26. Heine, V., Haydock, R. and Kelly, M.J. (1980) Electronic structure from the point of view of the local atomic environment, *Solid State Physics* **35**, 1.
27. Haydock, R. and Kelly, M.J. (1973) Surface densities of states in the tight-binding approximation, *Surf. Sci.* **38**, 139.
28. Andersen, O.K., Madsen, J., Poulsen, U.K., Jepsen, O. and Kollär, J. (1977) Magnetic and cohesive properties from canonical bands, *Physica* **B 86-88**, 249.
29. Bozorth, R. (1968), *Ferromagnetism*, Van Nostrand, Princeton.
30. Moruzzi, V.L., Janak, J.F., and Williams, A.R. (1978), *Calculated Electronic Properties of Metals*, Pergamon, New York.
31. Janak, J.F. (1977) Uniform susceptibilities of metallic elements, *Phys. Rev.* **B 16**, 255.
32. Megeghetti, D. and Sidhu, S.S. (1957) Magnetic Structures in Copper-Manganese Alloys, *Phys. Rev.* **105**, 130 ; Bacon, G.E., Dummur, I.W., Smith, J.H., and Street, R. (1957) The antiferromagnetism of manganese copper alloys, *Proc. Roy. Soc.* **A 241**, 223 ; Hick, T.J., Pepper, A.R., and Smith, J.H. (1968) Antiferromagnetism in γ-phase manganese-palladium and manganese-nickel alloys, *J.Phys.* **C 1**, 1683.
33. Oguchi, T. and Freemann, A.J. (1984) Magnetically induced tetragonal lattice distortion in antiferromagnetic fcc Mn, *J. Magn. Magn. Mater.* **46**, L1.
34. Antropov, V.P., Katsnelson, M.I., van Shilfgaarede, M., and Harmon, B.N. (1995) Ab Initio Spin Dynamics in Magnets, *Phys. Rev. Lett.* **75**, 729 ; Antropov, V.P., Katsnelson, M.I., Harmon, B.N., van Shilfgaarede, M., and Kusnezov, D. (1996) Spin dynamics in magnets: Equation of Motion and finite temperature effects, *Phys. Rev.* **B 54**, 1019.
35. Kennedy, S.J., and Hicks, T.J. (1986) Magnetic structure of γ-iron-manganese, *J. Phys. F: Met. Phys* **17**, 1599.
36. Bisanti, P., Mazzone, G., and Sacchetti, F. (1987) Electronic structure of FCC Fe-Mn alloys: II. Spin-density measurements, *J. Phys. F: Met. Phys.* **17**, 1425.
37. Andreani, C., Mazzone, G., and Sacchetti, F. (1987) Electronic structure of FCC Fe-Mn alloys: I. Charge-density measurements, *J. Phys. F: Met. Phys* **17**, 1419.
38. Kübler, J., Hock, K.H., Sticht, J. and Williams, A.R. (1988) Local spin-density functional theory of noncollinear magnetism, *J. Phys. F: Met. Phys.* **18**, 469.
39. Süss, F. and Krey, U. (1993) On the itinerant magnetism of Mn and its ordered alloys with Fe and Ni, *J. Magn. Magn. Mater.* **125**, 351.
40. Bacon, G.E., Dunmur, I.W., Smith, J.H. and Street, R. (1957) The antiferromagnetism of manganese copper alloys, *Proc. Roy. Soc.* **A 241**, 223.
41. Antropov, V.P., Anisimov, V.I., Lichtenstein, A.I., and Postnikov, A.V. (1988) Electronic structure and magnetic properties of 3d impurities in antiferromagnetic metals, *Phys. Rev.* **B 37**, 5603.

SPIN FLUCTUATION THEORY VERSUS EXACT CALCULATIONS

V. BARAR, W. BRAUNECK AND D. WAGNER
Ruhr-Universität Bochum, Theoretische Physik III
Universitätsstr. 150, 44780 Bochum, Germany

Abstract. The infinite dimensional simplified Hubbard or Falicov-Kimball model is used for a check of a generalized spin fluctuation theory which had been successfully applied to Invar systems. In particular we calculate the order parameter and the volume for this model from the exact free energy and from an application of the spin fluctuation theory which we adjust to this model. We find rather large discrepancies in the temperature behaviour.

1. Introduction

Spin fluctuation theory has become quite essential for our present understanding of itinerant ferromagnetism [1]. In the weak ferromagnetic limit it results in a Ginzburg-Landau like theory and becomes especially simple and tractable [2]. A generalized spin fluctuation theory in this limit has proven quite successful in describing the magnetomechanical effects of Invar [3, 4]. For a review of Invar behaviour with special attention to the experiments we refer to Wassermann [5]

A crucial point of the theory of Refs. [3, 4] is its connection with band structure calculations which are performed in the frame of the so-called fixed-spin-moment (FSM) method [6]. In contrast to conventional band structure calculations the FSM scheme treats the magnetic moment on the same footing as the volume. As a result, FSM calculations yield the total energy as a function of the volume per site as well as of the magnetic moment per site, i.e. the energy or binding surface $E = E(V, M)$. Such calculations have led to the conclusion that Invar effects should be intimately connected with certain properties of these energy surfaces: the existence of two stationary points, a ferromagnetic minimum and a saddle point with vanishing or small moment and with a smaller volume per site than that of the minimum, with only a slight energy difference between these points [7, 8, 9, 11]. On this ground the main Invar effect, the decrease of volume with increasing temperature, is then simply understood by the fact that the free energy of the ferromagnetic phase and the free energy of the paramagnetic phase are approaching each other when going from low temperatures to the transition point. Anti-Invar effects can be explained similarly [5]. In a sense, this picture of Invar reminds of the "low-moment state" and the "high-moment state" of the older phenomenological "2-state model" of Weiss [10].

To describe the behaviour at nonzero temperatures the energy surfaces $E(V, M)$ are chosen as the input of a generalized spin fluctuation theory in Ref. [3] and along similar lines also in Ref. [4]. The theory includes spin (moment) fluctuations as well as volume (density) fluctuations. As a starting point, an effective Hamiltonian of the fluctuations is defined [3] which reads

$$H(v(\mathbf{x}), \mathbf{m}(\mathbf{x})) =$$

$$\frac{1}{V} \int \left[E(V + v(\mathbf{x}), \mathbf{M} + \mathbf{m}(\mathbf{x})) + \frac{C}{2} \sum_{i,j} (\nabla_j m_i)^2 + \frac{D}{2} (\boldsymbol{\nabla} v(\mathbf{x}))^2 \right] d^3x \quad (1)$$

Here $\mathbf{M}$ is the thermodynamic magnetization (mean magnetic moment per site), V is the thermodynamic volume (per site), and $\mathbf{m}(\mathbf{x})$, $v(\mathbf{x})$ are the local fluctuations of the magnetic moment or the volume, respectively. The subscripts i,j denote components along space directions. The function $E(...,...)$ is taken from band structure calculations as indicated above. C and D are constants which can in principle be derived from experiment.

Considering a classical limit only, the fluctuations are interpreted as classical coordinates; the partition function Z is then given by a functional integral over the phase space of fluctuations Γ

$$Z = \int \exp\left[-\beta H \{v(\mathbf{x}), \mathbf{m}(\mathbf{x})\}\right] d\Gamma \quad (2)$$

$\beta = 1/k_B T$, and the free energy results from

$$F(V, M, T) = -k_B T \ln Z \quad (3)$$

Z is then calculated in Gaussian approximation. For the final numerical evaluations a fit of the energy surface at $T = 0$, $E(V, M)$, to a polynomial is used

$$E(V, M) = \sum_{k,l} a_{kl} M^{2k} V^l \quad (4)$$

Thereby the theory becomes Ginzburg-Landau like.

Here is not the proper place to go into further details and to give detailed results of this theory, instead the interested reader is referred to the original papers [3, 11, 12] and also to a review by Staunton [13].

In this paper we want to present a check of this theory. For this purpose we have looked for a model system which can be solved exactly even for nonzero temperatures and at the same time is nontrivial, exhibiting a phase transition from an ordered to a disordered phase. It should then be possible to compare the results from an exact treatment with the results from the application of the above spin fluctuation theory to this model. It turns out that this is indeed attainable by means of the so-called simplified Hubbard or spinless Falicov-Kimball model in the limit of infinite space dimensions.

2. The model system

The Hamiltonian of the simplified Hubbard model is given by

$$H = t \sum_{\langle i,j \rangle} d_i^+ d_j + E_f \sum_i f_i^+ f_i + U \sum_i f_i^+ f_i d_i^+ d_i \qquad (5)$$

where d_i (d_i^+) annihilates (creates) a "band electron" or "d particle", the band being generated by the hopping term t, f_i (f_i^+) annihilates (creates) a "localized electron" or "f particle" with atomic energy E_f, and U is the on-site interaction between d and f particles. The sums $\sum_i$ and $\sum_{\langle i,j \rangle}$ run over all lattice sites or all nearest neighbour pairs of sites, respectively. The Hamiltonian H in (5) results from the Hamiltonian of the usual Hubbard model

$$H_{Hubbard} = \sum_{\langle i,j \rangle} t_{ij,\sigma} c_{i\sigma}^+ c_{j\sigma} + U \sum_i c_{i\uparrow}^+ c_{i\uparrow} c_{i\downarrow}^+ c_{i\downarrow} \qquad (6)$$

where $\sigma = \uparrow, \downarrow$ denotes the spin states of the electrons, if one assumes that electrons of one kind of spin ($\sigma = \uparrow$ e.g.) do not hop: $t_{ij\uparrow} = E_f \delta_{ij}$. Then with $t_{ij\downarrow} = t$ (i,j nearest neighbours), $t_{ij\downarrow} = 0$ (else) and upon introducing

$$f_i = c_{i\uparrow} \qquad d_i = c_{i\downarrow}$$

one gets (5).

Since the f occupation number operator $f_i^+ f_i$ commutes with the Hamiltonian H, the interaction $U \sum_i f_i^+ f_i d_i^+ d_i$ in (5) looks like a disordered potential for d particles. But the system is not a disordered problem if taken not as a "quenched", but as an "annealed" system where the trace in the grand partition function has also to be performed over all f occupation numbers [15].

For a bipartite lattice - that is a lattice which can be subdivided into two sublattices, A and B say, as for example a hypercubic lattice - it has been proven that for all dimensions $d \geq 2$ the ground state of the "symmetric" case, where $E_f = 0$ and the number of particles (d and f together) is equal to the number of lattice sites, exhibits a "superstructure" in which half of the particles are f particles and all of them occupy only the sites of one sublattice ("chess-board" state) [15]. For large on-site interaction U the existence of long-range order even at finite temperature T has been proven for $d = 2$ (square lattice) [19, 20]. If T is large there is no long range order so that the system must show a phase transition to a disordered state at some finite temperature. In Refs. [19, 20] also the "unsymmetric" case is considered. From these results one is led to the conclusion that at least in the neighbourhood of the "symmetric" case the model should show a phase transition which is connected with a break of the sublattice symmetry in the ordered phase. The order parameter can therefore be taken as

$$\delta w_f = w_A - w_B \tag{7}$$

where $w_{A(B)}$ is the mean occupation number of a site of sublattice A (B).

The proofs mentioned above are based on some mathematical reasoning and do not mean that the model is exactly solved. But there is an exact solution for the free energy of this model on a hypercubic lattice in the limit of infinite space dimensions due to Brandt and Mielsch [21, 22, 23]. This solution will be the basis of our procedure.

One might think that the limit of infinite dimensions leads to a trivial model as is the case for lattice spin models where it is known that for $d \to \infty$ mean field approximation becomes exact. Following this reasoning this limit should lead to the usual Hartree-Fock approximation for the simplified Hubbard model. But this is not true. A diagrammatic analysis of a perturbation expansion for the simplified Hubbard model with respect to the interaction U shows that there is an enormous simplification in the limit $d \to \infty$ since momentum conservation at the vertices is irrelevant and therefore the self-energy becomes independent of momentum (diagonal in space), however, energy conservation has to be observed so that the self-energy depends on energy (time) [16, 17]. Therefore despite of the simplification the limit $d \to \infty$ retains many of the features of the finite dimensional model, in particular also the phase transition.

The exact solution leads to a rather complicated set of equations that have to be solved numerically. The details of the calculating procedure can be found in the Appendix. But here we want to mention one feature which will play a certain role in what follows.

It turns out that in order to retain a non-trivial model in the limit of infinite dimensions, the hopping element t has to be rescaled as

$$t = \frac{t^*}{\sqrt{Z}} \tag{8}$$

where Z is the number of nearest neighbours: $Z = 2d$. The need for rescaling should be obvious from the model Hamiltonian H. It can rather easily be shown that with the choice as in (8) both kinetic and potential energy are kept finite and the density of states $\rho(\varepsilon)$ of the non-interacting system $(U = 0)$ acquires a Gaussian form

$$\rho(\varepsilon) = \frac{1}{\sqrt{2\pi t^*}} \exp\left(-\frac{\varepsilon^2}{2t^{*2}} + O\left(d^{-1}\right)\right) \tag{9}$$

The parameter t^* obviously has to be considered as an effective band width [16, 17].

3. The model system and "Invar"

As intended, we want to use this model to describe "Invar" behaviour. It is quite natural to choose the order parameter δw_f to play the role of the magnetic moment M; but one has also to introduce the volume V into the model. This is only possible by assuming at least one model parameter to depend on V. We have chosen t^* since the band width should depend on the lattice constant a and therefore on V. Generalizing an old argument by Heine [18] to general dimensions one gets

$$t^* \sim a^{-(d+O(d^{-1}))} \tag{10}$$

Therefore we choose

$$t^* = \frac{1}{V} \tag{11}$$

It is left to calculate the ground-state energy $E(V, \delta w_f)$ and the free energy $F(V, \delta w_f, T)$ for a fixed number of particles per site n (d and f together), the distribution among d and f particles (as well as the distribution of the latter among A and B sites) being determined by thermodynamics (annealed case). Accordingly, the parameters left are E_f, U, n.

As outlined in the Introduction, the energy surface $E = E(V, \delta w_f)$ should exhibit specific features in order to describe "Invar" behaviour. But it turns out that surfaces with these features cannot be obtained irrespective of the values which are chosen for the three parameters, because the energy behaves monotonic with respect to the parameter t^*. However one really gets there, if in addition to the band width t^* a second parameter in the Hamiltonian is assumed to depend on the volume V (or t^*, respectively). We have chosen $U = U(t^*)$ in the following way. What is desired is an energy surface which has a minimum with nonzero order pameter δw_f ("high moment" state) and a saddle point with $\delta w_f = 0$ ("low moment" state), these two points being connected by a curve in the δw_f-V plane

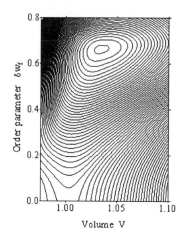
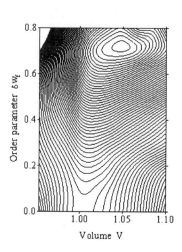

Figure 1. Energy surfaces $E = E(V, \delta w_f)$ for particle number value $n = 0.9$ and atomic energy values $E_f = -0.025$ and -0.05. The figures show contour lines. A minimum is at nonzero δw_f, a saddle point at $\delta w_f = 0$, $V = 1$.

with $\frac{\partial E}{\partial V}|_{\delta w_f} = 0$ (or $\frac{\partial E}{\partial t^*}|_{\delta w_f} = 0$) everywhere on the curve (zero "pressure" curve). With

$$\frac{\partial E}{\partial t^*}|_{\delta w_f} = \frac{\partial E}{\partial t^*}|_{U,\delta w_f} + \frac{\partial E}{\partial U}|_{t^*,\delta w_f} \cdot \frac{dU}{dt^*} = 0 \quad (12)$$

we have

$$\frac{dU}{dt^*} = -\frac{\frac{\partial E}{\partial t^*}|_{U,\delta w_f}}{\frac{\partial E}{\partial U}|_{t^*,\delta w_f}} \quad (13)$$

Since the right-hand side is a known function of t^*, U, δw_f, (13) can be used to fix the zero "pressure" line to some desired curve $\delta w_f = f(t^*)$ which can for example be taken from band structure calculations.

In Fig. 1 two examples of energy surfaces $E(V, \delta w_f)$ are shown which have been calculated for the two parameter sets $n = 0.9$, $E_f = -0.025$ and $n = 0.9$, $E_f = -0.05$. The figures show contour lines of these surfaces. As one can see, the surfaces have indeed the desired "Invar" features, moreover they are very similar to surfaces which are obtained from band structure calculations (see for example Refs [11, 9, 14]).

Fig. 2 shows the free energy surface for the parameter values $n = 0.9$, $E_f = -0.025$ at four different temperatures: $T = 0$, $0.5T_c$, $0.75T_c$, T_c, where T_c is the Curie temperature. The $T = 0$ surface is the same as the corresponding energy surface in Fig.1. From the figure it can be seen

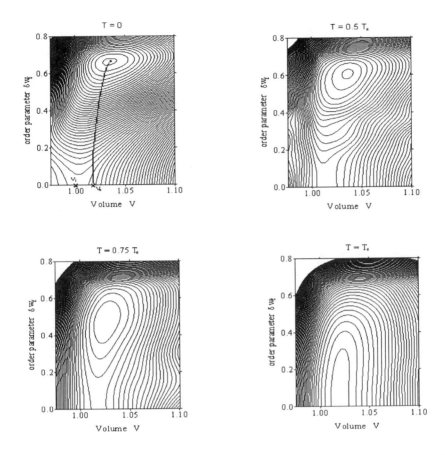

Figure 2. Free energy surface for particle number value $n = 0.9$ and atomic energy value $E_f = -0.025$ for temperatures $T = 0$, $T_c/2$, $3/4T_c$, T_c.

that with increasing temperature the equilibrium volume (= volume of the "high moment" state) is decreasing. At $T = T_c$ the minimum is at order parameter $\delta w_f = 0$ and volume $V = V_c$, the saddle point has disappeared. In the left top panel the line is sketched along which the volume decreases. It is obvious that the model shows "Invar" behaviour. Moreover, the surfaces are very similar to corresponding surfaces shown in Ref. [11] where the spin fluctuation theory, outlined in the Introduction, is applied to the system $Fe_3\text{-}Ni$.

Now we are in the position to perform a check of the spin fluctuation theory. From the exact free energy one gets all thermodynamic quantities precise (apart from numerical errors). In addition we apply the spin fluctuation theory to our model: as input we take the energy surface of our model and replace the moment M by the order parameter δw_f. In this

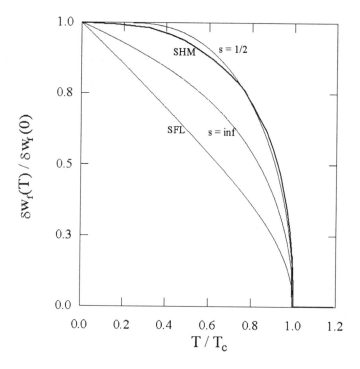

Figure 3. Order parameter as function of temperature (SHM: exact result, SFL: "spin" fluctuation theory result; also shown: Brillouin function curves for spins $s = 1/2$ and $s = \infty$)

way we have calculated the equilibrium values of the order parameter and of the volume as functions of temperature twice, viz. from the exact free energy and from the "spin" fluctuation theory. Fig. 3 shows the results for the order parameter δw_f. As can be seen the difference in temperature dependence turns out to be rather drastic. The curve for the spin fluctuation result even lies outside the region between the Brillouin function curves for the spin values $s = 1/2$ and $s = \infty$. Large differences are found also between the results for the volume as is shown in Fig. 4. It is true that the difference decreases with decreasing absolute values of the atomic energy E_f, but in any case the curvatures have opposite signs.

What might be the reason for this discrepancy? In spin fluctuation theory thermal fluctuations are treated as purely classical excitations and quantum excitations are completely left out. Since Invar behaviour is typically an effect at higher temperatures, this procedure seems to be justified, for classical excitations should dominate at higher temperatures, of course. But there is an abrupt transition from a quantum regime at zero temperature to a purely classical regime at all nonzero temperatures built in. It

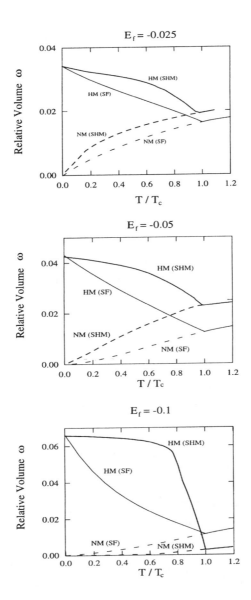

Figure 4. Relative volume $\omega = V - 1$ as function of temperature (HM: volume of the "high moment" state, NM: volume of the "low moment" state, SHM: simplified Hubbard model result, SF: "spin" fluctuation theory result, dashed curves: non-equilibrium results)

seems that it is this abrupt transition which influences the results also at higher temperatures. Morover, in the spin fluctuation theory the energy surface, that is a property at zero temperature, determines the whole thermodynamics of the system in a high degree. An alternative explanation might be that the transition point of the simplified Hubbard model lies within the quantum region, a comparison of both theories would be difficult then.

Thus through the above check questions arise which do not only refer to the spin fluctuation theory of Invar, but are of more general interest. To approach an answer to these questions we plan to apply our check to ferromagnetic non-Invar systems, for which the check should be easier.

4. Appendix

In this appendix we give a short outline of the procedure we used to calculate the free energy $F(V, \delta w_f, T)$ for the simplified Hubbard model in infinite space dimensions. For details see the papers by Brandt and Mielsch [21, 22, 23]. Our procedure is a slight generalization of their work.

It is convenient to turn over to the grand canonical ensemble where the grand potential is defined by the Hamiltonian

$$H = \frac{t^*}{\sqrt{2d}} \sum_{\langle i,j \rangle} d_i^+ d_j + E_f \sum_i f_i^+ f_i + U \sum_i f_i^+ f_i d_i^+ d_i - \mu \sum_i \left(d_i^+ d_i + f_i^+ f_i \right)$$

$$-2h \left(\sum_{i \in A} f_i^+ f_i - \sum_{i \in B} f_i^+ f_i \right) \qquad (14)$$

Here μ is the chemical potential and h is the field which is conjugate to the order parameter δw_f. $\sum_{i \in A(B)}$ denotes summation over the sites of sublattice A (B).

For infinite dimensions the grand potential $\Omega(\mu, h, V, T)$ per site is determined by the local self-energy

$$\Sigma_{ii}(i\omega_n) = \begin{cases} \Sigma_n^A & i \in A \\ \Sigma_n^B & i \in B \end{cases} \qquad (15)$$

where ω_n is the Matsubara frequency, $\omega_n = \pi/\beta \cdot (2n+1)$, and by the local Green's function G_n^v, $v = A, B$. Moreover in this limit local self-energy depends only on local Green's function. One has

$$\Sigma_n^v = \frac{U}{2} - \frac{1}{2G_n^v} \left(1 - \sqrt{1 + U^2 (G_n^v)^2 + 4UG_n^v \cdot \left(w_v - \frac{1}{2} \right)} \right) \qquad (16)$$

and
$$G_n^v = \frac{i\omega_n + \mu - \Sigma_n^v}{\tilde{z}_n^2} G_0(\tilde{z}_n) \quad (17)$$

Here the mean f occupation numbers w_A und w_B are given by
$$w_v = (1 + Y_v)^{-1} \quad (18)$$

$v = A, B$, with
$$\ln Y_v = \beta(E_f + h_v - \mu) - \sum_n \ln\left(1 - \frac{U}{\Sigma_n^v + (G_n^v)^{-1}}\right) e^{i\omega_n \delta} \quad (19)$$

and the free Green's function $G_0(z)$ is
$$G_0(z) = \frac{i\sqrt{\pi}}{t^*} w(z) \quad (20)$$

where $w(z)$ is the complex error function. The quantity $\tilde{z}_n$ is
$$\tilde{z}_n = \sqrt{(i\omega_n + \mu - \Sigma_n^A)(i\omega_n + \mu - \Sigma_n^B)} \quad (21)$$

In (19) $h_A = -2h$, $h_B = +2h$ and δ is a positive infinitesimal.

Equs. (16)-(21) form a complete set for the determination of the self-energy Σ_n^v and of the Green's function G_n^v as functions of μ, h, t^* (or V), and T. From these two quantities the grand potential $\Omega(\mu, h, V, T)$ can be calculated by the help of

$$\Omega(\mu, h, V, T) = \frac{1}{2\beta}\{\ln((1-w_A)(1-w_B)) - g(\mu, h, V, T)\} \quad (22)$$

where we have used the abbreviation

$$g(\mu, h, V, T) = \sum_n \left(\ln\left(X_n^A X_n^B\right) + \sum_\mathbf{k} \ln\left(\tilde{z}_n^2 - \epsilon(\mathbf{k})^2\right)\right) \quad (23)$$

Here
$$X_n^v = \Sigma_n^v G_n^v + 1 \quad (24)$$

$\epsilon(\mathbf{k})$ is the band energy and the summation over $\mathbf{k}$ has to be taken per site.

From $\Omega(\mu, h, V, T)$ one finally gets the free energy $F(V, \delta w_f, T)$ through the Legendre transformation

$$F(V, \delta w_f, T) = \Omega(\mu, h, V, T) + \mu n + h\delta w_f \quad (25)$$

where the chemical potential μ and the conjugate field h have to be expressed by the number of particles per site n and the order parameter δw_f by the help of

$$n = \frac{1}{2\beta} \sum_n \left(G_n^A + G_n^B\right) e^{i\omega_n \delta} + \frac{1}{2}\left(w_A + w_B\right) \qquad (26)$$

and

$$\delta w_f = w_A - w_B \qquad (27)$$

As can be imagined from this set of equations the numerical calculations are rather tedious and time-consuming [24].

Acknowledgement

The work has been supported within Sonderforschungsbereich 166 Duisburg-Bochum.

References

1. Moriya, T. (1985) *Spin Fluctuations in Itinerant Electron Magnetism*, Springer-Verlag, Berlin.
2. Murata, K. K. and Doniach, S. (1972) Theory of magnetic fluctuations in itinerant ferromagnets, *Phys. Rev. Lett.* **29**, 285-288.
3. Wagner, D. (1989) Fixed-spin-moment method and fluctuations, *J. Phys.: Condens. Matter* **1**, 4635-4642.
4. Schröter, M., Entel, P. and Mishra, S. G. (1990) Metallic magnetism and magnetic volume collapse, *J. Magn.Magn. Mater.* **87**, 163-176.
5. Wassermann, E. F. (1990) Invar: Moment-volume instabilities in transition metals and alloys in: K. H. J. Bushow and E. P. Wohlfahrt (eds.), *Ferromagnetic Materials*, Elsevier, Amsterdam, Vol. 5, p.237-322.
6. Williams, A. R., Moruzzi, V. L., Gelatt, C. D. Jr., Kübler, J. and Schwarz, K. (1982), Aspects of transition-metal magnetism, *J. Appl. Phys.* **53**, 2019-2023.
7. Moruzzi, V. L. (1990) High-spin and low-spin states in Invar and related alloys, *Phys. Rev. B* **41**, 6939-6946.
8. Moruzzi, V. L., Marcus, P. M., Kübler, J. (1989) Magnetovolume instabilities and ferromagnetism versus antiferromagnetism in bulk fcc iron and manganese, *Phys. Rev. B* **39**, 6957-6961.
9. Podgórny, M. (1992) Magnetic instabilities in $PtFe_3$ and in the fcc Ni-Fe system, *Phys. Rev. B* **46**, 6293-6961.
10. Weiss, R. J. (1963) The origin of the 'Invar' effect, *Proc. Phys. Soc. London* **82**, 281-288.
11. Mohn, P., Schwarz, K. and Wagner, D. (1991) Magnetoelastic anomalies in Fe-Ni Invar alloys, *Phys. Rev. B* **43**, 3318-3324.
12. Podgórny, M., Thon, M. and Wagner, D. (1992) Electronic structure and thermodynamic properties of Fe-Pt alloys, *J. Magn. Magn Mater.* **104-107**, 703-704.
13. Staunton, J. B. (1994) The electronic structure of magnetic transition metallic materials, *Rep. Prog. Phys.* **57**, 1289-1344.
14. Entel, P., Hoffmann, E., Mohn, P., Schwarz, K. and Moruzzi, V. L.(1993) First-principles calculations of the instability leading to the Invar effect, *Phys. Rev. B* **47**, 8706-8720.
15. Lieb, E.H. (1986) A model for crystallization: A variation on the Hubbard model, *Physica A* **140**, 240-250.

16. Metzner, W. and Vollhardt, D. (1989) Correlated lattice fermions in $d = \infty$ dimensions, *Phys. Rev. Lett.* **62**, 324-327.
17. Müller-Hartmann, E. (1989) Correlated fermions on a lattice in high dimensions, *Z. Phys. B* **74**, 507-512.
18. Heine, V. (1967) s-d Interactions in transition metals *Phys. Rev. B* **153**, 673-682.
19. Brandt, U. and Schmidt, R. (1986) Exact results for the distribution of the f-level ground state occupation in the spinless Falicov-Kimball model, *Z. Phys. B* **63**, 45-53.
20. Brandt, U. and Schmidt, R. (1987) Ground state properties of a spinless Falicov-Kimball model: Additional features, *Z. Phys. B* **67**, 43-51.
21. Brandt, U., Mielsch, C. (1989) Thermodynamics and correlation functions of the Falicov-Kimball model in large dimensions, *Z. Phys. B* **75**, 365-370.
22. Brandt, U., Mielsch, C. (1990) Thermodynamics of the Falicov-Kimball model in large dimensions II, *Z. Phys. B* **79**, 295-299.
23. Brandt, U., Mielsch, C. (1991) Free energy of the Falicov-Kimball model in large dimensions, *Z. Phys. B* **82**, 37-41.
24. Barar, V. (1994) Die Grundzustandsenergieflaeche des Falicov-Kimball-Modells im Grenzfall unendlich grosser Dimension (The ground-state energy surface of the Falicov-Kimball model in infinite dimensions) *Diplomarbeit, Bochum.*

MAGNETOVOLUME EFFECT AND LONGITUDINAL SPIN FLUCTUATIONS IN INVAR ALLOYS

A.Z.MENSHIKOV, V.A.KAZANTSEV, E.Z.VALIEV
AND S.M. PODGORNYKH
*Institute of Metal Physics, Ural Division RAS,
Ekaterinburg, RUSSIA.*

Abstract: The correlation between the large magnetovolume effect and longitudinal spin fluctuations in invar alloys is considered using γ-$(Ni_{0.9-x}Fe_x)Mn_{0.1}$ alloys and comparing their properties with γ-$Ni_{1-x}Fe_x$ alloys. We have performed thermal, magnetic and magnetoelastic measurements to calculate the spontaneous and forced volume magnetostriction, magnetization, and high-field magnetic susceptibility. The parameters of the magnetoelastic interaction are obtained and the relation between the spontaneous volume magnetostriction and longitudinal high-field susceptibility is discussed to shed more light on the invar problem.

1. Introduction

One hundred years ago Guillaume discovered the iron-nickel alloy $Fe_{65}Ni_{35}$ with a zero thermal expansion coefficient (invar alloy). Nowadays a lot of other alloys on the basis of γ-Fe with the similar physical properties are known. They are characterized by a very large spontaneous magnetostriction ω_s, as well as by the forced volume magnetostriction coefficient $d\omega/dH$. Moreover, a number of other properties show abnormal behavior in the invar region because of the mixed exchange interaction in γ-$Ni_{1-x}Fe_x$ alloys ($J_{NiNi}>0$, $J_{NiFe}>0$, $J_{FeFe}<0$) (see, e.g., reviews [1,2]). Therefore, it is very important to find some typical peculiarity, which could make the invars to be different from the conventional ferromagnetic alloys.

Now it is clear that the compensation of the thermal expansion in invar alloys has a magnetic nature due to a very large magnetovolume interaction. The magnetovolume effect in ferromagnets can be described by using the well known expression for the spontaneous magnetostriction (see e.g., [3])

$$\omega_s = \gamma_o B^{-1} M^2 = \gamma M^2, \qquad (1)$$

where γ_o is the constant of magnetovolume interaction, B is the bulk modulus, M is the magnetization, $\gamma = \gamma_o B^{-1}$.

Using formula (1) we can write the magnetic part of the thermal expansion coefficient α_M as a derivative of $d\omega_s/dT$,

$$\alpha_M(T) = \alpha_{exp} - \alpha_p = \gamma M (dM/dT), \quad (2)$$

where α_p is the thermal expansion coefficient in the paramagnetic state. This formula is suitable for the description of $\alpha_M(T)$ in the iron-nickel invar alloys, if the experimental M(T) dependencies are used [4,5], which however, doe not follow the Brillouin function. Moreover, the stiffness coefficient D in the spin-wave relation $E=Dq^2$ calculated from the low temperature dependence of the magnetization, is less by a factor of two, than the one determined from the inelastic neutron scattering measurements [6,7]. Therefore, besides the transverse spin-waves, some additional (or hidden) excitations must be suggested to describe the temperature dependence of the magnetization in invar alloys. Actually it was shown in [8] that the temperature dependence of the magnetization in iron-nickel invar alloys may be written as follows

$$M(T) = M(0)(1 - AT^{3/2} - ST^2), \quad (3)$$

where $AT^{3/2}$ term is caused by spin wave excitations and ST^2 must be related to some other magnetic excitations. The value of S is increased abruptly in the invar region of $Ni_{1-x}Fe_x$ alloys. To explain this experimental fact it was suggested [9] that the longitudinal spin fluctuations may be responsible for the abnormal behavior of M(T) in invar alloys. In this theory the temperature dependence of the magnetization in ferromagnets at temperatures $T/T_c \approx 0-0.5$ was written as follows [10]

$$M(T) \approx M(0)\left(1 - \frac{<m_\perp^2>}{M_0^2} - \frac{3}{2}\frac{<m_\parallel^2>}{M_0^2}\right), \quad (4)$$

where $M_0=M(0)$ is the magnetization at zero temperature, and $<m_\perp^2>$ and $<m_\parallel^2>$ are mean squared amplitudes of transverse and longitudinal magnetization fluctuations, respectively. The contribution of the temperature dependencies due to the transverse spin fluctuations may be described by the well-known formula of the spin-wave theory:

$$<m_\perp^2> = \mu_B M_0 \frac{(k_B T)^{3/2}}{(4\pi)^{3/2} D^{3/2}} F(3/2) = AT^{3/2} \quad (5)$$

where F(x) is the Rieman zeta function.

The contribution of the longitudinal fluctuations to the temperature dependence of the magnetization was written as follow [9]

$$<m_\parallel^2> = \mu_B M_0 \frac{(k_B T)^2}{2\pi D \hbar \Delta (r_c^{-1} + a)} +$$

$$+ \chi_\parallel \frac{\Delta B}{B} \left\{ \frac{(k_B T)^2}{4\pi l^2 \hbar \Delta (l^{-1} + a)} + \frac{(k_B T)^2}{l^2 \hbar c} \left[\frac{1}{12} - \frac{1}{2} u g(u)\right]\right\}, \quad (6)$$

where: $r_c = (\chi_\parallel \alpha)^{1/2}$ is the correlation radius of the longitudinal spin fluctuations,

$\alpha = D/2\mu_B M_0$; $\Delta = \alpha\beta^{-1}$ is the spin diffusion coefficient, β is a kinetic coefficient;
$a = \left[\left(\frac{\pi k_B T}{3} + \hbar\tau^{-1}\right) / \hbar\Delta\right]^{1/2}$; $\tau = \chi_{\parallel}\beta$ is the time of the homogeneous relaxation,

$u = \hbar\tau^{-1} / 2\pi k_B T$; $g(u) = 2u\left(\ln(u) - \frac{1}{2}u\psi(u)\right)$;

$\psi(u)$ - is Euler's psi function; l - is the radius of longitudinal fluctuations with account of the magnetoelastic interaction.

As we can see, the microscopic parameters of longitudinal spin fluctuations, r_c, Δ, and τ, depend on the longitudinal susceptibility $\chi_{\parallel}$. These dependencies may be obtained from the forced magnetostriction measurements. E.g., from (1) at a constant pressure, P = const., we get

$$\chi_p(T=0) = \left(\frac{d\omega}{dH}\right)(\gamma M_0)^{-1}, \tag{7}$$

where $(d\omega/dH)$ is the forced magnetostriction coefficient. Here $\chi_p \equiv \chi_{\parallel}$ is the susceptibility at a constant pressure, which coinsides with the longitudinal susceptibility. In order to determine the values of χ_p and γ we have to know the following values: ω_s, $d\omega/dH$, M_0 at T=0, as well as χ_{Hf} and χ_P since in the general case

$$\chi_{HF} = \chi_{\parallel} + \chi_{\perp} \tag{8}$$

The correlation between ω_s, $d\omega/dH$, M_0, and χ_{HF} depending on the iron concentration is very important for the γ-$Ni_{1-x}Fe_x$ system, where the invar effect takes place. However, we have limited information about concentrational dependencies of these values, because of the γ/α transformation at x=0.68. Manganese atoms are usually added to avoid this structural transformation [11,12], which does not affect the nature of an exchange interaction in ternary γ-FeNiMn alloys which is not changed basically due to the antiferromagnetic interaction between the manganese atoms ($J_{MnMn}<0$).

In the present work we consider the correlations between different properties of the invar γ-$(Ni_{0.9-x}Fe_x)Mn_{0.1}$ alloys, where the γ/α-transformation does not take place within the ferromagnetic region right up to the spin-glass transformation. We compare all the properties in this system with the similar ones in γ-$Ni_{1-x}Fe_x$ alloys to show the close relation between the magnetovolume effect and longitudinal spin fluctuations. In addition, we study the influence of Mn atoms on invar anomalies in γ-$Ni_{1-x}Fe_x$ alloys.

2. Experimental.

We have prepared the quasibinary γ-$(Ni_{0.9-x}Fe_x)Mn_{0.1}$ alloys with x varying from 0.2 to 0.5 by steps of 0.05, and from 0.5 to 0.65 by steps of 0.025 using the high purity constituents. All the prepared alloys were quenched from $1050°C$, and up to 4.2K had a face-centered cubic structure determined by x-ray and neutron diffraction measurements.

To determine the spontaneous magnetostriction, we have measured the thermal expansion of these alloys in the temperature range from 70 to 900K using the DL-1500 quartz dilatometer. The longitudinal magnetostriction was measured at 4.2K in the magnetic fields up to 9T, using the capacitance dilatometer.

To identify the magnetic state of these alloys we have measured the magnetization at 4.2K in the magnetic fields up to 9T, as well as the ac-susceptibility in the range from 4.2 to 300K. The Curie temperatures, T_c, of the alloys were determined from the magnetization measurements in a small magnetic field (kink-method).

3. Results

3.1. SPONTANEOUS AND FORCED MAGNETOSTRICTION.

Fig.1 shows both the experimental (points) and calculated (solid lines) temperature dependencies of the thermal expansion coefficient for some investigated alloys. One can see, that below the Curie temperature there exists a strong deviation of the experimental curves from the curves calculated from the Debye-Gruneisen equation. The difference between these two types of curves arises due to the spontaneous volume magnetostriction ω_s in the magnetic state [13]. Fig.2 shows the concentrational dependencies of ω_s for the investigated alloys compared to the similar data for γ-$Ni_{1-x}Fe_x$ alloys [14]. One can seen, that the magnetovolume effect in γ-$(Ni_{0.9-x}Fe_x)Mn_{0.1}$ alloys is lower than that in the binary alloys. Besides that, the iron concentrations at which the value ω_s has maxima in both systems, differ by about 15%.

The experimental field dependencies of the longitudinal magnetostriction $\lambda_\parallel = \Delta L/L$ are presented in Fig.3. The coefficient of the forced magnetostriction (dω/dH) calculated from the field dependencies of $\lambda_\parallel(H) = 1/3\omega(H)$ varies nonmonotonously with the iron concentration, passing through a maximum at x=0.55 (Fig.4). We can suppose that for γ-$Ni_{1-x}Fe_x$ alloys the maximum value of ω_s could occur at x=0.7- 0.72. However, this cannot be verified due to the absence of the γ-phase at low temperatures because of the γ/α transformation.

3.2. MAGNETIC PROPERTIES.

The magnetization curves, measured at 4.2K in the magnetic field up to 9T are shown for some alloys in Fig.5. These data allows us to calculate the average magnetic moment per atom $\overline{\mu}$, as well as the value of the high-field magnetic susceptibility χ_{HF}. For a comparison, the concentrational dependencies of $\overline{\mu}$ for γ-$(Ni_{0.9-x}Fe_x)Mn_{0.1}$ and γ-$Ni_{1-x}Fe_x$ alloys are shown in Fig.6. One can see that manganese strengthens decreasing of the average magnetic moment in the region of invar alloys.

In addition, we must notice, that the alloys with the concentration of iron from x=0.6 to 0.65 show the magnetization irreversibilities in high magnetic fields (Fig.5), which are characteristic of the reentrant spin-glass states (RSG). This experimental fact follows from the ac-susceptibility measurements as well. The very sharp maximum of the

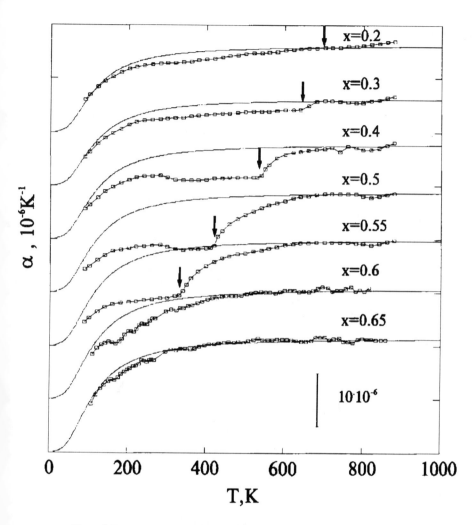

Figure 1. Temperature dependencies of the thermal expansion coefficient α for the γ-$(Ni_{0.9-x}Fe_x)Mn_{0.1}$ alloys. Arrows show the Curie temperatures of the alloys.

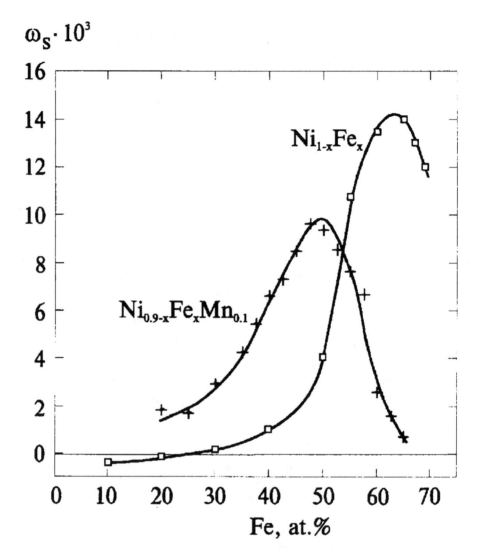

Figure. 2. Concentrational dependences of the spontaneous magnetostriction ω_s for the γ-$(Ni_{0.9-x}Fe_x)Mn_{0.1}$ alloys and for the γ-$Ni_{1-x}Fe_x$ alloys from [14].

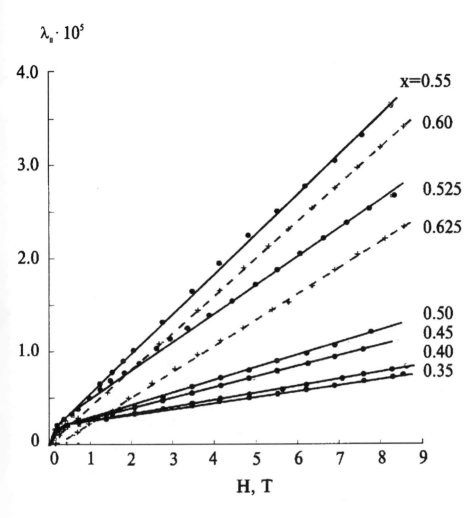

Figure.3. Longitudinal magnetostriction $\lambda_{\|}$ as a function of the external magnetic field for the γ-$(Ni_{0.9-x}Fe_x)Mn_{0.1}$ alloys at 4.2K.

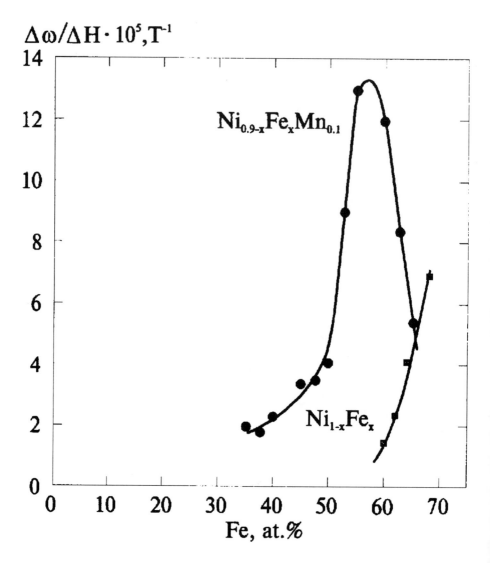

Figure.4. Concentrational dependencies of dω/dH at 4.2K for the γ-(Ni$_{0.9-x}$Fe$_x$)Mn$_{0.1}$ alloys and for the γ-Ni$_{1-x}$Fe$_x$ (• - data from [15]).

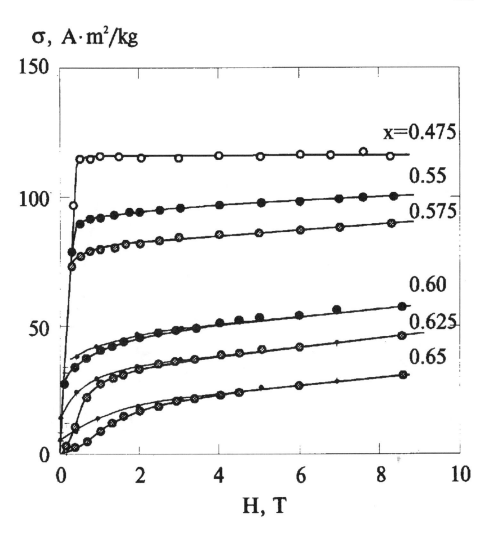

Figure.5. Magnetization curves for some γ-$(Ni_{0.9-x}Fe_x)Mn_{0.1}$ alloys. The solid and dashed lines are the curves which are measured in increasing and decreasing magnetic fields, respectively

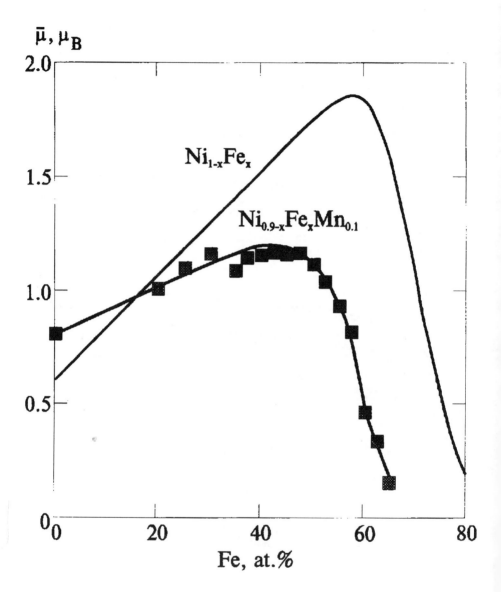

Figure 6. Concentrational dependence of the average magnetic moment per atom $\bar{\mu}$ in γ-(Ni$_{0.9-x}$Fe$_x$)Mn$_{0.1}$ alloys (this work) in comparison with the similar curve for γ-Ni$_{01-x}$Fe$_x$ alloys.

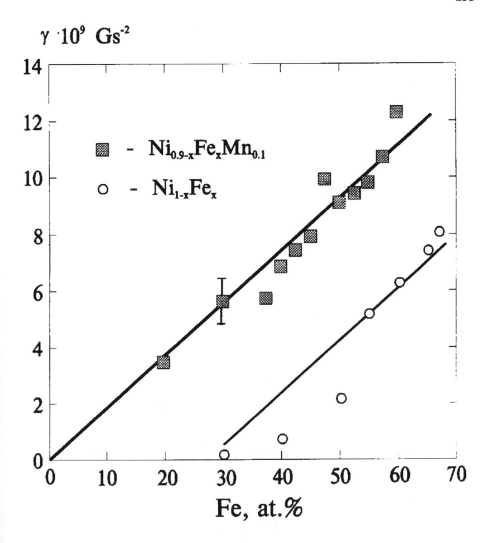

Figure 7. The constant of the magnetoelastic interaction $\gamma=\gamma_o B^{-1}$ as a function of the iron concentration in γ-$(Ni_{0.9-x}Fe_x)Mn_{0.1}$ and γ-$Ni_{1-x}Fe_x$ alloys.

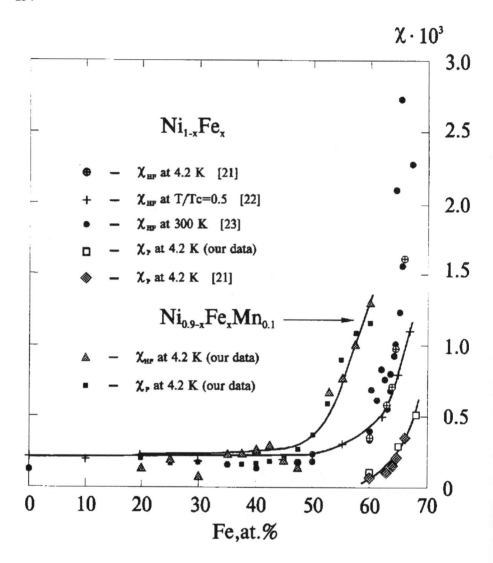

Figure 8. Concentrational dependencies of the measured high-field χ_{HF} and longitudinal χ_P susceptibilities calculated from the experimental data using formula (7).

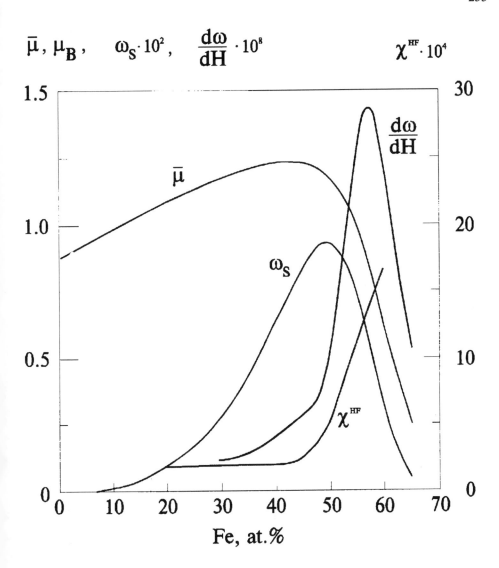

Figure 9. (a) The dependencies of $\bar{\mu}$, ωs, dω/dH, and χ_{HF} for γ-(Ni$_{0.9-x}$Fe$_x$)Mn$_{0.1}$ and γ-Ni$_{1-x}$Fe$_x$ on the iron concentration.

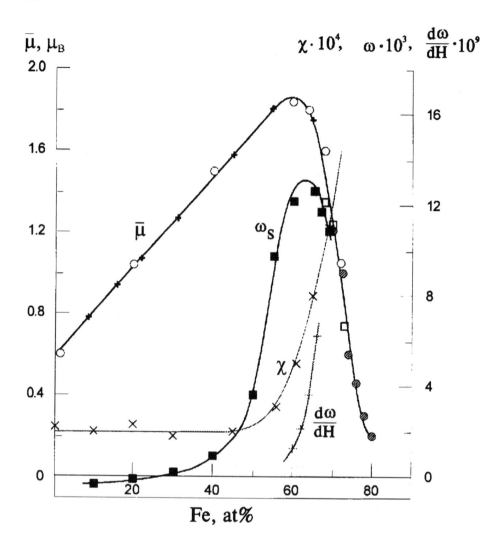

Figure 9. (b) The dependencies of $\bar{\mu}$, ω_s, $d\omega/dH$, and χ_{HF} on the iron concentration for the γ-$(Ni_{0.9-x}Fe_x)Mn_{0.1}$ alloys:
x – from [24], ° - from [25], - from [26], • - from [27].

ac-susceptibility occurs in the real χ' and imaginary χ'' parts of the dynamic susceptibility for the alloys with $x = 0.65$ and $x = 0.625$. Therefore, it is possible to conclude that the γ-$(Ni_{0.9-x}Fe_x)Mn_{0.1}$ alloys show a transition from a ferromagnetic to spin-glass state, as the iron concentration changes, within the same type of the crystal structure down to the low temperatures.

4. Discussion

The present experimental data allows us to calculate some parameters describing the magnetovolume effect in the invar alloys.

First of all, from $\omega_s(x)$ and $M(x)$ dependencies obtained at 4.2K we have determined the value of $\gamma = \gamma_0 B^{-1}$. It is seen from Fig.7, that γ increases linearly with the iron concentration in $(Ni_{0.9-x}Fe_x)Mn_{0.1}$ alloys, and its extrapolation to $x=0$ vanishes. Therefore, the magnetovolume effect in these alloys arises due to iron atoms only. An almost linear dependence of γ on x may be obtained for the binary γ-$Ni_{1-x}Fe_x$ alloys, if we take into account that the magnetovolume effect in these alloys vanishes at $x=0.25$. To our mind, these experimental facts support the theoretical results of [16,17], where the magnetovolume effect was caused by an instability of the iron local magnetic moment or by a coexistence of the low spin (LS) and high spin (HS) states of iron atoms in γ-$Ni_{1-x}Fe_x$ alloys.

Using the experimental data related to the forced magnetostriction $(d\omega/dH)$ at $T=4.2K$ we may calculate the susceptibility $\chi_p(T=0)$ from Eq. (7). The concentrational dependencies of χ_p are shown in Fig.8 for both types of alloys. The comparison of these susceptibilities with χ_{HF} obtained from the magnetic measurements in γ-$(Ni_{0.9-x}Fe_x)Mn_{0.1}$ and γ-$Ni_{1-x}Fe_x$ alloys are presented in Fig.8 as well. One can see, that in the first system χ_{HF} and χ_p are comparable. Taking into account the expression (8) we can conclude, that in γ-$(Ni_{0.9-x}Fe_x)Mn_{0.1}$ alloys $\chi_{HF} = \chi_p \equiv \chi_{||}$.

On the other hand, the longitudinal susceptibility in γ-$Ni_{1-x}Fe_x$ is much less than χ_{HF} because of the large transverse components of the total magnetic moment. The neutron polarization analysis [18] confirms this result. The transverse components of $\overline{\mu}$ arise due to the noncollinear magnetic structure in γ-$Ni_{1-x}Fe_x$ invar alloys [19]. As far as the iron magnetic moment and its direction depends on the nearest surroundings, it was suggested, that the ferromagnetic ground state of invar γ-$Ni_{1-x}Fe_x$ alloys is strongly affected by the magnetic inhomogeneities, which give rise to the transverse and longitudinal spin fluctuations. Probably, the centers of such fluctuations with a size from 10 to 15 $\overset{\circ}{A}$ are the iron atoms surrounded in the nearest coordination sphere only by iron atoms.

As to the magnetic state of the $(Ni_{0.9-x}Fe_x)Mn_{0.1}$ invar alloys we can conclude, that its magnetic structure is a collinear ferromagnet, since $\chi_{HF}=\chi_p$. However, in the region of $0.4<x<0.65$ it is inhomogeneous, because of the nonequivalent number of z-projections of total magnetic moments due to the strong antiferromagnetic interaction between Mn-Mn atoms. It was shown earlier [11,20], that in the ternary γ-FeNiMn alloys the exchange integral $J_{MnMn} \approx -300MeV$ is larger, than $J_{FeFe} \approx -9MeV$ in γ-$Ni_{1-x}Fe_x$ alloys. This means, that the probability of finding the iron atom in the zero molecular field is much more in γ-$Ni_{0.9-x}Fe_xMn_{0.1}$ alloys than in γ-$Ni_{1-x}Fe_x$.

5. Conclusion

To conclude, the properties of the γ-$Ni_{0.9-x}Fe_xMn_{0.1}$ alloys are investigated as functions of the iron concentration. The dependencies of $\overline{\mu}$, ω_s, $d\omega/dH$, and χ_{HF} are presented in Fig.9(a), where one can see a very distinct separation of different curves depending on the concentration scale. Anyhow, this physical picture does not hold for γ-$Ni_{1-x}Fe_x$ alloys, as it follows from Fig.9(b). The existence of different concentrational dependencies of ω_s, $d\omega/dH$ may be understood, if we take into account, that these quantaties depend on the magnetization in a different manner, $\omega_s \sim M^2$ and $d\omega/dH \sim M\chi_p$. As a result, the maximum for ω_s is close to the maximum of $\overline{\mu}$, but the maximal value of $d\omega/dH$ arises at a higher concentration due to a sharp increase of χ_p at a high iron concentration.

Finally, in the present work we did not use the χ_p experimental results for the calculations of the microscopic parameters of the longitudinal spin fluctuations, like Δ, r_c, and τ, due to the lack of the magnetization temperature dependencies analysis. We will do it carefully in the near future.

We gratefully acknowledge Mrs.. E.V. Scherbakova for magnetization measurements at low temperature. This work was supported by the Russian research program "Neutron Study of Condensed Matter".

References

1. Wasserman, E.F., (1990) Invar: Moment-Volume Instabilities in Transition Metals and Alloys, in K.H. Buchow and E.P. Wohlfarth (eds.), *Ferromagnetic Materials* **5**, North-Holland, Amsterdam, pp. 239-322.
2. Shiga, M. (1992) *Materials Science and Technology*, **3B**, p.II, 238-322.
3. Valiev, E.Z. (1991) Phenomenological Theory of Magnitoelastic Interaction in Invars and Elinvars,*Uspekhi Fiz. Nauk* **161(8)**, 87-128.
4. Menshikov, A.Z., (1989) On the Invar Problem, *Physica* B **161**, 1-8.
5. Zatoplyaev, A.K., Menshikov, A.Z. and Podgornykh, S.M. (1989) Magnetic Excitations and Invar Anomaly in γ-Fe-Ni-Pd Alloys, *Physica* **B161**, 25-28.
6. Ishikawa, Y. (1986) Neutron Studies of Hidden Magnetic Excitation inInvar Alloys, *Physica* **B136**, 451-454.
7. Lynn, J.W., Rosov, N., Acet, M. and Bach, H. (1994) Polarization Analysis of the Magnetic Excitations in $Fe_{65}Ni_{35}$ Invar, *J.Appl.Phys.* **75**, 6069-6071.
8. Yamada,O. and Nakai, I. (1981) The Magnetovolume Effect due to the Decrease of Local Magnetic Moment in Fe-Ni Invar Alloys, *J.Phys.Soc.Japan* **50**, 823-827.
9. Valiev, E.Z., Menshikov, A.Z. (1995) Longitudinal Spin Fluctuations (hidden magnetic excitations) in Invar Alloys, *JMMM* **147**, 189-191.
10. Lonzarich, G. and Taillefer, L. (1985) Effect of Spin Fluctuations on the Magnetic Equation of Stateof Ferromagnetic or near Ferromagnetic Metals,*J.Phys.C.:Sol.State Phys.* **18**, 4339-4371.
11. Menshikov, A.Z., Kazantsev, V.A., Kuzmin, N.N.and Sidorov, S.K. (1975) Exchange Interaction and Magnetic of Iron-Nickel-Manganese Alloys,*JMMM* **1**, 91-97.
12. Menshikov, A.Z., Kazantsev, V.A.,Kuzmin, N.N. (1976) Magnetic State of Iron-Nickel-Manganese Alloys, *Zh. Exsp. Theor. Fiz.* **71**, 648-656.
13. Grishkin, V.Yu., Kazantsev, V.A., Menshikov, A.Z. and Podgornykh, S.M. (1996)Magnetovolume Effects in Invar γ-$(Ni_{0.9-x}Fe_x)Mn_{0.1}$ Alloys, *Phys. Met. Metallogr.* **82**, 65-70.

14. Zakharov, A.I., Menshikov, A.Z.and Uralov, A.S. (1973) Thermal Expansion Coefficient of Fe-Ni Alloys with FCC lattice, *Fiz. Met. Metalloved.* **36**, 1306-1310.
15. Fujimori, Y.and Shiga, M. (1978) Thermal expansion and Magnetovolume Effects of Invar Alloys, *Physics and Applications of Invar Alloys*, Tokyo, pp. 80-122.
16. Moruzzi,V.L. (1990) High-Spin and Low-Spin States in Invar and Related Alloys, *Phys. Rev.* **B41**, 6939-6946; (1992) *Solid State Commun*, **83**, 739.
17. Mohn, P., Schwarz, K., Wagner, D. (1991) Magnetoelastic Anomalies in Fe-Ni Invar Alloys, *Phys. Rev.* **B43**, 3318-3324.
18. Menshikov, A.Z. and Schweizer, J. (1996) The Transverse Magnetization Components in the Ground State of Invar γ-$Ni_{1-c}Fe_c$ Alloys, *Solid State Commun.* **100**, 251-255.
19. Menshikov, A.Z., Sidorov, S.K.and Arkhipov, V.E. (1971) Magnetic Structure FCC Iron-Nickel Alloys, *Zh.Eksp. Teor. Fiz.* **61**, 311-319; Menshikov, A.Z., Shestakov, V.A. (1977) Magnetic Inhomogeneities in Invar Iron-Nickel Alloys, *Fiz.Met.Metalloved.* **43**, 722-733.
20. Menshikov, A.Z., Shestakov, V.A., Sidorov, S.K. (1976) The Critical Neutron Scattering in Invar Iron-Nickel Alloys, *Zh.Eksp.Teor.Fiz.* **70**, 163-171.
21. Hiroyoshi, H., Fujimori, H. and Saito, H. (1971) High-Field Susceptibility of Invar Alloys, *J.Phys.Soc.Japan.* **31**, 1278.
22. Romashov, L.N., Menshikov, A.Z.and Fakidov, I.G. (1970) The Susceptibility of Paraprocess of FCC Fe-Ni Alloys in High Magnetic Fields, *Fiz. Tverd. Tela* **12**, 2758-2761.
23. Shimizu, M. (1979) Origin of the Anomalies and Thermodynamic Aspects in Fe-Ni, *JMMM* **10**, 231-240.
24. Kondorskiy, E.I. and Fedotov, L.N. (1952) Magnetization Saturation Dependence on Temperature for Binary Iron-Nickel Alloys in Low Temperature Region,*Izv Acad.Nauk SSSR* **16**, 432-437.
25. Crangle, J.and Hallam, G.C. (1963) Temperature Dependence of the Invar Alloys Magnetization, *Proc.Roy.Soc.* **A272**, 119-128.
26. Asano, H. (1969) Magnetization of γ-Fe-Ni Invar Alloys, *J.Phys.Soc.Jap.* **27**, 542-549.
27. Bando, Y. (1964) The Magnetization of Face-Centered Cubic Iron-Nickel Alloys in the Vicinity of Invar Region, *J.Phys. Soc.Jap.* **19**, 237-246.

HIGH TEMPERATURE THERMAL EXPANSION OF RMn_2 INTERMETALLIC COMPOUNDS WITH HEAVY RARE EARTH ELEMENTS

I.S. DUBENKO
MIREA,
117454 Moscow, Russia.
I.YU. GAIDUKOVA, S.A. GRANOVSKY, R.Z. LEVITIN,
A.S. MARKOSYAN, A.B. PETROPAVLOVSKY, V.E. RODIMIN,
AND V.V. SNEGIREV.
M.V. Lomonosov Moscow State University,
119899 Moscow Russia.

Abstract. The temperature variation of the thermal expansion coefficient, α, of cubic and hexagonal Laves phase compounds RMn_2 (R = Gd ÷ Lu and Y) was measured in a wide temperature range 5-900 K by the X-ray diffraction and dilatometric methods. The value of α is substantially enhanced, up to 55×10^{-6} K^{-1}, in the paramagnetic region. At high temperatures, 600-800 K, α passes over a maximum and decreases with a further increase of temperature. Some change in $\alpha(T)$ depending on the type of the rare earth ion was discovered suggesting the instability of the Mn electronic structure related to the strong temperature dependence of the mean squared amplitude of spin fluctuations.

1. Introduction

Thermal expansion is one of the most important parameters widely used when discussing the features of 3d-magnetism in intermetallic compounds and metal alloys. In the magnetically ordered state the magnetovolume effect (volume magnetostriction) is proportional to the squared 3d moment. Hence, its magnitude, temperature and field dependencies can characterize the magnetic behavior of the 3d-electron system [1, 2]. The magnetic contribution to the thermal expansion is considered arising due to the strong dependence of the interatomic distances on the d-band width, which leads to a large positive magnetovolume effect, $\Delta V/V > 10^{-3}$ [3]. In the ordered state the magnetic part of the thermal expansion coefficient α_{mag} is proportional to the spontaneous magnetization of the d-electron system. In strongly correlated 3d-electron systems, where fluctuating local atomic moments persist above T_C, the importance of the magnetovolume effect in studying the temperature dependence of the spin-fluctuation

(SF) amplitude in the paramagnetic temperature region was emphasized by many authors [4 - 7]. The temperature variation of the mean squared amplitude of SF, $<m_{sf}^2>$, leads to a temperature dependent contribution to the thermal expansion coefficient in the paramagnetic temperature range [4, 6]. According to the SF theory, at moderate temperatures $<m_{sf}^2>$ is proportional to T and tends to saturate at higher temperatures; this region is determined by a characteristic SF temperature T_{sf}. With increasing temperature α_{mag} is temperature independent and then at higher temperatures asymptotically reduces to zero.

This concept was used to account for the giant thermal expansion observed in the Mn-rich pseudobinary alloys $Y(Mn_{1-x}Al_x)_2$ and to derive qualitatively the temperature variation of the mean squared SF amplitude in the undiluted compound YMn_2, which was found to be in agreement with the inelastic neutron scattering experiments [8, 9]. The value of α_{mag} in YMn_2 was found comparable to the lattice contribution α_{lat} in the paramagnetic temperature range. Recently strongly enhanced thermal expansion coefficients, up to 55×10^{-6} K^{-1}, were observed in the paramagnetic region in other binary Y-Mn intermetallic compounds (YMn_{12}, Y_6Mn_{23} and the $Y_6(Mn_{1-x}Fe_x)_{23}$ pseudobinary system), too [7, 10]. This value is significantly larger than α observed in other well-known spin fluctuation systems, such as YCo_2, Pd, FeSi, MnSi, etc., where α does not exceed $(18 \div 20) \times 10^{-6}$ K^{-1} [11, 2, 5]. This makes it possible to follow the anomalies of $\alpha(T)$ caused by the SF contribution.

The anomalously high thermal expansion observed in the binary Y-Mn intermetallics was attributed to the instability of the 3d-electron configuration of Mn-ions [7]. The strong temperature dependence of α can be treated as if the mean ionic radius of Mn is temperature dependent providing a new contribution to the thermal expansion coefficient due to the temperature variation of the Mn ionic radius (this situation could resemble the intermediate valence case observed in some 4f-compounds being its "itinerant" analog). The thermal expansion data show that the SF excitations in the Mn intermetallics have essentially different temperature dependence compared to the "normal" temperature induced ones predicted by the conventional SF theory [7].

The above suggestion is to be checked by experiments in other Mn compounds showing enhanced thermal expansion. The aim of the present work is to study the thermal expansion coefficient in the RMn_2 intermetallics with heavy rare earths, RE. The RMn_2 compounds crystallize in two different structures, $MgCu_2$-type (with RE= Gd - Ho, and Y) or $MgZn_2$-type (with Ho - Lu), cubic and hexagonal type structures, respectively [12]. The systematic study through this series permits to find out the influence of different RE metals and crystal structure on the thermal expansion in the paramagnetic high temperature range.

2. Experimental Part and Discussion

Polycrystalline samples of YMn_2 and RMn_2 with R= Gd, Tb, Dy, Er, Tm and Lu were prepared by induction melting under argon gas atmosphere and subsequently homogenized for one week at 720°C. The cubic modification of HoMn$_2$ was used in the present study, which crystallized when the ingot was quenched after homogenization.

Thermal expansion was measured by the X-ray diffraction (in the range 5 ÷ 900 K) and by dilatometric (300 ÷ 900 K) methods. The X-ray data for the hexagonal compounds were reduced to the effective cubic cell parameter using the formula $a_{eff} = \sqrt[3]{\sqrt{3}a^2 c}$, and all the dilatometric measurements were normalized to the room temperature lattice parameter in order to compare them with the X-ray measurements.

Figure 1 shows the temperature dependence of the cubic lattice parameter of the RMn$_2$ compounds. A large jump-like volume expansion observed for YMn$_2$, GdMn$_2$ and TbMn$_2$ is attributed to a formation of a long range order in the Mn-sublattice (due to a first-order type phase transition). The value of the magnetovolume effect and the characteristic temperature of the phase transition are in agreement with those referred in [13, 14, 6]. In the paramagnetic region all the temperature dependencies are essentially enhanced. This enhancement is more clearly seen in Fig.2, which shows the thermal expansion coefficients of the RMn$_2$ compounds as a function of temperature. As can be seen from Fig.2 the α-value is noticeably higher than those observed in most of the 3d-intermetallic compounds and alloys (in which it normally does not exceed 20×10^{-6} K^{-1}) [2, 15]. In all the compounds studied the thermal expansion coefficient increases with increasing temperature and passes over a maximum (with probably TmMn$_2$ left out). Although several factors can influence the magnitude of T_{max}, at which the maximal value of α is reached, a general tendency for increasing of T_{max} with increasing RE atomic number can be indicated (see the inset in Fig.2). At the temperature T_{max} the values of α are in the range between 37×10^{-6} K^{-1} (GdMn$_2$) and 56×10^{-6} K^{-1} (DyMn$_2$).

The total thermal expansion coefficient of metallic compounds can be written in an additive form as a sum of different contributions:

$$\alpha_{tot} = \alpha_{lat} + \alpha_{el} + \alpha_{mag}$$

Here the subscript indices in different terms denote "total", "lattice", "electronic", and "magnetic" contributions, respectively. In the Debye theory the lattice contribution α_{lat} is independent on the temperature roughly above the Debye temperature; the electronic contribution, α_{el}, is proportional to the temperature.

In the paramagnetic state α_{mag} is determined by the temperature dependence of the mean squared SF amplitude, and the non-linear behavior of $\alpha_{tot}(T)$ in the paramagnetic region as well as the existence of a characteristic temperature T_{max} should be attributed to this term. Following the arguments given in [7] for the Y-Mn intermetallics it can be assumed that the anomalously large thermal expansion in the binary RMn$_2$ intermetallics with magnetic RE arises due to an instability of the Mn electronic structure giving rise to a new contribution to the thermal expansion. Therefore, α_{mag} consists of two contributions, the first one caused by the temperature induced fluctuations of spin density in the hybridized (3d-5d)-band, and the second one being due to the temperature dependence of the effective Mn ionic radius.

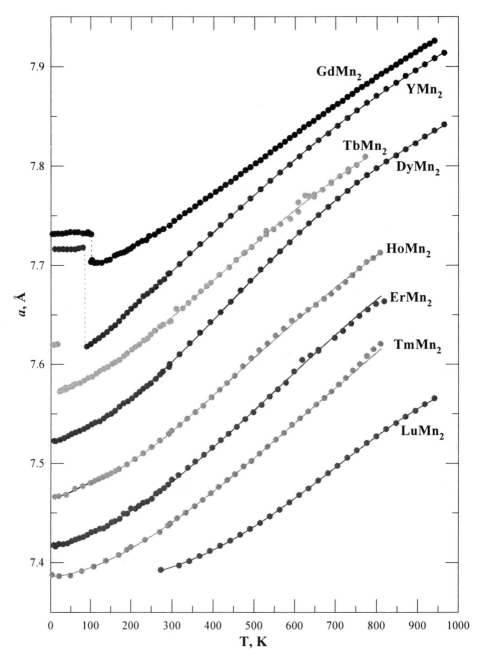

Figure 1. Temperature dependence of the linear thermal expansion of the RMn$_2$ compounds.

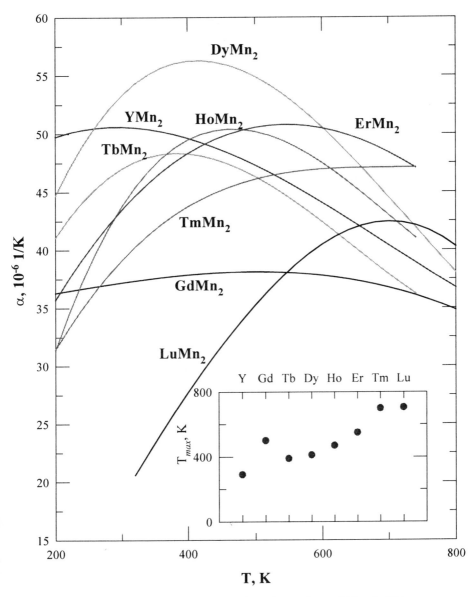

Figure 2. Temperature dependence of the thermal expansion coefficient for RMn_2. The inset shows the variation of T_{max} as a function of the RE element.

The situation in this "unstable" case must be different from the intermediate valence case arising due to the 4f-5d hybridization with two well-defined $4f^n$ and $4f^{n+1}$ valence states. In the case of Mn-intermetallics there shall exist a distribution of different 3d-electronic states which can easily exchange electrons with each other as well as with the d-band. The Mn electronic structure instability can account for the large low temperature SF amplitude of the order of 1 μ_B observed by inelastic neutron scattering experiments in some paramagnetic YMn_2-based intermetallics with small Y-substitutions [8]. The substantial increase of $<m_{sf}^2>$ up to about 2 μ_B at 300 K corresponds to additional two 3d-core electrons, which may not be excluded when considering the Mn ionic radii as a function of temperature or substitutions.

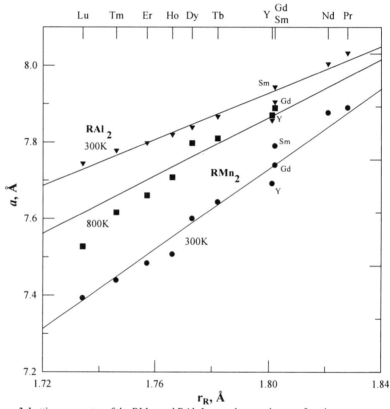

Figure 3. Lattice parameter of the RMn_2 and RAl_2 Laves phase series as a function of the RE atomic number at 300K and RMn_2 at T=800K.

By comparing the variation of the lattice parameters at the room temperature of different RM_2 compounds (M = Al, Mn, Fe, Co and Ni) vs ionic radius of R^{3+} [7] it was shown that the Mn ionic radius cannot be considered as a constant throughout the RMn_2 series. We assume that this dependence can be due to different sizes of the Mn coordination polyhedra in these compounds. Within this assumption T_{max} determined in this work means the temperature above which the electronic configuration of Mn becomes the same in different RMn_2 compounds. As it is shown in Fig.3, the variation of the lattice parameter vs RE radius in the RMn_2 series more resembles the normal behavior at 800 K and does not appreciably differ from that in the other RM_2 series at room temperatures.

Acknowledgments

This work was supported by RFBR, project No 96-02-16846, 96-02-16373 and INTAS foundation, project No 96-630

References

1. Shiga, M. (1993) Magnetic Ordering and Frustration in Itinerant Systems: RMn_2, *JMMM* **129**, 17-25.
2. Andreev, A.V. (1995) *Handbook on Magnetic Materials*, **8**, 59-175, Elsevier Sci. Publ. B.V., North-Holland, Amsterdam.
3. Shimizu, M. (1981) Intinerant Electron Magnetism, *Rep. Prog. Phys.*, **44**, 329-409.
4. Morya, T. (1985) *Spin Fluctuations in Itinerant Electron Magnetism*, Springer, Berlin.
5. Takahashi, Y. and Morya, T. (1983) On the Spin Fluctuations in Weak ItinerantFerromagnets, *J. Phys. Soc. Jap.*, **52**, 4342-4348.
6. Shiga, M. (1987) Magnetism and Spin Fluctuations of Laves Phase Manganese Compounds,*Physica B*, **BC 149**, 293-305.
7. Dubenko, I.S., Gaidukova, I.Yu., Granovsky, S.A., Gratz, E., Gurjazkas, D., Markosyan, A.S., and Muller, H. (1997) Strongly Enhanced Thermal Expansion in Binary Y-MnIntermetallic Compounds YMn_2, Y_6Mn_{23} and YMn_{12}, *Sol. State Commun.*, **103**, 495-499.
8. Shiga, M., Wada, H., Nakamura, H., Deportes, J., Ouladdiaf, B., andZiebeck, K.R.A. (1988) Giant Spin Fluctuations in YMn_2 and Related Compounds, *J. Phys. Col.*. **C8-49**, 241-246.
9. Shiga, M., Wada, H., Nakamura, H., Yoshimura, K., and Nakamura, Y., (1987) Characteristic Spin Fluctuations in $Y(Mn_{1-x}Al_x)_2$, *J. Phys. F.*, **17**, 1781-1793.
10. Dubenko, I.S., Granovsky, S.A., Gratz, E., Levitin, R.Z., Lindbaum, A., and Markosyan, A.S. (1996) Enhanced Paramagnetic Thermal Expansion of theIntermetallic Compounds $Y_6(Mn_{1-x}Fe_x)_{23}$, *J. Magn. Magn. Mater.*, **157/158**, 629-630.
11. Gratz, E. and Lindbaum, A. (1994) The Influence of the Magnetic State on the Thermal Expansion in 1:2 Rare Earth Intermetallic Compounds, *JMMM* **137**, 115-121.
12. Iandelli, A. and Palenzona, A. (1979) *Handbook on the Physics and Chemistry of Rare Earths*, **5**, Elsevier Science Publishers B.V., North-Holland, Amsterdam.
13. Gaidukova, I.Yu., Kovalev V.V., Markosyan, A.S., Menshikov A.Z., and Pirogov, A.N. (1988) On the Nature of the Phase Transition in YMn_2, *JMMM* **72**, 357-359.
14. Gaidukova, I.Yu., Krugliashov, S.A., Markosyan, A.S., Levitin, R.Z., Pastushenkov, Yu.G., and Snegirev V.V. (1983) Manganese Subsystem Metamagnetism in RMn_2 Intermetallic Compounds, *Zh. Eksp. Theor. Fiz.*, **84**, 1858-1867.
15. Wasserman, E.F. (1990) *Ferromagnetic Materials*, **5**, 237-322,Elsevier Science Publishers B.V., North-Holland, Amsterdam.

MAGNETOELASTICITY AND ISOTOPE EFFECT IN FERROMAGNETS

V.M. ZVEREV
*P.N. Lebedev Physical Institute
of the Russian Academy of Sciences,
117924 Moscow, Leninsky pr., 53, Russia*

Abstract. A phenomenological approach to the theory of the effects on the magnetic properties of ferromagnets due to acustic phonons is reviewed. The approach is based on the phenomenon of magnetoelasticity. The parameters of the approach are determined from the experimental data for Fe, Ni and for Fe-Ni, Fe-Ni-Mn, Fe-Pt Invar alloys. Quantitative calculations of the contributions of acoustic phonons to the Curie constant, Curie temperature and to the isotope shift of the Curie temperature of these ferromagnets are given. It is shown that thermal phonons have an anomalously strong effect on the Curie constant and the Curie temperature of the Invar alloys Fe-Pt due to shear magnetoelasticity. The isotope shift of the Curie temperature in Fe-Pt Invar alloys is comparatively small.

1. Introduction

Traditionally, the influence of the interaction of electrons with lattice vibrations on magnetism is considered to be relatively small (see, e.g., [1]). This conventional point of view bases on the fact that the contribution of the electron-phonon interaction to the magnetic susceptibility is relatively small, because the relevant parameter for this contribution, namely the ratio $\kappa_B \Theta / \varepsilon_F$, where Θ is the Debye temperature and ε_F is the Fermi energy, is small [1]. We note that the effect of ferromagnetism on the lattice properties is not taken into account in this consideration. A similar correction due to electron-phonon interaction appears in the Stoner exchange-enhancement factor

$$S = [1 + 2\nu(\psi + \psi_{\text{el-ph}})]^{-1}, \tag{1}$$

which defines the increase of the magnetic susceptibility in comparison with the susceptibility of the noninteracting electron gas. Here ν is the electron density of states at the Fermi level, ψ is the electron exchange coupling constant, and $\psi_{\text{el-ph}}$ results from the dependence of the energy of the zero-point lattice vibrations on the electron spin polarization [2],

$$\psi_{\text{el-ph}}/\psi \sim \kappa_B \Theta / \varepsilon_F \sim \sqrt{m_e/M_i}, \qquad (2)$$

Here m_e is the electron mass and M_i is the mass of the atoms in the crystal lattice. According to Hopfield [3] the electron-phonon interaction can play an important role in the case of very weak itinerant ferromagnets with strong exchange enhancement:

$$|S| \gg 1. \qquad (3)$$

The prediction [3] of a very large isotope effect on the Curie temperature of weak ferromagnets of $ZrZn_2$ type based specifically on this condition. The result in [3] becomes obvious from (1)-(3), if we take into account the following relation for the Curie temperature: $T_C \propto |S|^{-\gamma}$, where $\gamma = 1/2$ in the Stoner model [4] and $\gamma = 3/4$ in the paramagnon model of a ferromagnetic metal (see, e.g., Refs. [5] and [6]). We then obtain (c.f. [3])

$$I = \frac{d\ln T_C}{d\ln M_i} = \gamma S \frac{\psi_{\text{el-ph}}}{2\psi}. \qquad (4)$$

Hopfield [3] suggested to use the following parameters for $ZrZn_2$: $\gamma = 1/2$, $S \simeq -238$, and $\psi_{\text{el-ph}}/\psi \sim 0.1$. This leads to $|I| \gg 1$. However, experiments [7] did not confirm this prediction and gave the value $I = -0.1 \pm 0.3$ for the Zn isotopes and $I = -0.2 \pm 0.2$ for the Zr isotopes in $ZrZn_2$. This disagreement between theory and experiment can be attributed to the wrong use of the value $\psi_{\text{el-ph}}/\psi \sim 0.1$ to estimate the isotope coefficient I in $ZrZn_2$ in Ref. [3]. Moreover, according to Fay and Appel [2] and to Pickett [8], attempts to calculate $\psi_{\text{el-ph}}$ theoretically lead only to a rough estimate and do not allow to consider the contribution of the electron-phonon interaction to the Stoner factor (1). Therefore we use a phenomenological approach to the theory of the isotope effect in ferromagnets, in which the parameters of the theory can be related to experimentally measured quantities and can be estimated for real ferromagnets. The usefulness of this approach will become clear in connection with the recent experimental discovery of a giant isotope shift $\Delta T_C > 20$ K of the Curie temperature in the perovskite $La_{0.8}Ca_{0.2}MnO_{3+y}$, when the ^{16}O oxygen isotope is replaced by ^{18}O, for which the isotope coefficient is $I \simeq -0.85$ [9]. This discovery is a direct indication of the existence of such ferromagnets for which the electron-phonon interaction plays an important role.

A phenomenological approach to the theory of the influence of acoustic phonons in ferromagnets was proposed in [10, 11]. The theory bases on the phenomenon of magnetoelasticity. This phenomenon is manifested as a dependence of the elastic moduli of a ferromagnet on the magnetization, which leads to an analogous dependence of the Debye temperature. Therefore, the contribution of acoustic phonons to the free energy of a ferromagnet depends on the magnetization. This is the reason why acoustic phonons are manifested in the magnetic properties of ferromagnets. Within our approach it was possible to find a relatively small set of experimentally measurable parameters, which enable us to calculate the contribution of acoustic phonons to various magnetic properties of ferromagnets and to study the role of phonons in magnetism in general.

In this communication we present a review of the main results of the phenomenological approach to the theory of the influence of acoustic phonons on the magnetic properties of ferromagnets. The parameters of the approach are determined from an analysis of the available experimental data for the metals Fe, Ni and for the Fe-Ni, Fe-Ni-Mn, and Fe-Pt Invar alloys. We present quantitative calculations of the contributions of acoustic phonons to the Curie constant, the Curie temperature and to the isotope shift of the Curie temperature of these ferromagnets . We also analyze the conditions for observing the isotope shift of the Curie temperature in ferromagnets with high and low Curie temperatures.

2. General relations

Here we focus on magnetoelasticity which plays a fundamental role below. For numerous ferromagnets (see, for example, Ref. [12]) the dependence of the elastic moduli on the magnetization has a very simple form. The model of an isotropic elastic medium which is characterized by a bulk modulus K and a shear modulus G is frequently used to describe ferromagnets. For K and G at constant magnetization M, we can write [12]

$$K_M(T, M) = K(T) + K'M^2, \qquad G_M(T, M) = G(T) + G'M^2, \quad (5)$$

Here $K(T)$ and $G(T)$ are the elastic moduli of the paramagnetic state, which depend on temperature due to the anharmonicity of phonons. The values of the elastic moduli $K(T)$ and $G(T)$ in the ferromagnetic phase are found by extrapolation from the paramagnetic region, where they exhibit a normal temperature dependence (see, e.g., [13]). It is important to stress that, according to [12], the magnetoelastic coefficients K' and G' for most of ferromagnets are independent of temperature.

In experimental investigations of the elasticity of ferromagnets the elastic moduli are usually measured at constant magnetic field H. The bulk modulus at constant magnetic field K_H is related to K_M by the well-known thermodynamic relation [13]

$$\frac{1}{K_H(P,T)} - \frac{1}{K_M(P,T)} = \frac{1}{\chi_P}\left(\frac{\partial \omega}{\partial H}\right)_{P,T}^2. \qquad (6)$$

Here $\chi_P = (\partial M/\partial H)_{P,T}$ is the isothermal paraprocess magnetic susceptibility at constant pressure P, and $(\partial \omega/\partial H)_{P,T} = V^{-1}(\partial V/\partial H)_{P,T}$ is the forced volume magnetostriction, where V is the volume of the ferromagnet. Within the phenomenological approach relations (5) and (6) can be used to determine the dependence of the bulk modulus K_M on the magnetization and to find the bulk magnetoelastic coefficient K' (see, e.g., [10] and [13]). The experimental data for the remaining quantities appearing in (5) and (6) are used for this purpose. We stress that the shear modulus at constant magnetic field, G_H, does not differ from the shear modulus at constant magnetization G_M. This simplifies the determination of the shear magnetoelastic coefficient G' on the basis of the experimental data [10].

As was shown in [14] existing microscopic theories of the elasticity of ferromagnetic metals [15] do not distinguish between different bulk moduli. We also note that since the elasticity of a lattice is associated with the Debye screening of the Coulomb field in the itinerant-electron model, the Debye screening radius at constant magnetic field r_H and the analogous radius at constant magnetization r_M are different as follows from the microscopic treatment [16].

Magnetoelasticity makes the contribution of acoustic phonons to the free energy of a ferromagnet dependent on the magnetization. In the Grüneisen corresponding-states model we write the following expression for this contribution [17]:

$$F_{\text{ph}}(V,T,M) = \Theta_M^l f_l\left(\frac{T}{\Theta_M^l}\right) + 2\Theta_M^t f_t\left(\frac{T}{\Theta_M^t}\right). \qquad (7)$$

Here we have introduced the partial Debye temperatures for the longitudinal modes of the acustic phonons

$$\Theta_M^l = \frac{\hbar}{\kappa_B \rho^{1/2}}\left(\frac{6\pi^2}{v}\right)^{1/3}\sqrt{K_M + \frac{4}{3}G_M}, \qquad (8)$$

and for their transverse modes

$$\Theta_M^t = \frac{\hbar}{\kappa_B \rho^{1/2}}\left(\frac{6\pi^2}{v}\right)^{1/3}\sqrt{G_M}, \qquad (9)$$

where ρ is the density and v is the unit-cell volume of the crystal. In our treatment we shall neglect the possible dependence of Θ_M^l and Θ_M^t on temperature, i.e., we shall neglect the anharmonicity of the phonons. At the same time, according to (5), we shall take into account the dependence of the Debye temperatures (8) and (9) on magnetization caused by magnetoelasticity. Bearing in mind the smallness of the magnetoelasticity contribution to the moduli in (5), we can write (8) and (9) in the form

$$\Theta_M^l = \Theta_l + \Theta_l' M^2, \qquad \Theta_M^t = \Theta_t + \Theta_t' M^2, \qquad (10)$$

where

$$\frac{\Theta_l'}{\Theta_l} = \frac{3K' + 4G'}{2(3K + 4G)}, \qquad \frac{\Theta_t'}{\Theta_t} = \frac{G'}{2G}. \qquad (11)$$

The magnetization-dependent phonon contribution to the free energy (7) of a ferromagnet can be written in accordance with expressions (10) in the form

$$\Delta F_{\rm ph}(V, T, M) = \sum_{s=l,t} (1 + \delta_{s,t}) \Theta_s' \varphi_s\left(\frac{T}{\Theta_s}\right) M^2, \qquad (12)$$

where

$$\varphi_s(x) = f_s(x) - x f_s'(x), \qquad (13)$$

and $\delta_{s,t} = 1$ for $s = t$ and $\delta_{s,t} = 0$ for $s \neq t$.

Let us now consider the temperature range near the ferromagnetic transition. Then for (12) we can write the following expansion with respect to temperature:

$$\Delta F_{\rm ph}(V, T, M) = V \sum_{s=l,t} (1+\delta_{s,t}) \frac{\Theta_s'}{\Theta_s} \left[\frac{\Theta_s}{V}\varphi_s\left(\frac{T_C}{\Theta_s}\right) + C_{\rm ph}^s(T_C)(T - T_C)\right] M^2, \qquad (14)$$

where $C_{\rm ph}^s(T) = -(T/V\Theta_s) f_s''(T/\Theta_s)$ is the partial contribution of the longitudinal ($s = l$) and transverse ($s = t$) modes of the acoustic phonons to the lattice specific heat [18]. We note here that a linear temperature dependence appears in (14) not only near T_C, but also over a broad temperature range $T \geq T_C$ for ferromagnets with high Curie temperatures ($T_C \gg \Theta_{l,t}/4$), where the phonon specific heat $C_{\rm ph}^s$ is constant [18].

Of course, apart from the phonon contribution to the free energy (14) of a ferromagnet, the theory should take into account magnetism of electrons caused by their exchange interaction as a primary factor. For the magnetization-dependent electronic contribution to the free energy not very far from the Curie temperature, where the magnetization is small, we can use the expansion

$$\Delta F_{\rm el}(V, T, M) = V \left\{\frac{1}{2}[a_1(T_C) + \alpha(T - T_C)] M^2 + \frac{1}{4} a_3 M^4\right\}. \qquad (15)$$

Formulae (14) and (15) enable us to write the expression for the magnetic part of the free energy of a ferromagnet in the form of a Landau theory of phase transitions [18]:

$$\Delta F_M(V,T,M) = V\left[\frac{1}{2C}(T-T_C)M^2 + \frac{1}{4}a_3 M^4\right],\qquad(16)$$

where the Curie temperature is defined by

$$a_1(T_C) + \frac{2}{V}\sum_{s=l,t}(1+\delta_{s,t})\Theta'_s\varphi_s\left(\frac{T_C}{\Theta_s}\right) = 0,\qquad(17)$$

and the Curie constant C is given by the relation

$$\frac{1}{C} = \alpha + 2\sum_{s=l,t}(1+\delta_{s,t})C^s_{\text{ph}}(T_C)\frac{\Theta'_s}{\Theta_s}.\qquad(18)$$

According to (17) and (18) acoustic phonons affect the Curie temperature and the Curie constant. It is convenient to rewrite Eq. (18) in the following manner:

$$C = \frac{1 - X(T_C)}{\alpha},\qquad(19)$$

where the magnitude of the dimensionless parameter

$$X(T_C) = X_l(T_C) + 2X_t(T_C)\qquad(20)$$

quantitatively determines the contribution of the acoustic phonons to the Curie constant, and the parameters

$$X_s(T_C) = 2CC^s_{\text{ph}}(T_C)\frac{\Theta'_s}{\Theta_s},\qquad s=l,t\qquad(21)$$

characterize the partial contributions of the respective acoustic modes. It is clear that only thermal phonons make a contribution to the Curie constant (19). Now, in the limit of high Curie temperature ($T_C \gg \Theta_{l,t}/4$) the parameters

$$X_s = \frac{2\kappa_B C}{v}\frac{\Theta'_s}{\Theta_s},\qquad s=l,t\qquad(22)$$

do not depend on T_C, and for low Curie temperature ($T_C \ll \Theta_{l,t}/4$) they decrease as

$$X_s(T) \sim C^s_{\text{ph}}(T_C) \sim (T_C/\Theta_s)^3,\qquad s=l,t.\qquad(23)$$

Therefore, metals and alloys which have comparatively high Curie temperatures will be used below as examples for the quantitative determination of

$X(T_C)$ in real ferromagnets. Relations (16)-(18) enable us to study effects of acoustic phonons in various thermodynamic properties of ferromagnets.

Let us examine the dependence of the Curie temperature on the atomic mass, i.e., the isotope effect. Differentiating (17) with respect to the atomic mass M_i and using Eqs. (18) and (21), we find the following expression for the dimensionless isotope coefficient:

$$\frac{d\ln T_C}{d\ln M_i} = -\frac{1}{2}\sum_{s=l,t}(1+\delta_{s,t})X_s(T_C)\left[1 - \frac{\Theta_s\varphi_s(T_C/\Theta_s)}{VT_C C_{\text{ph}}^s(T_C)}\right]. \quad (24)$$

Here we assumed that according to (11) the ratios $\Theta'_{l,t}/\Theta_{l,t}$ do not depend on the atomic mass. The main difference between (24) and the result in [19] is the presence of the factors $X_{l,t}(T_C)$ in (24), which determine the magnitude of the isotope effect to a significant extent. This difference is attributed to the model used in [19], where the Curie temperature is completely determined by the contribution of the thermal phonons. In addition, Eq. (24) unlike the results in Ref. [19], takes into account the contribution of both the longitudinal and transverse modes of the acoustic phonons, as well as the contribution of the zero-point lattice vibrations.

Let us present the expressions, which can be derived from Eq. (24) for ferromagnets with high or low Curie temperature and by using the Debye model. For high Curie temperature ($T_C \gg \Theta_{l,t}/4$), using the expansion

$$\varphi_s(x) = f_s(0) + \kappa_B N x\left(1 - \frac{\alpha_s}{x} + \frac{\beta_s}{x^2}\right), \qquad s=l,t, \quad (25)$$

where N is the number of unit cells in the crystal, from (24) the following expression for the isotope coefficient is found:

$$\frac{d\ln T_C}{d\ln M_i} = \frac{1}{2}\sum_{s=l,t}(1+\delta_{s,t})X_s\frac{\Theta_s}{T_C}\left(\frac{f_s(0)}{\kappa_B N} - \alpha_s + 2\beta_s\frac{\Theta_s}{T_C}\right). \quad (26)$$

Here the parameters X_s do not depend on T_C and are given by Eq. (22). Applying the Debye model where $f_s(0) = 3\kappa_B N/8$, $\alpha_s = 3/8$, and $\beta_s = 1/20$ from (26) we obtain

$$\frac{d\ln T_C}{d\ln M_i} = \frac{1}{20}\left[X_l\left(\frac{\Theta_l}{T_C}\right)^2 + 2X_t\left(\frac{\Theta_t}{T_C}\right)^2\right]. \quad (27)$$

In the case of low Curie temperature ($T_C \ll \Theta_{l,t}/4$) where

$$\varphi_s(x) = f_s(0) + O(x^4), \qquad s=l,t, \quad (28)$$

formula (24) leads to an isotope effect which is completely determined by zero-point lattice vibrations:

$$\frac{d\ln T_C}{d\ln M_i} = \frac{1}{2}\sum_{s=l,t}(1+\delta_{s,t})X_s\frac{\Theta_s}{T_C}\left(\frac{f_s(0)}{\kappa_B N}\right). \tag{29}$$

For the Debye model Eq. (29) takes the form

$$\frac{d\ln T_C}{d\ln M_i} = \frac{3}{16}\left(X_l\frac{\Theta_l}{T_C} + 2X_t\frac{\Theta_t}{T_C}\right). \tag{30}$$

Comparing (26) and (27) with (29) and (30) for the same values of $X_{l,t}$, we can conclude that the absolute value of the isotope effect is larger in ferromagnets with low Curie temperature ($T_C \ll \Theta_{l,t}/4$). This conclusion differs qualitatively from the result obtained above, according to which the contribution of the acoustic phonons to the Curie constant (19) should be larger for high Curie temperature (high in comparison with the Debye temperature of the acoustic phonons).

Equation (17) enables us to study the renormalization of the Curie temperature due to phonons. Assuming this renormalization is small we find

$$T_C = T_C^{(0)}\left[1 - \sum_{s=l,t}(1+\delta_{s,t})X_s\frac{\Theta_s\varphi_s(T_C^{(0)}/\Theta_s)}{\kappa_B N T_C^{(0)}}\right], \tag{31}$$

where the Curie temperature $T_C^{(0)}$ is found from the equation $a_1(T_C^{(0)}) = 0$. For high Curie temperature ($T_C \gg \Theta_{l,t}/4$) we find from (25) and (31) the following expression for the phonon renormalization of the Curie temperature:

$$T_C = T_C^{(0)}(1 - X), \tag{32}$$

where $X = X_l + 2X_t$, and the parameters $X_{l,t}$ (see 22) do not depend on T_C.

In the case of low Curie temperature ($T_C \ll \Theta_{l,t}/4$), formulae (28) and (31) lead to a phonon renormalization of the Curie temperature which is completely determined by the zero-point lattice vibrations:

$$T_C = T_C^{(0)}\left[1 - \sum_{s=l,t}(1+\delta_{s,t})X_s\frac{\Theta_s}{T_C^{(0)}}\left(\frac{f_s(0)}{\kappa_B N}\right)\right]. \tag{33}$$

Finally, for the Debye model Eq. (33) takes the form

$$T_C = T_C^{(0)}\left[1 - \frac{3}{8}\left(X_l\frac{\Theta_l}{T_C^{(0)}} + 2X_t\frac{\Theta_t}{T_C^{(0)}}\right)\right]. \tag{34}$$

3. Experimental data

The treatment performed above shows that for the quantitative determination of the effects of acoustic phonons on various thermodynamic properties of real ferromagnets we need the dimensionless parameters $X_{l,t}(T_C)$ or (22). We can find them from experimental data. We need experimental values for the Curie constant C, the phonon specific heat $C^s_{\text{ph}}(T_C)$, the elastic moduli K and G, and the magnetoelastic coefficients K' and G'. Below we report a set of these parameters which have been determined on the basis of experimental data for the pure metals Fe and Ni, the iron-nickel Invar alloys $\text{Fe}_{1-x}\text{Ni}_x$ ($0.30 < x \leq 0.45$), the ternary alloys $\text{Fe}_{0.65}(\text{Ni}_{1-x}\text{Mn}_x)_{0.35}$ ($0 \leq x \leq 0.13$), and the iron-platinum Invar alloys $\text{Fe}_{1-x}\text{Pt}_x$ ($x = 0.28, 0.25$) with various degrees of order s of the atoms on the crystal-lattice sites. These ferromagnets correspond to the case of high Curie temperature ($T_C \gg \Theta_{l,t}/4$).

Table 1 gives the experimental data obtained for the Curie constant C from Refs. [20] -[24] and the data obtained for the phonon specific heat $C_{\text{ph}} = C^l_{\text{ph}} + 2C^t_{\text{ph}} = 3\kappa_B/v^3$, where for the case of ferromagnets with a cubic lattice the relation $v = a^3$ was used, where a is the lattice constant. In Table 1 we also give the experimental data for the product $2CC_{\text{ph}}$.

TABLE 1. Experimental data for lattice constant a, phonon specific heat $C_{\text{ph}} = 3\kappa_B/a^3$, Curie constant C, and for the product $2CC_{\text{ph}}$ in some metals and alloys.

Metals, alloys	a (Å)	C_{ph} ($10^7 \frac{\text{ergs}}{\text{cm}^3\text{K}}$)	C (K)	$2CC_{\text{ph}}$ ($10^7 \frac{\text{ergs}}{\text{cm}^3}$)
Ni	3.52	0.95	0.05	0.095
Fe	2.866	1.76	0.17	0.60
$\text{Fe}_{0.72}\text{Pt}_{0.28}$	3.749	0.79	0.31	0.49
$\text{Fe}_{0.75}\text{Pt}_{0.25}$	3.73	0.80	0.29	0.464
$\text{Fe}_{1-x}\text{Ni}_x$	3.585	0.90	0.28	0.50
$\text{Fe}_{0.65}(\text{Ni}_{1-y}\text{Mn}_y)_{0.35}$	3.585	0.90	0.28	0.50

$0.30 \leq x \leq 0.45$, $\quad 0 \leq y \leq 0.13$.

Let us now find the magnetoelastic coefficients. The shear magnetoelastic coefficient G' can be determined directly from the experimental data of Refs. [13] and [25] -[31] for the magnetoelastic contributions $\Delta C'$ and ΔC_{44} to the moduli C' and C_{44} of cubic crystals, presented in the form $\Delta C' = C'' M^2$ and $\Delta C_{44} = C'_{44} M^2$. By Voigt averaging one gets

$$G' = \frac{2C'' + 3C'_{44}}{5}. \tag{35}$$

TABLE 2. Elastic moduli K, G and magnetoelastic coefficients C'', C'_{44} of metals and alloys determined on the basis of experimental data.

Metals, alloys	K (10^2 kbar)	G (10^2 kbar)	C'' (10^5)	C'_{44} (10^5)
Ni	17.5	8.2	0.88	0.69
Fe	13.5	6.8	0.034	0.010
$Fe_{0.685}Ni_{0.315}$	11.7	7.8	-0.75	-0.97
$Fe_{0.668}Ni_{0.332}$	12.1	7.6	-0.81	-0.72
$Fe_{0.662}Ni_{0.338}$	12.5	7.4	-0.84	-0.72
$Fe_{0.647}Ni_{0.353}$	12.4 – 14.6	7.4	-0.88	-0.72
$Fe_{0.623}Ni_{0.377}$	12.6	7.2	-0.60	-0.38
$Fe_{0.60}Ni_{0.40}$	12.9	7.2	-0.50	-0.29
$Fe_{0.575}Ni_{0.425}$	12.9	7.2	-0.41	-0.24
$Fe_{0.548}Ni_{0.452}$	13.6	7.1	-0.30	-0.18
$Fe_{0.65}Ni_{0.35}$	11.8 – 15.0	7.4	-0.78	-0.75
$Fe_{0.65}(Ni_{0.96}Mn_{0.04})_{0.35}$	13.3 – 13.7	7.6	-0.78	-0.75
$Fe_{0.65}(Ni_{0.91}Mn_{0.09})_{0.35}$	12.6 – 13.4	7.9	-0.78	-0.75
$Fe_{0.65}(Ni_{0.87}Mn_{0.13})_{0.35}$	11.9 – 13.1	8.2	-0.78	-0.75
$Fe_{0.72}Pt_{0.28}$ ($s=0$)	12.1 – 15.8	7.0 – 7.1	-1.63	-1.47
$Fe_{0.75}Pt_{0.25}$ ($s=0.85$)	12.6 – 14.1	6.9 – 7.0	-1.42	-1.22
$Fe_{0.75}Pt_{0.25}$ ($s=0.70$)	12.6 – 14.1	6.9 – 7.0	-2.09	-2.12
$Fe_{0.75}Pt_{0.25}$ ($s=0.60$)	12.6 – 14.1	6.9 – 7.0	-1.91	-1.69
$Fe_{0.75}Pt_{0.25}$ ($s=0.55$)	12.6 – 14.1	6.9 – 7.0	-2.06	-1.78
$Fe_{0.75}Pt_{0.25}$ ($s=0.40$)	12.6 – 14.1	6.9 – 7.0	-1.96	-1.67

The bulk modulus K_M at constant magnetization cannot be measured directly. For this reason, to determine the bulk magnetoelastic coefficient K', the modulus K_M must be calculated from experimental data and the thermodynamic relations (6) which can be written as follows [13]

$$\frac{1}{K_H} - \frac{1}{K_M} = 2\gamma M \left(\frac{\partial \omega}{\partial H}\right)_{P,T} \tag{36}$$

Here γ is the spontaneous magnetostriction coefficient for which the experimental data were taken from Refs. [13, 32]. Using the temperature dependence of the bulk modulus K_H from [13] and [25]-[31], the data for the forced volume magnetostriction $(\partial \omega/\partial H)_{P,T}$ from [32, 33] and the data for the spontaneous magnetization M from [34]-[36], we calculate the magnetoelastic coefficients C''', C'_{44}, G', and K' and present them in Tables 2 and 3, as well as the data for the paramagnetic elastic moduli K and G.

TABLE 3. Magnetoelastic parameters of metals and alloys determined from the experimental data.

Metals, allous	G' (10^5)	K' (10^5)	$(\Theta'/\Theta)_G$ $(10^{-8}G^{-2})$	$(\Theta'/\Theta)_K$ $(10^{-8}G^{-2})$
Ni	0.77	0.5 – 0.9	3.8	0.3 – 0.5
Fe	0.02	0.04	0.12	0.03
Fe$_{0.685}$Ni$_{0.315}$	-0.88	1.3 – 1.5	-4.6	1.0 – 1.1
Fe$_{0.668}$Ni$_{0.332}$	-0.76	1.3 – 1.5	-4.1	1.0 – 1.1
Fe$_{0.662}$Ni$_{0.338}$	-0.77	1.3 – 1.5	-4.3	1.0 – 1.1
Fe$_{0.647}$Ni$_{0.353}$	-0.78	1.3 – 1.5	-(4.2 – 4.3)	0.9 – 1.1
Fe$_{0.623}$Ni$_{0.377}$	-0.47		-2.6	
Fe$_{0.60}$Ni$_{0.40}$	-0.37		-2.0	
Fe$_{0.575}$Ni$_{0.425}$	-0.31		-1.7	
Fe$_{0.548}$Ni$_{0.452}$	-0.23		-1.3	
Fe$_{0.65}$Ni$_{0.35}$	-0.76	1.3 – 1.5	-(4.1 – 4.2)	0.9 – 1.2
Fe$_{0.65}$(Ni$_{0.96}$Mn$_{0.04}$)$_{0.35}$	-0.76	1.3 – 1.5	-4.0	0.9 – 1.1
Fe$_{0.65}$(Ni$_{0.91}$Mn$_{0.09}$)$_{0.35}$	-0.76	1.3 – 1.5	-3.9	0.9 – 1.1
Fe$_{0.65}$(Ni$_{0.87}$Mn$_{0.13}$)$_{0.35}$	-0.76	1.3 – 1.5	-3.8	0.9 – 1.1
Fe$_{0.72}$Pt$_{0.28}$ ($s=0$)	-1.53	1.0 – 1.3	-(8.6 – 8.9)	0.7 – 1.0
Fe$_{0.75}$Pt$_{0.25}$ ($s=0.85$)	-1.30	1.0 – 1.3	-(7.4 – 7.5)	0.7 – 1.0
Fe$_{0.75}$Pt$_{0.25}$ ($s=0.70$)	-2.11	1.0 – 1.3	-(12.0 – 12.2)	0.7 – 1.0
Fe$_{0.75}$Pt$_{0.25}$ ($s=0.60$)	-1.78	1.0 – 1.3	-(10.2 – 10.3)	0.7 – 1.0
Fe$_{0.75}$Pt$_{0.25}$ ($s=0.55$)	-1.89	1.0 – 1.3	-(10.8 – 11.0)	0.7 – 1.0
Fe$_{0.75}$Pt$_{0.25}$ ($s=0.40$)	-1.79	1.0 – 1.3	-(10.2 – 10.4)	0.7 – 1.0

In the limit of high Curie temperature it is convenient to introduce the new dimensionless parameters

$$X_G = 2CC_{\rm ph}\left(\frac{\Theta'}{\Theta}\right)_G, \quad X_K = 2CC_{\rm ph}\left(\frac{\Theta'}{\Theta}\right)_K, \tag{37}$$

instead of $X_{l,t}$. Here the magnetoelastic parameters

$$\left(\frac{\Theta'}{\Theta}\right)_G = \frac{2(K+2G)}{(3K+4G)}\frac{\Theta'_t}{\Theta_t} = \frac{(K+2G)}{(3K+4G)}\frac{G'}{G},$$

$$\left(\frac{\Theta'}{\Theta}\right)_K = \frac{1}{3}\left(\frac{\Theta'_l}{\Theta_l} - \frac{4G}{3K+4G}\frac{\Theta'_t}{\Theta_t}\right) = \frac{K'}{2(3K+4G)} \tag{38}$$

are completely determined by the shear magnetoelastic coefficient G' and the bulk magnetoelastic coefficient K' respectively. It is obvious, that

$$X = X_G + X_K \tag{39}$$

and the parameters $X_{l,t}$ are connected with the new ones by the relations:

$$X_l = X_K + \frac{2G}{3(K+2G)}X_G, \qquad X_t = \frac{3K+4G}{6(K+2G)}X_G. \qquad (40)$$

The data in Tables 1,2 and, 3 make it possible to determine both $(\Theta'/\Theta)_G$ and $(\Theta'/\Theta)_K$, as well as the dimensionless parameters X_G, X_K, and X, which are given in Tables 3 and 4.

TABLE 4. Parameters determining the contribution of acoustic phonons to the magnetic properties of ferromagnets.

Metals, alloys	X_G	X_K	X
Ni	0.036	$(3-5) \times 10^{-3}$	0.04
Fe	0.007	0.002	0.009
$Fe_{0.685}Ni_{0.315}$	-0.23	0.05 − 0.06	-(0.17 − 0.18)
$Fe_{0.668}Ni_{0.332}$	-0.20	0.05 − 0.06	-(0.14 − 0.15)
$Fe_{0.662}Ni_{0.338}$	-0.22	0.05 − 0.06	-(0.16 − 0.17)
$Fe_{0.647}Ni_{0.353}$	-(0.21 − 0.22)	0.04 − 0.06	-(0.15 − 0.18)
$Fe_{0.623}Ni_{0.377}$	-0.13		
$Fe_{0.60}Ni_{0.40}$	-0.10		
$Fe_{0.575}Ni_{0.425}$	-0.08		
$Fe_{0.548}Ni_{0.452}$	-0.06		
$Fe_{0.65}Ni_{0.35}$	-(0.20 − 0.21)	0.04 − 0.06	-(0.14 − 0.17)
$Fe_{0.65}(Ni_{0.96}Mn_{0.04})_{0.35}$	-0.20	0.04 − 0.06	-(0.14 − 0.16)
$Fe_{0.65}(Ni_{0.91}Mn_{0.09})_{0.35}$	-0.20	0.04 − 0.06	-(0.14 − 0.16)
$Fe_{0.65}(Ni_{0.87}Mn_{0.13})_{0.35}$	-0.19	0.04 − 0.06	-(0.13 − 0.15)
$Fe_{0.72}Pt_{0.28}$ ($s=0$)	-(0.42 − 0.44)	0.03 − 0.05	-(0.37 − 0.41)
$Fe_{0.75}Pt_{0.25}$ ($s=0.85$)	-(0.34 − 0.35)	0.03 − 0.05	-(0.29 − 0.32)
$Fe_{0.75}Pt_{0.25}$ ($s=0.70$)	-(0.56 − 0.57)	0.03 − 0.05	-(0.51 − 0.54)
$Fe_{0.75}Pt_{0.25}$ ($s=0.60$)	-(0.47 − 0.48)	0.03 − 0.05	-(0.42 − 0.45)
$Fe_{0.75}Pt_{0.25}$ ($s=0.55$)	-(0.50 − 0.51)	0.03 − 0.05	-(0.45 − 0.48)
$Fe_{0.75}Pt_{0.25}$ ($s=0.40$)	-(0.47 − 0.48)	0.03 − 0.05	-(0.42 − 0.45)

4. Discussion of the results and conclusions

By the help of the data in Table 4 and the formulae (19) and (32) we can formulate quantitative statements regarding the role of the acoustic phonons in the renormalization of the Curie constant and the Curie temperature for numerous metals and alloys.

The main contribution of acoustic phonons to the magnetic properties of Fe, Ni, and Fe-Ni, Fe-Ni-Mn, Fe-Pt Invar alloys comes from shear magnetoelasticity and not from bulk magnetoelasticity, since $|X_G| \gg |X_K|$.

Acoustic phonons have a relatively weak effect on the magnetic properties of Fe and Ni, since the parameter X does not exceed several percent for these metals. For comparison, we call attention to [37], where by using model calculations the effect for iron is estimated to be 20 %.

Acoustic phonons play a much larger role in the magnetic properties of Fe-Ni and Fe-Ni-Mn Invar alloys for which depending on the composition of the alloy the parameter X reaches the value $-(13-18)$ %.

An especially strong influence of acoustic phonons on the magnetic properties was found in Fe-Pt Invar alloys. For example, $X \simeq -40$ % for the disordered alloy $Fe_{0.72}Ni_{0.28}$ and in the case of the alloy $Fe_{0.75}Ni_{0.25}$ the parameter X ranges from -30 % to -50 %, depending on the degree of ordering of the alloy. It may be said that this large effect is anomalous.

Let us discuss the isotope shift of the Curie temperature caused by magnetoelasticity in the ferromagnets considered above. Because of an anomalously large value for the dimensionless parameter $X \simeq -(0.51-0.54)$ found in the Invar alloy $Fe_{0.75}Pt_{0.25}$ with a degree of order $s = 0.70$ (see Table 4), it is natural to estimate the isotope shift of the Curie temperature in reference to this alloy. For other Fe-Pt, Fe-Ni, and Fe-Ni-Mn Invar alloys the isotope effect will be smaller than in the $Fe_{0.75}Pt_{0.25}$ alloy.

The data in Tables 2 and 4, and the relations (40) allow us to estimate the dimensionless parameters $X_l \simeq -(0.04-0.07)$ and $2X_t \simeq -0.47$, which characterize the partial contributions of the longitudinal-acoustic mode and the two transverse-acoustic modes to the isotope effect. Next, we determine the Debye temperature of the acoustic phonons, $\Theta_{l,t}$, from formulae (8) and (9) using the data presented in Table 2 for the elastic moduli K and G and taking for the density a value of $\rho = 11.7$ g/cm^3 from [29] and for the unit-cell volume $v = 51.9$ Å^3 in accordance with Table 1. This gives $\Theta_l \simeq 350$ K and $\Theta_t \simeq 194$ K. Since the condition $T_C \gg \Theta_{l,t}/4$ holds for the Curie temperature $T_C \simeq 386$ K of this alloy, as determined from Table 3 in [29], Eq. (26) can be used for numerical estimates of the isotope effect in this alloy. Assuming $f_l(0) = f_t(0) = f(0)$, $\alpha_l = \alpha_t = \alpha$, and $\beta_l = \beta_t = \beta$ for the coefficients in (26), we find

$$\frac{d\ln T_C}{d\ln M_i} \simeq -(0.14-0.15)\left(\frac{f(0)}{\kappa_B N} - \alpha\right) - (0.16-0.17)\beta. \qquad (41)$$

We next consider the case where $|f(0)/\kappa_B N - \alpha| \sim \alpha < 1$ and $\beta \ll \alpha$. From (41) we then obtain the estimate

$$\left|\frac{d\ln T_C}{d\ln M_i}\right| \simeq 0.14\alpha \qquad (42)$$

and for the absolute value of the isotope shift of the Curie temperature we have

$$|\Delta T_C| \simeq (54 \text{ K}) \left(\alpha \frac{\Delta M_i}{M_i}\right), \qquad (43)$$

where ΔM_i is the isotope change of the atomic mass. For the replacement of the ^{54}Fe isotope by ^{56}Fe or by ^{58}Fe we find $\Delta M_i/M_i \simeq (3.7 - 7.4)$ %. Then, formula (43) gives $|\Delta T_C| \simeq (2-4)\alpha$ K. Since $\alpha < 1$, for an upper estimate of the isotope shift of the Curie temperature in the Invar alloy Fe$_{0.75}$Pt$_{0.25}$ we obtain $|\Delta T_C| < (2-4)$ K. Finally, we present an estimate of the isotope shift of the Curie temperature for this alloy according to the Debye model. Then we have $f(0)/\kappa_B N = \alpha = 3/8$ and $\beta = 1/20$. From (41) we obtain in this case

$$\Delta T_C \simeq -(0.1 - 0.2) \text{ K}. \qquad (44)$$

Thus, in contrast to the anomalously large phonon renormalization of the Curie constant and the Curie temperature for this alloy discovered above, the contribution of the acoustic phonons to the isotope shift of the Curie temperature is comparatively small. This finding is not unexpected since according to the phenomenological approach developed in Sec. 2, the Invar alloy Fe$_{0.75}$Pt$_{0.25}$ is a ferromagnet with a relatively high Curie temperature ($T_C \gg \Theta_{l,t}/4$). On the contrary, the estimates of the dimensionless parameters X_l and $2X_t$ presented above allow us to conclude that acoustic phonons can cause a large isotope shift of the Curie temperature when the equality $|d\ln T_C/d\ln M_i| \sim 1$ holds in ferromagnets with relatively large parameters $X_{l,t}$ (as, e.g., in the Invar alloy Fe$_{0.75}$Pt$_{0.25}$) and the Curie temperature is low compared with the Debye temperature. We should also mention the work of Karchevsky et al. [38], who first observed an isotope shift of the Curie temperature of $\Delta T_C = 4.0 \pm 0.5$ K in going from uranium hydride, UH$_3$, to uranium deuteride, UD$_3$, for which the isotopic coefficient is relatively small, $I \simeq -2 \times 10^{-2}$.

In conclusion, we would like to stress that the phenomenological approach reviewed here indicates a set of parameters which can be derived from experiment and permit a quantitative calculation of the contributions of the acoustic phonons to various properties of real ferromagnets caused by magnetoelasticity.

The author acknowledges support by the Russian Foundation for Basic Research (project 96-02-17318-a) and the State Council for Support of Leading Scientific Schools (project 96-15-96750).

References

1. Herring, C. (1966) Exchange interactions among itinerant electrons, in G.T. Rado and H. Suhl (eds.), *Magnetism*, Academic Press, New York and London, Vol. IV, p. 290.
2. Fay, D. and Appel, J. (1979) Phonon contribution to the Stoner enhancement factor: Ferromagnetism and possible superconductivity of $ZrZn_2$, *Phys. Rev.* **B20**, 3705-3708; Fay, D. and Appel, J. (1980) Comment on the isotope effect in weak itinerant ferromagnets, *Phys. Rev.* **B22**, 1461-1463.
3. Hopfield, J.J. (1968) The isotope effect in very weak itinerant ferromagnets, *Phys. Lett.* **A27**, 397-399.
4. Edwards, D.M. and Wohlfarth, E.P. (1968) Magnetic isotherms in the band model of ferromagnetism, *Proc. Roy. Soc. London*, **A303**, 127-137.
5. Dzyaloshinsky I.E. and Kondratenko P.S. (1976) On the theory of weak ferromagnetism in a fermi fluid, *Zh. Eksp. Teor. Fiz* **70**, 1987-2005. [*Sov. Phys. JETP* **43**, 1036].
6. Moriya, T. (1985) *Spin fluctuations in Itinerant Electron Magnetism*, Springer-Verlag, Berlin.
7. Knapp, G.S., Corenzwit, E., and Chu, C.W. (1970) Attemp to measure the isotope effect in the weak itinerant ferromegnet $ZrZn_2$, *Solid State Commun.* **8**, 639-641.
8. Pickett, W.E. (1982) Renormalized thermal distribution function in an interacting electron-phonon system, *Phys. Rev. Lett.* **48**, 1548-1551. Generalization of the theory of the electron-phonon interaction: Thermodynamic formulation of superconducting- and normal- state properties, *Phys. Rev.* **B 26**, 1186-1207.
9. Zhao, G.M., Conder, K., Keller, H., and Müller, K.A. (1996) Giant oxygen isotope shift in the magnetoresistive perovskite $La_{1-x}Ca_xMnO_{3+y}$, *Nature* **381**, 676-678.
10. Zverev, V.M. and Silin, V.P. (1996) Magnetoelasticity and effect of thermal phonons on the magnetic properties of ferromagnets, *JETP Lett.* **64**, 37-41.
11. Zverev, V.M. (1997) Isotope effect in ferromagnets, *Zh. Eksp. Teor. Fiz.* **112**, 1863-1872. [*JETP* **85**, 1019-1023].
12. Hausch, G. (1973) Magnetic exchange energy contribution to the elastic constants and its relation to the anomalous elastic behaviour of Invar alloys, *phys. stat. sol. (a)* **15**, 501-510.
13. Shiga, M., Makita, K., Uematsu, K., Muraoka, Y,. and Nakamura, Y. (1990) Magnetoelasticity of $Fe_{65}(Ni_{1-x}Mn_x)_{35}$ Invar alloys: I. Temperature and field dependences of the elastic constants, *J. Phys.: Condens. Matter* **2**, 1239-1252.
14. Zverev, V.M. and Silin, V.P. (1988) On the self-consistent fluctuation-phonon theory of metallic magnetism. Comments on Kim's articles, *Fiz. Met. Metalloved.* **65**, 895-906.
15. Kim, D.J. (1988) The electron-phonon interaction and itinerant electron magnetism, *Phys. Rept.* **171**, 129-229.
16. Zverev, V.M. and Silin, V.P. (1989) Dynamic elasticity theory of ferromagnetic metals under constant-magnetization conditions, *Fiz. Tverd. Tela (Leningrad)* **31**, 123-128. [*Sov. Phys. Solid State* **31**, 788-791].
17. Zverev, V.M. and Silin, V.P. (1988) Phase transitions in a self-consistent phonon-fluctuation model of magnetic materials, *Fiz. Tverd. Tela (Leningrad)* **30**, 1989-1998. [*Sov. Phys. Solid State* **30**, 1148-1152].
18. Landau, L.D. and Lifshitz, E.M. (1980) *Statistical Physics*, Vol. 1, Pergamon Press, Oxford-New York.
19. Zverev, V.M. and Silin, V.P. (1987) Fluctuation-phonon approach to the theory of magnetism, *Zh. Eksp. Teor. Fiz.* **93**, 709-722. [*Sov. Phys. JETP* **66**, 401-407].
20. Arajs, S. and Miller, D.S. (1960) Paramagnetic susceptibilities of Fe and Fe-Si alloys, *J. Appl. Phys.* **31**, 986-991.
21. Arajs, S. and Colvin, R.V. (1963) Paramagnetism of polycrystalline nickel, *J. Phys. Chem. Solids* **24**, 1233-1237.

22. Kouvel, J.S. and Wilson, R.H. (1961) Magnetization of iron-nickel alloys under hydrostatic pressure, *J. Appl. Phys.* **32**, 435-441.
23. Shiga, M. (1967) Magnetic properties of $Fe_{65}(Ni_{1-x}Mn_x)_{35}$ ternary alloys, *J. Phys. Soc. Japan* **22**, 539-546.
24. Sumiyama, K., Shiga, M., and Nakamura Y. (1980) Paramagnetic susceptibility in Fe-Pt Invar alloys, *J. Phys. Soc. Japan* **48**, 1393-1394.
25. Alers, G.A., Neighbours, J.R., and Sato, H. (1960) Temperature dependent magnetic contributions to the high field elastic constants of nickel and an Fe-Ni alloy, *J. Phys. Chem. Solids* **13**, 40-55.
26. Dever, D.J. (1972) Temperature dependence of the elastic constants in α-iron single crystals: relationship to spin order and diffusion anomalies, *J. Appl. Phys.* **43**, 3293-3300.
27. Hausch, G. and Warlimont, H. (1973) Single crystalline elastic constants of ferromagnetic face centered cubic Fe-Ni Invar alloys, *Acta metall.* **21**, 401-414.
28. Hausch, G. (1974) Elastic constants of Fe-Pt alloys. I. Single crystalline elastic constants of $Fe_{72}Pt_{28}$, *J. Phys. Soc. Japan* **37**, 819-823.
29. Ling, H.C. and Owen, W.S. (1983) The magneto-elastic properties of the Invar alloy, Fe_3Pt, *Acta metall.* **31**, 1343-1352.
30. Kawald, U., Schulenberg, P., Bach, H., and Pelzl, J. (1991) Elastic anomalies in iron-platinum Invar alloys, *J. Appl. Phys.* **70**, 6537-6539.
31. Mañosa, Ll., Saunders, G.A., Rahdi, H., Kawald, U., Pelzl, J., and Bach, H. (1992) Acoustic-mode vibrational anharmonicity related to the anomalous thermal expansion of Invar iron alloys, *Phys. Rev.* **B 45**, 2224-2236.
32. Sumiyama, K., Shiga, M., Morioka, M.,and Nakamura, Y. (1979) Characteristic magnetovolume effects in Invar type Fe-Pt alloys, *J. Phys. F: Metal Phys.* **9**, 1665-1677.
33. Ishio, S. and Takahashi, M. (1985) Temperature dependence of forced volume magnetostriction in fcc Fe-Ni alloys, *J. Magn. Magn. Mater.* **50**, 271-277.
34. Crangle, J. and Hallam, G.C. (1963) The magnetization of face-centered cubic and body-centered cubic iron+nickel alloys, *Proc. Roy. Soc.* **272**, 119-132.
35. Crangle, J. and Goodman, G.M. (1971) The magnetization of pure iron and nickel, *Proc. Roy. Soc. London.*, Ser. A **321**, 477-491.
36. Sumiyama, K., Shiga, M. and Nakamura, Y. (1976) Magnetization, thermal expansion and low temperature specific heat of $Fe_{72}Pt_{28}$ Invar alloy, *J. Phys. Soc. Japan* **40**, 996-1001.
37. Turov, E.A. and Grebennikov, V.I. (1989) The transition metal properties in the spin fluctuation theory, *Physica* **B59**, 56-60.
38. Karchevsky, A.I., Artyushkov, E.V. and Kikoin, L.I. (1959) Isotope shift of the Curie point in the uranium hydride and uranium deuteride, *Zh. Eksp. Teor. Fiz.* **36**, 636-637. [*Sov. Phys. JETP* **9**, 442-443].

SPIN–FLOP AND METAMAGNETIC TRANSITIONS IN ITINERANT FERRIMAGNETS

A. K. ZVEZDIN AND I. A. LUBASHEVSKY
Instite of General physics, Russian Academy of Science,
Moscow, Russia

R. Z. LEVITIN
Moscow State University,
Moscow, Russia

G.M.MUSAEV
Dag GU,
Dagestan, Russia

AND

V. V. PLATONOV AND O. M. TATSENKO
VNIIEF,
Sarov, Russia

Abstract. The present paper discusses magnetic field-induced phase transitions and the H–T phase diagrams of itinerant metamagnets with an unstable d-sublattice. In particular, the effect of the magnetic anisotropy on the field induced phase transitions is studied in detail. The theoretical results are compared with the available experimental data for the compounds $(YR)(CoAl)_2$ and RCo_2, where R is a heavy rare-earth element.

1. Introduction

Spin-flop transitions in ferrimagnets have been considered in detail both theoretically and experimentally. The concept of these transition was proposed by Néel in 1936 for antiferromagnets and they were found later in $CuCl_2H_2O$ by the Dutch group of Poulis *et al.* in 1951. In these transitions the collinear phase turns into a canted phase and further into a ferromagnetic one as the magnetic field increases. Later similar phase transitions were investigated in ferrimagnets which behavior turned out to be more sophisticated and interesting (see, e.g., Zvezdin et al. [1]). The ferrimagnetic structure may happen to be non-collinear in a certain range of magnetic fields and temperatures. Such structures arise due to the competition between the negative exchange interaction within the sublattices which forces their magnetic moments to be antiparallel to each other.

The common materials, which have been setting for most of the research of these phase transitions, are compounds of rare-earth elements and transition metals of the iron group: the iron garnets $R_3Fe_5O_{12}$ and the intermetallic compounds R_nT_m, where R is a rare-earth element, and T is a transition metal. Such systems are generally described in the approximation of a "rigid" d-sublattice in which the magnetization of the d-sublattice is assumed to be independent of the external magnetic field and of the exchange field exerted on it by the rare-earth sublattice. From the microscopic point of view the d-sublattice can be described in terms of the Heisenberg model of coupled local spin moments. The spin moments of the d-sublattice are nearly saturated and can fluctuate in orientation but not in magnitude.

Much interest has been attracted by itinerant f-d metamagnets which are opposite to the materials mentioned above relating microscopic properties. In these compounds the magnetic moment of the d-sublattice is unsaturated. Itinerant ferromagnets were recognized in 1936 by Slater and Stoner. Presently, an extensive literature exists on this subject (see, e.g., [2, 3] and references therein). Among various itinerant ferromagnets the particular interest is associated with the materials in which the d-subsystem is a weak itinerant ferromagnet and can undergo a metamagnetic phase transition with a substantial jump in the magnetization. Typical members of this family are YCo_2, $LuCo_2$, and RCo_2, and the alloys $FePt$, $FeRh$, etc., which exhibit metamagnetic transitions in the electron d-subsystem from a paramagnetic state to a ferromagnetic one in the magnetic field.

Magnetic field-induced transitions of this sort were predicted by Wohlfarth and Rhodes in 1962 [2]. According to the theory of Bloch et al. [3], Cyrot and Lavagna [4], and Yamada et al. [5] such a transition is possible in YCo_2. It was indeed observed by Goto et al. [6] and Murata et al. [7] in fields of the order 10^6 Oe and is explained by peculiarities in the energy dependence of the density of states $N(E)$ of the d-electrons near the Fermi level.

Qualitatively the magnetization process in the RCo_2 ferrimagnets looks as follows.

For the compounds $R_xY_{1-x}FeO_3$ the critical field is decreased in magnitude (see the papers of Wohlfarth and Rodes [2], Cyrot and Lavagna [4], Duc et al. [8], Steiner et al. [9], and Ballou et al. [10]). This is because the metamagnetic transition arises under the effective field which is the vector sum of the external field $\mathbf{H}$ and the molecular field exerted on the d-ions by the rare-earth ions:

$$\mathbf{H}_{\text{eff}} = \mathbf{H} - \lambda \mathbf{M}, \qquad (1)$$

where λ is the constant of the molecular field ($\lambda > 0$ for the heavy rare earth ions) and $\mathbf{M}$ is the magnetization of the f-sublattice. In all compounds RCo_2 the magnetization of the d-subsystem m does not exceed $2\mu_B$ per formula unit and $M > m$.[1] Therefore, during magnetizing these compounds the magnetic moment of the d-subsystem is initially oriented antiparallel to the external field and the effective field H_{eff} is negative (taking the direction of the external field as positive). As the external magnetic field increases, the effective field H_{eff} decreases in magnitude and after attaining the metamagnetic transition field $-H_p$ causes the d-subsystem to transit from the ferromagnetic state to the paramagnetic one. The external field at which this transition takes place is given by the formula

$$H_{c1} = \lambda M - H_p.$$

This means that in the itinerant metamagnets RCo_2 a reentrant phase transition of the Co-subsystem from a ferromagnetic ordered into a disordered paramagnetic state is possible [12]. Obviously, in this case the spin gap in the d-electron spectrum becomes zero. The physical mechanism of this transition is similar to the mechanism of the induced superconducting state in the strong magnetic field discovered in the rare-earth Chevrel phase superconductors (($SmEu)Mo_6S_8$).

As the external magnetic field increases further, the magnetic moment of the paramagnetic d-subsystem is reoriented parallel to the field, because the effective field H_{eff} becomes positive. Under the external field H_{c2} when H_{eff} approaches the value H_p a transition of the d-subsystem to a ferromagnetic state occurs again. The critical field of this transition is

$$H_{c2} = H_p + \lambda M.$$

Therefore, during magnetizing of the ferrimagnetic compounds RCo_2 at least two successive jump-like metamagnetic transitions should be observed due to an abrupt demagnetization and reentrant magnetization of the d-subsystem.

In this paper we discuss magnetic field induced phase transitions and the H–T phase diagrams in itinerant metamagnets with an unstable d-sublattice. This subject is of general interest for physics of f-d intermetallic

---

[1] Another possibility for a decrease in the critical field of a metamagnetic transition is realized in compounds of the type $(RY)Co_{2-x}Al_x$ (Aleksandryan et al. [11]).

compounds, because according to the present understanding of itinerant ferromagnetism the d-subsystem in such compounds can be usually treated as an unsaturated (weak) ferromagnet.

2. Theory

2.1. GROUND STATE

We first consider the ground state of an f-d itinerant ferrimagnet at $T = 0$. The thermodynamic potential of the d-subsystem of a weak itinerant ferrimagnetic is conventionally written as an expansion in powers of the magnetization m [2]:

$$F^d(m) = \frac{1}{2}am^2 - \frac{1}{4}bm^4 + \frac{1}{6}cm^6, \qquad (2)$$

where a, b, and d are coefficients which temperature dependence is determined by the particular band structure and/or by spin fluctuations. As a first approximation, these parameters may be considered to be independent on the magnetic field H. Let us assume, as is usually done in the theory of itinerant metamagnetism, that the following relations hold: $a > 0$, $b > 0$, $c > 0$. The parameters b and c are weakly dependent on temperature, and $a = a(T)$ varies approximately quadratically with the temperature T at low temperatures and linearly at high temperatures. The equation of state can be, evidently, written as

$$\frac{H}{m} = a - bm^2 + cm^4.$$

As follows from (2) the potential $F^d(m)$ has two minima at $m = 0$ and $m = m_1$, and one local maximum at $m = m_*$, where

$$m_1^2 = \frac{b + \sqrt{b^2 - 4ac}}{2c}, \qquad m_*^2 = \frac{b - \sqrt{b^2 - 4ac}}{2c},$$

provided the inequality $b^2 - 4ac > 0$ holds. In the case where no magnetic field is applied to the d-subsystem, the function (2) describes the first order paramagnetic–ferromagnetic phase transition, when $3b^2 = 16ca_*$ at a certain temperature T_* (where $a_* = a(T_*)$). Let us regard the local maximum of $F^d(m)$ at $m = m_*$ as the characteristic energy

$$F^* = \frac{1}{4}cm_*^4(m_1^2 - \frac{1}{3}m_*^2)$$

of the d-subsystem self-interaction and treat it in the present model as a large parameter. Then in the vicinity of this phase transition the difference

$\Delta = F^d(m_1)$ between the values of the thermodynamic potential at $m = m_1$ and $m = 0$ ($F^d(0) = 0$) can be written as

$$\Delta \approx \frac{27}{4} F^* \frac{(a - a_*)}{a_*}.$$

Let us confine our consideration to the system behavior in the neigborhood of the temperature T_* assuming $|a - a_*| \ll a_*$. It should be noted that this inequality seems to hold in a wide temperature interval down to $T = 0$. Under these conditions the applied magnetic field can induce the phase transitions of the d-subsystem not affecting substantially its magnetization in the ferromagnetic state. So, when the paramagnetic state of the d-subsystem is stable itself, i.e. $a > a_*$, the applied magnetic field will induces the phase transition to the ferromagnetic state if it exceeds the critical value

$$H_p \simeq \frac{\Delta}{m_1} \simeq \frac{27 F^*}{4 m_1} \frac{(a - a_*)}{a_*}. \qquad (3)$$

Further we will consider the case in which the d-subsystem is paramagnetic down to $T = 0$. It would seem at the first glance that in this case the whole f-d system should also be paramagnetic, since the molecular field created by the d-subsystem at the f-ions must be zero. However, this is not the case. Actually, the paramagnetic state (i.e., the state with the zero spontaneous magnetization) of such a subsystem is unstable in the presence of an f-ion with a degenerate ground state and the spontaneous magnetization arises in the f-d systems.

The onset of spontaneous magnetization below the threshold, i.e., formally, in the paramagnetic phase, can be seen directly on the magnetization curves reported by R. Ballou et al. [10]. The physical meaning of this phenomenon can be explained as follows. Let us consider a spin fluctuation in a system of d-ions surrounding a certain f-ion. This fluctuation causes the ground state of the f-ion to split, making this f-ion magnetized. The latter in turn gives rise to the further magnetization of the surrounding ions due to the f-d interaction. In this way a spontaneous magnetization arises in the d-f system. In other words, after a certain time the symmetry has to be spontaneously broken. (In the case of an isolated ion, of course, quantum fluctuations should lead to a restoration of the symmetry under time reversal, but for the case of the cooperative instability discussed below this symmetry breaking has a real meaning.)

A more detailed microscopic description of this interaction goes beyond the scope of the present paper, but we would like to point out that this picture of an isolated f-ion in a d-matrix may be observed at a low concentration of f-ions, at which local spin modes do not overlap. A mode overlap determines an interaction of centers, and if the concentration of centers is sufficiently high, it will lead to cooperative effects. The threshold value t^* of the relative f-ion concentration t can be estimated with the help of the percolation model. If the distance between two nearest f-ions is smaller than the correlation length ρ of perturbations in the d-subsystem due to the f-ions, then these ions affect each other substantially. As a result, the directions of spins are correlated. The percolation threshold can then be found approximately from $N^*\rho^3 = q$, where q is a numerical factor of the order of unity; for our purposes we can set it equal to one. Using the estimate of the correlation length ρ [13, 14, 15, 16] the concentration threshold is:

$$t^* \approx (a/\rho)^3 \approx 0.1,$$

where a is the average interatomic distance in the compound.

2.2. FIELD-INDUCED PHASE TRANSITIONS AND PHASE DIAGRAMS

We now turn to phase transitions induced by the magnetic field and to phase diagrams at $T = 0$, assuming that the concentration of f-ions is large enough, so, that the behavior of the system can be regarded as cooperative. In other words, we assume $t > t^*$. In this case we can ignore the nonuniform–exchange energy and can write the thermodynamic potential in the form [1]

$$\Phi = F^d(m) - mH_{\text{eff}} - \mathbf{MH} - TS(M). \qquad (4)$$

Here the function $F^d(m)$ is given by (2), the effective magnetic field acting on the d-subsystem is specified by (1), thus

$$H_{\text{eff}} = \sqrt{H^2 + \lambda^2 M^2 - 2\lambda M H \cos\psi},$$

where ψ is the angle between the magnetic field H and the magnetization M of the f-sublattice, $\mathbf{MH} = MH\cos\psi$ is their scalar product, and $S(M)$ is the entropy of the f-subsystem.

Under the adopted assumptions we may consider the value of $\mathbf{M}$ fixed and M is constant. Then minimizing (4) with respect to m and ψ we find

$$\frac{dF^d(m)}{dm} - H_{\text{eff}}(M, \psi) = 0, \qquad (5)$$

$$(H_{\text{eff}} - \lambda m)\sin\psi = 0. \qquad (6)$$

The former equation specifies the dependence of the magnetization m of the d-sublattice on the magnetic field $h = H_{\text{eff}}$ acting on it, $m(h)$. Taking into account the results of the previous subsection let us use the following approximation for the function $m(h)$,

$$m(h) = \begin{cases} \chi_d h, & h < H_p, \\ m_1, & h > H_p, \end{cases} \qquad (7)$$

where H_p is the threshold field for the metamagnetic transition in the d-subsystem. It should be noted that the magnetization of the d-subsystem is not constant in the ferromagnetic phase and slightly increases with the field increasing. This point is simple to deal with mathematically, but taking it into account makes the analysis more complicated without leading to qualitatively new facts. Since this feature tends to obscure the overall picture, we will ignore it. Besides, for simplicity we will also ignore a possible hysteresis of the real $m(h)$ curve.

Eqs. (5) and (6), along with the stability conditions ($\delta^2 \Phi > 0$), determine the following solutions (phases) and the regions where they exist:

$$
\begin{aligned}
&W: & \psi &= 0, & m &= \chi_d(H - \lambda M_f); \\
&F_{s1}: & \psi &= 0, \theta = \pi, & m &= m_1; \\
&F_{s2}: & \psi &= \pi, \theta = 0, & m &= m_1; \\
&C: & 0 &< \psi < \pi, & m &= m_1; \\
&F_e: & \psi &= 0, \theta = 0, & m &= m_1,
\end{aligned}
$$

where θ is the angle between the magnetic field $\mathbf{H}$ and $\mathbf{m}$. The phases $W(F_{s1}$ and $F_{s2})$ can be weak (strong) ferrimagnetic collinear phases, F_e is a ferromagnetic phase, while the phase C is the canted (angular) phase.

To plot H–M phase diagrams we need to find the energies of the coexisting phases and find the curves of first-order phase transitions from these equations. In this way we get the phase diagrams shown in Fig. 1 that gives a general picture of the magnetization curves and critical fields of itinerant metamagnets. The nature of the H–M phase diagrams depends on the relation between H_p and λm_1. If $H_p < \lambda m_1/2$, the phase diagram will have, in addition to the "ordinary" lines of the second-order phase transitions between the collinear and angular phases (which are determined by the well-known expressions for the critical fields $H_1 = \lambda|m_1 - M|$ and $H_2 = \lambda|m_1 + M|$), the lines of the first-order transitions, which are determined by the equation (under the condition $\lambda \chi_d \ll 1$)

$$H_M = \frac{H_p - \lambda M}{1 - 2M/m}. \tag{8}$$

In the interval $\lambda m_1/2 < H_p < \lambda m_1$ the phase diagram (Fig. 1b) has two interesting features. Not only $W \to F_{s2}$ phase transitions can occur here, but there are also first-order transitions from the weak ferrimagnetic phase A_1 to the canted phase C. Transitions of this sort have not previously been seen in isotropic systems. Lines QL and Q^*L^* are determined by Eq. (8), and lines QN and Q^*N^* by

$$H = \lambda|M + \epsilon m_1|, \quad \text{and} \quad H = \lambda|M - \epsilon m_1|, \tag{9}$$

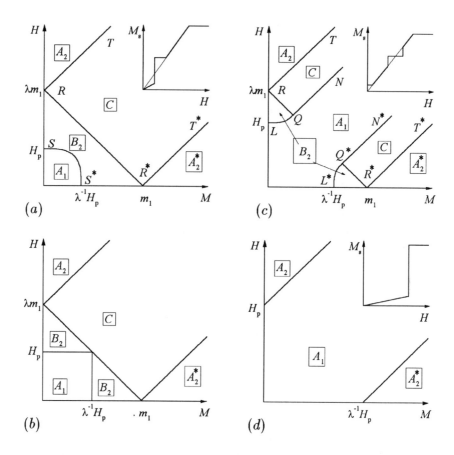

Figure 1. Phase diagrams of an itinerant ferrimagnet with an unstable d-subsystem. a) $H_p < \lambda m_1/2$; b) $H_p = \lambda m_1/2$; c) $\lambda m_1/2 < H_p < \lambda m_1$; d) $H_p > \lambda m_1$. The inserts are schematic diagrams of the corresponding magnetization curves. (after Zvezdin [26])

where $\epsilon = ((2H_p/\lambda m_1) - 1)^{1/2}$. The points Q and Q^* have the coordinates $M = \frac{1}{2}m_1(1 \pm \epsilon)$, $H = \frac{1}{2}\lambda m_1(1 \mp \epsilon)$.

Another unusual feature of this diagram is that there can be an "inverse transition" from the strong ferrimagnetic phase to the weak ferrimagnetic phase. This transition would occur as the magnetic field is increased [12]. The same feature is seen, even more clearly, when $H_p > \lambda m_1$ (Fig. 1c), where the inverse transition occurs abruptly, without an intermediate canted phase.

Fig. 2 shows the $H-T$ phase diagrams of itinerant ferrimagnets deduced from the minimization of the thermodynamic potential (4) where the temperature dependence $a(T)$ is taken into account [1].

The results presented in the given subsection consider properties of such materials assuming the f-subsystem to be isotropic. This is, however, a

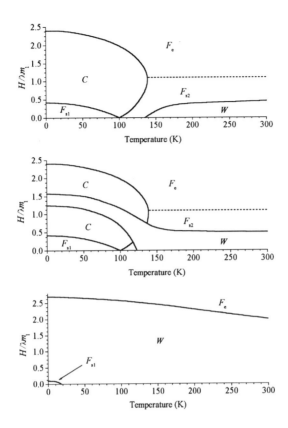

Figure 2. H–T phase diagrams of an itinerant ferrimagnetics with unstable d-subsystem. a) $H_p < \lambda m_1/2$; b) $\lambda m_1/2 < H_p < \lambda m_1$; c) $\lambda m_1 < H_p$ (after Evangelista and Zvezdin [1]).

special case and there are materials where the f-subsystem is extremely anisotropic. The properties of the field induced phase transitions in this case are the subject of the next subsection.

2.3. EFFECT OF THE MAGNETIC ANISOTROPY ON THE FIELD INDUCED PHASE TRANSITIONS

Let us consider the case where this effect is most pronounced. Namely, we study the field induced phase transition assuming the magnetization of the f-subsystem to be governed by the Ising model. In this model the f-subsystem possesses an easy direction **n** and its magnetization **M** does not practically deviate from it. So, should the materials under consideration be homogeneous the external magnetic field H can induce two first order spin-flop transitions as it increases. This is due to the fact that the field H, at first, supresses the effective field H_{eff} below the critical value H_p, and the system passes to the paramagnetic state. Then, as the field H increases, the effective field H_{eff} exceeds again the threshold H_p and the d-subsystem has to transit by a jump to the ferromagnetic state.

However, such materials are heterogenous and involve a large number of small grains randomly oriented in space. Therefore, the corresponding local phase transitions in various grains can occur at different values of the external magnetic field H. So, in these heterogenous materials the field induced phase transitions shall be fuzzy and, at least qualitatively, may be treated as second order ones. Exactly this problem is studied in detail in the present subsection.

Following the previous section we specify the local dependence of the d-subsystem magnetization $\mathbf{m}$ on $\mathbf{H}$ and $\mathbf{M} = \mathbf{n}M$ in the form

$$\mathbf{m} = m_1 \Theta \{|\mathbf{H} - \lambda \mathbf{n}M| - H_p\} \frac{\mathbf{H} - \lambda \mathbf{n}M}{|\mathbf{H} - \lambda \mathbf{n}M|}, \tag{10}$$

where $\Theta\{\zeta\}$ is the stepwise Heaviside function ($\Theta = 0$ if $\zeta < 0$ and $\Theta = 1$ for $\zeta > 0$) and the values of M and m_1 are regarded as fixed. In addition, we assume that the inequality $\lambda M > H_p$ holds. Then averaging (10) over random orientations $\mathbf{n}$ and choosing the direction of the z-axis along the external field $\mathbf{H} = \{0, 0, H\}$ we get the following formula for the dependence of the mean magnetization $\langle \mathcal{M} \rangle_z$ on the filed H:

$$\langle \mathcal{M} \rangle_z = \frac{1}{2} M + m_1 \int_0^1 d\zeta\, \Theta\{H_{\text{eff}}(\zeta) - H_p\} \frac{H - \lambda M \zeta}{H_{\text{eff}}(\zeta)}, \tag{11}$$

where

$$H_{\text{eff}}(\zeta) = \sqrt{H^2 + \lambda^2 M^2 - 2\lambda H M \zeta}.$$

The obtained dependence is illustrated by Fig. 3. As it follows from it, the phase transitions that occur at $H_l = \lambda M - H_p$ and $H_u = \lambda M + H_p$ may be regarded as the second order ones, the former being much more pronounced than the latter. Besides, it should be mentioned, that the less is the difference $\lambda M - H_p$, the more pronounced is the phase transition.

3. Experiment

3.1. $(YR)(COAL)_2$ INTERMETALLIC ALLOYS

Metamagnetic transitions in the d-subsystem almost never have been observed in pure compounds RCo_2, since estimates show that in most compounds they should occur in fields above 100 T. Nevertheless, metamagnetic transitions have been realized experimentally and studied in the substituted

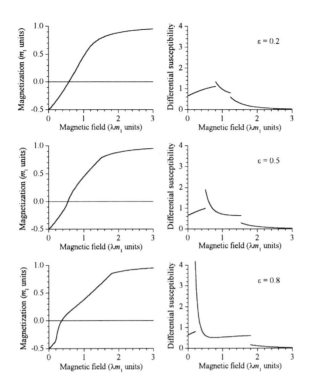

Figure 3. Magnetization and differential susceptibility of the d-sublattice of highly anisotropic itinerant ferrimagnetics vs external magnetic field for different values of $\varepsilon = H_p/(\lambda M) < 1$.

compounds $(R, Y)(Co, Al)_2$ and $(R, Lu)(C0, Al)_2$, where they are displaced to weaker fields [17]. However, the phase diagrams of the substituted ferrimagnets are more complicated, since in these compounds the metamagnetic transitions can "interfere" with transitions to the noncollinear ferrimagnetic phase. Moreover, now it is not known how the magnitude of the f-d exchange depends on the type and degree of a substitution in RCo_2.

Substituted intermetallic compounds $Y_{1-t}R_t(Co_{1-x}Al_x)_2$ (where R is a heavy rare-earth element) with negative intersublattice exchange interaction may serve as representatives of the itinerant ferrimagnets with one unstable magnetic sublattice. One stable magnetic subsystem of these two-sublattice ferrimagnets is formed by the localized moments of the $4f$-shells of rare-earth ions. The second itinerant magnetically unstable subsystem is formed by the magnetic $3d$-electrons of Co hybridized with the $5d(4f)$ electrons of rare-earth ions. For this subsystem, the density of d-states at the Fermi level and the magnitude of the d-d exchange are such that the Stoner criterion for the appearance of itinerant ferromagnetism is not fulfilled [3],

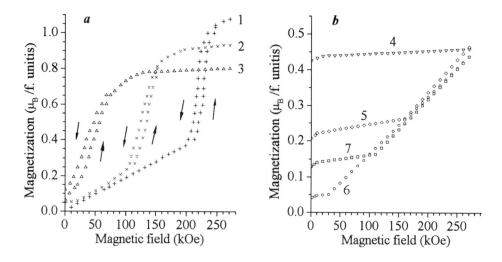

Figure 4. Magnetization curves of the compounds $(Y_{1-t}Gd_t)(Co_{0.95}Al_{0.05})_2$ at 4.2 K and $t = 0.0$ (1), 0.04 (2), 0.1 (3), 0.12 (4), 0.15 (5) 0.18 (6), 0.2 (7) (after Ballou et al. [10]).

[5]. Because of this YCo_2 and $LuCo_2$ compounds are exchange-enhanced itinerant paramagnets [18].

In RCo_2 compounds with magnetic rare-earth ions, both magnetic subsystems are magnetically ordered [19], [20]). In this case the magnetic ordering of the itinerant subsystem is extrinsic and is due to a magnetizing molecular field acting on the d-subsystem. The investigations of the f-d exchange effect on the properties of the d-subsystem have been carried out on the $(RY)Co_2$ compounds for $R = Gd, Tb, Ho$, and Er [18], [8].

In order to study the influence of the f-d interaction on the magnetic order of the unstable d-subsystem Ballou et al. [10] investigated the magnetic properties of the intermetallic compounds $Y_{1-t}Gd_t(Co_{1-x}Al_x)_2$, where $0 \leq t \leq 0.2$ and $0 \leq x \leq 0.105$. Partial replacement of the cobalt by aluminum in YCo_2 leads to a decrease of the field of the metamagnetic transition and to the appearance of the itinerant ferromagnetism in the $Y(Co_{1-x}Al_x)_2$ compounds for $x \geq 0.12$ [11].

Fig. 4 shows the field dependence of the magnetization at 4.2 K for certain compounds belonging to the system $Y_{1-t}Gd_t(Co_{0.095}Al_{0.05})_2$. It is clear that for small substitutions of Y by gadolinium ($t < 0.12$) there is no spontaneous magnetization. Increasing the gadolinium content leads to an increased weak-field susceptibility; the magnetization curves of the compounds with Gd become nonlinear and exhibit a tendency towards saturation in strong fields. Compounds with Gd content $t \geq 0.12$ possess a spontaneous moment. The value of this spontaneous moment de-

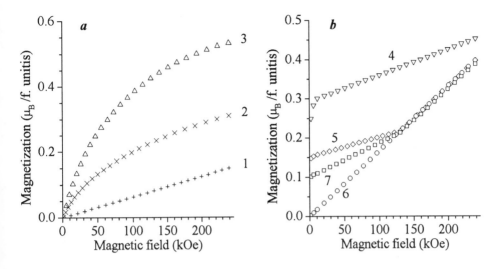

Figure 5. Magnetization curves for the compounds $(Y_{1-t}Gd_t)(Co_{0.915}Al_{0.085})_2$ at 4.2 K and $t = 0.0$ (1), 0.02 (2), 0.04 (3), 0.10 (4), 0.15 (5), 0.18 (6), 0.20 (7) (after Ballou et al. [10]).

creases as the gadolinium content increases, passing through a minimum at $t_{comp} \approx 0.17 - 0.18$ and then increasing once more. In compounds near t_{comp} kinks in the magnetization curves take place, which are characteristic of a transition from a collinear ferrimagnet to the noncollinear phase.

Fig. 5 shows the magnetization curves of several compounds of the $(Y_{1-t}Gd_t)(Co_{0.915}Al_{0.085})_2$ system at 4.2 K. It is clear that the original compound $Y(Co_{0.915}Al_{0.085})_2$ is an itinerant metamagnet with a critical magnetic transition field $H_M = 225$ kOe. As the gadolinium concentration increases, the metamagnetic transition field H_M decreases, and for concentrations $t \geq 0.04$ these compounds possess a spontaneous magnetization. At relatively small gadolinium concentrations ($0.04 \leq t \leq 0.06$) the spontaneous magnetization is small in the magnetically ordered region, and application of a field leads to a metamagnetic transition from a weakly ferrimagnetic to a strongly ferrimagnetic state. For a larger gadolinium content ($t \geq 0.06$) metamagnetic transitions are not observed: these compounds are in the strongly ferrimagnetic state even in the zero field. The saturation magnetization decreases with increasing t, and this causes it to increase once more. As with the compounds with low aluminum content, near this concentration a transition from the collinear ferrimagnetic phase to the noncollinear phase is observed in the external magnetic field.

A decrease in the metamagnetic transition field is also observed in other systems, e.g., in $Y_{1-t}Gd_t(Co_{1-x}Al_x)_2$ with high aluminium content ($x = 0.07$ and 0.105), as the gadolinium content increases. This decrease

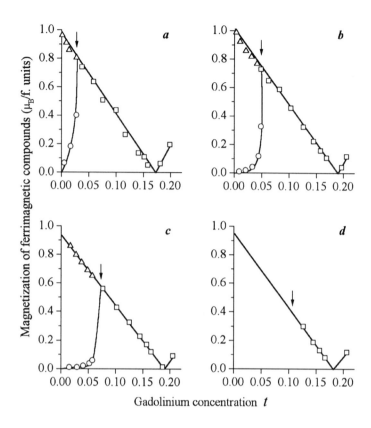

Figure 6. Dependence of the magnetization on gadolinium content t for various compounds of the system $(Y_{1-t}Gd_t)(Co_{1-x}Al_x)_2$ at 4.2 K. ○ is spontaneous magnetization of the weakly ferrimagnetic compounds, △ is magnetization of the weakly ferrimagnetic compounds in the field of 270 kOe, □ is spontaneous magnetization of the strongly ferrimagnetic compounds. $x = 0.105$ (a), 0.075 (b), 0.07 (c), 0.05 (d) (after Ballou *et al.* [10]).

is followed by the appearance of a weakly ferrimagnetic phase, which is then replaced by the strongly ferrimagnetic phase. The only difference is that increasing of the aluminium content in these compounds causes the metamagnetic transition field to increase. It also causes the concentration region where the paramagnetic and weakly ferromagnetic metamagnetic phases can exist to shrink. At the same time, the values of the magnetization and gadolinium concentrations at which a compensation of the magnetic moments of the f- and d-subsystems is observed are close for all these systems, although with certain differences. Furthermore, the spontaneous magnetization of the low-aluminium-content system with $x = 0.05$ is also close to the magnetization of those compounds corresponding to systems with larger amounts of aluminium ($x = 0.07, 0.085$, and 0.105).

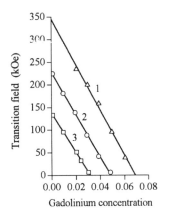

Figure 7. Dependence of the metamagnetic transition field on gadolininium concentration t for the system $(Y_{1-t}Gd_t)(Co_{1-t}Al_x)_2$: $x = 0.07$ (1), 0.085 (2), 0.105 (3). The straight line is calculated using Eq. (8) and the points are experimental data after Ballou et al. [10].

All of these features are easy to see in Figs. 6 and 7, where the basic characteristics of all the discussed systems are plotted. Fig. 6 presents the data on the magnetization as a function of the gadolinium concentration for systems with various values of x. Here the spontaneous magnetization of weakly ferrimagnetic and strongly ferrimagnetic samples are shown as well as the magnetization in the field 270 kOe for samples with metamagnetic transitions. Fig. 7 shows the measured dependence of the metamagnetic transition field on gadolinium content for systems with various aluminium contents. A comparison of the experimental data with calculations for f-d magnetic systems shows that they agree in most of the cases, at least qualitatively.

3.2. RCO_2 IN THE MEGAGAUSS RANGE FIELDS

We measured the critical fields of metamagnetic transitions in an itinerant d-subsystem in compounds RCo_2 with most of the heavy rare-earths. The measurements were performed at 4.2 K by an induction method in the pulsed magnetic fields up to 300 T, produced by an explosive method [21], on powders of the compounds RCo_2. The rise time of the field in a pulse was 15 μs. These measurements have the drawback that the signal from the magnetic field cannot be completely compensated, since the measurements are performed only once: the measuring coils and the samples are destroyed after each pulse. Therefore, the signal induced in the measuring coils can

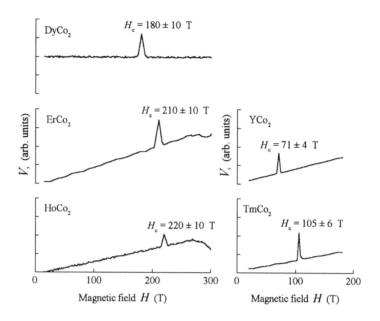

Figure 8. Field dependence of the signal induced in the measuring coil of a pulsed solenoid for the compounds RCo_2 at 4.2 K (after R. Z. Levitin et al. [27]).

be written in the form:

$$V(H) \sim K\frac{DH}{dt} + \frac{dM}{dt} = \frac{dH}{dt}(K + \frac{dM}{dH}).$$

Since dH/dt is a monotonic function of the field, we can determine the magnitude of the critical field according to the maximum of the signal $V(H)$. The experiments were performed several times in order to eliminate random noise arising in the megagauss fields.

Fig. 8 displays the signals induced in the measuring coils for some intermetallics RCo_2 as a function of the magnetic field. Peaks indicating a sharp increase in magnetization are clearly seen in all curves. The fields corresponding to these resonances can be regarded as the critical fields of the metamagnetic transitions. For YCo_2 and $LuCo_2$ where there are only two itinerant subsystem the values of these fields agree within the experimental error with the value of H_M obtained by Goto et al. [6].

The critical fields obtained for the intermetallics with the magnetic rare earths are given in Table 1. The Table also gives the critical fields calculated from the values obtained from the indirect data: the Curie temperatures of the RCo_2 compounds [22] and the magnetization curves of the substituted compounds $(R,Y)(Co,Al)_2$ from [17]. By comparing these values with the experimental ones for the critical fields we were able to identify the ob-

TABLE 1. Critical fields of the metamagnetic transitions in RCo_2

	H_{c_1}, T			H_{c_2}, T	
Experiment	Theory [22]	Theory [23]	Experiment	Theory [22]	Theory [23]
Tm		4	3; -12 [25]	105	144	143; 128 [25]
Er	55 [24]	83	123	210	223	263
Ho	220	160	190		300	330
Dy	180	220	300		360	440

served transitions as metamagnetic accompanied by the demagnetization (reentrant transitions) or magnetization of the d-sublattice. It is obvious that the agreement between the values of the critical fields determined experimentally in the present work is only qualitative. The discrepancies could be due to the fact that the main formulae (9) for the critical fields were obtained in the exchange approximation, i.e. the effects of the crystal fields were neglected. Moreover, it was assumed that the itinerant d-subsystem is identical in all compounds RCo_2, and, therefore, the critical field H_M in the d-subsystem does not depend on the atomic number of the rare earth element. Probably, this simple model is too rough, since, the features of the d-electron density of states are different in different intermetallics RCO_2, which could affect the value of H_M, as shown in [17].

References

1. Evangelista, L. R. and Zvezdin, A. K. (1995) Phase Transitions and Critical Points in Rare Earth-Transition Metal Ferrimagnets, *J. Magn. Magn. Mater.*, **140-144**, pp. 1569-1570.
2. Wohlfarth, E. P. and Rhodes, P. (1962) Collective Electron Metamagnetism, *Phil. Mag.*, **7**, pp. 1817–1824.
3. Bloch, D, D., Edwards, M., Shimizu, M., and Voiron, J. (1975) First Order Phase Transitions in ACo_2 compounds, *J. Phys.*, **F 5**, pp. 1217–1226.
4. Cyrot, M. and Lavagna, M. (1979) Density of States and Magnetic Properties of Rare-Earth Compounds RFe_2, RCo_2, RNi_2, *J. Phys. (Paris)*, **40**, pp. 763–771.
5. Yamada, H., Jnoue, J., Terao, K., Kanda, S., and Shimizu, M. (1984) Electron Structure and Magnetic Properties of YM_2 Compounds ($M = Mn, Fe, Co, Ni$), *J. Phys.*, **F 14**, pp. 1943–1960.
6. Goto, T., Fukamichi, K., Sakakibara, T., and Komatsu, H. (1989) Itinerant Electron Metamagnetism in YCo_2, *Solid State Commun.*, **72**, pp. 945–947.
7. Murata, M., Fukamichi, K., Komatsu, A., Sakakigura, E., and Goto, T. (1991) Itinerant Electron Metamagnetic Transitions in Exchange Enhanced Paramagnetic Compounds $Lu(CO_{1-x}Sn_x)_2$, *J. Phys.: Conden. Matt.*, **3**, pp. 2515–2521.
8. Duc, N. H., Hien, T. D., Brommer, P. E., and Franse, J. J. M., (1988) Electronic and Magnetic Properties of $Er_xY_{1-x}Co_2$, *J. Phys.*, **F 18**, pp. 275–284.

9. Steiner, W., Gratz, E., Ortbauer, H., and Camen, W. (1978) Magnetic Properties, Electric Resistivity and Thermal Expansion of $(Ho, Y)Co_2$ Compounds, J.Phys., F **8**, pp. 1525-1527.
10. Ballou, R., Gamishidze, Z. M., Lemaire, R., Levitin, R. Z., Marcosyan A. S., and Snegirev, V. V. (1992) The Effect of f-d Interaction on the Magnetic State of d-Subsystems in the Itinerant Magnets $Y(Co_{1-x}Al_x)_2$: Investigation of the Compounds $Y_{1-x}Gd_x(Co_{1-x}Al_x)_2$, Sov. Phys. JETP, **75**, pp. 1041-1048.
11. Aleksandryan, V. V., Lagutin, A. S., Levitin, R. Z., Marcosyan A. S., and Snegirev V. V. (1985) Itinerant Metamagnetism of d-Electrons in $ItCo_2$: Study of Metamagnetic Transitions in $It(Co, Al)_2$, Sov. Phys. JETP, **62**, pp. 153-157.
12. Utochkin, S. N. and Zvezdin, A. K. (1992) Itinerant Metamagnetism in f-d Systems, J. Mag. Magn. Mat., **104-107** pp. 1479-1480.
13. Belov, K. P., Zvezdin, A. K., Kadomtseva, A. M., and Levitin, R. Z. (1976) Spin-Reorientation Transitions in Rare-Earth Magnets, Uspechi Fiz. Nauk (Sov. Adv. Phys.)), **119**, pp. 447-486.
14. Zvezdin, A. K., Muhin, A. A., and Popov, A. I. (1976) Magnetic Structure Instability due to the Energy Level Intersection, JETP Letters, **23**, pp. 267-271.
15. Belov, K. P., Zvezdin, A. K., Kadomtseva, A. M., and R. Z. Levitin (1979) Spin-reorientation Transitions in Rare-Earth Magnets. Nauka, Moscow (in Russian).
16. Zvezdin, A. K., Matveev, A. A., Muhin, A. A., and Popov, A. I. (1985) Rare-Earht Ions in magnetically ordered crystals. Nauka, Moscow.
17. Dubenko, I. S., Levitin, R. Z., Markosyan, A. S., Snegirev, V. V., and Sokolov, A. Yu (1995) f-d Exchange Interaction in the RCo_2 Type Intermetallic Compounds with Heavy Rare-Earthes, J. Magn. Magn. Mater., **140-144**, pp. 825-826.
18. Lemaire, R. and Schweizer, J. (1966) Changes in the Cobalt Magnetic Moment in the $Cd_uY_{1-u}Co_2$ compounds, Phys. Lett., **21**, pp. 366-368.
19. Bloch, D. and Lemaire, R. (1970) Metallic Alloys and Exchange Enhanced Paramagnetism. Application to RE Cobalt Alloys, Phys. Rev., **B 2**, pp. 2648-2650.
20. Levitin, R. Z. and Markosyan, A. S. (1988) Itinerant Metamagnetism, Sov. Phys. Usp., **155**, pp. 623-657.
21. Pavlovskii, A. I., Kolokol'chikov, N. P., and Tatsenko, O. M. (1980) Megagauss physics and Techniques, edited by Turchi P. Plenum Press, New York.
22. Duc, N. H., Hien, T. D., and Givord, D. (1992) Magnetic Coupling in the CdT intermetallics $(T = Fe, Co)$, J. Magn. Magn. Mater., **104-107**, pp. 1343-1344.
23. Dubenko, I. S., Levitin, R. Z., Markosyan, A. S., Sokolov, A. Yu, and Snegirev, V. V. (1995) Exchange of f-d Interaction in Rare-Earth Compounds with RCo_2, Sov. Phys. JETP, **80**, pp. 296-305.
24. Goto, T. Private communication.
25. Brommer, P. E., Dubenko, I. S., Franse, J. J., Levitin, R. Z., Markosyan, A. S., Radvansky, R. J., Sokolov, A. Yu, and Snegirev, V. V. (1993) Field Induced Noncollinear Magnetic Structures in Al Stabilized RCo_2 Lavis Phases, Physica **B 183**, pp. 364-368.
26. Zvezdin, A. K. (1993) Spin-Flop Transitions in Itinerant Metamagnets JETP Letters, **58**, pp. 719-725.
27. Dubenko, I. S., Zvezdin, A. K., Lagutin, A. S., Levitin, R. Z., Markosyan, A. S., Platonov, V. V., and Tatsenko, O. M. (1996) Investigation of Metamagnetic Transitions in the Itinerant d-Subsystem of the Intermetallics RCo_2 in Superstrong Magnetic Field up to 300 T, JETP Lett., **64**, pp. 202-206 (Pis'ma ZhETF **64**, pp. 188-192, 1996).

THE PHASE DIAGRAM OF THE KONDO LATTICE

C. LACROIX
*Laboratoire Louis Néel, CNRS,
BP 166, 38042 Grenoble-Cedex 9, France*

1. The Kondo Lattice

Heavy Fermions systems have received much attention from both experimental and theoretical points of view because they exhibit many unusual phenomena due to the competition between the long range RKKY interactions, $I(R_i-R_j)$, and the local negative s-f exchange, J, responsible for the Kondo effect: the ground state can be either magnetic or nonmagnetic depending on the relative strength of these interactions, $\lambda = T_K/I$ ($T_K \approx 1/\rho \exp(1/\rho J)$ is the characteristic Kondo energy) [1], which can be varied by applying pressure or alloying. At T=0°K a critical value of λ, λ_C, separates the two phases and non-Fermi liquid behavior has been observed around λ_C in many cases [2]. Also in several heavy fermions compounds superconductivity can be stabilized close to this critical point [3]. Since the first paper of Doniach [1] there have been a large number of articles devoted to this problem. The purpose of the present paper is not to give a review on this problem but rather to stress a few points such as the importance of the band filling, the role of intersite exchange interactions and the essential difference between ferro- and antiferromagnetic intersite exchange.

The description of heavy fermions involves two types of electrons: conduction electrons (s, p, or d electrons) and localized 4f or 5f electrons, which hybridize with the conduction band. The number of conduction electrons, n, is an important parameter for determining the ground state of these systems [4]: λ_C depends on n, and a semiconducting ground state is usually obtained if n=1. On the other hand, it was proposed by Nozieres that in a lattice, screening of all Kondo ions is not possible because it requires too many conduction electrons. In Section 2 the stability of the magnetic ground state will be discussed, and we emphasize the role of the conduction electrons concentration. In Section 3 we will show that screening is more easily realized in the presence of antiferromagnetic interactions between the magnetic ions. We also discuss in this paper the applicability of the present model to some magnetic and non-magnetic Cerium compounds.

We start from the Hamiltonian

$$H = \sum_{k\sigma} \varepsilon_k c_{k\sigma}^+ c_{k\sigma} + E_0 \sum_{i\sigma} n_{i\sigma}^f - J \sum_i \mathbf{s}_i \cdot \mathbf{S}_i - I \sum_{i\delta} \mathbf{S}_i \cdot \mathbf{S}_{i+\delta}, \tag{1}$$

where J is the local negative exchange interaction between the f-electron spin, $\mathbf{S}_i$, and the conduction electron spin, s_i; I is the intersite exchange between f-spins, and we consider here only the nearest neighbor interaction. In principle, the RKKY exchange interaction is induced by the local exchange, and I is long ranged and proportional to ρJ^2 (ρ is the conduction electrons density of states), but here we introduce both parameters.

The competition between the RKKY interaction and Kondo effect has been extensively studied for the two impurities problem [5, 6, 14] and an exact solution of the 1D Kondo lattice has been recently proposed [7]. For the 3D Kondo lattice model no exact results exist and we present here the results within the mean field approximation, for the Kondo interaction this approximation was introduced in [8]: the local exchange term of Eq. (1) is linearized by introducing a fictitious hybridization calculated self-consistently:

$$V_{i\sigma} = 2J \left\langle f_{i\sigma}^+ c_{i\sigma} \right\rangle. \qquad (2)$$

It is usually supposed that $V_{i\sigma}$ is independent on i and σ. In the following we define $V = V_{i\sigma}$.

This approximation has been used by several authors [8, 9, 10] and it is quite similar to the slave boson method [11]. Both methods give a simple explanation of the origin of the Heavy Fermion behavior: they yield a renormalization of the density of states for the f-quasiparticles, which is proportional to $1/T_K$ leading to a large enhancement of the effective mass.

2. The Magnetically Ordered Phase

In order to describe the consequences of the Kondo interaction in the magnetic phase it is necessary to introduce several self-consistent parameters: the Kondo parameter V introduced in the preceding Section (Eq. (1)), the magnetic moment of the f-electrons, μ_i, and eventually of the conduction electrons, μ'_i. The resulting T=0 phase diagram is shown on Figure 1.

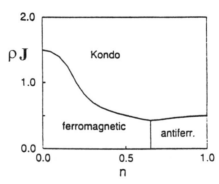

Figure 1. Phase diagram obtained in the mean field approximation as a function of n [8]

It is found that the number of conduction electrons, n play an important role because it determines the type of magnetic ordering which is stabilized: for small values of n the interaction is ferromagnetic and a quite large value of J is necessary to destroy the ferromagnetic order; in [8] it was found that the ferromagnetic state is stable up to $\rho|J| \leq 1.5$ for small n. In the 1D case the ferromagnetic state is stable for all values of J if n is small [7]. On the other hand if the band is nearly half-filled the RKKY interaction is antiferromagnetic and the critical value of J above which ordering is destroyed is much smaller: it is found that Kondo state is stable for $\rho|J| > 0.5$. There are two reasons for this difference: (i) the Kondo effect is more effective in the antiferromagnetic phase since the ground state in the absence of the Kondo interaction is already a singlet; in the ferromagnetic case the effective moment is large and complete screening requires a larger value of J. This is in agreement with the two-impurities results [5, 6]; (ii) in the half-filled case more conduction electrons can participate in the screening. As it is pointed out in [4] that this factor is important in the Kondo lattice since the number of conduction electrons per the Kondo impurity is limited, which is different from the two impurities case.

Figure 2 shows the schematic behavior of the Néel temperature T_N as a function of the parameter ρJ : it exhibits a maximum and vanishes at the critical value $(\rho J)_c$ above which the ground state is non-magnetic; the critical value is reached when the one-impurity Kondo temperature T_K^0 is approximately equal to the intersite exchange energy I or ρJ^2. On this figure T_{N0} represents the Néel temperature in the absence of the Kondo effect ($T_{N0} \approx |I|$). This diagram was proposed initially by Doniach [1] and is known as the "Doniach diagram".

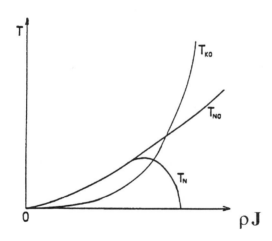

Figure 2. Schematic behavior of the Néel temperature as a function of J.

Such a behavior of the ordering temperature has been observed in many Cerium compounds in which a variation of J can be induced by pressure or chemical pressure. Some well known examples are CeAg, $CePd_2Al_3$, Ce(Ni,Pt), $CeSi_x$...(for a review see [15]). In the magnetic phase, the f-magnetic moments are often reduced compared to their ionic value due to the Kondo effect; such a reduction of the Cerium magnetic moments is also present in several compounds.

In the vicinity of the critical point the non-Fermi liquid behavior has been observed [16, 17] and described by several authors [2, 13]. Different mechanisms have been proposed for this behavior, such as large spin fluctuations near the critical point, intrinsic disorder or the multi-channel Kondo effect. Another kind of a ground state can be stabilized near the critical point in the presence of frustrated interactions: in the compounds UNi$_4$B [18] , CeSb [19] or CePdAl [20] magnetic and Kondo sites coexist in the ordered state giving rise to peculiar magnetic structures. In a system which is close to the magnetic Kondo instability frustration of the intersite interactions may be avoided if some sites are non-magnetic. Such frustrated exchange interactions are always present either due to the crystallographic structure or to the long-range nature of the RKKY mechanism.

3. The Non-Magnetic Phase

In the mean field approximation the Kondo temperature of the lattice T_K is found to be equal to the one-impurity Kondo temperature, i.e. $T_K^0 = 1/\rho \exp(1/\rho J)$. However there is an experimental evidence that T_K should be reduced compared to T_K^0: (i) compounds which exhibit the largest effective masses are close to the critical point while T_K^0 is expected to increase continuously with ρJ [21]; (ii) in many Cerium compounds two characteristic temperatures appear, and the lowest one, being often called the coherence temperature, is much smaller than T_K^0 [22]. This behavior can be easily understood if intersite correlations are taken into account in the non-magnetic phase. In fact such short range magnetic correlations have been observed by neutron diffraction in several Cerium compounds such as CeCu$_6$, CeRu$_2$Si$_2$ [23].

Starting from the Hamiltonian (1), it is possible to describe the magnetic correlations by decoupling the intersite interaction term and introducing a new mean field parameter as was proposed in [10], similar to the RVB decoupling:

$$\Gamma_{ij} = \left\langle f_{i\sigma}^+ f_{j\sigma} \right\rangle \quad (3)$$

The calculation at a finite temperature yields the following results [21] for the antiferromagnetic exchange I: (i) two characteristic temperatures are obtained: T_K is defined by the temperature at which the Kondo parameter V given by (2) vanishes and T_c is defined as the temperature at which the intersite correlations Γ vanish; (ii) T_c is always found to be larger (or equal) than T_K; (iii) T_K is substantially reduced compared to T_K^0. These results are summarized on Figure 3 where we have supposed that J and I are not independent, $I = -\alpha \rho J^2$ ($\alpha = 2$ in Fig. 3).

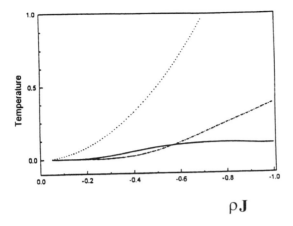

Figure 3. Variation of T_K and T_c as a function of ρJ [21] (solid, dashed, and dotted curves are related to T_K, T_K^0, and T_c, respectively)

It can be seen on Figure 3 that T_K has a much weaker dependence on ρJ than T_K^0 and exhibits a weak maximum around $\rho |J| \approx 0.5$ which is very close to the critical value at which the magnetic - non magnetic transition takes place. Thus, this model provides an explanation for the large effective masses observed near the transition. Moreover, such a behavior of T_K has been observed in several Ce compounds like $CeRh_2Si_2$ under pressure [24]. This model also gives a clear explanation of the occurence of short range antiferromagnetic correlations observed at temperatures larger than T_K [23].

It is also interesting to look at the case of ferromagnetic intersite exchange (I>0): in this case the Kondo state is not stabilized by intersite magnetic correlations. As discussed in Section 2, this difference is again due to the fact that the Kondo effect is not favored by ferromagnetic intersite correlations since screening requires too many conduction electrons as predicted by Nozières [4]. This could explain the fact that the heavy fermion behavior is much more often observed in nearly antiferromagnetic systems than in nearly ferromagnetic ones.

References

1. Doniach, S. (1977) The Kondo lattice and weak antiferromagnetism, *Physica* **B91**, 231-234.
2. Moriya, T. and Takimoto, T. (1995) Anomalous properties around magnetic instability in Heavy electron systems, *J. Phys. Soc. Jpn.* **64**, 960-969.
3. Steglich, F., Geibel, C., Modler, R., Lang, M., Hellmann, P., and Gegenwart, P. (1995) Classification of strongly correlated f - electrons systems, *J. Low Temp. Phys.* **99,** 267-281.
4. Nozieres, P. (1985) Magnetic impurities and Kondo effect, *Ann. Phys. (Paris)* **10**, 19-35.
5. Jones, B.A. and Varma, C.M. (1989) Critical point in the solution of the magnetic impurity problem, *Phys. Rev.* **B40**, 324-329.
6. Fye, R.M. and Hirsch, J.E. (1989) Quantum Monte Carlo study of the two impurity Kondo Hamiltonian, *Phys. Rev.* **B40**, 4780-4796.

7. Tsunetsugu, H., Sigrist, M., and Ueda, K. (1997) The ground state phase diagram of the one-dimensional Kondo lattice model, *Rev. Mod. Phys.* **69**, 809-863.
8. Lacroix, C. and Cyrot, M. (1979) Phase diagram of the Kondo lattice, *Phys. Rev.* **B20**, 1969- 1976.
9. Irkhin, V.Y. and Katsnelson, M.I. (1990) On the mean field theory of magnetically ordered Kondo lattices, *J. Phys.: Condens. Matter* **2**, 8715-8719.
10. Coleman, P. and Andrei, N. (1989) Kondo stabilized spin liquids and heavy fermion superconductivity, *J. Phys.: Condens. Matter* **1**, 4057-4080.
11. Coleman, P. (1984) New approach to the mixed valence problem, *Phys. Rev.* **B29**, 3035-3044.
12. Read, N. and Newns, D.M. (1983) On the solution of the Coqblin-Schrieffer hamiltonian by the large-N expansion technique, *J. Phys.* **C16**, 3273-3295.
13. Millis, A.J. (1993) Effect of a nonzero temperature on quantum critical points in itinerant fermion systems, *Phys. Rev.* **B48**, 7183-7196.
14. Gorkov, L.P. and Kim, J.H. (1996) Interplay between the Kondo effect and the RKKY interaction in the two-impurity model, *Phil. Mag.* **B74**, 447-456.
15. Coqblin, B., Le Hur, K., Iglesias, J.R., Lacroix, C., and Arispe, J. (1997) On the competition between magnetism and the Kondo effect in Cerium compounds, *Mol. Phys. Rep.* **17** (to be published).
16. Maple, M.B., de Andrade, M.C., Herrmann, J., Dalichaouch, Y., Gajewski, D.A., Seaman, C.L., Chau, R., Movshovich, R., Aronson, M.C., and Osborn, R. (1995) Non Fermi liquid ground states in strongly correlated f-electron materials, *J. Low Temp. Phys.* **99**, 223-249.
17. Lohneysen, H.V., Pietrus, T., Portisch, G., Scchlager, H.G., Schroder, A., Sieck, M., and Trappman, T. (1994) Non-Fermi liquid behavior in a heavy fermion alloy at a magnetic instability, *Phys. Rev. Lett.* **72**, 3262-3265.
18. Lacroix, C., Canals, B., and Nunez-Regueiro, M.D. (1996) Magnetic ordering and Kondo screening in frustrated UNi_4B, *Phys. Rev. Lett.* **77**, 5126-5129.
19. Coqblin, B., Arispe, J., Iglesias, J., Lacroix, C., and Le Hur, K. (1996) Competition between the Kondo effect and Magnetism in Heavy Fermion Compounds, *J. Phys. Soc. Jpn.* **65** Suppl. B, 64-77.
20. Nunez-Regueiro, M.D., Lacroix, C., and Canals, B. (1997) Magnetic ordering in the frustrated Kondo lattice compound CePdAl, *Physica* **C282-287**, 1885.
21. Iglesias, J.R., Lacroix, C., and Coqblin, B. (1997) Revisited Doniach diagram: influence of short-range antiferromagnetic correlations in the Kondo lattice, *Phys. Rev.* **B56**, 11820-11826.
22. Mielke, A., Rieger ,J.J., Scheidt, E.W., and Stewart, G.R. (1994) Important role of coherence for the heavy fermion state in $CeCu_2Si_2$, *Phys. Rev.* **B49**, 10051-10053.
23. Rossat-Mignod, J., Regnault ,L.P., Jacoud, J.L., Vettier, C., Lejay, P., Flouquet ,J., Walker, E., Jaccard, D., and Amato, A. (1988) Inelastic neutron scattering study of Cerium heavy fermion compounds , *J. Magn. Magn. Mat.* **76-77**, 376-384.
24. Graf, T., Thompson, J.D., Hundley, M.F., Movshovitch, R., Fisk , Z., Mandrus, D., Fischer, R.A., and Phillips, N.E. (1997) Comparison of $CeRh_2Si_2$ and $CeRh_2-xRu_xSi_2$ near their magnetic-nonmagnetic boundaries, *Phys. Rev. Lett* **78**, 3769-3772.

PRESSURE EFFECT ON THE MAGNETIC SUSCEPTIBILITY OF THE YBINCU$_4$ AND GDINCU$_4$ COMPOUNDS

I.V. SVECHKAREV, A.S. PANFILOV AND S.N. DOLJA
*B. Verkin Institute for Low Temperature Physics
and Engineering, Kharkov, 310164, Ukraine*

AND

H. NAKAMURA AND M. SHIGA
*Department of Material Science and Engineering,
Kyoto University, Kyoto 606-01, Japan*

Abstract. Pressure effect on the magnetic susceptibility of YbInCu$_4$ and GdInCu$_4$ compounds has been studied at fixed temperatures of 78, 150 and 300 K under helium gas pressure up to 2 kbar. The data for the YbInCu$_4$ compound at $T \geq 100$ K are well described in terms of the characteristic Kondo temperature change $dT_K/d\ln V = -640$ K, which is typical for other Kondo systems and far exceeds the corresponding derivative values for usual mechanisms of the indirect $f - f$ interaction. Moreover, the last contribution reveals a negligible pressure dependence for the reference GdInCu$_4$ compound. For the YbInCu$_4$ compound at $T = 78$ K, the pressure effect is significantly enhanced and is explained, in line with the temperature dependence of the susceptibility itself near the transition temperature T_V, in terms of spatial dispersion of its pressure dependence due to some atomic disorder.

1. Introduction

The integration of local f-level of rare-earth ions into quasi-continuous states of the band spectrum in metal systems has been of great interest to researchers over many years but the problem is still far from being solved . In some cases information on the electron parameters that characterize this integration can be obtained from the temperature dependence of the magnetic susceptibility $\chi(T)$. A pressure effect on the susceptibility $d\ln\chi/dP$

(or the magnetovolume effect $d\ln\chi/d\ln V$) provides, along with the improvement of reliability of the parameters to be obtained, their dependence on atomic volume. This dependence is of importance in elucidating both a microscopic origin of the parameters and their role in determination the type of the phase transition between the modes (or phases) with different extent of the interaction between f-levels and band states.

In particular, under change of different thermodynamic variables or a composition, the transition from the Kondo or heavy fermion state (HFS) to the intermediate valence state (IVS) may proceed gradually as, for instance, in the $CeIn_{3-x}Sn_x$ system [1] or it may be realized through the first-order phase transition followed by a considerable change in the volume with the lattice symmetry remaining unchanged. Examples of the latter type of valence transition are pure cerium in which the transition with volume change of about $\Delta V/V \sim 0.1$ [2] is initiated by high pressure and the $YbInCu_4$ compound in which the IVS-HFS phase transition is realized with increasing of the transition temperature up to $T_V \simeq 40$ K ($\Delta V/V=0.05$ [3, 4]. The transition temperature in this compound is very sensitive to pressure and magnetic field [4, 5], and variations in the composition result in a change from the jump-like first order transition to a smooth one [6]. Hence, the $YbInCu_4$ compound is particularly attractive and convenient for study with the above magnetovolume effect applied.

In this paper we present the results of the study of the pressure effect up to 2 kbar at $78 \leq T \leq 300$ K on magnetic susceptibility of the unstable-valence compound $YbInCu_4$ and its isostructural analog with the stable valence $GdInCu_4$ as a reference one. Up to now, this effect has been studied for the nonstoichiometric compound $Yb_{0.8}In_{1.2}Cu_4$, resulting in the $T_V(P)$ dependence only [5]. The magnetostriction data for $YbInCu_4$ [4] cover the range of phase transition and are more illustrative in revealing a shift of T_V in the high magnetic fields. There are no reliable data on the magnetovolume effect beyond the transition region.

The second section of the paper describes the experimental procedure and results, the third one concerns the role of disorder in the magnetic susceptibility of the $YbInCu_4$ sample and the behavior of the paramagnetic Curie temperature under pressure. Summary presents the basic conclusions.

2. Experimental Procedure and Results

Polycrystalline ingots of the $YbInCu_4$ and $GdInCu_4$ compounds were prepared from a stoichiometric composition of the elements (Yb, Gd – 3N; Cu, In – 5N purity) in an argon arc furnace. The ingots were then annealed in evacuated quartz tubes for a week at 750-850°C. The samples of appropriate sizes were spark-cut from the ingots.

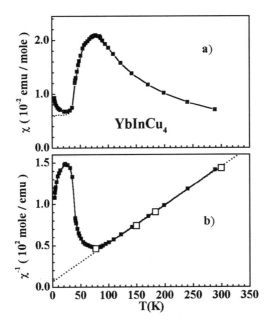

Figure 1. Temperature dependence of the magnetic susceptibility for YbInCu$_4$: (a) the correction for low-temperature impurity "tail" is shown by the dashed curve, and its reciprocal value, (b) the dashed straight line corresponds to the Curie–Weiss law (for explanation of the symbols, see the text).

The temperature dependence of the magnetic susceptibility was measured by the Faraday method at $T = 4.2 - 300$ K in the magnetic field of 0.8 T. The effect of helium-gas-produced pressure (up to 2 kbar) on the susceptibility was studied at temperatures 78, 150 and 300 K and the magnetic field up to 2 T by using a pendulum magnetometer placed directly in the high-pressure chamber [7]. In both types of measurements the absolute value of the magnetic susceptibility appears to be almost identical (see Fig. 1 where the pendulum magnetometer data at zero pressure are marked by the squares). The relative error of measurements under pressure was not more than 0.05%. In all the cases the effect of pressure was linear, and the hysteresis was not observed up to $T = 78$ K.

The temperature dependence of the susceptibility for the stable valence compound GdInCu$_4$ of in the region studied obeys the Curie-Weiss law:

$$\chi(T) = C/(T - \Theta) \qquad (1)$$

with the Curie temperature Θ and constant C given in Table 1.

TABLE 1. Magnetic parameters and their pressure derivatives in RInCu$_4$ compounds

Compound	T (K)	$10^3\chi$ (emu/mole)	$d\ln\chi/dP$ (Mbar^{-1})	Θ (K)	μ_{eff} (μ_B/f.u.)	$d\Theta/dP$ (K/Mbar)
GdInCu$_4$	300	22.2	0.15±0.2	−47	7.86	30±30
	77.5	61.9	0.25±0.3	−45 [8]	8.15 [8]	
YbInCu$_4$	300	6.97	2.2±0.3	−14±3	4.17	640±50
	150	13.5	4.0±0.3	−7.2 [9]	4.37 [9]	
	77.5	21.6	25.5±1.0	−15 [10]		

These quantities are in good agreement with the data presented in [8, 11], so the dependence $\chi(T)$ for GdInCu$_4$ is not shown here. The measured value of the magnetic moment corresponds to the ground state of a free Gd^{3+} ion ($\mu = 7.94\mu_B$).

The pressure effect on the susceptibility of this compound (Fig. 2) and hence in the paramagnetic Curie temperature,

$$d\ln\chi/dP = d\ln C/dP + (\chi/C)d\Theta/dP = (\chi/C)d\Theta/dP, \qquad (2)$$

($d\ln C/dP=0$ for the pure ion moment) appeared to be lower then the experimental resolution: $d\ln\Theta/dP=-0.7 \pm 0.7$ Mbar^{-1} (Table 1).

The temperature dependence of the magnetic susceptibility of YbInCu$_4$ is shown in Fig. 1. As a whole it is consistent with the results of other works [6, 9, 10, 12, 13, 14]: a low temperature plateau with an impurity "tail", a sharp rise at $T \sim 40$ K, a smooth maximum in the vicinity of $T \sim 80$ K with a subsequent hyperbolic decay. The low temperature Curie law "tail" is described by the free Yb^{3+} ions contribution in amounts of about 0.5% of the total number of the Yb atoms. They are supposed to be stabilized on the structural imperfections. Above the phase transition temperature ($100 \leq T \leq 300$ K) the magnetic susceptibility obeys the Curie-Weiss law (Fig. 1(b)) with the parameters which are comparable to those obtained for the samples similar in quality (Table 1).

For YbInCu$_4$ the total splitting of the multiplet $J = 7/2$ of the Yb ion ground state by the crystal electric field (CEF) is modest ($\Delta = 44$ K [15]). Therefore, there is no need to take it into consideration for the HF phase susceptibility at $T \geq 100$ K [12, 14] (as well as the non-f-state contributions based on the weak susceptibility of "nonmagnetic" analogs YInCu$_4$ and LuInCu$_4$ [16]). Thus, a certain discrepancy between the observed magnetic moment μ and its value for the free Yb^{3+} ion ($\mu/\mu_B = 4.54$) may be a consequence of the Kondo demagnetization [17].

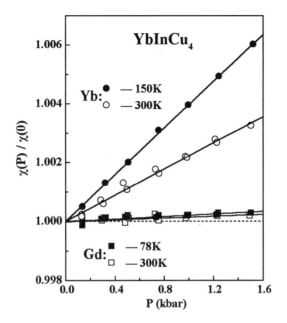

Figure 2. Pressure dependence of the magnetic susceptibility for YbInCu$_4$ and GdInCu$_4$.

The pressure effect on the susceptibility of YbInCu$_4$ (see Fig. 2) is characterized by high values of derivatives $d\ln\chi/dP$ presented in Table 1. In the high temperature region, where the Curie–Weiss law takes place, these derivatives fit Eq. (2) with $d\ln C/dP=0$ -(a linear part of the $d\ln\chi/dP$-versus-χ dependence and a dashed extension in Fig. 3b).Thus, we obtain $d\Theta/dP=640 \pm 50$ K/Mbar.

But at $T = 78$ K that corresponds to the peak in the $\chi(T)$ curve (Fig. 3a) the derivative $d\ln\chi/dP$ is several times higher that the expected one (a dashed line in Fig. 3b). Such behavior of $d\ln\chi/dP$ as well as the $\chi(T)$ dependence itself in this temperature region may be affected by the phase transition and needs a special analysis (see below).

3. Discussion of Results

3.1. YBINCU$_4$ COMPOUND: THE VICINITY OF T_V

The high pressure sensitivity of the phase transition temperature T_V [3, 4, 5], together with the large jump of the susceptibility at the transition point, results in giant values of the magnetovolume effect in the neighborhood of T_V [4]. With the appropriate dispersion of T_V, the effect of pressure may

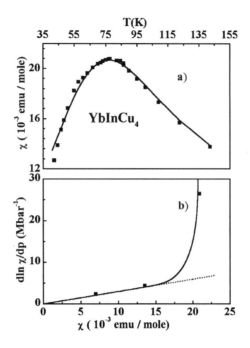

Figure 3. Temperature dependence of the magnetic susceptibility for YbInCu$_4$ in the vicinity of the phase transition temperature (a) and the pressure effect value d$\ln\chi$/dP vs χ (b). The model descriptions are given by the solid lines, the dashed straight line results from the Curie–Weiss behavior.

remain high even at $T = 78$ K, which is far away from the transition temperature. On the other hand, the deviation of the susceptibility at $T = 78$ K from the Curie-Weiss one at high temperatures (of about several percents) seems to be insufficient for this mechanism, and needs a quantitative consideration.

To do this, we state that the phase transition temperature in the sample is spatially inhomogeneous [18]. This may arise due to different reasons, for example, due to incomplete ordering of the atoms in sublattices. We shall set the concrete definition of the T_V inhomogeneity at a microscopic level for the time being. We assume that the temperature-independent susceptibility, χ_0, of the IVS Fermi liquid for the i-th volume element of the sample makes a jump at T_V to the Curie-Weiss value, $\chi = C/(T - \Theta)$, for the HFS localized moments. Then using the distribution function $W(T)$ for T_V and assuming that other characteristics of the sample susceptibility are identical, the expression for the susceptibility of an imperfect sample is of the form:

$$\chi(T) = \chi_0 + [C/(T - \Theta) - \chi_0]W(T). \tag{3}$$

A similar equation is used in [5, 12] where the susceptibility is treated in terms of the model of valence phase transition. In this equation χ_0 stands for the susceptibility of the Yb^{2+} state, $C = C(T)$ denotes the Curie constant of the Yb^{3+} state, and W (p$_3$ in [5]) means the Yb^{3+} state occupation. The latter is determined by the excitation energy of this state which is assumed to be dependent on the occupation, $E_{ex} = E_0(1 - \alpha W)$. As a consequence a positive feedback, the self-consistent solution for W changes abruptly from $W \approx 0$ to $W \approx 1$ at some temperature that corresponds to the first order phase transition Yb^{2+} →Yb^{3+}. According to Eq. (3), the magnetic susceptibility undergoes change in a jump-like manner too. The smearing of the susceptibility jump observed with varying of the basic compound composition is realized in [5, 12] by dragging out the phase transition through a choice of the constant α and additional Lorentzian broadening of the E_0 level.

Unfortunately, the reasonable treatment of some aspects of the susceptibility behavior of YbInCu$_4$ in terms of the model given in [5, 12] is more likely to be a good simulation. In this model neither the ground state (Yb^{2+} instead of IVS) nor the value of valence jump ($\sim$ 1 instead of 0.1 [3, 4]) and the accepted CEF splitting ($\Delta = 135$ K instead of $\Delta = 44$ K [15]) corresponds to the reality. Moreover, the recent paper [6] reports that the magnetic susceptibility of a perfect single-crystal YbInCu$_4$ exhibits an extremely abrupt jump at the phase transition point and obeys Eq. (3) with $W(T < T_V) = 0$ and $W(T \geq T_V) = 1$ (the other parameters are listed in Table 1). Hence, there is no other mechanism for smearing of $\chi(T)$ dependence, except the structure disorder, at least for YbInCu$_4$.

The paper [6] is an important argument in support of our model (Eq. (3)) which calls attention to the role of the system disorder defining the mechanism of the phase transition. The main problem in the model is the choice of an adequate distribution function of T_V. For simplicity, we apply the normal Gaussian distribution, although *a priori* it is hard to present convincing arguments in favor of this choice. It corresponds to a symmetric distribution of T_V^i with respect to some average value of T_V:

$$W(T) \equiv W(T, T_V, \sigma) = (1/2)[1 + \mathrm{erf}\Big(\frac{T - T_V}{\sqrt{2}\sigma}\Big)] \qquad (4)$$

To describe the susceptibility in the vicinity of the phase transition by Eqs. (3), (4) it is sufficient to choose the values of the distribution function parameters T_V and σ using the asymptotic values of χ_0, μ and Θ from Table 1.

The derivative $d\ln\chi/dP$ follows obviously from Eqs. (3), and (4) and does not require any new unknown quantities. Among the derivatives $d\chi_0/dP$, dT_V/dP and $d\Theta/dP$, the first one can be neglected considering the contribution of χ_0 to be small at $T > 78$ K. The dT_V/dP derivative is estimated in [4, 5] to fall in the range from -2.3 to -1.9 K/Mbar depending on the property under study and the method of identifying the phase transition. Finally, the value $d\Theta/dP$ is given above (see Table 1). Note, that one or more derivatives may be included in the set of parameters, together with T_V and σ.

Fig. 3 shows the results of simultaneous description of the temperature dependence of the magnetic susceptibility and magnetovolume effect by Eqs. (3) and (4) with the parameters:
$$T_V = 57\ K, \quad \sigma = 22\ K, \quad dT_V/dP = -1.9\ \text{K/Mbar}.$$

As it is seen, the distribution (4) describes adequately the behavior of magnetic susceptibility in the rather wide temperature range around the phase transition point.

Although the derivative dT_V/dP was estimated far from the transition temperature it coincides with those found in [5]. The value of σ appears to be rather high for our sample of ordinary quality. An increase of the existing dispersion by the unavoidable inhomogeneity in alloys may result in significant transformations of their $\chi(T)$ dependences.

The sharpness of the valence phase transition in perfect crystals YbInCu$_4$ [6] makes this compound suitable for comprehensive analysis of the effects of different imperfections and chemical disorder on the phase transition. The model proposed in combination with the magnetovolume effect may be a useful tool to perform the analysis. Our paper illustrates such a possibility rather than realizes it, pursuing quite a different goal, namely, to prove a reliability of the pressure dependence of the Curie-Weiss law parameters. As seen from Fig. 3, the shift in T_V under pressure does not affect the derivatives $d\ln\chi/dP$ at $T \geq 100$ K and the estimated value of $d\Theta/dP$ (Table 1) needs no correction.

As we approach the phase transition temperature, we may expect a sharp increase of the magnetovolume effect (see Fig. 3b) qualitatively similar to that observed in magnetostriction data [4]. To avoid the possible built-up of hysteresis effects and accumulation of defects in the sample, we did not perform measurements below 78 K. A pronounced non-linearity of the pressure effects in the magnetic susceptibility at 78 K is expected at $P > 5$ kbar. This limit was not reached in our experiment.

3.2. YBINCU$_4$ COMPOUND: KONDO PHASE

As it is evident from the previous section, at $T \geq 100$ K the compound YbInCu$_4$ is assumed to be in a single phase of HFS. For this state one may expect that the paramagnetic Curie temperature in the region of the Curie-Weiss law validity is close to the characteristic Kondo temperature

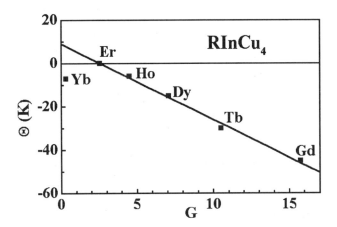

Figure 4. Experimental value of the Curie temperature from [6,20] against the de Gennes factor for heavy rare earth RInCu$_4$ compounds.

T_K ($\sim |\Theta|$) [19]. The contribution of the indirect exchange interaction between f-moments to Θ is proportional to the de Gennes factor $G = (g_J - 1)^2 J(J+1)$. As it follows from Fig. 4, the antiferromagnetic indirect interaction is dominant in the heavy rare earth RInCu$_4$ compounds. It should be negligible for YbInCu$_4$.

Nevertheless, the straight line $\Theta(G)$ does not pass through the origin. It is likely that in the compounds under consideration there exists a latent positive contribution of another type of an interaction. Although based on the data for GdInCu$_4$, this contribution, along with the indirect exchange, would be expected to be pressure independent, it introduces an uncertainty in the value of T_K. Taking the extrapolated value $\Theta(G=0)$ to be a reference point for T_V and using the value $\Theta = -7.2$ K [6] shown in Fig. 4, one obtains the estimate $T_K \sim 16$ K. It is close to the value $T_K \sim 20$ K [6] which follows from the theoretical dependence $\chi(T)$ for the Kondo system with $J = 7/2$ [17] fitted to the experimental data. The latter value for T_K combined with relation $dT_K/dP \approx -d\Theta/dP$ and bulk modulus $B(300\ K) = 1.0$ Mbar [4] results in the Grüneisen parameter G_K:

$$G_K \equiv -d\ln T_K/d\ln V = (d\Theta/d\ln V)/T_K = -32. \quad (5)$$

The sign of the pressure effect on the magnetic susceptibility of the other known Yb compounds with an unstable valence [21, 22] corresponds to the negative Grüneisen parameter value, whereas its sign for the unstable valence Ce compounds is positive [21, 23, 24]. With account of the electron-hole symmetry this fact implies a common tendency for both systems: the f-state energy increases under pressure with respect to the Fermi level.

The absolute value of G_K in YbInCu$_4$ is also typical for other Kondo systems. It is twice as much as its maximum value at the boundary between HF and IV states in the system CeIn$_{3-x}$Sn$_x$ [23] in which the valence transition is continuous. Similar pressure effects were observed in the systems which do not exhibit the valence phase transition (e.g. in CeAl$_2$ [24]). The observed correlation between pressure dependences of T_V and T_K - both values approach zero in the pressure range 25-30 kbar - is an argument in favor of the Kondo volume-collapse model of the phase transition, which is based on the strong pressure dependence of T_K [25]. However behavior of the same parameters under chemical pressure in Yb$_{1-x}$Y$_x$InCu$_4$ [10] does not support this correlation. Therefore, the large G_K value in YbInCu$_4$ would not seem to be convincible for the first order valence phase transition. It may be due to a special feature of the individual microscopic parameters which determine the magnitude and properties of the Kondo temperature. Thus for a conduction band of width D the characteristic temperature

$$T_K = D\exp(-1/8\,|\,s\,|) \qquad (6)$$

relates through the parameter $s = N(E_F)J_{bf}$ to the microscopic characteristics such as the density of the band states at the Fermi level $N(E_F)$ and the parameter of the effective exchange interaction between band and f-electrons J_{bf}. Here all degrees of freedom for $J = 7/2$ octet are assumed to be excited because the condition $T > \Delta$ is fulfilled [15]. Hence the pressure effect on Θ is determined by the dependence $s(P)$ which follows from Eq. (6),

$$\mathrm{dln}(T_K/D)/\mathrm{dln}s = -\ln(T_K/D) \qquad (7)$$

The variation of D in the range from $8 \cdot 10^3$ to $8 \cdot 10^4$ K (its pressure derivative $\mathrm{dln}D/\mathrm{dln}V \approx -1$ may be neglected) has no essential effect on $\mathrm{dln}s/\mathrm{dln}V$ which is determined from Eqs.(5) and (7),

$$\mathrm{dln}s/\mathrm{dln}V = 4.6 \pm 0.6.$$

This derivative is almost identical to those obtained, for example, in the CeCu$_6$ (–6 [26]) and CeAl$_2$ (–5.5 [24]) compounds. For a comprehensive analysis of this derivative the calculated volume dependences of the band characteristics are required before the dominant contribution can be extracted from the observed effect. The necessity of such calculations for

YbInCu$_4$ arises from a semimetallic character of its electronic structure [27] which probably also stimulates the phase instability. The atomic volume change in this compound could lead to unusual results which cannot be described by the simple renormalization of the band width. Unfortunately, no direct calculations of the pressure effect on the YbInCu$_4$ band structure were performed up to now.

As for the phase diagram for YbInCu$_4$, the data obtained here suggests that for HFS the external pressure stabilizes the trivalent state and the antiferromagnetic Kondo interaction decreases. The critical pressure for its disappearing is estimated to be $P_c \sim 30$ kbar using the experimental derivative dΘ/dP= 640 K/Mbar. Then for a perfect samples YbInCu$_4$ at hydrostatic pressure $P > P_c$, one would expect the occurrence of a ferromagnetic phase at liquid helium temperatures caused by the above-mentioned latent interaction of positive sign. According to the estimates of dT_V/dP the IVS phase is assumed to be no longer present. The proximity of the ferromagnetic and Kondo phases is possible [28] and is observed in cerium compounds [29].

The CEF effect should significantly suppress the characteristic Kondo temperature through the reduction of the electron degrees of freedom under condensation onto the ground state level Γ_8 with four-fold degeneracy [15]. Since for this case the behavior of dΘ/dP is unclear it would make no sense to comment on the ferromagnetic phase. Nevertheless it is of interest to search for this phase, which may reveal the latent positive interaction between magnetic moments in YbInCu$_4$ and shed some light on its origin.

The negligible dΘ/dP derivative for GdInCu$_4$ needs a special analysis. For rare earth– sp metal compounds, the origin of the paramagnetic Curie temperature is still not clear, but its behavior is very sensitive to the band structure peculiarities [30]. We emphasize once again the necessity of the band structure calculations for RInCu$_4$ compounds at different lattice parameters for further progress in this interesting field.

4. Summary.

The study performed made it possible:

- to obtain the Grüneisen parameter for the characteristic temperature in the HF phase of the YbInCu$_4$ compounds;
- to conclude that large Grüneisen parameter for the characteristic Kondo temperature does not probably lead to the first order phase transition in YbInCu$_4$;
- to predict a possible occurrence of the ferromagnetic phase in YbInCu$_4$ at low temperatures under high pressure;

- to attribute the dramatic enhancement of the magnetovolume effect far from the valence phase temperature to the spatial dispersion of T_V due to the lattice disorder; the latter can be described by a simple model with the normal Gaussian distribution of the T_V value;
- to propose the magnetovolume effect to be employed as useful tool for the further study of the phase diagrams and nature of the interactions in $YbInCu_4$-based alloys.

References

1. Lawrence, J. (1979) Scaling behavior near a valence instability: The magnetic susceptibility of $CeIn_{3-x}Sn_x$, *Phys. Rev. B* **20**, 3770–3782.
2. Lawrence, J.M., Riseborough, P.S., and Parks, R.D. (1981) Valence fluctuation phenomena, *Repts. Progr. Phys.* **44**, 1–84.
3. Felner, I., Nowik, I., Vaknin, D., Potzel, U., Moser, J., Kalvius, G.M., Wortmann, G., Schmiester, G., Hilscher, G., Gratz, E., Schmitzer, C., Pillmayr, N., Prasad, K.G., de Waard, H., and Pinto, H. (1987) Ytterbium valence transition in $Yb_xIn_{1-x}Cu_2$, *Phys. Rev. B* **35**, 6956–6963.
4. Teresa, J.M. de, Arnold, Z., Moral, A. del, Ibarra, M.R., Kamarád, J., Adroja, D.T., and Rainford, B. (1996) Pressure and magnetic field effects on the volume anomaly associated with first-order valence change in $YbInCu_4$, *Solid State Commun.* **99**, 911–915.
5. Nowik, I., Felner, I., Voiron, J., Beille, J., Najib, A., du Tremolet de Lacheisserie, E., and Gratz, E. (1988) Pressure, substitution, and magnetic-field dependence of the valence phase transition in $Yb_{0.4}In_{0.6}Cu_2$, *Phys. Rev. B* **37**, 5633–5638.
6. Sarrao, J.L., Immer, C.D., Benton, C.L., Fisk, Z., Lawrence, J.M., Mandrus, D., and Thompson, D. (1996) Evolution from first-order valence transition to heavy-fermion behavior in $YbIn_{1-x}Ag_xCu_4$, *Phys. Rev. B* **54**, 12207–12211.
7. Panfilov, A.S. (1992) Pendulum magnetometer for measurements of magnetic susceptibility under pressure, *Phys. Tech. High Pressure (in Russian)* **2**, 61–66.
8. Nakamura, H., Ito, K., and Shiga, M. (1994) Magnetic and transport properties in the low-carrier system $Gd_{1-x}Lu_xInCu_4$, *J. Phys.: Condens. Matter* **6**, 6801–6813.
9. Sarrao, J.L., Benton, C.L., Fisk, Z., Lawrence, J.M., Mandrus, D., and Thompson, D. (1996) $YbIn_{1-x}Ag_xCu_4$: Crossover from first-order valence transition to heavy Fermion behavior, *Physica B* **223-224**, 366–369.
10. Nakamura, H., Shiga, M., Kitaoka, Y., Asayama, K., and Yoshimura, K. (1996) Nuclear resonance study of Yb–based compounds with C15b–type structure, *J. Phys. Soc. Japan* **65**, Suppl. B, 168–180.
11. Nakamura, H., Ito, K., Wada, H., and Shiga, M. (1993) Anomalous magnetism of the frustrated compound $GdInCu_4$, *Physica B* **186-188**, 633–635.
12. Felner, I. and Nowik, I. (1986) First-order valence phase transition in cubic $Yb_xIn_{1-x}Cu_2$, *Phys. Rev. B* **33**, 617–619.
13. Yoshimura, K., Nitta, T., Shimizu, T., Mekata, M., Yasuoka, H., and Kosuge, K. (1990) Valence transition in $YbIn_{1-x}Ag_xCu_4$, *Technical Report of ISSP Ser. A, No 2264*, p.p. 1–10.
14. Altshuler, T.S., Bresler, M.S., Elschner, B., Schlott, M., and Gratz, E. (1996) Evolution of the magnetic state in $Yb_{0.5}In_{0.5}Cu_2$ under first-order phase transition, *Zh. Eksp. Teor. Fiz.* **109**, 1359–1369.
15. Severing, A., Gratz, E., Rainford, B.D., and Yoshimura, K. (1990) Study of the valence transition in $YbInCu_4$ by inelastic neutron scattering, *Physica B* **163**, 409–411.

16. Nakamura, H., Ito, K., Uenishi, A., Wada, H., and Shiga, M. (1993) Anomalous transport and thermal properties of YInCu$_4$, *J. Phys. Soc. Japan* **62**, 1446–1449.
17. Rajan, V.T. (1983) Magnetic susceptibility and specific heat of the Coqblin-Schrieffer model, *Phys. Rev. Lett.* **51**, 308–311.
18. Lawrence, J.M., Kwei, G.H., Sarrao, J.L., Fisk, Z., Mandrus, D., and Thompson, J.D. (1996) Structure and disorder in YbInCu$_4$, *Phys. Rev. B* **54**, 6011–6014.
19. Bauer, E. (1991) Anomalous properties of Ce–Cu- and Yb–Cu-based compounds, *Adv. Phys.* **40**, 417–534.
20. Abe, S., Atsumi, Y., Kaneko, T., and Yoshida, H. (1992) Magnetic proper ies of RCu$_4$In (R=Gd–Er) compounds, *J. Magn. Magn. Mater.* **104-107**, 1397–1398.
21. Häfner, H.U. (1985) Volume magnetostriction of intermediate valence systems, *J. Magn. Magn. Mater.* **47-48**, 299–301.
22. Zell, W., Pott, R., Roden, B., and Wohlleben, B. (1981) Pressure and temperature dependence of the magnetic susceptibility of some ytterbium compounds with intermediate valence, *Solid State Commun.* **40**, 751–754.
23. Grechnev, G.E., Panfilov, A.S., Savchenko, N.V., Svechkarev, I.V., Czopnik, A., and Hackemer, A. (1996) Magnetovolume effect in paramagnetic alloys of CeIn$_{3-x}$Sn$_x$, *J. Magn. Magn. Mater.* **157-158**, 677–678.
24. Panfilov, A.S., Svechkarev,, I.V., and Fawcett E. (1993) The magnetovolume effect in CeAl$_2$, in: J.Kübler and P.Oppeneer (eds.), *Physics of Transition Metals*, Vol. 2, World Scientific, Singapore, p.p. 703–707.
25. Allen, J.W. and Martin, R.M. (1982) Kondo volume collapse and $\gamma \to \alpha$ transition in cerium, *Phys. Rev. Lett.* **49**, 1106–1110.
26. Shibata, A., Oomi, G., Ōnuki, Y., and Komatsubara, T. (1986) Effect of pressure on electrical resistivity and lattice spacing of dense Kondo material CeCu$_6$, *J. Phys. Soc. Japan* **55**, 2086–2087.
27. Nakamura, H., Ito, K., and Shiga, M. (1994) Semimetallic behaviour of YInCu$_4$ and LuInCu$_4$, *J. Phys.: Condens. Matter* **6**, 9201–9210.
28. Lacroix, C. and Cyrot, M. (1979) Phase diagram of the Kondo lattice, *Phys. Rev. B* **20**, 1969–1976.
29. Loewenhaupt, M. and Fischer, K.H. (1993) Valence–fluctuation and heavy–fermion 4f systems, in: K.A. Gschneidner,Jr. and L. Eyring (eds.), *Handbook on the Physics and Chemistry of Rare Earths*, Vol. 16, Elsevier Science Publisher B.V., Amsterdam, p.p. 1–105.
30. Buschow, K.H.J., Grechnev, G.E., Hjelm, A, Kasamatsu, Y., Panfilov, A.S., and Svechkarev, I.V. (1996) Exchange coupling in GdM compounds, *J. Alloys and Compounds* **244**, 113–120.

ATOMIC VOLUME EFFECT ON ELECTRONIC STRUCTURE AND MAGNETIC PROPERTIES OF UGA$_3$ COMPOUND

G.E. GRECHNEV, A.S. PANFILOV AND I.V. SVECHKAREV

B. Verkin Institute for Low Temperature Physics and Engineering, Lenin's pr. 47, Kharkov, 310164, Ukraine

A. DELIN, O. ERIKSSON AND B. JOHANSSON

Condensed Matter Theory Group, Department of Physics, University of Uppsala, Box 530, S-751 21 Uppsala, Sweden

AND

J.M. WILLS

Center for Materials Science and Theoretical Division, Los Alamos National Laboratory, Los Alamos, New Mexico 87545

Abstract. The magnetic susceptibility χ of the itinerant antiferromagnetic compound UGa$_3$ has been studied experimentally under pressure up to 2 kbar in the temperature range 64-300 K. This study reveals a pronounced pressure effect on magnetic properties of UGa$_3$ and the measured pressure derivative of the Néel temperature is found to be $dT_N/dP = -1.1$ K/kbar. In order to analyze the experimental magnetovolume effect, to be specific $d\ln\chi/d\ln V$, the volume dependent electronic structure of UGa$_3$ has been calculated *ab initio* in the paramagnetic phase by employing a relativistic full-potential LMTO method. The effect of the external magnetic field was included self-consistently by means of the Zeeman operator, as well as orbital polarization. The calculations have brought out a predominance of itinerant uranium 5f states at the Fermi energy, as well as large and competing orbital and spin contributions to χ. The calculated field-induced magnetic moment of UGa$_3$ and its volume derivative compare favorably with our experimental results.

Introduction

The delocalization of 5f electrons in uranium compounds and the related quenching of f-magnetic moment are usually attributed to the direct f–f overlap, or to f–spd hybridization [1, 2]. The uranium intermetallic compounds UX_3, where X is a non-transition element from the group-III or group-IV series, crystallize in the cubic $AuCu_3$-type structure, and the U-U spacing in all compounds is far above the critical Hill limit [2, 3]. Therefore direct 5f–5f interactions are weak, and these systems provide an exceptional opportunity to study the role of f–spd hybridization in magnetic properties, ranging from Pauli-like paramagnetism (UAl_3, USi_3, UGe_3) to spin-fluctuation behavior (USn_3) and local-moment ordering (UPb_3). It was demonstrated in Refs.[3] and [4], that f-ligand hybridization is apparently responsible for the development of magnetic properties in UX_3 system.

The compound UGa_3 is located between the isovalent temperature-dependent paramagnet UAl_3 and the antiferromagnetically ordered UIn_3. The gallide exhibits a magnetic susceptibility χ that is only weakly temperature dependent with a maximum at 67 K [5, 6, 7]. This maximum has been attributed to an antiferromagnetic transition [5], which was later proved by neutron diffraction [8, 9] and electrical resistivity [7, 10] studies, yielding the Néel temperature $T_N=67$ K. The simple antiferromagnetic structure was found [8, 9] with uranium moments of (0.8–0.9) μ_B coupled ferromagnetically within (111) planes and antiparallel to the moments on the adjoining (111) planes. No other magnetic phase transitions were found in UGa_3 by χ_{ac} measurements [10] down to 20 mK. The investigations of the pseudobinary $U(Ga_{1-x}Al_x)_3$ [11] and $U(Ga_{1-x}Ge_x)_3$ [12] systems have shown a rapid decrease in the Néel temperature upon substitution of Al (or Ge) for Ga, and for $x \cong 0.2$ the magnetic order was found to disappear in both systems. The attempt to fit the high temperature susceptibility of UGa_3 to the Curie-Weiss law, $\chi(T) = C/(T + T^*)$, has given unrealistic parameters [6, 12]: $T^* \cong 2000$ K and a paramagnetic moment $m_{\text{eff}} \cong 1.5\mu_B$, which appeared to be inconsistent with the neutron diffraction data [8, 9]. At the same time the value of m_{eff} is much smaller than the expected moment for localized 5f electrons with any reasonable valence configuration. The in Ref.[7] estimated reduced magnetic entropy $\Delta S_m = 0.14R\ln 2$ indicates the presence of the itinerant magnetism in UGa_3. Also, the obtained electronic specific heat coefficient, γ, appeared to be large (about 50 mJ/mole K^2) [7, 13], which suggests the existence of a narrow 5f band at the Fermi level (E_F).

Therefore, the analysis of the above-mentioned experimental data points to the itinerant character of 5f-electron magnetism in UGa_3. This suggestion is also supported by the results of recent band structure calculations: a scalar-relativistic FP-LAPW [14] and a fully relativistic LMTO-ASA [15]. These calculations, however, provided only a rough picture for the electronic structure, and were unable to explain the magnetic properties of UGa_3. In this connection, the main objective of the present work is to shed light on a nature of the itinerant magnetism in UGa_3 and related features of its electronic structure. For this reason the pressure effect on the magnetic susceptibility has been studied over a wide range of temperatures (Sec. 1). The pressure derivatives of χ and T_N, derived from these measurements, are of particular interest, owing to their assumed sensitivity to the nature of the magnetism. In particular, the local-moment magnetism of UPb_3 leads to a positive pressure derivative of T_N, whereas for itinerant systems this derivative is expected to be negative (see Ref.[2] for examples).

The experimental study is complemented by *ab initio* relativistic calculations of the volume dependent band structure and field-induced magnetization of UGa_3 in the paramagnetic phase. The considerably reduced value of the magnetic moment with respect to that in UPb_3, as well as the lack of Curie-Weiss behavior give some evidence that the magnetic state may be spin-degenerate above T_N, in which case the corresponding calculations are meaningful. The details of these calculations are described in Sec. 2. The calculated band structure, as well as atomic volume effects on spin and orbital contributions to the induced magnetization are discussed and compared with the experimental data in Sec. 3, and conclusions are given in Sec. 4.

1. Experiment

The polycrystalline sample of UGa_3 was prepared by arc-melting of the constituent metals in an argone atmosphere and annealed under vacuum at 600° C. The sample was checked by x-ray diffractometry and appeared to be single phase, crystallized in the expected $AuCu_3$ crystal structure.

The magnetic susceptibility was studied under helium gas pressure up to $P=2$ kbar in the temperature range 64–300 K, i.e. at and above the Néel temperature. The measurements were carried out by the Faraday method, using a pendulum magnetometer placed into the pressure cell [16]. The relative errors of our measurements did not exceed 0.05% for magnetic fields used ($H \cong 1$ T).

It can be seen from Fig. 1 that the pressure dependence of the susceptibility, $\chi(P)$, is linear. The resulting values of $d\ln\chi/dP$ are -6.85 ± 0.5 and -5.4 ± 0.5 Mbar^{-1} at $T=77.5$ K and 300 K, respectively. These derivatives are also listed in Table I.

Also, the $\chi(T)$ dependencies were studied thoroughly in the vicinity of the Néel temperature for two different pressures (see Fig. 2). The re-

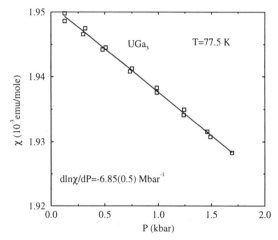

Figure 1. Pressure dependence of the magnetic susceptibility of UGa$_3$ at 77.5 K. The upper boxes correspond to the return change of pressure.

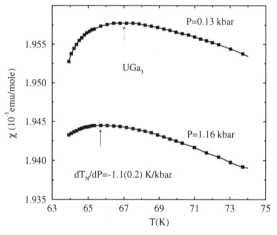

Figure 2. Temperature dependence of the magnetic susceptibility of UGa$_3$ in the vicinity of T_N (indicated by arrows) at $P=0.13$ kbar (curve 1) and $P=1.16$ kbar (curve 2). The solid line is a guide for the eye.

sulted pressure derivative, $dT_N/dP=-1.1\pm0.2$ K/kbar, appeared to be in close agreement with the value reported in Ref.[17], $dT_N/dP=-1.4$ K/kbar, derived from the resistivity measurements for a single-crystalline sample of UGa$_3$. Extrapolation gives that a critical pressure of about 50 kbar can be expected for disappearance of the antiferromagnetic order.

The corresponding logarithmic volume derivative, $d\ln T_N/d\ln V=16\pm3$, evaluated using the experimental bulk modulus value $B=0.99$ Mbar [18], is

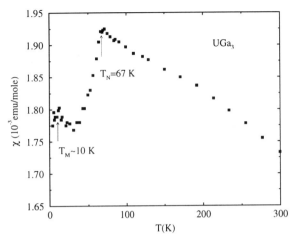

Figure 3. Temperature dependence of the magnetic susceptibility of UGa$_3$ in the range $4.2 < T < 300$ K at ambient pressure.

comparable to the values found for other itinerant antiferromagnetic systems: chromium ($d\ln T_N/d\ln V = 26$, Ref.[19]) and UN ($d\ln T_N/d\ln V = 19$, Ref.[1]).

In order to check the magnetic properties of UGa$_3$ at ambient pressure over the range 4.2–300 K, the $\chi(T)$ measurements were carried out on a high quality single crystal similar to that used in Ref.[17].

The measurements were made in a magnetic field applied along the [001] axis. As a result, in addition to a peculiarity on the $\chi(T)$ dependence at T_N, a conspicuous peak was observed at 10 K (see Fig. 3). Presumably, this peak can be attributed to another magnetic phase transition.

2. Details of calculations

In order to analyze the observed $d\ln\chi/dP$ values, the electronic structure of the paramagnetic phase of UGa$_3$ was calculated *ab initio* by employing the standard LMTO-ASA method [20], as well as the full-potential (FP) LMTO method [21]. The exchange-correlation potential was treated in the local spin density approximation (LSDA) using the von Barth-Hedin parameterization [22]. In the present calculations the spin-orbit coupling was included at each variational step, as well as the orbital polarization correction. The orbital polarization term has been adopted in the form [23] $\Delta\mathcal{H}_{\ell,m_\ell} = -R_{\ell,\sigma}\mu_{\ell,\sigma}\hat{l}_z$, where $\mu_{\ell,\sigma}$ is the orbital moment. The Racah parameter, $R_{\ell,\sigma}$, usually called E^3 for f states and B for d states, was calculated at each iteration step by using the radial d and f wave functions for each spin channel, energy set, and atom type.

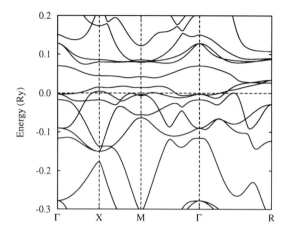

Figure 4. Energy bands of UGa$_3$ calculated at the equilibrium volume along the major symmetry directions. Energies are in Ry, and the Fermi level is marked by a horizontal dashed line.

In addition to this, the Zeeman term, $B \cdot (2\hat{s} + \hat{l})$, has been incorporated in the Hamiltonian for first-principle calculations of field-induced spin and orbital moments on the same footing [24, 25]. In the spin-polarized calculations the external magnetic field was assigned to 10 Tesla in the [001] direction. In this case the local symmetry is reduced for the gallium atoms in the AuCu$_3$ structure, and the unit cell contains three atom types, including two different Ga atoms. The muffin-tin radii where chosen to be of equal size and 3% from touching. The integration over the Brillouin zone was performed using the special point sampling [26] and with a Gaussian smearing of 10 mRy. After achievement of self-consistency, the tetrahedron method [20] was employed to get the density of states (DOS) on a fine energy mesh, and the spin and orbital moments were calculated as described in Ref.[27].

3. Results and discussion

The electronic structure calculations were performed for a number of lattice parameters close to the experimental one ($a \cong 8.0$ a.u.). The calculated equilibrium lattice spacing, $a_{\text{theor}} = 7.7$ a.u., and bulk modulus, $B_{\text{theor}} = 1.1$ Mbar, appear to be in reasonable agreement with experimental data ($B_{\text{exp}} = 0.99$ Mbar [18]). The calculated energy bands of paramagnetic UGa$_3$ are presented in Fig. 4 along high symmetry directions in the Brillouin zone.

As can be seen from this figure, there are two relatively flat bands crossing the Fermi level. The parts of these bands at Γ and M points are

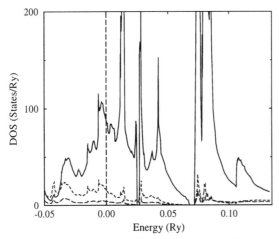

Figure 5. The partial DOS of paramagnetic UGa$_3$. The uranium 5f and 4d states, and gallium 4p states are indicated by solid line, long-dashed line, and short-dashed line, respectively. The Fermi level is marked with a vertical dashed line.

presumably responsible for the magnetic instability in UGa$_3$. For the next empty band above E_F the dispersion at the Γ point is expected to be crucial for creating an energy gap. Also, a family of extremely flat bands is seen up to 0.1 Ry above the Fermi level. The corresponding partial densities of states, evaluated on a fine energy mesh, are shown in Fig. 5. As can be seen from Figs. 4 and 5, there are two subbands originating mainly from the spin-orbit split 5f-states, divided by a "pseudogap" situated at $\sim$0.07 Ry above the Fermi level. According to the calculated partial DOS, the main contributions to the DOS at the Fermi level come from uranium 5f-states and gallium 4p-states, which overlap in energy. This hybridization is sufficiently strong to form a narrow band of itinerant 5f states which has most of their intensity in the energy range -0.1 to 0.1 Ry.

The values of the DOS at E_F, $N(E_F)$ are given in Table I together with the evaluated volume derivative, $d\ln N(E_F)/d\ln V$. The estimations for the electronic specific heat coefficient, $\gamma = 2\pi^2 k_B^2 N(E_F)/3$, made with calculated $N(E_F)$ values, appeared to be consistent with $\gamma_{\exp}$ (50 and 52 mJ/mole K^2, according to Refs. [7] and [13], respectively), providing a many-body enhancement factor, λ, which is about 1.5. It should be noted, however, that the experimental electronic specific heat coefficients were actually obtained at low temperatures for a spin-polarized phase of UGa$_3$, and can only be compared with the calculated γ, if there is no large change in the DOS at E_F at the magnetic ordering.

In order to estimate the spin magnetic susceptibility, the Stoner model was employed. In this model, the electron-electron interactions manifest themselves in magnetic properties through the enhancement of the Pauli spin susceptibility, proportional to $N(E_F)$, by the Stoner factor, $S = (1 - IN(E_F))^{-1}$, where the parameter I originates from intra-atomic Coulomb interactions. This multi-band Stoner parameter, representing the exchange interaction for all conduction electrons, has been calculated within the LMTO-ASA method, in line with Ref.[28]. With the calculated parameter of $I = 0.01$ Ry·cell, the Stoner criterion, $IN(E_F)=1$, is just about fulfilled for the theoretical lattice spacing. It indicates a possibility of magnetic instability in UGa$_3$ at low temperatures.

On the other hand in actinide systems the Stoner approach itself cannot explain the observed magnetic properties, because spin-orbit coupling in connection with spin polarization induces large orbital moments [27, 29]. In case the spin-orbit energy dominates over the Zeeman spin polarization energy, the orbital magnetic moment is larger than, and anti-parallel to, the spin moment [24, 25, 29]. Also, as suggested in Ref.[24], in systems with less than a half filled f-electronic shell (i.e. in uranium ones) the crucial parameter determining whether the spin and orbital moments are parallel or not, is the relation between the magnitude of the magnetic susceptibility (a small susceptibility favors a parallel alignment) and the spin-orbit parameter (a large spin-orbit coupling favors an anti-parallel alignment). Thus, the resulting moments are determined by the interaction between the Zeeman operator, exchange effects and spin-orbit coupling. In a general case the magnitudes and directions of these moments are hardly predictable from simple arguments, but can only be evaluated from first-principles calculations.

The calculated magnetic moments of UGa$_3$, induced by an external field of 10 T, are listed in Table I.

In agreement with Hund's third rule, the induced spin moment is antiparallel to the orbital moment and to the applied magnetic field. In line with the considerations presented in Ref.[24], the uranium susceptibility is sufficiently strong to make the spin and orbital moments antiparallel in UGa$_3$. In accordance with the calculated partial DOS at E_F, the uranium moment is almost exclusively of 5f character, and the gallium moment is mainly of 4p character. As a whole, the total magnetic moment of UGa$_3$ is dominated by contributions coming from the uranium site and composed of large orbital and spin moments, which are coupled in an antiparallel way.

Due to the close proximity of the induced spin-polarized state to the spontaneous magnetic ordering for lattice parameters $a \geq 7.9$ a.u., the calculated total magnetic moment rises gradually to a large value ($\simeq 1\mu_B$), close to the moment observed in the magnetically ordered state [9] ($0.9\mu_B$). This represents a spontaneous ferromagnetic phase transition and should

TABLE 1. Lattice spacing effect on the calculated electronic structure and magnetic parameters of UGa$_3$. $N(E_F)$ is DOS at the Fermi level in units of states/Ry·cell. The orbital (m_{orb}) and spin (m_{spin}) moments, induced in a magnetic field of 10 T, are given in units of $10^{-3} \mu_B$. The magnetic susceptibility χ is presented in units of 10^{-3} emu/mole. The volume logarithmic derivatives of these entities are given in the second part of the table. In order to transform the experimentally obtained pressure derivatives, $d\ln \chi/dP$, into the volume derivatives, we made use of the experimental value for the bulk modulus (0.99 Mbar[18]).

a (a.u.)	$N(E_F)$	Magnetic moments		χ
		m_{orb}	m_{spin}	
7.5	91.3	44.49	-27.33	0.96
7.6	96.1	57.36	-34.31	1.29
7.7	102.8	70.02	-40.13	1.67
7.8	110.0	85.27	-47.55	2.10
7.9	117.6			
8.0	125.0			1.95[a]
	$d\ln N/d\ln V$	$d\ln m_{orb}/d\ln V$	$d\ln m_{spin}/d\ln V$	$d\ln \chi/d\ln V$
	1.73	5.1	4.2	6.3
				6.8[b]
				5.4[c]

[a])Extrapolation of the experimental $\chi(T)$ to $T=0$.
[b])Experimental value at 77.5 K.
[c])Experimental value at 300 K.

not be compared to the experimental data above T_N.

The contributions to the magnetic susceptibility, presented in Table I, were derived from the calculated spin and orbital magnetic moments. The diamagnetic contributions to the susceptibility, coming from core and conduction electrons, were assumed to be negligible in comparison to the large paramagnetic contributions. It can be deduced from the moments in Table I that the orbital paramagnetic susceptibility is almost twice as large as the value of the spin susceptibility. The resulting total χ matches favourably the experimental value of $\cong 1.95 \cdot 10^{-3}$ emu/mole, extrapolated from the paramagnetic region to zero temperature. With the orbital polarization correction omitted our self-consistent spin-polarized calculations yield very close values of the magnitude of the anti-parallel orbital and spin moments,

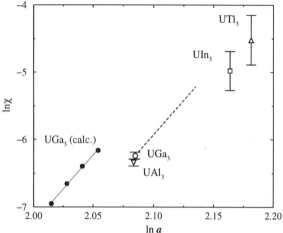

Figure 6. Magnetic susceptibility vs lattice spacing in UX$_3$ compounds. Filled circles are related to the calculated susceptibility of UGa$_3$. Other symbols indicate the experimental $\chi(T\to 0)$ for UGa$_3$ ($\circ$, present work), UIn$_3$ ($\square$)[6], and UTl$_3$ ($\triangle$)[6]. The dashed line represents the experimental $d\ln\chi/d\ln a$ derivative for UGa$_3$.

which results in a too small total susceptibility, suggesting that this correction is important in f-electron systems.

The obtained lattice spacing dependence of the magnetic susceptibility of UGa$_3$ is presented in Fig. 6. We observe that the agreement between the calculated and experimental $d\ln\chi/d\ln a$ derivatives is quite good, except for a shift of for the calculated $\chi(a)$ to smaller lattice parameters. In the same figure we also display magnetic susceptibilities of isovalent uranium compounds, UAl$_3$, UIn$_3$, and UTl$_3$, obtained by extrapolating the experimental $\chi(T)$ dependencies in the paramagnetic regions [5, 6] to zero temperature:

These data appear to be consistent with the trend derived for UGa$_3$, pointing out the dominant role that the interatomic spacing has on the magnetic properties of UX$_3$ compounds. The analogous effect has been also observed in UCu$_{5-x}$Au$_x$ alloys [30].

The volume derivatives of the calculated magnetic moments and magnetic susceptibility are listed in Table I. As can be seen from this table, the volume effect is more pronounced for the orbital part of the magnetic moment. Since the orbital moment is a van Vleck-type moment, originating from the mixing of states across the energy bands [27, 29], the magnitude of this moment is expected to be inversely proportional to the band width W. Simple canonical band theory [20] gives that $d\ln(1/W)/d\ln V=(2l+1)/3$. For the pure unhybridized f-band, not affected by spin-orbit coupling, this gives $d\ln(1/W)/d\ln V=7/3$. On the other hand, estimations based on the present band structure calculations yield $d\ln(1/W)/d\ln V \simeq 4$. This value is in reasonable agreement with the volume derivatives of the calculated orbital moment and the experimental magnetic susceptibility, 5.1 and 6.8, respectively, presented in Table I. Therefore the observed large magneto-volume effect in UGa$_3$ is apparently related to the rapid quenching of the induced orbital moment with increasing width of the 5f-band under applied pressure.

Also, it is clear from Table I that a large logarithmic volume derivative of the calculated spin magnetic moment can not be solely attributed to the volume dependence of $N(E_F)$. In the framework of Stoner theory, the volume dependence of the exchange interaction parameter, $I = 1/N - \mu_B^2/\chi_{spin}$, is assumed. Using the values of χ_{spin} derived from the calculated induced spin magnetic moments for different lattice spacings, and also $N(E_F)$ from Table I, the volume derivative of the interaction parameter has been evaluated and found to be $d\ln I/d\ln V = -1.42$. This surprisingly large derivative is actually close to the corresponding values obtained for a number of strongly enhanced itinerant systems [31] like vanadium, palladium alloys, TiCo and Ni$_3$Al compounds, where $d\ln I/d\ln V$ lies between -0.7 and -1.3.

4. Conclusions

The detailed experimental study of magnetic susceptibility has revealed a pronounced pressure effect on the magnetic properties of UGa$_3$, as well as a new magnetic phase transition at 10 K. Our calculations point to a strong hybridization of uranium 5f-states and gallium 4p-states, giving a narrow band of itinerant states in the energy range $\sim$0.2 Ry at E_F. The calculated field induced magnetic moments of UGa$_3$ and their volume derivatives appeared to be in a fair agreement with the experimental data obtained. It was also demonstrated that the magnetism in this systems is dominated by an orbital contribution. Also, the spin and orbital moments appeared to be anti-parallel in UGa$_3$ and thus this system obeys the Hund's third rule. The volume effect was found to be more pronounced for the orbital contribution to the magnetic moment, due to the rapid quenching of the induced orbital moment with increasing width of the 5f-band under pressure. A relatively large volume effect on the spin magnetic moment can not be entirely attributed to the volume dependence of DOS at the Fermi level. This effect is apparently enhanced by the strong volume dependence of the exchange interaction parameter. The calculated lattice spacing dependence of χ in UGa$_3$ was compared with the experimental data available for isovalent UX$_3$ compounds: UGa$_3$, UAl$_3$, UIn$_3$, and UTl$_3$. This comparison allows to conclude that magnetic properties of UX$_3$ compounds are predominantly determined by the interatomic spacing variations.

Acknowledgment

We are grateful to A. Czopnik and D. Kaczorowski for giving samples. Valuable discussions with M.S.S. Brooks and A. Hjelm are acknowledged. This work has been supported by the Swedish Natural Science Research Council and the Swedish Royal Academy of Sciences.

References

1. Fournier, J.-M. and Troć, R. (1985) Bulk properties of the actinides, in: A.J. Freeman and G.H. Lander (eds.), *Handbook on the Physics and Chemistry of the Actinides*, North-Holland, Amsterdam, vol. **2**, pp. 29–173.
2. Sechovsky, V. and Havela, L. (1988) Intermetallic compounds of actinides, in: E.P. Wohlfarth and K.H.J. Buschow (eds.), *Ferromagnetic materials*, vol. **4**, North-Holland, Amsterdam, pp. 309–492.
3. Koelling, D.D., Dunlap, B.D. and Crabtree, G.W. (1985) f-electron hybridization and heavy-fermion compounds, *Phys. Rev. B*, **31** pp. 4966–4971.
4. Norman, M.R. and Koelling, D.D. (1993) Electronic structure, Fermi surfaces, and superconductivity in f electron metals. in: K.A. Gschneidner, Jr., L. Eyring, G.H. Lander, C.R. Choppin (eds.), *Handbook on the Physics and Chemistry of the Rare Earths*, vol. **17**, North-Holland, Amsterdam, pp. 1–86.
5. Buschow, K.H.J. and van Daal, H.J. (1972) Comparison of anomalies observed in U- and Ce-intermetallics, in: D.C. Graham and J.J. Rhyme (eds.) *Magnetism and Magnetic Materials. Proceedings of the 17th Annual Conference on Magnetism and Magnetic Materials*, AIP. Conf. Proc. **No. 5**, AIP, New York, pp. 1464–1477.
6. Misiuk, A., Mulak, J. and Czopnik, A. (1972) Magnetic properties of UGa_3, UIn_3 and UTl_3. Discussion in terms of the paramagnetic susceptibility maximum model, *Bull. Acad. Pol. Sci. Ser. Sci Chim.*, **20**, pp. 891–896.
7. Kaczorowski, D., Troć, R., Badurski, D., Bohm, A., Shlyk L. and Steglich F. (1993) Magnetic-to-nonmagnetic transition in the pseudobinary system $UGa_{3-x}Sn_x$, *Phys. Rev. B*, **48**, pp. 16425–16431.
8. Murasik, A., Leciejewicz, J., Ligenza S. and Zygmunt, A. (1974) Antiferromagnetism in UGa_3, *Phys. Status Solidi A*, **23**, pp. K147–149.
9. Lawson, A.C., Williams, A., Smith, J.L, Seeger, P.A., Goldstone, J.A., O'Rourke, J.A. and Fisk, Z. (1985) Magnetic neutron diffraction study of UGa_3 and UGa_2, *J. Magn. Magn. Mater.*, **50**, pp .83–87.
10. Ott, H.R., Hulliger, F., Rudiger, H. and Fisk, Z. (1985) Superconductivity in uranium compounds with Cu_3Au structure, *Phys. Rev. B*, **31**, pp. 1329–1333.
11. Jee, C., Yuen, T., Lin, C.L. and Crow, J.E. (1987) Thermodynamic and transport properties of $U(Ga,Al)_3$, *Bull. Am. Phys. Soc.*, **32** p. 720.
12. Zhou, L.W., Lee, C.S., Lin, C.L., Crow, J.E., Bloom, S. and Guertin, R.P. (1987) Magnetic to nonmagnetic transition in a highly hybridized f-electron system: $UGa_{3-x}Ge_x$, *J. Appl. Phys.*, **61**, pp. 3377–3379.
13. van Maaren, M.H., van Daal, H.J. and Buschow, K.H.J. (1974) High electronic specific heat of some cubic UX_3 intermetallic compounds, *Solid State Commun.*, **14**, pp. 145–147.
14. Diviš, M. (1994) Electronic structure of UGa_3 calculated by tight binding and LDA methods, *Phys. Status Solidi B*, **182**, pp. K15–18.
15. Grechnev, G.E., Panfilov, A.S, Svechkarev, I.V., Kaczorowski, D., Troć, R. and Czopnik, A. (1996) Effect of pressure on magnetic properties of $UGa_{3-x}Sn_x$ alloys, *J. Magn. Magn. Mater.*, **157/158**, pp. 702–703.

16. Panfilov, A.S. (1992) Pendulum magnetometer for measurements of magnetic susceptibility under pressure, *Physics and Techniques of High Pressure (in russian)*, **2**, pp. 61–66.
17. Kaczorowski, D., Czopnik, A., Jezowski, A., Misiorek, H., Zaleski, A.J., Klamut, P.W., Wolcyrz, M., Troć, R. and Hauzer, R. (1996) Single-crystal study of itinerant 5f-electron antiferromagnetism in UGa$_3$, in: 26^{iemes} *Journees des Actinides, Program and Abstracts*, Szklarska Poreba, Poland, pp. 132–133.
18. Le Bihan, T., Heathman, S., Darracq, S., Abraham, C., Winand, J.M. and Benedict, U. (1996) High pressure X-ray diffraction studies of UX$_3$ (X=Al,Si,Ga,Ge,In,Sn), *High Temp. - High Press.*, **27/28**, pp. 157–162.
19. McWhan, D.B. and Rice, T.M. (1967) Pressure dependence of itinerant antiferromagnetism in chromium, *Phys. Rev. Lett.*, **19**, pp. 846–849.
20. Skriver, H.L. (1984) *The LMTO Method*. Springer, Berlin.
21. Wills, J.M. and Cooper, B.R. (1987) Synthesis of band and model Hamiltonian theory for hybridizing cerium systems, *Phys. Rev. B*, **36**, pp. 3809–3823;
    Price, D.L. and Cooper, B.R. (1989) Total energies and bonding for crystallographic structures in titanium-carbon and tungsten-carbon systems, *Phys. Rev. B*, **39**, pp. 4945–4957.
22. von Barth, U. and Hedin, L. (1972) A local exchange-correlation potential for the spin polarized case. I, *J. Phys. C*, **5**, pp. 1629–1642.
23. Eriksson, O., Johansson, B. and Brooks, M.S.S. (1990) Orbital polarization in narrow-band systems: application to volume collapses in light lanthanides, *Phys. Rev. B*, **41**, pp. 7311–7314.
24. Hjelm, A., Trygg, J., Eriksson, O., Johansson, B. and Wills, J.M. (1994) Field-induced magnetism in itinerant f-electron systems: U, Pu, and Ce, *Phys. Rev. B*, **50**, pp. 4332–4340.
25. Trygg, J., Wills, J.M., Johansson, B. and Eriksson, O. (1994) Field-induced magnetism in uranium compounds: UGe$_3$ and URh$_3$, *Phys. Rev. B*, **50**, pp. 9226–9234.
26. Chadi, D.J. and Cohen, M.L. (1973) Special points in the Brillouin zone, *Phys. Rev. B*, **8**, pp. 5747–5753;
    Froyen, S. (1989) Brillouin-zone integration by Fourier quadrature: special points for superlattice and supercell calculations, *Phys. Rev. B*, **39**, pp. 3168–3172.
27. Brooks, M.S.S. and Kelly, P.J. (1983) Large orbital-moment contribution to 5f band magnetism, *Phys. Rev. Lett.*, **51**, pp. 1708–1711.
28. Eriksson, O., Brooks, M.S.S. and Johansson, B. (1989) Relativistic Stoner theory applied to PuSn$_3$, *Phys. Rev. B*, **39**, pp. 13115–13119.
29. Brooks, M.S.S. (1993) Pressure effects upon the magnetization of actinide compounds, *Physica B*, **190**, pp. 55–60.
30. Panfilov, A.S., Svechkarev, I.V., Troć, R. and Tran, V.H. (1995) Effect of pressure on magnetic susceptibility of UCu$_{5-x}$Au$_x$ alloys, *J. Alloys and Compounds*, **224**, pp. 39–41.
31. Panfilov, A.S., Pushkar, Yu.Ya. and Svechkarev, I.V. (1989) Effects of pressure on the exchange-enhanced band paramagnetism of palladium alloys, *Sov. Phys. JETP*, **68**, pp. 426–431;
    Brommer, P.E., Grechnev, G.E., Franse, J.J.M., Panfilov, A.S., Pushkar, Yu.Ya. and Svechkarev, I.V. (1995) The pressure effect on the enhanced itinerant paramagnetism of Ni$_3$Al and TiCo compounds, *J. Phys.: Condens. Matter*, **7**, pp. 3173–3180.

ON THE TEMPERATURE DEPENDENCE OF THE ELECTRICAL RESISTIVITY OF $Er_{0.55}Y_{0.45}Co_2$

A.N. PIROGOV
Institute of Metal Physics RAS,
620219 Ekaterinburg, Russia
N.V. BARANOV AND A.A.YERMAKOV
Ural State University,
620083 Ekaterinburg, Russia
C. RITTER
Institute Max von Laue - Paul Langevin,
F-38042 Grenoble, France
AND
J. SCHWEIZER
CEA/Department de Recherche Fondamentale sur la Matiere Condensee,
SPSMS/MDN, 17, rue des Martyrs, 38054 Grenoble Cedex 9, France

Abstract. At low temperatures the electrical resistivity $\rho(T)$ dependencies for some $R_{1-x}Y_xCo_2$ (R = Er, Ho, Dy) in the concentration regions 0.2<x<0.7; 0.3<x<0.8 and 0.6<x<0.8, respectively, exhibit a pronounced minimum. The detailed neutron diffraction study performed on $Er_{0.55}Y_{0.45}Co_2$ shows that the behavior of the critical scattering of neutrons with the temperature and in an applied magnetic field correlates with the behavior of the electrical resistivity. The presence of the minimum on the $\rho(T)$ dependence results from the appearance of localized spin density fluctuations with the decreasing temperature below T = 18K. This temperature can be associated with the magnetic ordering temperature of the R-subsystem (T_C^R). An abrupt decrease of the resistivity with the further decrease of the temperature accompanies the appearance of the long range magnetic order in the itinerant d-subsystem at $T_C^{Co} \approx$ 11 K.

1. Introduction

Rare earth intermetallic compounds RCo_2 have two magnetic subsystems with the different origin [1]. The indirect exchange RKKY interaction between the localized 4f-electrons is responsible for the magnetic ordering in the R-subsystem. The exchange interaction between itinerant 3d-electrons of Co is not sufficient for the d-band splitting. Magnetic ordering in the Co-subsystem can be induced by an effective molecular field H_{mol} acting from the R-ions. When the value of H_{mol} is more than the critical one ($H_c \approx$ 70 T) the magnetic moment per Co atom (μ_{Co}) may reach a value of 1 μ_B. The $\mu_{Co}(H_{mol})$

dependence shows a metamagnetic character. H_{mol} can be changed by the substitution of R-atoms for the nonmagnetic yttrium or by an applied magnetic field.

An increase of the Y content in the $R_{1-x}Y_xCo_2$ compounds up to the critical value x_c leads to the disappearance of the d-band splitting. For x is just above x_c one can expect a first order phase transition in the d-subsystem in a relatively small applied magnetic field. Such a field-induced increase of μ_{Co} at $x \approx x_c$ was found in $(R_{1-x}Y_x)Co_2$ (R=Er, Ho) by means of thermal expansion, specific heat, neutron diffraction and electrical resistivity measurements [2,3].

The temperature dependencies of the electrical resistivity $\rho(T)$ of all the $R_{1-x}Y_xCo_2$ compounds reveal a common feature at high temperatures (T > 150 K). They show a tendency to the saturation with increasing the temperature due to the predominant contribution from the s-electron scattering on the thermally activated spin fluctuations of the itinerant d-electrons. At low temperatures the $\rho(T)$ dependencies for some $R_{1-x}Y_xCo_2$ (R = Er, Ho, Dy) in the concentration regions 0.2<x<0.7; 0.3<x<0.8, and 0.6<x<0.8, respectively, exhibit a pronounced minimum [3-5]. Moreover, a sharp decrease in the electrical resistivity at a further decrease of the temperature is observed in these compounds when the yttrium content is less than the critical values ($x_c \approx 0.5$; 0.58; and 0.7 for above mentioned systems). The presence of such a minimum was discussed in terms of the s-electron scattering on spin density fluctuations localized in a small region of the sample (clusters) [3]. These spin density fluctuations appear with decreasing temperature owing to the fluctuations of the effective molecular field H_{mol} acting from the diluted R-subsystem. The drastic decrease of the electrical resistivity at the further decrease of the temperatures was related to the suppression of spin fluctuations with the appearance of the long range magnetic order.

In order to study the magnetic state of $R_{1-x}Y_xCo_2$ compounds and the nature of the low temperature anomalies of the resistivity we have performed the detailed neutron diffraction study of the $Er_{0.55}Y_{0.45}Co_2$ compound.

2. Experimental

The $Er_{0.55}Y_{0.45}Co_2$ compound was prepared by arc melting and homogenized at 1220 K for 50 h. The powder neutron diffraction measurements were performed using the diffractometer D1B of the ILL high-flux reactor at a wavelength of 2.524 A in the temperature range 2 ÷25 K. The external magnetic field up to 1.05 T was applied vertically. The electrical resistivity was measured by the four-probe method on the 1 x 1 x 6 mm^3 sample in the magnetic fields up to 6 T.

3. Results and discussion

Fig.1 shows the magnetic neutron diffraction patterns of $Er_{0.55}Y_{0.45}Co_2$ at T=2 K obtained at different conditions: i) the sample was cooled at μ_0H=0 (virgin-state); ii) the sample was cooled from T = 50 K down to 2 K at μ_0H=1.05 T, after that the field was turned off ("on-off"-state); iii) the sample was cooled from T = 50 K down to 2 K at $\mu_0 H = 1.05$ T ("on"-state). For the virgin ("vgn")-state the pattern shows the magnetic diffusive maximum of a lorenzian shape in the vicinity of the Bragg peak positions.

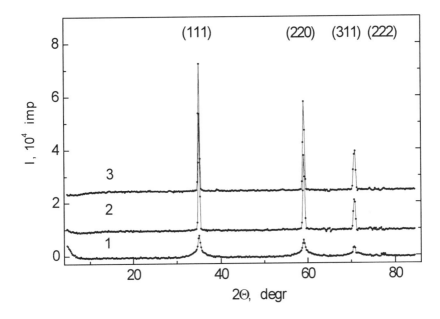

Figure 1. The magnetic neutron diffraction patterns of $Er_{0.55}Y_{0.45}Co_2$ at 2 K for the "vgn" (a) and "on-off" (b) states.

The magnetic diffusive scattering can be associated with the presence of localized spin density fluctuations. An estimation of a distance of the short-range correlations gives the value about 3 nm, i. e. about seven lattice spacing in the [111] direction. The Bragg reflections result from a long-range magnetic order and can be well described by the Gauss function. An application of the magnetic field increases significantly the intensity of the Bragg reflections and decreases the intensity of diffusive scattering. This indicates an enhancement of the long-range magnetic order and a suppression of the spin density fluctuations. The patterns for the "on" and "on-off" states are practically the same. This shows that the sample remains in the field-induced long-range magnetic state after the removal of the field. An irreversibility of the field-induced magnetic phase transition in this compound was observed in our earlier works [3,6] by the measurements of the electrical resistivity and specific heat and was connected with the presence of the hysteresis in the $\mu_{Co}(H)$ dependence in the vicinity of the critical molecular field owing to the first order phase transition in the d-subsystem.

Using the neutron diffraction patterns obtained at various temperatures for the sample with the different magnetic prehistory (i, ii, iii) we have observed the change of the magnetic state (long-range and short-range order, magnetization of R- and Co-subsystems) with increasing the temperature. The calculation of the μ_R and μ_{Co} values at each temperature was made using the experimental intensity of the Bragg reflections (111), (220), and (310). Heating of the sample in the "vgn"-state leads to the decrease of the intensity of the Bragg reflections (see Fig. 2) and to a diffusive maximum (Fig.3).

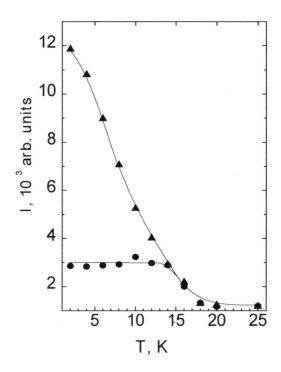

Figure 2. Temperature dependencies of the intensity of the Bragg reflections (111) in the "vgn" (circles) and "on-off" (triangles) states.

The long-range and short-range orders disappear at the same temperature about 18 K. As it follows from Fig.4, which shows the temperature dependencies of the magnetization of the R-subsystem in "vgn"- and in "on-off" states, this temperature corresponds to the ordering temperature of the R-subsystem (T_C^R). We have also found that the Co-magnetic moment disappears at the lower temperature T_C^{Co} of about 11 K (see Fig.5). The presence of the remarkable difference between the ordering temperatures of Er- and Co subsystems in $Er_{1-x}Y_xCo_2$ is in agreement with results of our study of the concentrational dependencies of μ_R and μ_{Co} in this system. As it was shown in [3] the Co-magnetic moment vanishes at the lower Y-concentration ($x = 0.5$) than the ordered magnetic moment in Er-subsystem ($x = 0.6$). Moreover, some indications on the presence of the separate ordering temperatures in $Er_{1-x}Y_xCo_2$ with $x \approx x_c$ were obtained by means of the specific heat measurements [6,7].

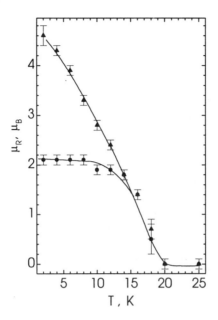

Figure.3. Temperature dependencies of the intensity of the diffusive maximum (111) in the "vgn" (circles) and "on-off" (triangles) states.

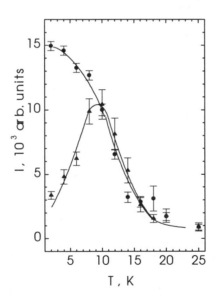

Figure 4. Temperature dependencies of the magnetization of the R-subsystem in the "vgn" (circles) and "on-off" (triangles) states.

It should be noted that Figs. 2 and 3 demonstrate the significant difference in the behavior of the intensity of the coherent and diffusive magnetic scattering for the samples in "vgn" and "on-off " states. As one can see, the intensity of the Bragg reflection (111) decreases with increasing the temperature for both the "vgn"and "on-off" states, but the temperature dependence of the intensity of the diffusive maximum (111) in "on-off " state shows the nonmonotonous character contrary to its monotonous decrease in the "vgn" state. After an irreversible suppression of spin density fluctuations in the field applied at T = 2 K, heating of the sample increases the intensity of the diffusive scattering I_D up to a maximal value reached at about 11 K (see Fig.3). At this temperature the intensities of the diffusive maximum for the "vgn" and "on-off" states become equal. Therefore, we can conclude that spin density fluctuations in these states become similar at T > 11 K.

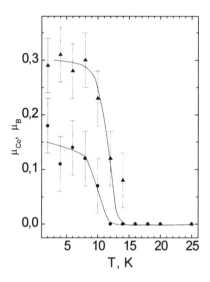

Figure5. Temperature dependencies of the magnetization of the Co-subsystem in the "vgn" (circles) and "on-off" (triangles) states.

An increase of the μ_R value under an ordering influence of the applied magnetic field results in an irreversible increase of μ_{Co} due to the metamagnetic transition in the itinerant d-electron subsystem [3]. This transition is accompanied by the suppression of localized spin density fluctuations. This results in the decrease of the intensity of magnetic diffusive scattering (Fig.3), of the coefficient γ in the linear term of the low temperature specific heat ($\Delta\gamma/\gamma$ = -44 %) [6], and of the electrical resistivity ($\Delta\rho/\rho \approx$ -40 %, see below).

The difference of the "vgn" and "on-off" magnetic states in $Er_{0.55}Y_{0.45}Co_2$ manifests itself in the temperature dependence of the small angle diffusive neutron scattering as well as in the temperature dependence of the electrical resistivity presented in Figs. 6a and 6b.

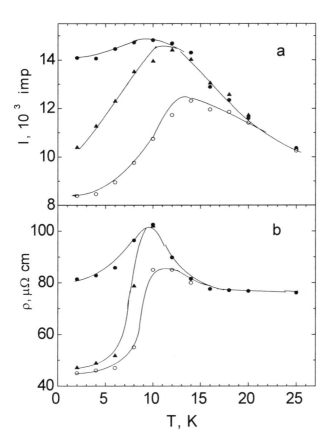

Figure 6. Temperature dependencies of the intensity of small angle neutron scattering (a) and electrical resistivity (b) for different magnetic states: full circles, full triangles, and open circles describe the "vgn", "on-off", and "on" states, respectively.

As one can see from Figs.6a and 6b the $I_D(T)$ and $\rho(T)$ dependencies are quite similar. Therefore, we can suggest that the critical scattering of both neutrons and electrons have the same origin. To our mind, this is caused by the change of localized spin density fluctuations attributed to the Co-subsystem with the temperature in the applied magnetic field. An increase of ρ with increasing temperature after application of the magnetic field (in the "on-off" state) can be related to the disappearance of the long-range magnetic order in the Co-subsystem which is accompanied by the growth of spin density fluctuations when approaching $T_C^{Co} \approx 11$ K. The minimum of ρ is observed at a temperature about 18 K, which corresponds to the ordering temperature of the R-subsystem T_C^R. In the temperature interval $11 \div 18$ K the molecular field H_{mol} is not sufficient for the appearance of the Co-magnetic moment in the whole volume of the sample. H_{mol} induces the Co-magnetic moment $\mu_{Co} \neq 0$ only in the small regions of the sample. The number and volume of these regions decrease with increasing temperature.

Therefore, at T > 11 K the contribution to the electrical resistivity due to the inhomogenous distribution of the spin density in the Co-subsystem decreases with increasing temperature. On other hand, an increase of the temperature leads to the increase of the contributions to the electrical resistivity from the scattering on phonons and on thermally activated spin fluctuations. The change of all these contributions with the temperature results in the presence of the minimum in the $\rho(T)$ dependence for $Er_{0.55}Y_{0.45}Co_2$ as well as for other $R_{1-x}Y_xCo_2$ (R= Dy, Ho, Er) compounds with the yttrium concentration near the critical value.

Acknowledgment

This work was partly supported by the Program "Neutron research of matter" of RAS (projects N 96-104, 96-305).

References

1. Bloch, D., Edwards, D.M., Shimizu, M. and Voiron, J. (1975) First order transition in ACo_2 compounds, *J.Phys. F: Metal Phys.* **5**, 1217-1226.
2. Levitin, R Z. .and Markosyan, A. (1988) Itinerant metamagnetism, *Sov. Phys. - Usp.* **155**, 623-657.
3. Baranov, N.V., Kozlov, A.I. Pirogov, A.N. and Sinitsyn, E.V. (1989) Itinerant metamagnetizm and the features of the magnetic structure of $Er_{1-x}Y_xCo_2$ compounds, *Sov.Phys.- JETP.* **96**, 674-683.
4. Steiner, W., Gratz, E. .Ortbauer, H. and Camen, H.W. (1978) Magnetic properties, electrical resistivity and thermal expansion of $(Ho,Y)Co_2$, *J. Phys.F: Metal. Phys.* **8**, 1525-1537.
5. Gratz, E, Pillmayr, N. Bauer, E. and Hilscher, G. (1987) Temperature and concentration dependence of the electrical resistivity in $(RE,Y)Co_2$ (RE- rare earth element), *JMMM* **70**, 159-161..
6. Baranov, N.V., Andreev, A.V,. Nakotte, H., de Boer, F.R. and Klasse, J.C.P., Irreversible suppression of spin fluctuations at the metamagnetic phase transition in $Er_{0.55}Y_{0.45}Co_2$, *J. Alloys and Comp.* **182**, 171-174.
7. Hauser, R., Bauer, E. Gratz, E, Rotter, E.M., Hilsher, G. Michor, H. and Markosyan, A.S. (1997) Evidence for separate magnetic ordering in the rare earth and the dsublattice of $Er_{0.6}Y_{0.4}Co_2$, *Physica* **B237-238**, 577.-578.

ELECTRICAL RESISTIVITY AND PHASE TRANSITIONS IN FeRh BASED COMPOUNDS: INFLUENCE OF SPIN FLUCTUATIONS

N.V. BARANOV AND S.V. ZEMLYANSKI
Institute of Physics and Applied Mathematics, Ural State University, 620083 Ekaterinburg, Russia
K. KAMENEV
Department of Physics, University of Warwick, Coventry CV4 7AL, England

Abstract. $(Fe_{1-x}Co_x)_{49}Rh_{51}$ compounds with $x = 0 \div 0.1$ have been studied by means of the electrical resistivity and magnetization measurements. The substitution of Fe for Co atoms up to 3.5 at. % decreases the critical ferro-antiferromagnetic (F-AF) transition temperature T_t. At $x \geq 0.035$ only the ferromagnetic order exists in these compounds as well as in the $(Fe_{1-x}Ni_x)Rh$ system. The field induced AF-F transitions at $T < T_t$ in $(Fe_{1-x}Co_x)_{49}Rh_{51}$ are accompanied by the giant magnetoresistance effect ($\Delta\rho/\rho \sim -80\%$). The temperature dependencies of the electrical resistivity at low temperatures show the T^2 law in both the AF and F states. The coefficient A of the T^2-term depends nonmonotonously on the Co-content reaching the maximal value near the critical boundary of antiferromagnetism owing to the contribution from spin fluctuations.

1. Introduction

The spin dependent electrical resistivity of antiferromagnetic metals below the ordering temperature is determined by several contributions including the gap effect of the conduction band and contributions from the scattering of conduction electrons on spin waves and spin fluctuations. The effect of the energy gap is essential for both the localized and itinerant metallic antiferromagnets. As it was shown by Ueda [1] spin fluctuations can play a predominant role in the electrical resistivity of the antiferromagnetic metal with s- and d-bands. Due to the scattering of s-electrons on spin fluctuations the low temperature electrical resistivity of the antiferromagnetic metal changes with the temperature as T^2 as well as the resistivity of weakly or nearly ferromagnetic metals. According to Ueda the coefficient in the T^2- term diverges at the critical boundary of the AF state.

In order to study the role of spin fluctuations in the electrical resistivity of antiferromagnets we have chosen the well known FeRh compound [2,3] where the antiferromagnetic order at low temperatures can be changed for a ferromagnetic one by means of the partial substitution of Fe or Rh by other transition metals [4,5]. The parent FeRh compound with the cubic structure of the CsCl type is ordered ferromagnetically below the Curie temperature $T_c \approx 670$ K. At the critical temperature $T_t=320 \div 340$ the first order phase transition from the ferromagnetic (F) to antiferromagnetic (AF) state was observed in this compound. The AF-F transition in FeRh based compounds is accompanied by a large volume expansion of about 0,9 % [6], giant magnetoresistance [4,6], and by a giant increase of the electronic specific heat coefficient γ (up to 150 % [4,5,7]). These experimental date are in good agreement with the band structure calculations [8,9] which show the significant difference in the electronic structure of FeRh in AF and F states and an itinerant character of the d-electrons.

The aim of the present paper is to study how the change of the magnetic order in FeRh arising due to a substitution of Fe for Co atoms influences the electrical resistivity.

2. Experimental details

The $(Fe_{1-x}Co_x)_{49}Rh_{51}$ compounds with $x = 0 - 0.1$ were prepared by melting in arc furnace in a helium atmosphere and homogenized at 1000 °C for 150 h. Using X-ray analysis it was found that the samples have CsCl-type structure. In the Co-substituted compounds the small traces of the bcc-phase were observed.

The electrical resistivity was measured by means of the four-probe method in the temperature range from 2 to 400 K in a magnetic field up to 7 T using prismatic specimens of about 1 x 1 x 6 mm^3 in size. The measurements of the magnetization were made using a vibrating sample magnetometer in the same magnetic field in the temperature range from 4.2 to 300 K.

3. Results and discussion

Fig.1 shows the temperature dependencies of the electrical resistivity of $(Fe_{1-x}Co_x)_{49}Rh_{51}$ compounds with different content of Co. The $\rho(T)$ dependence for the parent FeRh compound exhibits an abrupt increase of ρ on lowering the temperature to $T_t = 325$ K. The higher value of ρ in the AF state can be related to the appearance of the energy gap in the electron spectrum of FeRh, because the unit cell volume of the AF structure has a doubled period compared with the ferromagnetic one [10]. An existence of such an energy gap results in the much lower value of the electronic specific heat coefficient γ and, consequently, of the density of the electronic states at the Fermi level in the AF state compared with the F state [4,5,7]. An increase in the Co content up to $x = 0.035$ leads to the decrease of the T_t value. At $x > 0.035$ the AF state is not observed and the F state is stable in the whole temperature range below the Curie temperature T_c. The change of the magnetic state of FeRh due to the substitution manifests itself in the change of the electrical resistivity. We have not found any anomalies of the $\rho(T)$ dependencies for the ferromagnetically ordered samples.

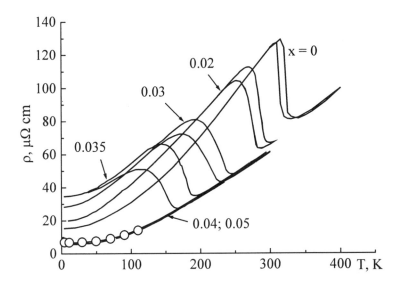

Figure1. Temperature dependencies of the electrical resistivity for the sample $(Fe_{1-x}Co_x)_{49}Rh_{51}$ with x = 0.035. Open circles correspond to the ferromagnetic state (B = 7 T).

An application of the magnetic field to the antiferromagnetically ordered samples (x ≤ 0.035; T < T_t) leads to the abrupt decrease in ρ. As can be seen from Fig.1, the electrical resistivity of the sample with x = 0.035 in the F-state at B = 7 T has the same value as well as the electrical resistivity of the F-ordered samples (x > 0.035) measured without the field. This behaviour of ρ can be attributed to the significant change of the electronic structure of the FeRh-based compounds due to the disappearance of the energy gap in the magnetic field rather than to the critical AF-F transition field.

Using the measurements of the electrical resistivity and magnetization we have obtained the magnetic phase diagram of $(Fe_{1-x}Co_x)_{49}Rh_{51}$ (see Fig.2a).

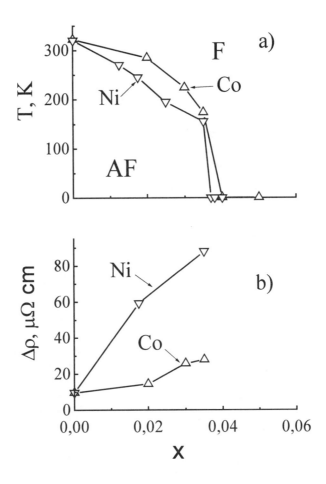

Figure 2. a) magnetic phase diagrams for $(Fe_{1-x}M_x)_{49}Rh_{51}$ (M = Co, Ni); b) $\Delta\rho = \rho_{AF} - \rho_F$ versus x.

The change of the magnetic order in this system takes place practically at the same critical concentration as in the $(Fe_{1-x}Ni_x)Rh$ system [4]. However, we have found that the Co-substitution results in the smaller change in the difference of the electrical resistivity in AF and F states ($\Delta\rho_0 = \rho_{AF} - \rho_F$) at $T \to 0$. As it follows from Fig. 2b the substitution of 3.5 at. % Co results in the $\Delta\rho_0$ value about 30 $\mu\Omega$ cm in comparison with $\Delta\rho_0 = 90$ $\mu\Omega$ cm for the Ni-substituted compound. This difference in $\Delta\rho$ can be related to the different change of 3d-electron concentration during the Co- and Ni-substitution because we have not found the difference in the lattice parameters of $(Fe_{0.95}Co_{0.035})_{49}Rh_{51}$ and $(Fe_{0.95}Ni_{0.035})_{49}Rh_{51}$ at the room temperature where both compounds are in the F-state.

The magnetoresistance effect in $(Fe_{0.95}Co_{0.035})_{49}Rh_{51}$ at $T < T_t$ reaches the giant values, about 80% (see Fig.3) as well as in the other FeRh-based compounds.

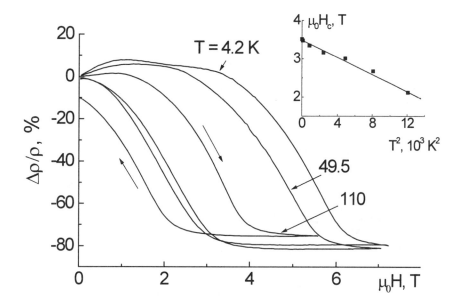

Figure 3. Field dependencies of the magnetoresistance for $(Fe_{1-x}Co_x)_{49}Rh_{51}$ at various temperatures. The inset shows the average critical field as a function of T^2.

The average critical field H_c which corresponds to the center of the AF-F transition changes with the temperature as T^2 (shown in the inset in Fig. 3). Using the thermodynamic relation $\Delta\gamma = 2\alpha\Delta M$ [11], where ΔM is the jump of the magnetization at the AF-F transition and α is the coefficient in the T^2-term in the temperature dependence of the average critical field we have estimated the change of γ. The estimation gives the value about 30 mJ kg^{-1} K^{-2} which is in the reasonable agreement with the experimental data for FeRh where Fe is substituted by 3.5% Ni ($\Delta\gamma = 44$ mJ kg^{-1} K^{-2} [4]).

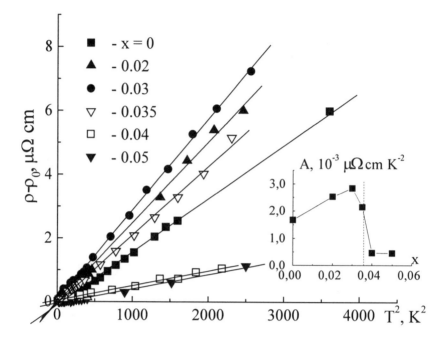

Figure 4. ρ - ρ₀ versus T^2 for $(Fe_{1-x}Co_x)_{49}Rh_{51}$ compounds. Inset shows the concentration dependence of the coefficient A in the T^2- temperature dependence of the electrical resistivity: $\rho = \rho_0 + AT^2$.

The low temperature measurements of the electrical resistivity of $(Fe_{1-x}Co_x)_{49}Rh_{51}$ presented in Fig. 4 show that the ρ(T) dependencies have the T^2 character in both the AF (x≤ 0.035) and F (x > 0.035) states. The coefficient A in the T^2-term depends nonmonotonously on the Co-content with a maximum near the critical concentration of the AF-F transition. An analogous behaviour was observed in the Co-doped NiS_2 compound where the coefficient A in the T^2- temperature dependence of the electrical resistivity tends to diverge as one approaches the critical boundary of antiferromagnetic ordering under hydrostatic pressure [12]. This change in the ρ(T) dependencies for the modified Ni_2S compound was explained by Ueda [1] using the self-consistent theory of spin fluctuations. According to this theory the coefficient in the T^2 term diverges as $|1-\alpha|^{-1/2}$, where α is a coupling constant.

The dramatic change of the physical properties of $(Fe_{1-x}Co_x)_{49}Rh_{51}$ at the critical concentration allows us to suggest that the change of magnetic ordering in $(Fe_{1-x}Co_x)_{49}Rh_{51}$ with increasing of the Co content occurs through the first order phase transition. An increase of the coefficient in the T^2-term with the concentration increase can be related to the increase of the contribution from spin fluctuations due to the instability of the d-electron subsystem in the vicinity of the AF-F phase transition.

Acknowledgement

We are indebted to D.V. Morozov for the assistance in measurements. This work was partly supported by the Russian Foundation for Basic Research (Project 97-02-16504) and by the Ministry of Education (Project 97-7.3-179).

References

1. Ueda, K. (1977) Electrical resistivity of antiferromagnetic metals, *J.Phys. Soc. Jap.*, **3**, 1497-1508.
2. Bergevin, F. and de Muldaver, L. (1961) Antiferromagnetic-ferromagnetic transformations in FeRh, *J. Ghem. Phys.* **35**, 1904-1905.
3. Kouvel, J.S. and Hartelius, C.C. (1962) Anomalous magnetic moments and transformations in the ordered alloy FeRh, *J. Appl. Phys.* **33**, S1343-1344.
4. Baranov, N.V. and Barabanova, E.A. (1995) Electrical resistivity and magnetic phase transitions in modified FeRh compounds, *J. Alloys and Compounds* **219**, 139-148.
5. Baranov, N.V., Markin, P.E., Zemlyanski, S.V., Michor, H., Hilsher, G. J. (1996) Giant magnetoresistance in antiferromagnetically ordered FeRh and Mn_2Sb based alloys, *J. Magn.Magn Mater.* **157-158**, 401-402.
6. Zakharov, A.I., Kadomtseva, A.M., Levitin, R.Z., and Ponyatovski, E.G. (1964) Magnetic and magnetoelastic properties of a metamagnetic iron-rhodium alloy, *Sov. Phys.: JETP* **19**, 1348-1353.
7. Kreiner, K., Michor, H., Hilsher, G., Baranov, N.V., and Zemlyanski, S.V. (1998) Evolution of the electronic specific heat and magnetic order in $(Fe_{1-x}Ni_x)Rh$, *JMMM* **177-181** (to be published).
8. Koening, C. (1982) Self-consistent band structure of paramagnetic, ferromagnetic and antiferromagnetic ordered FeRh, *J.Phys. F: Metal. Phys.* **12**, 1123-1137.
9. Hasegava, H. (1987) Electronic structure and local magnetic moments in ferromagnetic and antiferromagbetic Fe_xRh_{1-x} alloys, *JMMM* **66**, 175-186.
10. Cren, E., Pal, E., and Scabo, P. (1964) Neutron diffraction investigation of the antiferromagnetic-ferromagnetic transformation in the Fe-Rh alloy, *Phys.Lett.* **9**, 297-298.
11. Baranov, N.V., Khrulev, Yu. A., Bartashevich, M.I., Goto, T., and Katori, H.A. (1994) High-field magnetization process in Mn1.9Cr0.1Sb, *J. Alloys and Comp.* **210**, 197-200.
12. Vatanabe, T., Mori, N., and Mitsui, T. (1976) Effect of spin fluctuations on the electrical resistivity in the pressure-induced metallic phase of 7 at. % Co-doped NiS_2, *Solid State Commun.* **19**, 837-839.

RESISTIVITY, MAGNETORESISTANCE AND HALL EFFECT IN $Co_{(100-x)}(CuO)_x$ ($10 \leq x \leq 70$ wt.%) COMPOSITES

V. PRUDNIKOV, A. GRANOVSKY AND M. PRUDNIKOVA
Department of Physics, Lomonosov University,
119899 Moscow, Russian Federation
AND H. R. KHAN
Material Physics, Forschungsinstitut für Edelmetalle und Metallchemie,
73525 Schwäbisch Gmünd, Germany
Department of Physics, University of Tennessee,
Knoxville, Tennessee, USA

Abstract. The experimental data on morphological, structural, transport, and magnetic properties of sintered $Co_{100-x}(CuO)_x$ composites are reported. The Mooij correlation is not valid for these composites because the resistivity increases with the temperature up to 320 µOhm·cm and does not saturate below the percolation threshold. The positive transverse magnetoresistance was observed at low fields. This effect is due to the strong spin-dependent scattering of conduction electrons on oxidized surface layers of Co particles. Hall resistivity behavior unusual for ferromagnets was observed and explained in the framework of the effective medium approach.

1. Introduction

The discovery of the giant magnetoresistance (GMR) [1-3] and giant Hall effect (GHE) [4,5] in ferromagnetic granular alloys has caused the increased interest relating the investigations of structural, magnetic and transport properties of these materials. The experimental GMR and GHE data in granular systems have revealed a series of factors which cannot be explained even qualitatively in the frameworks of earlier developed theories for homogeneous magnetic materials. As a rule, the GMR in magnetic multilayers and granular alloys "ferromagnetic metal - nonmagnetic metal" is negative and the resistivity decreases in the magnetic field. However, recently it has been shown that in some multilayers it may be positive as well (called the inversed GMR) [6]. A rather large negative magnetoresistance (MR) was also observed [2,3] in granular "ferromagnetic metal - insulator" composites near the percolation threshold.

Another unusual magnetic transport phenomena in granular alloys is the extraordinary Hall effect (EHE) [4,5,7,8]. For "metal-metal" granular alloys exhibiting the GMR the well-known correlation between the EHE coefficient R_S and electrical resistivity ρ does not exist [7], and there arises a nonmonotonical field dependence of R_S [8]. In the $(Co_{70}Fe_{30})_xAg_{100-x}$ granular alloys the EHE shows an opposite sign compared to the homogeneous $Co_{70}Fe_{30}$ alloy [9]. In "metal - insulator" nanocomposites the Hall resistivity $\rho_H(H)$ increases sharply for compositions close to the percolation threshold and reaches the values of (100 to 200)·10^{-6} Ohm·cm which is larger by a factor of 10^3 than the Hall resistivity in homogeneous ferromagnetic materials [4,5].

This paper reports the structural, magnetic and transport (electrical resistivity, magnetoresistance, and Hall effect) properties of the sintered $Co_{100-x}(CuO)_x$ granular alloys which consist of ferromagnetic metal and nonmagnetic semiconducting granules.

2. Experimental details

The granular alloys $Co_{100-x}(CuO)_x$ with x between 10 and 70 wt.% Co were prepared using the following techniques. Fine powders of CuO and cobalt with the particle sizes smaller than 20 μm were mixed in proper ratios and pressed in the form of tablets of 15 mm diameter and 2 mm thickness with a pressure 0.8 ton using an isostatic press. The tablets were annealed at a temperature of 900 °C for a period of 1 hour in a vacuum of ~10^{-5} mbar. The $Co_{(100-x)}(CuO)_x$ samples were polished for the morphological investigations.

The microstructure of all the composites was studied by the optical and scanning electron microscopy techniques. The chemical composition of the different grains formed after the heat treatment was determined by the Energy Dispersive X-ray Analysis (EDXA). The formation of the phases of the composites after the annealing was investigated by the X-ray diffraction technique using the Siemens diffractometer with CuK_α-radiation of 0.154178 nm wavelength and Si monochromator.

The magnetization was measured using the conventional vibration sample magnetometer with the magnetic field oriented perpendicular to the sample plane.

The magnetoresistance of the composites in the current in plane (CIP) geometry with the magnetic field perpendicular to the sample plane as well as the Hall effect were measured by the four-point technique in magnetic fields up to 18 kOe. The magnetic and electrical properties investigations were performed in the temperature range 77 - 300 K

3. Results and discussion

3.1. MICROSTRUCTURE

The results of the structural research show that the formation of Co, CoO, Cu_2O, and Cu phases occurs after annealing at 900 °C.

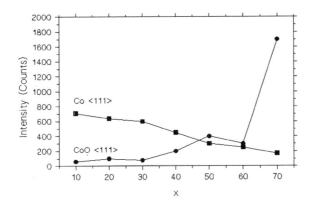

Figure 1. Plot of the intensities of the X-ray diffraction lines Co <111> and CoO <111> as a function of CuO concentration

The intensities of the X-ray diffraction lines of Co <111> and CoO <111> lines from the X-ray diffractograms are displayed as a function of the CuO concentration (Fig. 1). The intensities of these lines roughly correspond to the volume fraction of Co and CoO phases in the composite. As can be seen in the figure, the volume fraction of the metallic cobalt decreases and that of the CoO phase increases with increasing of the CuO concentration.

After 60 wt.% of CuO a large fraction of Co is oxidized to CoO. This confirms that the composite CuO transforms to Cu_2O and the oxygen is released after annealing at 900 °C. The released oxygen oxidizes the metallic cobalt particles to CoO, and with increasing of the CuO concentration more oxygen is available to oxidize the metallic cobalt particles.

3.2. RESISTIVITY

The temperature dependence of the electrical resistivity ρ for various CuO concentrations is shown in Fig.2a. For all samples the electrical resistivity increases with the temperature and the temperature coefficient of the resistivity is positive. The percolation threshold is observed in the concentration range from 30 to 32 wt.% Co. The anomaly in the temperature dependence of $\rho(T)$ can be related to the antiferromagnet - paramagnet transition of CuO phase (Fig.2a). The concentration dependence of the electrical resistivity is shown on Fig.2b. It is unusual for the mixture of low (Co) and high (CuO) resistivity particles and cannot be explained by the mean-field approach. Probably, it is due to the phase transformations discussed above. As one can see from Figs.2a and 2b, the resistivity increases with the temperature for x from 10 to 60 but does not exceed 320 µOhm·cm and 100 µOhm·cm at 300 K and 77 K, respectively. Therefore, these samples are metallic, and conduction electrons move along metal links and are strongly scattered by the impurities, grain boundaries on the surface of ferromagnetic particles and on the contacts between them.

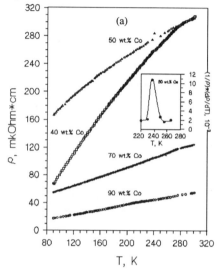

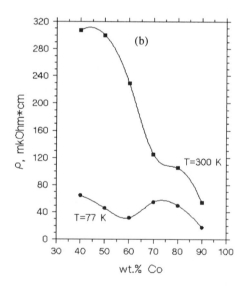

Figure 2. (a) Specific electrical resistivity as a function of temperature for the composition with the cobalt concentration 40, 50, 70,and 90 wt.%. (b) Concentration dependencies of specific electrical resistivity at 77K and 300 K

It should be pointed out that the observed temperature dependence of the resistivity is extremely strong and the Mooij correlation between ρ and $d\rho/dT$ is not valid for these high resistivity composites [10].

3.3. TRANSVERSE MAGNETORESISTANCE

The strong spin-dependent scattering leads to the complex behavior of the transverse MR which is positive for all the samples with x = 10 to 60 at 300 K up to 12 kOe. At 77 K the MR is negative for x = 60 but positive for other samples in low fields. The field dependence of the transverse MR, $\Delta\rho/\rho$, for $Co_{40}(CuO)_{60}$ is presented in Fig.3.

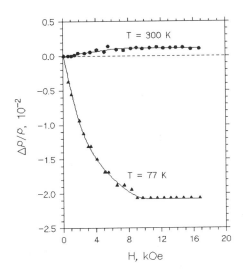

Figure 3. Plot of the transverse magnetoresistance as a function of applied field for the annealed composites with cobalt concentration 40wt.% at 300K and 77 K

As shown in [11] the unusual positive (inverse) sign of the MR can be related to the enhanced spin-dependent surface scattering. It follows from the structural data that Co particles are oxide-covered and surrounded by CuO, CoO, and Cu_2O. All these oxides are antiferromagnets. So, the surface of Co particles can be viewed as a disordered alloy with enhanced spin-dependent scattering on localized magnetic impurities.

This spin-dependent scattering can provide the positive (inverse) MR with the amplitude up to 5% [11]. It is easy to explain the experimental data on the transverse MR as a sum of two contributions – the positive (inverse) MR and negative anisotropic MR [11]. However, the origin of the large negative MR for $Co_{40}(CuO)_{60}$ composite at 77 K (Fig.3) is still unclear. Probably, near the percolation threshold at low temperatures the negative tunneling MR should be taken into account.

3.4. HALL EFFECT

All samples were magnetized to a saturation in magnetic the field about 8 kOe (Fig.4). The difference between the magnetization curves at 77 and 300 K is less than ~ 1%.

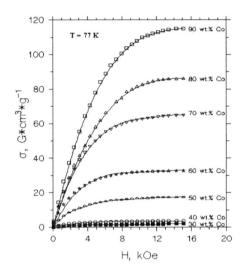

Figure 4. Magnetization curves at 77 K in the magnetic field perpendicular to the sample plane

For all compositions the saturation magnetization is smaller than one can expect from the nominal Co fraction. It is due to the Co particles oxidation that is in agreement with the structural data. It follows from Figs.2a, 2b and 4 that the granular alloys $Co_{100-x}(CuO)_x$ with x between 10 and 60 wt.% are high resistivity ferromagnetic metals. Therefore, it is quite reasonably to suppose that in these composites the EHE should be rather large, not so large as the GHE in the ferromagnetic nanocomposites [4,5], but at least larger than the ordinary Hall effect contribution. However, this is not the case (see Figs. 5a, 5b, and 6a and 6b)

The Hall resistivity of composites was measured at T = 77 K and T = 300 K. The isotherms of the Hall resistivity field dependencies for all investigated concentrations at T = 77 K are shown in Fig.5a. One can see that no saturation and deviation from the linear field dependence take place up to 18 kOe. Thus, we can conclude that the EHE contribution is extremely small at 77 K, and the Hall resistivity is defined by the ordinary Hall effect.

Essentially different is the $\rho_H (H)$ dependence at T = 300 K (Fig.5b), where at $H > H_S$ (H_S > 8 to 10 kOe) the slope of the curves changes, and $d\rho_H (H)/dH$ becomes positive for all Co fraction, except for the $Co_{40}(CuO)_{60}$ sample. It should be mentioned that the Hall resistivity is negative at T = 300 and T = 77 K for all composites.

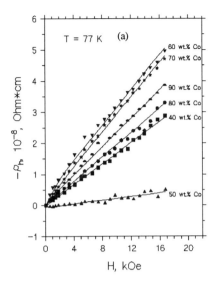

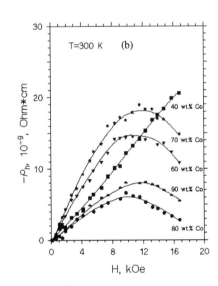

Figure 5. Hall resitivity as function of magnetic field at (a) 77 K and (b) 300 K

Within the framework of the effective-medium approach (EMA) the Hall resistivity for systems containing ferromagnetic particles with the volume fraction x, resistivity ρ_1, ordinary Hall effect coefficient R_{01}, EHE coefficient R_{S1}, magnetization M_1 and second type of particles with concentration y and parameters ρ_2, R_{02}, R_{S2}, M_2 are given by [12]:

$$\rho_H = \frac{x}{x+y\left(\frac{\rho\,\rho_1 + 2}{\rho\,\rho_2 + 2}\right)^2} \cdot [R_{01}B_{z1} + 4\pi R_{s1} M_{z1}]\left(\frac{\rho}{\rho_1}\right)^2 +$$

$$+ \frac{y}{x\left(\frac{\rho\,\rho_1 + 2}{\rho\,\rho_2 + 2}\right)^2 + y} \cdot [R_{02}B_{z2} + 4\pi R_{s2} M_{z2}]\left(\frac{\rho}{\rho_2}\right)^2 =$$

$$= DR_{01}B_{z1}\left(\frac{\rho}{\rho_1}\right)^2 + D4\pi R_{s1} M_{z1}\left(\frac{\rho}{\rho_1}\right)^2 + CR_{02}B_{z2}\left(\frac{\rho}{\rho_2}\right)^2 ,$$

(where B_z is the magnetic induction) provided the second type of particles are nonmagnetic with the typical for semiconductors values of ρ_2 and R_{02}.

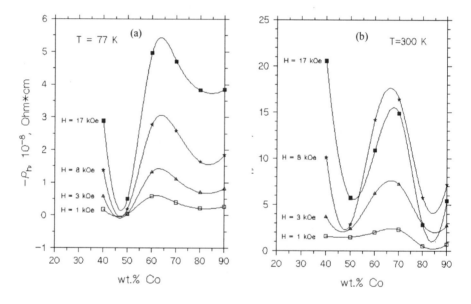

Figure 6. Hall resistivity as a function of Co concentration at (a) 77 K and (b) 300 K

For pure ferromagnetic metals like Ni and Fe the first term in the last line of this formula is by an order of magnitude less than the second one. For bulk scattering in pure Co particles R_0 and R_S are of the same order and $R_S > 0$ at room temperatures. However, R_S is rather small and negative at 77 K [13]. The strong impurity and interface scattering can change both the sign and magnitude of R_S [7,10]. The third term in the Hall resistivity is the ordinary contribution from the second type particles like CuO, Cu_2O, and CoO oxides or from the high resistivity material of intergranular contacts. Due to impurity doping these semiconductors might have not very large resistivity and the third term cannot be neglected.

Let us now explain the experimental data shown in Figs.5a, 5b, and 6a and 6b basing on the expression for the Hall resistivity. It is evident that the linear part of the Hall resistivity curves at $H < H_S$ at room temperatures (Fig.5b) is defined by the EHE contribution (second term) because the slope of the curves changes at $H = H_S$, where this term does not depend on the magnetic field. In the fields $H > H_S$ the Hall resistivity is caused by the ordinary Hall effect (first and third terms), and $d\rho_H /dH$ is rather large and positive (Fig.5b). The positive sign of R_0 is in agreement with the hole-type conductivity of CuO, Cu_2O, and CoO oxides at 300 K. It should be pointed out that the magnitude of $d\rho_H /dH$ slightly depends on the Co volume fraction and, therefore, the first and third terms in the expression for the Hall resistivity are of the same order.

The behavior of ρ_H at 77 K is quite different (Fig.5a). The second term significantly decreases at low temperatures, which is typical for Co-based alloys [10], and one can observe only the ordinary Hall effect (the first and third terms) with an effective negative Hall effect coefficient. The change of the sign of the ordinary Hall effect at low temperatures is probably due to the strong impurity doping of semiconducting particles. Since the Hall resistivity of composites "ferromagnetic metal-semiconductor" is a sum of three (very sensitive to resistivity of ferromagnetic and semiconducting components) contributions, ρ_H is a very complex and nonmonotonic function of the ferromagnetic volume fraction (Figs.6a and 6b).

The authors hope that this first experimental study of transport properties in composites "ferromagnetic metal - semiconductor" will stimulate more experimental interest of the curious phenomena in composites in a wide range of compositions and temperatures.

Acknowledgments

This work was partially supported by Russian Foundation for Basic Research, INTAS (grant N 93-0718) and NATO (grant HTECH.LG 95152). We want to thank Profs. F. Brouers and J.P.Clerc for the fruitful discussions.

References

1. Berkovitz, A.E., Mitchell, J.R., Carey, M.J., Yong, A.P., Zhang, S., Spada, F.E., Parker, F.T., Hutten, A. and Thomas, G. (1992) Giant magnetoresistance in heterogeneous Co-Cu alloys,*Phys.Rev.Lett.* **68**, 3745-3748.
2. Gerber, A., Milner, A., Groisman, B., Karpovsky, M. and Gladkikh, A. (1997) Spin-dependent electron transport in granular ferromagnets, *Phys. Rev.* **B55**, 6446-6458.
3. Inoue, J. and Maekawa, S. (1996) Theory of tunneling magnetoresistance in granular magnetic films, *Phys. Rev.* **B53**, 11, 927-944
4. Pakhomov, A., Yan, X., Xu, Y. (1996) Observation of giant Hall effect in granular magnetic films, *J. Appl. Phys.* **79**, 6140-6144
5. Brouers, F., Granovsky, A., Sarychev, A. and Kalitsov, A. (1997) The influence of boundary scattering on transport properties of metal-dielectric nanocomposites, *Physica* **A241**, 284-287.

6. Hsu, S.Y., Barthelemy, A., Holody, P., Loloee, R., Schoeder, P. and Fert, A. (1997) Towards a unified pictures of spin-dependent transport and perpendicular giantmagnetoresistance in bulk alloys, *Phys. Rev. Lett.* **78**, 2652-2655.
7. Granovsky, A., Brouers, F., Kalitsov, A. and Chshiev, M. (1997) Extraordinary Hall effect in magnetic granular alloys, *JMMM* **166**, 193-198
8. Granovsky, A., Kalitsov, A. and Brouers, F. (1997) Field dependence of extraordinary Hall effect coefficient of granular alloys with giant magnetoresistance, *JETP Lett.* **65**, 511-513
9. Prudnikova, M., Granovsky, A. and Prudnikov, V. (1996) Hall effect in granular $(Co_{70}Fe_{30})_xAg_{1-x}$ alloys, *Proceeding of the Russian-Japanese joint seminar 'The Physics and modeling of intelligent materials and their applications'*, Moscow, MSU, 85.
10. Vedyaev, A., Granovsky, A. and Kotelnikova, O. (1992) *Transport phenomena in disordered ferromagnetic alloys*, Moscow, MSU, 158.
11. Khan, H.R., Granovsky, A., Prudnikova, M., Prudnikov, V., Brouers F., Vedyaev, A., and Radkovskaya, A. Positive transverse magnetoresistance, magnetic and structural properties of $Co_{100-x}(CuO)_x$ ($10 \leq x \leq 70$ wt. %) composites, *JMMM* (in press).
12. Granovsky, A., Vedyaev, A. and Brouers, F. (1994) Extraordinary Hall effect of ferromagnetic composites in the effective medium approximation, *JMMM* **136**, 229-236
13. Vasilieva, R., Cheremushkina, A., Yazliev, S., Kadirov, Y. (1974) Hall effect and electrical resistance of Fe-Ni alloys, *Sov. Phys.: Phys. Met. Metallurgy* **38**, 289-296

THEORY OF ITINERANT-ELECTRON SPIN-GLASS IN AMORPHOUS Fe

T. UCHIDA AND Y. KAKEHASHI
Hokkaido Institute of Technology
Maeda, Teine-ku, Sapporo 006, Japan

Abstract. The physics of itinerant-electron spin-glass and the ferromagnetism in the vicinity of amorphous Fe has been clarified on the basis of the finite-temperature theory of amorphous metallic magnetism. The theory is extended to describe the noncollinear magnetism in amorphous transition metals. It is demonstrated that the transverse spin degrees of freedom yield the isotropic spin-glass and noncollinear ferromagnetism around amorphous Fe. These structures are shown to be stabilized by the higher-order biquadratic exchange energies, and are consistent with the experimental data of recent neutron scatterings and Mössbauer measurements.

1. Introduction

Recent experimental studies on amorphous transition metal alloys have shown that the magnetic properties of amorphous systems are quite different from those of the crystalline counterparts. In the case of Fe-rich amorphous transition metal alloys, the ferromagnetism completely collapses and the spin-glass (SG) phase appears beyond 90 at.% Fe, as shown in Fig.1 [1]. A remarkable feature of the magnetic phase diagram in Fig.1 is that the SG transition temperatures beyond 90 at.% Fe are the same irrespective of the second elements. This suggests that the spin-glass beyond 90 at.% Fe is caused by the structural disorder intrinsic to the pure amorphous Fe, instead of the configurational disorder between Fe and the second element. In addition, the Fe-rich amorphous alloys show a variety of peculiar magnetism which is not observed in ordinary local moment SG, such as the resistivity minimum [1], anomalous thermal expansion [1], large forced

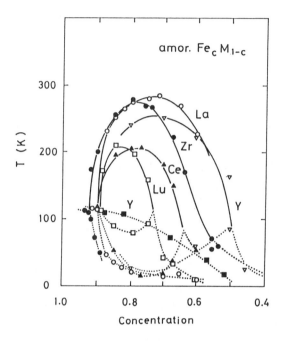

Figure 1. Magnetic phase diagram showing the Curie temperatures (soild curves) and the spin-glass transition temperatures (dotted curves) in amorphous Fe_cM_{1-c} alloys (M = La, Zr, Ce, Lu, and Y) [1,10].

magnetostriction at the reentrant SG transition temperatures [2], and large pressure effect [3]. Therefore it has been a fundamental issue in the theory to clarify the origin of the SG around amorphous Fe and the related peculiar magnetism from the microscopic point of view. In order to solve this problem, we developed the finite-temperature theory of amorphous metallic magnetism and have clarified the various aspects of magnetism observed in amorphous transition metal alloys [4]. Moreover, we have recently extended the finite-temperature theory to include the transverse spin degrees of freedom leading to the isotropic itinerant-electron SG realized in amorphous Fe [5-7].

In the present paper, we review the recent development of the finite-temperature theory of amorphous metallic magnetism, and present the physical picture for a variety of magnetism around amorphous Fe. In particular, we discuss the magnetic phase diagram obtained from the new theory with transverse spin degrees of freedom.

2. Development of theory

In the past twenty years, there has been much progress in the theory of electronic structure calculations for the structurally disordered systems. The recursion method [8] made possible the electronic structure calculations without assuming translational symmetry. Fujiwara applied the method to the electronic structure calculations of amorphous Fe on the basis of the first principles tight-binding Hamiltonian, and found that the main peak in the nonmagnetic density of states (DOS) shifts from the Fermi energy to the higher energy region when the structure is changed from the bcc to the amorphous one, so that the Stoner condition is not satisfied for amorphous Fe [9]. This suggests that the amorphous Fe does not show the uniform ferromagnetism.

Kakehashi developed the finite-temperature theory of amorphous metallic magnetism to investigate the problem of structure vs magnetism in transition metals [10]. The theory takes into account the thermal spin fluctuations by means of the functional integral method [11], and the fluctuations of the local moments due to structural disorder by means of the distribution function method [12]. Numerical studies based on this approach showed that the itinerant-electron SG is realized in pure amorphous Fe because of the nonlinear magnetic couplings between neighboring local moments (LM's) and the local environment effects on LM's due to the structural disorder. The result is consistent with the experimental data on Fe-rich amorphous transition metal alloys in Fig.1. The theory also explained the peculiar magnetism observed around 90 at.% Fe, such as the large pressure effect and the anomalous forced volume magnetostriction by means of the variable amplitude of LM, which is characteristic of itinerant-electron magnetism. The theory was extended to the case of amorphous alloys [13], so that the basic feature of the magnetic phase diagram shown in Fig.1 was reproduced by taking into account the alloying effect. It was verified that the SG beyond 90 at.% Fe is caused by the structural disorder as in the case of pure amorphous Fe, although the ferromagnetism is stabilized below 90 at.% Fe because of the atomic size effect of the second elements.

There are, however, some experimental results which support the ferromagnetism of amorphous Fe, instead of SG. For example, in Fe-B alloys, there are no signs of vanishing ferromagnetic moments around 90 at.% Fe [14], and in the thin film of Y/amorphous Fe/Y layered structure, the ground-state magnetization 1.2 μ_B is reported [15]. In the amorphous Fe powder containing 2 wt.% H, 3 wt.% C, 1 wt.% O etc., the magnetization 1.4 μ_B is obtained [16]. Although the ferromagnetism in these amorphous Fe might be ascribed to the strong effect of the second elements, it is also probable that the magnetism is greatly influenced by the change of the volume and microscopic structure in amorphous Fe. Numerical calculations based on the finite-temperature theory revealed that the phase transition from the ferromagnetism to SG and then from the SG to the paramagnetism occurs in amorphous Fe with decreasing the volume [17]. The phase diagram shows that the volumes of 90 at.% Fe alloys in Fig.1 correspond to the SG phase near the boundary between the SG and the ferromagnetic phases. Therefore, the small volume expansion expected in the Y/amorphous Fe/Y layered structure could induce the ferromagnetism.

In order to clarify the influence of the degree of structural disorder on the magnetism of amorphous Fe, we proposed the finite-temperature theory which interpolates between the crystal and the amorphous structure using microscopic parameters such as the fluctuation of atomic distance $[(\delta R)^2]_s^{1/2}/[R]_s$ and the average coordination number z^*. Here, $[\ ]_s$ denotes the structural average and R denotes the nearest-neighbor interatomic distance. The magnetic phase diagram obtained from the interpolation theory shows that the SG phase in amorphous Fe appears in the region of $[(\delta R)^2]_s^{1/2}/[R]_s \approx 0.06$ and $z^* \approx 11.5$, while the ferromagnetism is realized in the region of $[(\delta R)^2]_s^{1/2}/[R]_s \approx 0.06$ and $z^* \approx 10.5$ [18]. Therefore, the ferromagnetism of amorphous Fe powder containing 2 wt.% H etc. and Fe-B alloys can be explained by the latter region of the structural parameters and thus does not contradict with the SG in Fe-rich amorphous alloys shown in Fig.1.

3. Theory of amorphous metallic magnetism with vector spins

The finite-temperature theory of amorphous metallic magnetism mentioned in the previous section explained the SG and various aspects of magnetism observed in Fe-rich amorphous alloys. However, it does not describe the isotropic SG and noncollinear ferromagnetism, since the transverse spin degrees of freedom were neglected in the theory. Recent neutron scattering experiment [19] and the Mössbauer measurement [20] suggest the existence of noncollinear magnetism in Fe-rich amorphous alloys. In order to treat more realistic case, we have recently extended the theory to the noncollinear case [5-7].

In the theory of the noncollinear magnetism, we introduce locally rotated coordinates into the degenerate-band Hubbard model with Hund's rule coupling. We then apply the functional integral method to take into account the thermal spin fluctuations within the static approximation. The free energy reduces to the generalized Hartree-Fock one at the ground state and leads to Anderson's super-exchange interactions for the half-filled band in the insulator limit. By making use of the molecular field approximation, the thermal average of the vector LM on the central site is given by

$$\langle m_0 \rangle = \frac{\int d\xi \, (1 + \frac{4}{\beta \tilde{J} \xi^2}) \xi \, e^{-\beta E(\xi)}}{\int d\xi \, e^{-\beta E(\xi)}}. \quad (1)$$

Here β is the inverse temperature and $\tilde{J}$ is the effective exchange energy parameter.

The energy functional $E(\xi)$ in Eq. (1) consists of the single-site energy $E_0(\xi)$, the atomic pair energies $\Phi^{(a)}_{0j}(\xi)$ between the central site 0 and the neighboring site j, and the three types of exchange energies $\Phi^{(e)}_{0j\alpha}(\xi)$, $\Phi^{(b)}_{0j\delta}(\xi)$, and $\Phi^{(c)}_{0j}(\xi)$:

$$\begin{aligned} E(\xi) = & \, E_0(\xi) \\ & + \sum_{j \neq 0} \left[\Phi^{(a)}_{0j}(\xi) - \sum_\alpha \Phi^{(e)}_{0j\alpha}(\xi) \frac{\langle m_{j\alpha} \rangle}{\tilde{a}_{j\alpha}} \right. \\ & \left. + \sum_{(\alpha,\gamma)} \Phi^{(b)}_{0j\delta}(\xi) \frac{\langle m_{j\alpha} \rangle \langle m_{j\gamma} \rangle}{\tilde{a}_{j\alpha} \tilde{a}_{j\gamma}} + \Phi^{(c)}_{0j}(\xi) \frac{\langle m_{jx} \rangle \langle m_{jy} \rangle \langle m_{jz} \rangle}{\tilde{a}_{jx} \tilde{a}_{jy} \tilde{a}_{jz}} \right]. \quad (2) \end{aligned}$$

Here the subscripts α, γ, and $\delta(\neq \alpha, \gamma)$ in Eq.(2) take x, y, and z components, respectively. $\sum_{(\alpha,\gamma)}$ means the summation over different pairs chosen from x, y, and z components. The energies $\Phi^{(e)}_{0j\alpha}$, $-\Phi^{(b)}_{0j\delta}$, and $-\Phi^{(c)}_{0j}$ describe the magnetic energy gain of the central LM when the nearest-neighbor (NN) local moments $\langle m_{j\alpha} \rangle$, $\langle m_{j\alpha} \rangle \langle m_{j\gamma} \rangle$, and $\langle m_{jx} \rangle \langle m_{jy} \rangle \langle m_{jz} \rangle$ take positive values, respectively. $\tilde{a}_{j\alpha}$ denotes the α component of the amplitude of LM on site j, which is defined by $\tilde{a}_{j\alpha} = (1 + 4/\beta \tilde{J} \langle \xi^2_j \rangle_0) \langle \xi^2_{j\alpha} \rangle_0^{1/2}$, $\langle \, \rangle_0$ being the thermal average with respect to the single-site energy on site j.

The energies at the right-hand side of Eq.(2) are expressed by a one-electron Green's function with structural disorder as well as the diagonal disorder due to the random exchange fields $\{\xi_i\}$. After making use of the Bethe-type approximation to the Green function, the central LM $\langle m_0 \rangle$ is given by the coordination number z on the NN shell, the neighboring average LM's $\{\langle m_j \rangle\}$, the square of transfer integrals $\{t^2_{0j}\}$, the effective self-energy $\{S_\sigma\}$ for electrons with spin σ describing the structural disorder outside the NN shell, and the effective medium $\{\mathcal{L}^{-1}_\sigma\}$ describing the thermal spin fluctuations. The first three quantities change randomly due to the structural disorder. Introducing the distribution functions $g(\langle m_j \rangle)$ for the surrounding LM's $\{\langle m_j \rangle\}$, the probability $p(z)$ of finding a coordination number z and the probability $p_s(y_j)$ for the squares of transfer integrals $\{y_j = t_{0j}^2\}$, we obtain the distribution of the central LM from $g(\langle \boldsymbol{m}_j \rangle)$, $p(z)$, and $p_s(y_j)$ via Eq.(1). Since it is identical with the surrounding ones, we obtain an integral equation for the distribution function as follows:

$$g(M) = \sum_z p(z) \int \delta(M - \langle m_0 \rangle) \prod_{j=1}^{z} [p_s(y_j) dy_j g(m_j) dm_j]. \qquad (3)$$

By making use of the decoupling approximation at the right-hand side of Eq.(3), we obtain the self-consistent equations for the magnetization $[\langle m_z \rangle]_s$ and the spin-glass order parameters for each direction $[\langle m_\alpha \rangle^2]_s^{1/2}$ ($\alpha = x, y, z$), where $[\ ]_s$ denotes the structural average.

In the present approximation, the local environment is described by a plus or minus direction of each component of LM with the magnitude $[\langle m_{j\alpha} \rangle^2]_s^{1/2}$ on the surrounding sites and a contraction or a stretch of the NN pair by $[(\delta R)^2]_s^{1/2}$. The structural average of LM's was calculated by means of 8000 Monte-Carlo sampling since the number of configurations is of the order of 10^6. The effective media $\mathcal{S}_\sigma$ and $\mathcal{L}_\sigma^{-1}$ are also obtained self-consistently by using the average DOS in the nonmagnetic state. The variables z and $\{t_{0j}^2\}$ are described by the average coordination number z^* and the fluctuation of atomic distance $[(\delta R)^2]_s^{1/2}/[R]_s$ in the present approximation.

4. Noncollinear magnetism and isotropic SG in amorphous Fe

The application of the theory with transverse spin fluctuations to amorphous transition metals leads to the noncollinear magnetism in the phase diagram [5-7]. Figure 2 shows the calculated distribution of LM's when the d electron number N is decreased from the value around amorphous Co ($N \sim 8.0$) towards the value ($N \sim 7.0$) of amorphous Fe. A strong ferromagnetism is realized around amorphous Co since the peak in the nonmagnetic DOS shifts to the Fermi energy due to the structural disorder. With decreasing the d electron number N, the strong ferromagnetism begins to collapse as the Fermi energy approaches the top of the DOS for up-spin electrons. The LM's then show a broad distribution as shown in Fig.2(a) due to the strong local environment effects. At lower d electron numbers, the LM's antiparallel to the magnetization appear due to the antiferromagnetic interaction (Fig.2(b)). Further decrease of the d electron number leads to the frustrated LM's with weak molecular fields due to the increased antiparallel LM's on the surrounding sites. The higher-order

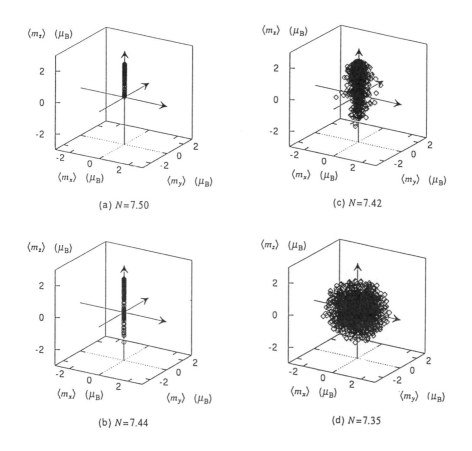

Figure 2. Distributions of local moments at $T = 35K$ obtained by 4000 Monte-Carlo sampling for the d electron numbers (a)$N = 7.50$, (b)$N = 7.44$, (c)$N = 7.42$, and (d)$N = 7.35$. Here, N denotes the d electron number and the the magnetic polarization is assumed be in the z direction [7].

bilinear exchange interactions stabilize these frustrated LM's, leading to the noncollinear ferromagnetism (Fig.2(c)). After the disappearance of the ferromagnetism, the isotropic SG is finally realized, as is clearly shown by the spherically symmetric distribution of Fe LM's (Fig.2(d)). The magnetic phase diagram around amorphous Fe is shown in Fig.3. It is seen that the isotropic SG phase extends over the region $N \lesssim 7.4$. Amorphous Fe ($N \sim 7.0$), therefore, forms an isotropic SG. The obtained SG transition temperatures $50 \sim 100K$ are comparable to the value $110K$ obtained from the extrapolation of experimental data on Fe-rich alloys as shown in Fig.1.

The analysis of the single-site and pair energies which determine the

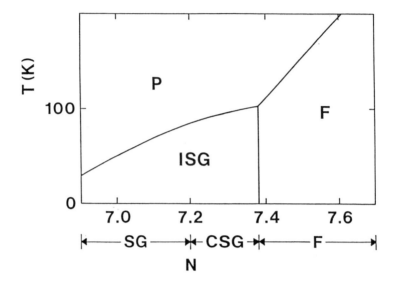

Figure 3. Magnetic phase diagram around amorphous Fe showing the ferromagnetic (F), paramagnetic (P), and the isotropic SG (ISG) states. The isotropic SG is classified into the normal SG (SG) and the cluster SG (CSG). The noncollinear ferromagnetism is realized near the boundary between the ferromagnetic and the SG phases [7].

thermal average of the central LM reveals the existence of the nonlinear magnetic couplings such that the Fe LM's with large amplitudes ferromagnetically couple with the neighboring LM's, while the Fe LM's with small amplitudes antiferromagnetically couple with the neighboring ones. Since the amplitude of LM depends strongly on the surrounding environments, the structural disorder causes the competition between the ferro- and antiferromagnetic interactions via the nonlinear couplings. This leads to the itinerant-electron SG in amorphous Fe. The SG formation mechanism is essentially the same as that proposed for the collinear SG in the finite-temperature theory with Ising type spins [10]. The bilinear exchange interactions in the vector spin space stabilize the noncollinear magnetic structures leading to the isotropic SG.

On the other hand, the nonlinear magnetic couplings between the nearest-neighbor (NN) LM's disappear in the d electron number region $N \gtrsim 7.25$; the NN couplings are ferromagnetic in any local environment. The isotropic SG in this region is caused by the competition of the short-range ferromagnetic couplings and the long-range antiferromagnetic couplings. Since the NN couplings are ferromagnetic, the resulting SG is accompanied by ferromagnetic clusters. Thus it is called the cluster SG (see the region $7.25 \lesssim N \lesssim 7.40$

in Fig.3). Although the present theory cannot treat the clusters larger than the NN shell, the size of the ferromagnetic clusters in the SG should increase with increasing the d electron number, finally realizing the bulk ferromagnetism after the second-order phase transition. In fact, the recent neutron scattering experiment on amorphous $Fe_{90}Zr_{10}$ alloy suggests the coexistence of the spin wave excitations and the SG in the reentrant SG region, being consistent with the above mentioned picture [21].

5. Summary

We have presented the physical picture for the itinerant-electron SG in the vicinity of amorphous Fe on the basis of the finite-temperature theory of amorphous metallic magnetism. In particular, we have demonstrated that the theory with the transverse spin degrees of freedom leads to the isotropic SG and noncollinear ferromagnetism around amorphous Fe, which are consistent with the recent experimental data on Fe-rich amorphous transition metal alloys.

Recently, the spin polarized electronic structure calculations have been performed towards quantitative understanding of the magnetism of amorphous Fe at the ground state [22-26]. Earlier calculations using the supercell [22-24], however, led to the ferromagnetism for amorphous Fe, in contradiction with the results of the finite-temperature theory. They also disagree with the recent experimental facts that the amorphous $Fe_{93.5}Zr_{6.5}$ alloy exhibits a clear phase transition from the paramagnetism to the SG as a function of temperature [27], and that the application of small pressures of about 2 GPa induces a clear transition from the ferromagnetism to the SG in the Fe-rich amorphous alloys around 90 at.% Fe [1]. The discrepancy probably originates in the small number of atoms ($\sim$ 60 atoms) in the unit cell, the difference in the structural model, and the difference in the boundary condition. In particular, the periodic boundary condition with use of the same unit cell as the crystalline one may overestimate the magnetization in competing systems, because the competing interactions generally lead to larger magnetic unit cell. The finite-temperature theory presented here, on the other hand, does not give rise to such a problem since the magnetization and the SG order parameters are determined self-consistently without using the periodic boundary condition. In fact, Lorenz et al. [25] have recently obtained a magnetization of about 0.5 μ_B for amorphous Fe in their revised calculation scheme in which the outside of the unit cell consisting of 64 atoms is treated by an average potential instead of using the periodic boundary condition. According to the recent first-principle calculation by Liebs et al. [26], the calculated magnetization decreases as the size of the unit cell is increased, suggesting the existence of SG in amorphous Fe. To obtain more solid conclusion on the ground-state magnetism in amorphous Fe, however, it is desired to develop a calculation scheme finding automatically the most stable magnetic structure among various metastable structures.

References

1. Fukamichi, K., Goto, T., Komatsu, H., and Wakabayashi, H. (1989) Spin glass and Invar properties of iron-rich amorphous alloys, in W. Gorkowski, H. K. Lachowicz, and H. Szymczak (eds.) *Proceedings of the Fourth International Conference on the Physics of Magnetic Materials, Poland, 1988*, World Scientific, Singapore, 354–381.
2. Tange, H., Tanaka, Y., Goto, M., and Fukamichi, K. (1989) Saturation and forced volume magnetostriction of Fe-rich FeZr amorphous alloys, *J. Magn. Magn. Mater.* **81**, L243–246.
3. Goto, T., Murayama, C., Mori, N., Wakabayashi, H., Fukamichi, K., and Komatsu, H. (1988) Pressure effect on the magnetic properties of Fe-La amorphous alloys, *J. Phys. Colloq.* (Paris) **49**, 1143–1144.
4. Kakehashi, Y. and Tanaka, H. (1995) Theory of magnetism in amorphous transition metals and alloys, in J. A. Fernandez-Baca and W. Y. Ching (eds.) *The Magnetism of Amorphous Metals and Alloys*, World Scientific, Singapore, 1–84.
5. Uchida, T. and Kakehashi, Y. (1997) Finite temperature theory of noncollinear magnetism in amorphous Fe, *J. Appl. Phys.* **81**, 3859–3861.
6. Uchida, T. and Kakehashi, Y. (1997) Finite temperature theory of amorphous metallic magnetism with vector spins, *Physica* **B237-238**, 504–505.
7. Uchida, T. and Kakehashi, Y. (1997) Theory of isotropic spin-glass and noncollinear ferromagnetism in amorphous Fe, *Phys. Rev.* **B**, to be submitted.
8. Haydock, R., Heine, V., and Kelly, M. J. (1975) Electronic structure based on the local atomic environment for tight-binding bands II, *J. Phys.* **C8**, 2591–2605.
9. Fujiwara, T. (1984) Electronic structure calculations for amorphous alloys, *J. Non-Cryst. Solids* **61-62**, 1039–1048.
10. Kakehashi, Y. (1989) Single-site theory of finite-temperature magnetism in amorphous and liquid alloys, *Phys. Rev.* **B40**, 11063–11069; Kakehashi, Y. (1990) Finite-temperature theory of local environment effects in amorphous and liquid magnetic alloys, *Phys. Rev.* **B41**, 9207–9220.
11. Hubbard, J. (1979) The magnetism of iron *Phys. Rev.* **B19**, 2626–2636; Hubbard, J. (1979) Magnetism of iron II *Phys. Rev.* **B20**, 4584–4595; Hasegawa, H. (1979) Single-site functional approach to itinerant-electron ferromagnetism, *J. Phys. Soc. Jpn.* **46**, 1504–1514; Hasegawa, H. (1980) Single-site spin fluctuation theory of itinerant-electron systems with narrow bands, *J. Phys. Soc. Jpn.* **49**, 178–188.
12. Matsubara, F. (1974) Theory of the random magnetic mixture II Ising system, *Prog. Theor. Phys.* **52**, 1124–1134; Katsura, S., Fujiki, S., and Inawashiro, S. (1979) Spin-glass in the site Ising model, *J. Phys.* **C12**, 2839–2846.
13. Yu, M., Kakehashi, Y., and Tanaka, H. (1994) Finite-temperature theory of amorphous magnetic alloys, *Phys. Rev.* **B49**, 352–367.
14. Hasegawa, R. and Ray, R. (1978) Iron-boron metallic glassses, *J. Appl. Phys.* **49**, 4174–4179.
15. Handschuh, S., Landes, J., Köbler, U., Sauer, Ch., Kister, G., Fuss, A., and Zinn, W. (1993) Magnetic properties of amorphous Fe in Fe/Y layered structures, *J. Magn. Magn. Mater.* **119**, 254–260.
16. Bellissent, R., Galli, G., Grinstaff, M. W., Migliardo, P., and Suslick, K. S. (1993)

Neutron diffraction on amorphous iron powder, *Phys. Rev.* **B48**, 15797–15800.
17. Yu, M. and Kakehashi, Y.(1994) Existence of the spin-glass state in amorphous Fe , *Phys. Rev.* **B49**, 15723–15729.
18. Kakehashi, Y., Uchida, T., and Yu, M. (1997) Metallic magnetism from crystals to amorphous structures in Fe, Co, and Ni, *Phys. Rev.* **B56**, 8807–8818.
19. Cowley, R. A., Cowlam, N., and Cussens, L. D. (1988) A non-collinear magnetic structure for the amorphous ferromagnet $Fe_{83}B_{17}$, *J. Phys. Colloq.* (Paris) **49**, C8 1285–1286.
20. Vincze, I., Kaptás, D., Kemény, T., Kiss, L. F., and Balogh, J. (1995) Temperature and external magnetic field dependence of the spin freezing in amorphous $Fe_{93}Zr_7$, *J. Magn. Magn. Mater.* **140-144**, 297–298.
21. Fernandez-Baca, J. A., Rhyne, J. J., Fish, G. E., Hennion, M,, and Hennion, B. (1990) Spin dynamics of amorphous $Fe_{90-x}Ni_xZr_{10}$, *J. Appl. Phys.* **67**, 5223–5228.
22. Krey, U., Krauss, U., and Krompiewski, S. (1992) Itinerant spin glass states and asperomagnetism of amorphous Fe and iron-rich Fe/Zr alloys, *J. Magn. Magn. Mater.* **103**, 37–46.
23. Bratkovsky, A. M. and Smirnov, A. V. (1993) Amorphous magnetism in iron-boron systems: First-principles real-space tight-binding LMTO study, *Phys. Rev.* **B48**, 9606–9610.
24. Liebs, M., Hummler, K., and Fähnle, M. (1995) Influence of structural disorder on magnetic order: An ab initio study of amorphous Fe, Co, and Ni, *Phys. Rev.* **B51**, 8664–8667.
25. Lorenz, R. and Hafner, J. (1995) Noncollinear magnetic structures in amorphous iron and iron-based alloys , *J. Magn. Magn. Mater.* **139**, 209–227.
26. Liebs, M. and Fähnle, M. (1996) Amorphous iron revisited: An ab initio study, *Phys. Rev.* **B53**, 14012–14015.
27. Nicolaides, G. K. and Rao, K. V. (1993) Melt-spun $Fe_{93.5}Zr_{6.5}$: a 112K 'spin-glass' alloy in the vicinity of the multicritical point, *J. Magn. Magn. Mater.* **125**, 195–198.

THE FORMATION OF THE MAGNETIC PROPERTIES IN DISORDERED BINARY ALLOYS OF METAL-METALLOID TYPE

A.K.ARZHNIKOV AND L.V.DOBYSHEVA
Physical-Technical Institute
Ural Branch of Russian Academy of Sciences
Kirov str, 132, Izhevsk, 426001, RUSSIA

Abstract. A theoretical description of the magnetic properties in disordered alloys Fe-M (M=Al, Si, Sn, Ga) on the basis of a two-band model of the Hubbard type is presented. It is shown that at T=0 the local magnetic moment is determined by the number of metalloid atoms in the nearest environment and weakly depends on the concentration, which allows to use the Jaccarino-Walker model for the description of the experimental results. The thermodynamical Stoner excitations have essential effect at high concentration of the metalloid. The spin density fluctuations should be taken into account in a calculation of the concentration dependence of the Curie temperature at low concentration. So, at all concentrations, the itinerant character of the magnetism of these alloys does not allow to use localized models.

For a long time the disordered alloys of metal-metalloid type ($Fe - Al$, $Fe - Si$, $Fe - P$, $Fe - Sn$) have been attracting the attention of scientists. This interest is mainly due to the fact that these alloys have a wide concentration range for the homogeneous disordered state and this constancy of structure makes them a good model object for studying the magnetism in disordered systems. The successful interpretation of the experimental results by the modified Jaccarino-Walker ($J - W$) models gives the grounds for the usage of these models [1, 2]. However, the $J - W$ models are conceptually closer to the localized models whereas magnetism in iron is of itinerant character. In the alloys, the magnetism has even more collectivized character, which follows from the increase in the Rhodes-Wohlfarth parameter (RW) with concentration (this parameter is the relation of the magnetic moment, found from the Curie constant, divided by the saturation moment). For $Fe-Si$ alloys RW increases from 1.42 to 1.82 when concentration changes from 0 to 24 $at.\%Si$. So, it is unclear why the $J-W$ models are so good in describing the experiments. Consequent self-consistent calculations of the magnetic properties of itinerant electron models including thermodynamical excitations are not possible today. Thus we believe it is necessary to use simple models which take into account main details of the above-mentioned alloys.

The available experimental data of the concentration dependencies which are qualitatively similar for different metalloid atoms let us suggest that the description of these alloys requires only a few number of physical parameters connected with the crystal structure and the electron structure of the metalloid atoms. Really, an X-ray structure analysis [3, 4, 5] shows that these alloys in a wide concentration range may be considered as substitutional alloys with body centered crystal structure, the possible topological disorder and the corresponding change of wave functions overlap being insignificant, which may be proved by the coincidence of the magnetic characteristics for the amorphous and crystalline samples [3].

The common feature of the metal-metalloid alloys is the presence of narrow d-like and wide sp-like bands near the Fermi level, the metalloid d-band being considerably far away in energy from the metal d-band unlike the sp-bands which may be regarded as common for the iron and metalloid atoms. Starting from these assumptions, these alloys can be described by a two-band Hamiltonian of the Hubbard type which on the one hand takes into account the peculiarities mentioned above, and on the other hand is most suitable for the description of the itinerant magnetic properties:

$$H = H_s + H_d + H_{sd} + H_U \qquad (1)$$

$$H_s = \sum_{k,\sigma} E_{sk} a^+_{sk\sigma} a_{sk\sigma} \qquad H_d = \sum_{j,\sigma} E_{dj} a^+_{dj\sigma} a_{dj\sigma} + \sum_{j,j',\sigma} t_{ij} a^+_{dj\sigma} a_{dj'\sigma}$$

$$H_{sd} = \sum_{j,\sigma} \gamma_j [a^+_{dj\sigma} a_{sj\sigma} + a^+_{sj\sigma} a_{dj\sigma}] \qquad H_U = U \sum_i n_{i\uparrow} n_{i\downarrow}$$

where j, j' are the sites of the crystal lattice, $a^+_{jm\sigma}, a_{jm\sigma}$ are the creation and destruction operators of the Wannier states with index m at a site j and spin σ, k is a wave vector, E_{sk} is the dispersion law of sp-electrons, $n_{i\sigma} = a^+_{di\sigma} a_{di\sigma}$, U is the intraatomic interaction of electrons. The parameters in the Hamiltonian are chosen in the following way: $\left|\frac{E^B - E^A}{W_d}\right| \gg 1$, $W_s \gg W_d$, $U \simeq W_d$ and $\gamma \simeq W_d$, W_d is the d-band width, W_s is the sp-band width, E^α is the d-level position of the atom of α type. In the following we will regard the component B as magnetic and A as nonmagnetic. U_j and γ_j are on-site interaction values and, besides, $U_j = U$, because this parameter is mainly determined by the screening of the d-electrons by the electrons of the inner levels. The sp-bands are wide and close to each other in energy for all kinds of atoms. The overlapping integral (t_{ij}) is taken independent of the type of atoms i and j. For wide sp-band this approximation, as well as the approximation of $E_{si} = Const$, is natural by virtue of self-averaging. For d-electrons this approximation is not essential, since the metalloid d-band is considerably higher than E_F. The orbital degeneracy of the d-states is taken into account by a subsequent renormalization of the physical quantities [6].

The Hamiltonian (1) in Hartree-Fock approximation was repeatedly used to describe the magnetic properties of d-metal alloys. The distinctive feature of the Hamiltonian used in this paper is that E_{dj}, γ_j and M_j (M_j is the magnetization at site j) are considered to depend not only on the kind of the atom at site j, but also on the configuration of the nearest-environment cluster Ω_j. We assume that γ_j depends linearly on the number of nonmagnetic atoms Z in Ω_j ($\gamma_{\Omega_j} = \gamma^0 + \gamma^1 Z$). $E_{d\Omega_j}$ is determined by the condition of electroneutrality of the cell. We set $t_{ij} = t$ for nearest neighbors and $t_{ij} = 0$ otherwise.

For the calculation of the local magnetic moment in the ground state we use the Hartree-Fock approximation for H_U:

$$H_U \simeq \frac{U}{2} \sum_j M_j [a^+_{dj\downarrow} a_{dj\downarrow} - a^+_{dj\uparrow} a_{dj\uparrow}] \qquad (2)$$

and the coherent potential approximation (CPA) for the effective medium, in which the cluster of the specified configuration Ω_j is inserted, and we make a self-consistent procedure over the parameters $E_{d\Omega_j}$ and M_{Ω_j}. We neglect the fluctuations of $M_{j'}$, $\gamma_{j'}$ and $E_{dj'}$ arising from the difference in the number of nonmagnetic impurities in the environment at the sites j' neighboring j and set them equal to the average (for a given atom type) values, that is, the values with which the effective medium is calculated.

The self-consistent calculation gives us the matrix element of the resolvent G^Ω of the Hamiltonian of the cluster Ω centered at j embedded in an effective medium. Fig.1 displays the variation of the average magnetic moment of an atom B with concentration.

Fig.2 shows the result of the self-consistent calculation of the local magnetic moment of an atom B as a function of the number Z of nonmagnetic atoms A in the nearest environment for different concentrations.

First of all we should note that the local magnetic moment for a given configuration Ω is only weakly dependent on concentration, which is the main peculiarity of the $J - W$ models; the dependence of the local moment on Z agrees qualitatively with experiments. So, in spite of the band character of the model we obtain results which are similar to the results of

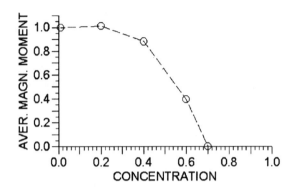

Figure 1. Average magnetic moment of the atom B divided by the magnetic moment at $c = 0$ versus concentration c

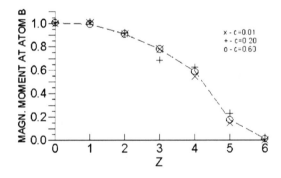

Figure 2. The local magnetic moment of the atom B divided by the magnetic moment at $c = 0$, as a function of Z

the J-W model. This certainly justifies the frequent usage of such models in interpreting the experiments. Also remarkable is the fact that at high Z ($Z = 3, 4, 5$) the local magnetic moment is directed opposite with respect to the total magnetization.

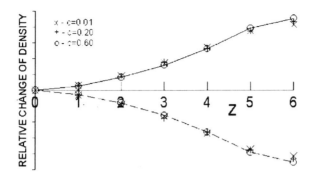

Figure 3. The integrated density of states of sp- and d-electrons at atom B as a function of Z. Dashed line indicates the sp-electrons, solid line is for d-electrons.

At low impurity concentration the number of atoms with negative local magnetic moment is naturally small. Near the critical concentration for ferromagnetism the number becomes large enough for experimental detection (see [7, 8]).

Fig.3 displays our results for the change of the integrated density-of-states of the *sp*- and *d*-electrons.

They agree qualitatively with the known experiments on the isomer shift obtained from Mossbauer spectroscopy [9, 10].

It is common knowledge that the Stoner theory does not consistently describe the transition metals, i.e. the influence of Stoner type excitations (STE) on the magnetic properties is negligible. To obtain the observed temperature dependence of physical properties it is necessary to take into account the spin-wave excitations. Usually this statement about the STE role is extended to the disordered alloys. Any magnetic alloy on the basis of $3d$-metals has undoubtedly a temperature below which the magnetization is determined by the spin-wave excitations, but this range narrows as the concentration of non-magnetic impurities increases. The ratio of the intensities of the Stoner and the spin-wave excitations can then change with increasing concentration in favor of the former. Experimental indication of the possible enhancement of the STE influence in the disordered alloys is the increase in the Rhodes-Wolfarth parameter.

To study the influence on the magnetization from the Stoner excitations only we have to insert the Fermi distribution function into the self-consistent equation for the local magnetic moment, considering M_{Ω_j} and $E_{d\Omega_j}$ as dependent on temperature:

$$M_{\Omega_j}(T) = -\frac{1}{\pi} Im \int_{-\infty}^{+\infty} n(E)[G_{jj}^{\Omega\uparrow}(E, E_{d\Omega_j}, M_{\Omega_j}) - G_{jj}^{\Omega\downarrow}(E, E_{d\Omega_j}, M_{\Omega_j})] \, dE$$

together with the equation

$$n_{\Omega_j}(T) = -\frac{1}{\pi} Im \int_{-\infty}^{+\infty} n(E)[G_{jj}^{\Omega\uparrow}(E, E_{d\Omega_j}, M_{\Omega_j}) + G_{jj}^{\Omega\downarrow}(E, E_{d\Omega_j}, M_{\Omega_j})] \, dE$$
$$= const.,$$

where $n(E) = \left[\exp\left(\frac{E-\mu}{kT}\right) + 1\right]^{-1}$, μ is the chemical potential. The shift of $E_{d\Omega_j}$ occurs due to the difference of the energy spectrum of electrons in the cell surrounded by the different number of metalloid atoms. The shift of the chemical potential results in a rescreening of the d-electrons by the mobile sp-electrons. This rescreening is taken into account by requiring electroneutrality of the cell at all temperatures. It should be noted that in a consistent consideration of the many-electron phenomena the sd-hybridization should also undergo a change with temperature. However we believe that since

$$\left|\frac{E_{d\Omega_j}(T=0) - E_{d\Omega_j}(T)}{E_F}\right| \simeq \left|\frac{\gamma(T=0) - \gamma(T)}{\gamma(T=0)}\right|,$$

the following inequality holds

$$\left|E_{d\Omega_j}(T=0) - E_{d\Omega_j}(T)\right| \gg |\gamma(T=0) - \gamma(T)|.$$

So, the largest changes in the parameters of the Hamiltonian with temperature occur due to the shift of the d-level.

It is natural to expect that the effect of the Stoner excitations should be largest for the local magnetic moments of atoms with many non-magnetic atoms in their nearest environment, because the d-band becomes flatter due to the higher hybridization, which resembles the situation in weak ferromagnets. Fig.4 shows the temperature dependence of the local magnetic moment of an atom with $Z = 5$ non-magnetic impurities in its nearest environment.

According to our results, the local magnetic moment at $Z = 5$ is opposite to the total magnetization of the alloy. One can see that the influence of the STE is essential even at low temperatures 0.015-0.025 eV. It should be mentioned that the temperatures characteristic of STE in the non-impurity

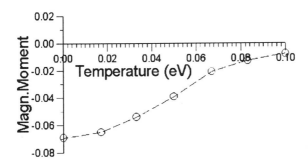

Figure 4. The temperature dependence of the local magnetic moment of an atom with Z=5 non-magnetic impurities in nearest environment with the concentration c=0.20.

case are higher by an order-of-magnitude, i.e. $kT = 0.5 - 0.6$ eV. It is clear that configurations with such a large Z are rare at low concentration of non- magnetic impurities and do not essentially affect the magnetization behavior. At concentrations near the percolational threshold their influence may be noticeable.

The configurations with $Z = 3$ and $Z = 4$ are more probable. Fig.5 gives the local moment of a configuration with $Z = 3$ for the concentration $c = 0.20$.

As should be expected for lower hybridization the influence of the STE now manifests itself at higher temperatures ($kT = 0.07 - 0.08eV$). But at these temperatures a second solution appears which is positive relative to total magnetization and belongs to a higher energy. This energy relation holds up to the temperature $\simeq 0.12eV$. A different situation with respect to the energies of the two solutions arises at higher concentration ($c = 0.60$). A ferromagnetic solution still exists at this concentration. The results of the calculations for $c = 0.60$ and $Z = 3$ are shown in Fig.6 and demonstrate that the energies of the two solutions become equal at $kT = 0.05 - 0.06$ eV. The local moment with $Z = 4$ behaves in the same way (Fig.7).

So, at this temperature a change of the magnitude and the direction of the local magnetic moment must occur and cosequently an increase of the total magnetization should be observed.

It should be noted that there indeed exists experimental evidence for an increase of the magnetization at concentrations near the critical one [11].

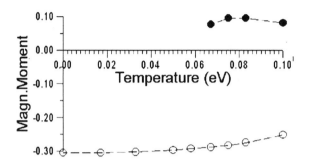

Figure 5. The temperature dependence of the local moment of the configuration Z=3 at the concentration c=0.20.

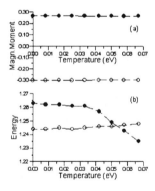

Figure 6. The temperature dependence of the local moment (a) and energy (b) of the configuration Z=3 at the concentration c=0.60 (dashed circles are for the solution with positive magnetic moment).

But as a rule these changes are explained by a transition from the reentrant spin glass state into the ferromagnetically ordered state, which is mainly characterized by the presence of frustrations. Though spin disordering is

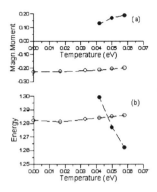

Figure 7. The same as in Fig.6 at Z=4.

present in our case, there are no frustrations, and the ground state of this model can be better classified as a Mattis glass.

Assuming only that the d-states of the metalloid atoms are far away in energy from the Fe d-band and that the s-d hybridization at a Fe atom is determined by the number of metalloid atoms in its nearest environment, the main laws governing the formation of magnetism in these alloys were described qualitatively. An itinerant character of the magnetism appeared to be found at high metalloid concentration. This is reflected in a decrease in the amplitudes of the magnetic moments with concentration and in a high influence of the Stoner type excitations at relatively low temperatures. These excitations cause both changes in the magnitude of the individual local magnetic moments and changes in their directions. Therefore it became clear that at high metalloid concentration the description of the magnetic states of these alloys on the basis of localized models like the Ising and Heisenberg models is not justified. However at low metalloid concentration the situation should be seemingly different (the Stoner type excitations are insignificant, and the local magnetic moments do not vary with the concentration).

Figs.8 and 9 show the experimental results for the Curie temperature and the magnetization of alloys of Fe with Al, Si, Ga, Sn.

This set of experimental data clearly demonstrates that the concentration dependencies of the magnetic disordering temperatures for different alloys (Fe with Al, Si, Ga, Sn) do not differ quantitatively and are very

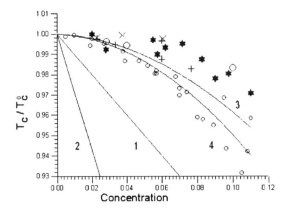

Figure 8. The concentrational dependence of the Curie temperature of disordered alloys: ($\star$) - $Fe - Al$ [1, 12-14], (o) - $Fe - Si$ [12-13, 15-18], ($\bigcirc$) - $Fe - Ga$ [14], ($\times$) - $Fe - Sn$ [12, 19]; curve 1 - the Heisenberg model; 2 - the model $T_c(x) \sim J_{eff}(x)$; 3 - this paper calculation without taking into account the charge fluctuations; 4 - with taking into account the charge fluctuations.

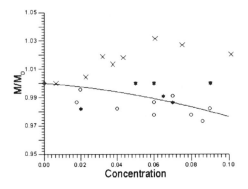

Figure 9. The concentrational dependence of the average magnetic moment at Fe atom in disordered alloys: ($\star$) - $Fe - Al$ [5, 12-13], (o) - $Fe - Si$ [12-13, 18], ($\times$) - $Fe - Sn$ [12, 19]; curve - this work calculation.

close to each other. The same can be said about the concentration behavior of the magnetic moments (Fig.9). So, there are no qualitative differences in spite of the essential dissimilarities of the metalloid atoms added to Fe. The electron configuration in the row Al, Si, Ga, Sn changes as $3p^1$, $3p^2$, $4p^1$, $5p^2$. The size of the impurity atoms also varies. And the fact, that magnetic characteristics do not qualitatively differ, shows that they are determined by a limited number of factors. At least, the metalloid p-electrons, the size of the impurity atoms and the distance between atoms do not affect them essentially. In spite of the weak concentration dependence of the local magnetic moment at the iron atom B (Fig.9) and a negligible influence of the STE, an estimation of the Curie temperature through localized models of the Heisenberg type turned out to be impossible even at low concentrations of metalloid atoms (curve 1, Fig.8). This shows the necessity of taking into account the itinerant character of the electrons forming the magnetic state. The above analysis of experimental data allows us to carry out an additional restriction of the factors having an influence on the magnetism. We keep the main peculiarities: the presence of a d-band for iron and $\left|\frac{E^B - E^A}{W_d}\right| \gg 1$. Then the Hamiltonian can be rewritten as follows:

$$\begin{aligned} H &= \sum_{i\sigma} E_i a_{i\sigma}^+ a_{i\sigma} + \sum_{ij} t_{ij} a_{i\sigma}^+ a_{j\sigma} + U \sum_i n_{i\uparrow} n_{i\downarrow} \\ &= H_0 + H_{int} \\ H_{int} &= \sum_{i\sigma} E_i a_{i\sigma}^+ a_{i\sigma} + U \sum_i n_{i\uparrow} n_{i\downarrow} \end{aligned} \quad (3)$$

The randomly distributed energy parameter E_i takes the value E^A for the metalloid atoms and E^B for the metal atoms. As before, we take t_{ij} and U identical for metal and metalloid atoms.

This representation of the Hubbard model is chosen, because we do not consider the transverse fluctuations of the spin density, i.e., we do not take into account the hopping of electrons with a spin-flip.

Following the common procedure [20] the partition function of the system in the interaction representation

$$Z = Tr \left\{ \exp(-\beta \tilde{H}_0) T_\tau \exp\left[-\int_0^\beta d\tau H_{int}(\tau)\right] \right\}, \quad (4)$$

can be rewritten with a functional integration:

$$\begin{aligned} Z = \int &\left(\Pi_l \frac{1}{i} \delta w_l(\tau) \delta v_l(\tau) \right) \\ &\exp\left[-\frac{\pi}{\beta} \int_0^\beta d\tau \sum_l [v_l^2(\tau) - w_l^2(\tau)] \right] Z\{w_l(\tau), v_l(\tau)\}, \end{aligned} \quad (5)$$

$$Z\ \{w_l(\tau), v_l(\tau)\} =$$
$$= Tr \exp\left(-\beta \tilde{H}_0\right) T_\tau \exp\left\{-\int_0^\beta d\tau \sum_l Tr_\sigma \left(V_l(\tau)\rho_l(\tau)\right)\right\} \quad (6)$$

The transition from (5) to (6), and (7) is possible, since all the operators in H_{int} commute with one another, including the disorder term. Here an effective potential at a site l appears:

$$V_l = \begin{pmatrix} \sqrt{\frac{\pi U}{\beta}}(w_l + v_l) + E_l 0 \\ 0 \sqrt{\frac{\pi U}{\beta}}(w_l - v_l) + E_l \end{pmatrix}$$

This result is diagonal over the spin matrix. It includes the fluctuating fields v_l and w_l, they correspond to the thermodynamical fluctuations of spin and charge densities. It also includes the random parameter E_l which depends on the type of atom at site l. With the exception of E_l this is a commonly used expression for representing the Hubbard model by a two-field functional integral. Here the field w_l which describes the charge fluctuations has been calculated in two ways. In the first case we neglected the charge fluctuations, which are rapid variables with respect to the spin fluctuations. We chose them equal to a constant potential according to the condition that the site is electroneutral. This approximation means that the sp-electrons have enough time to rearrange and to compensate the changes of charge due to the spin density fluctuations. In the second case the charge fluctuations were taken into account within the saddle point method.

The functional integration was done in static approximation which can be used because we are not interested in the low-energy excitations, our aim is to obtain the Curie temperature, which usually belongs to the higher range of temperature. Here the temperature is much higher than the characteristic energy of collective spin excitations.

Minimization of the free energy gives us a self-consistent system of equations for the self energy Σ:

$$g^\sigma = (z - H_0 - \Sigma^\sigma(z))_{ll}^{-1}$$

$$\int_{-\infty}^{+\infty} dv \int_{-\infty}^{+\infty} dE_l\, t_l W(v, E_l) = 0, \quad (7)$$

where $t_l = (\tilde{V}_l - \Sigma)[1 - g(\tilde{V}_l - \Sigma)]^{-1}$. $W(v, E_l)$ is the distribution probability density of $v = \sqrt{\pi U T} v_l$ and E_l at a site l,

$$W(v, E_l) = C_l \left[x\delta(E^A - E_l) + (1-x)\delta(E^B - E_l)\right] \times$$
$$\exp\left\{-\frac{1}{UT}\left(v^2 - w^2 + \frac{U}{\pi}\int_{-\infty}^{\mu} dz\, Tr_\sigma Im\, \ln\left[1 - g(\tilde{V}_l^\sigma - \Sigma)\right]\right)\right\}, \quad (8)$$
$$\tilde{V}_l^\sigma = \sigma v + w + E_l.$$

The normalizing constants C_l are found from

$$\int_{-\infty}^{+\infty} dv \int_{-\infty}^{+\infty} dE_l\, W(v, E_l) = 1$$

This approach of calculation corresponds to the quenched disorder case. The equation (7) is the coherent potential approximation for the self-energy of the Green function. While obtaining these equations we also used the fact that the Stoner excitations, as we mentioned earlier, are insignificant at low concentration, and replaced the Fermi distribution function by the step function with integration up to the Fermi energy. More detailed description of the calculation is given in Ref.[19]

The precision of the approximation made is here proportional to $(a_1 c^2 + a_2 c <v^2> + a_3 <v^2>^2)$. This estimation allows us to neglect the terms with non-diagonal Green function and so our approximation is valid only at low concentrations.

A simple two-peak model density of states of Fe was used for the calculations (Fig.10).

It is rather simple to calculate the effective exchange interaction in the Hartree-Fock approximation (2),

$$J_{eff}(c) = \int_{-\infty}^{\mu} dz\, Tr_\sigma Im\, \ln\left[1 - g(\tilde{V}_l - \Sigma)\right]\Big|_{v=MU/2, E_l=E^B} - \int_{-\infty}^{\mu} dz\, Tr_\sigma Im\, \ln\left[1 - g(\tilde{V}_l - \Sigma)\right]\Big|_{v=-MU/2, E_l=E^B}, \quad (9)$$

which is most often used in the first-principle calculations of the Curie temperature $T_c \sim J_{eff}(c)$. The concentration dependence $T_c(c)$ is shown in Fig.8, curve 2. It is clearly seen that such an interpolation of $J_{eff}(c)$ from the ground state is not satisfactory and one should take into account the concentration dependency of the thermodynamical fluctuations of the charge and the spin densities.

In Figs.8 and 9 the results of the calculation of the magnetic moment and T_C are shown for the cases with and without accounting for the charge density fluctuations along with the experimental data. As can be seen,

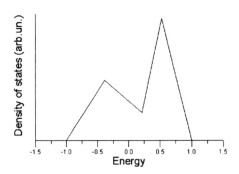

Figure 10. The model density of states, for Fe

taking into account the spin density fluctuations allows us to describe correctly the magnetic characteristics at low concentrations. A more accurate description depending on the type of metalloid requires an account of the sp-d hybridization, i.e. the consideration of the second band. As our calculations showed, the sp-d hybridization, first, increases the number of d-electrons (Fig.3) and, second, flattens the d-band. Both mechanisms affect the slope of the T_c curve.

Thus, the thermodynamic magnetic characteristics cannot be described by localized models such as Ising and Heisenberg models, despite the fact that the local magnetic moments remain constant and the Stoner fluctuations are negligibly small. The usage of the effective exchange interaction calculated for the ground state also fails because it varies as the temperature increases due to the spin density fluctuations and their influence is different at different concentrations of a metalloid. So, the itinerant character of magnetism in disordered metal- metalloid-alloys manifests itself both at low and at high concentration of metalloid atoms.

This work was supported by the Russian Foundation for Basic Research, Grant No. 97-02-16270.

References

1. Yelsukov, E.P., Voronina, E.V. and Barinov, V.A. (1992) Mossbauer study of magnetic properties formation in disordered Fe-Al alloys, *J. Magn. Magn. Mater.* **115**, 271-280.

2. Perez Alcazar, G.A., Plascak, J.A. and Galvao de Silva, E. (1986) Site - diluted Izing model for the magnetic properties of $Fe_{1-q}Al_q$, $0 < q < 0.5$ alloys in disordered phase, Phys.Rev.B **34**, 1940-1943.
3. Elsukov, E.P., Barinov, B.A. and Konygin, G.U. (1986) The effect of order-disorder transition on the structural and magnetic properties of the BCC Fe-Si alloys Fis. Met. i Met. (Russian) **62**, 719-723.
4. Eymery, J.P., Fnipiki, A., Denenot, M.F. and Krishuan, R. (1988) Magnetic properties of $Fe_{60}Al_{40}$ films prepared by coevaporation, IEE Trans. Magn. **24**, 1697-1700.
5. Besnus, M.J., Kerr, A. and Meyer, A.J.P. (1975) Magnetization of disordered cold- worked Fe-Al alloys up to 51 at.% Al, J.Phys.F: Met. Phys. **5**, 2138-2147.
6. Kakehashi, Y. (1986) Degeneracy and quantum effects in Hubbard model, Phys.Rev.B **34** 3243-3253.
7. Shukla, P. and Wortis, M. (1980) Spin-glass behavior in iron-aluminum alloys. A microscopic model, Phys.Rev. B **21**, 159-164.
8. Grest, G.S. (1980) Monte Carlo study of the transition from a ferromagnet to a spin glass in Fe-Al alloys, Phys.Rev. B **21**, 165-168.
9. Chacham, H., Silva, E.G.da, Guenzburger, D. and Ellis, D.E. (1987) Electronic structure, magnetic properties, Mossbauer isomer shifts, and hyperfine fields of disordered Fe-rich Fe-Al alloys, Phys.Rev. B **35**, 1602-1608.
10. Dubiel, S.M. and Zinn, W. (1983) Spin and charge density changes in $\alpha - Fe$ on substituting its atoms by Al, Si, Sn and V, J. Magn. Magn. Mater. **31-34**, 530-532.
11. Shiga, M. and Nakamura, Y. (1976) Mossbauer study of bcc Fe-Al alloys near the critical concentration, J.Phys.Soc.Jap. **40**, N5, 1295-1299.
12. Fallot, M. (1936) Ferromagnetisme des alliages de fer, Ann. Phys. **6**, 305-387.
13. Parsons, D., Sucksmith, W. and Thompson J.E. (1958) The magnetization of cobalt - aluminium, cobalt - silicon, iron - aluminium and iron - silicon alloys, Phil. Mag. **3**, 1174-1184.
14. Vincze, I. and Cser, L. (1972) Phys. Stat. Sol. (b) **50**, 709.
15. Arais, S. (1965) Ferromagnetic Curie temperatures of iron solid solutions with germanium, sylicon, molybdenium and manganese, Phys. Stat. Sol. **11**, 121-126.
16. Pepperhoff, W. and Ettwig, H.-H. (1967) Uber die spezifischen Warmen von Eisen - Silizium - Legierungen, Z. Phys. **22**, 496-499.
17. Meinhardt, D. and Krisement, O. (1965) Fernordnung im System Eisen - Silicium, Arch. Eisenhuttenw. **36**, 293-297.
18. Elsukov, E.P., Konygin, G.N., Barinov, V.A. and Voronina, E.V. (1992) Local atomic environment parameters and magnetic properties of disordered crystalline and amorphous iron-silicon alloys, J. Phys.: Cond. Matt. **4**, 7597-7606.
19. Arzhnikov, A.K., Dobysheva, L.V., Yelsukov, E.P. and Zagaynov, A.V. (1996) Concentration dependence of the magnetization and the Curie temperature in disordered Fe-M alloys (M=Al, Si, Sn), JETP **83**, 623-627.
20. Moriya, T. (1987) *Spin Fluctuations in Itinerant Electron Magnetism*, Springer-Verlag, New York.

ITINERANT ELECTRONS AND SUPERCONDUCTIVITY IN EXOTIC LAYERED SYSTEMS

V. A. IVANOV, E. A. UGOLKOVA, AND M.YE. ZHURAVLEV
N. S. Kurnakov Institute of General & Inorganic Chemistry of the Russian Academy of Sciences,
31 Leninskii prospect, Moscow, 117 907 Russia

Abstract. In this paper our recent studies of organic salts are reviewed including new results. We analysed an electronic structure and superconductivity of layered organic materials on the basis of bis(ethylenedithio)tetrathiafulvalene molecule (BEDT-TTF, hereafter ET) with essential intraET correlations of electrons. Taking into account the Fermi-surface topology the normal and superconducting electronic density of states are calculated in the explicit form for a realistic model of κ-ET$_2$X salts. For an electronic pair in an empty triangular ET$_2$ lattice the critical binding energy is evaluated. The d-symmetry of superconducting order parameter is obtained and interplay between its nodes on the Fermi-surface and superconducting phase characteristics is found. The results are in satisfactory agreement with known band parameters of the normal phase and measured nonactivated temperature dependencies of superconducting specific heat and NMR-relaxation rate of central carbon nuclei ^{13}C in ET. A comparison of the present results with available data is made. The developed approach is applicable to other layered organic salts parent to κ-ET$_2$X ones.

1. Introduction

The condensed organics is a branch of the condensed matter science since the discovery of conductivity [1-3] and superconductivity [4] in organic matter. The electron donor bis(ethilenedithio)tetrathiafulvalene, i. e. BEDT-TTF molecule $C_{10}S_8H_8$ ($\equiv$ ET hereafter) can form a wide class of salts [5] with the most attractive for fundamental science and applications κ-ET$_2$X family. The κ-ET$_2$X breakthrough has been stimulated by the discovery of superconductivity in the β-ET$_2$I$_3$ salt [6] under ambient pressure. Irrespectively of the similarity of electronic and crystal structure and the same carrier concentration of half a hole per ET molecule the κ-ET$_2$X family includes semiconductors (*e. g.*, for X = Cu[N(CN)$_2$]Cl, d$_8$-Cu[N(CN)$_2$]Br), normal metals and superconductors (*e. g.*, 10 K-class superconductors under ambient pressure for X = Cu[N(CN)$_2$]Br, Cu[N(CN)$_2$]CN, Cu (NCS)$_2$). Up to now the κ-ET$_2$X superconductors have the highest critical superconducting temperature T_c~13 K among organic superconductors without metallic atoms. At the same time the κ-ET$_2$X metals are the early organic ones in which the Fermi-surface topology was found experimentally from the Shubnikov oscillations [7,8]. In recent years systematic nuclear magnetic resonance (NMR), electrodynamic, thermodynamic and transport measurements under pressure (see reviews [9-11]) have shed some light on the problem of the the κ-phase salts of ET-based compounds. Its crystal motif is made by dimers, ET_2^+ arranged in a crossed-dimer manner in ET-layers (so-called κ-packing), separated by alternating polymerised anion sheets, X^-, with a sheet periodicity of about 15 Å. The ET_2 dimers are fixed in sites of plane lattice closed to triangular one and its unit cell is the rectangular a-by-$\sqrt{3}a$ including two dimers ET_2 (hereafter the lattice constant $a=1$). According to the quantum chemical calculations the charge density profile of the hole within the ET is concentrated around the central fragment C_2S_4. The intermolecular distance within the ET_2 dimer is 3.2Å whereas the separation between the neighbouring dimers ET_2 is about 8Å. The electron localisation on centers of ET molecules and the small overlap of the molecular π-orbitals means that carriers prefer to stay at ET or ET_2 than to travel. This leads to the enhancement of intraET Coulomb repulsion of π-electrons, U_{ET} [12-15], and to the narrowness of the energy band in κ-ET_2X materials. The highest occupied molecular orbitals (HOMOs) of the ET molecules are π-orbitals which are arranged perpendicular to the molecule ET planes and are oriented nearly parallel to the conduction ET_2 layers, providing quasi two-dimensional character of κ-ET_2X salts. Noteworthy, that nowdays the κ-ET_2X organic salts represent the condensed matter example with much more stronger electron correlations than such inorganic solids as high-T_c superconductors or transition metal compounds. In early Refs. [16,17] the importance of consideration of electron-electron correlations have been noticed for layered organic salts based on the ET molecule. Authors [18] discussed the magnetic properties of κ-ET_2X, α-ET_2I_3, $ET_2MHg(SCN)_4$ in the light of the correlation effects restricting the consideration to the frame of the Hartree-Fock approximation.

The strong electron correlations suppress charge fluctuations near the top of the band filling, and the long range interaction of electrons with lattice phonons is screened. It leads to a reduction of the electron-phonon coupling. Along with it C-S and C-C vibrational ET phonon modes do not influence the origin of the isotopic shift, ΔT_c, in ^{34}S and ^{13}C substituted ET-superconductors [10] and the normal properties of ET-salts. So, the role of the central intraET stretching motions and electron-phonon interaction may not be crucial for a formation of κ-ET_2X normal and superconducting properties. Due to these reasons we neglect here the role of phonons.

In the present review the we present the analysis of electronic structure of κ-ET_2X salts and superconductivity based on the assumption that κ-ET_2X properties are governed by the scale of $U_{ET} \geq 1eV$ (intraET electron-electron repulsion), $t_0 \sim 0.2eV$ (intraET_2 carrier hopping between ET molecules) and $t_{1,2,3} \sim 0.1eV$ (interET_2 carrier hoppings between nearest molecules of neighbouring dimers) [12, 13]. The diagram technique for generalised Hubbard-Okubo operators [12, 19] makes it possible to take into account electron correlations and this scale of characteristic energies contrary to the Hartree-Fock approximation.

The normal electronic density of states and dispersion relations in the explicit form are derived in Sec. 2 and 3 for realistic κ-ET_2X lattice symmetry in the approximation of the effective dimer Hamiltonian. On the basis of the Hubbard model with two ET_2^+ sites per a unit cell the insulating state is obtained for the κ-ET_2X family and a phase transition to a metallic state is observed in Sec. 4 and 5. For a discussion of the metallic state the ET_2^+ dimers are considered as entities with two degenerate energy levels. The employed X-operator machinery of the generalised Okubo-Hubbard operators is briefly discussed in Sec. 4 and 6. Besides the discussion of the nature of the superconducting mechanism the anisotropic pairing of different symmetries is studied and in the BCS approximation coupling constants are calculated in Sec. 7 and 8 taking into account correlated narrowing of energy bands. In Sec. 9 for an electronic pair in the empty triangular lattice of ET_2 dimers the critical binding energy is evaluated. Strong electron correlations manifests themselves in renormalisation of the hopping integrals and chemical potential due to the correlation factor. In Sec. 10 we discuss a link between the derived superconducting density of electronic states and the topology of the Fermi surface. In Sec. 11 it is shown that nodes of superconducting order parameter are responsible for the description of such quantities as nonactivated superconducting specific heat, ^{13}C NMR, London penetration depth. A relation between the spin-lattice relaxation rate and electronic specific heat is suggested for an experimental test of the κ-ET_2X superconductors.

2. Tight-binding electron energy dispersions

The tight-binding carrier energy dispersion relations have been derived and applied in [12,13,18-20] under the assumption of the square ET_2-layer lattice. However, the real ET molecular arrangement has a rhombic distortion. For κ-$(ET)_2Cu(NCS)_2$ at 104K an interdimer azimuthal distance to the pair of nearest neighbours is 8.4 Å while other four neighbours are at distances 7.7 Å [21]. We suppose that ET_2 site positions are fixed by lattice vectors $x_{1,2}$ in such a way that azimuthal nearest neighbours are located at $\pm x_1$, and the nonazimuthal nearest neighbours at $\pm x_2$ and $\pm(x_1+x_2)$. Then a pair of angles between nonazimuthal directions $\pm x_2 \wedge \pm(x_1+x_2)$ is around 66^0 while the other four angles between the azimuthal and others directions, namely $\pm x_1 \wedge \pm(x_1+x_2)$, $x_1 \wedge -x_2$ and $-x_1 \wedge x_2$, are equal to 57^0. Other κ-ET_2X solids possess similar crystal geometries. So, the arrangement of ET_2 dimers in κ-ET_2X is closer to a triangular lattice than to a square lattice as is the case of [22].

For an ideal triangular ET_2-layer the vectors of lattice are

$$\boldsymbol{x}_1 = (0,1)a, \boldsymbol{x}_2 = (\frac{\sqrt{3}}{2}, -\frac{1}{2})a$$

Below we set the lattice constant $a = 1$. For clarity all of the nonazimuthal hopping integrals are assumed to be the same. In reality, they are slightly split [23] owing to the non-symmetry of the dimer position in some salts as was shown by a number of quantum chemical computations (see, e.g., [21]). An azimuthal hopping t_2 in the crystal direction b is set along the y axis in Fig. 1.

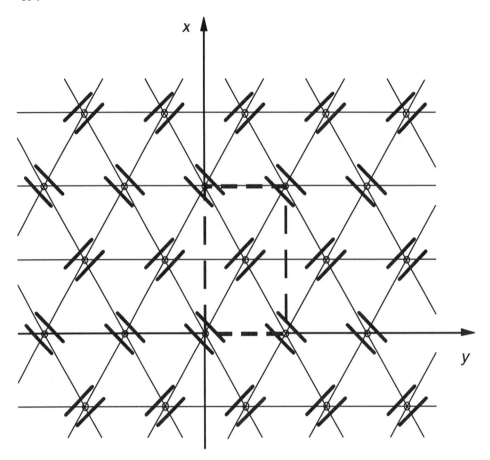

Figure 1. The scheme of positions of the ET$_2$ dimers in the sites of the triangular lattice. Lattice constants of the unit cells (dashed line) are related as $\sqrt{3}:1$.

The single ET$_2$ layer Hamiltonian corresponding to Fig. 1 can be written in the form

$$H = \sum_{i,j,\sigma}[-t_0(a^+_{\sigma,i,j}b_{\sigma,i,j} + b^+_{\sigma,i,j}a_{i,j} + c^+_{\sigma,i,j}d_{\sigma,i,j} + d^+_{\sigma,i,j}c_{\sigma,i,j})$$
$$-t_2(a^+_{\sigma,i,j}b_{\sigma,i,j+1} + b^+_{\sigma,i,j}a_{\sigma,i,j-1} + c^+_{\sigma,i,j}d_{\sigma,i,j+1} + d^+_{\sigma,i,j}c_{\sigma,i,j-1})$$
$$-t_1(a^+_{\sigma,i,j}d_{\sigma,i-1,j-1} + d^+_{\sigma,i-1,j-1}a_{\sigma,i,j} + a^+_{\sigma,i,j}d_{\sigma,i-1,j} + d^+_{\sigma,i-1,j}a_{\sigma,i,j})$$
$$-t_3(b^+_{\sigma,i,j}c_{\sigma,i,j} + c^+_{\sigma,i,j}b_{\sigma,i,j} + b^+_{\sigma,i,j}c_{\sigma,i,j-1} + c^+_{\sigma,i,j-1}b_{\sigma,i,j})] \qquad (1)$$

where a, b, c, d are HOMO orbitals of ET molecules in the cell, $a_{i,j}^{(+)}, b_{i,j}^{(+)}, c_{i,j}^{(+)}, d_{i,j}^{(+)}$ are electron annihilation (creation) operators in the site (i,j), and $\mathbf{R} = i \cdot \mathbf{x} + j \cdot \mathbf{y}$.

In the momentum representation it can be written as

$$H = \sum_\sigma \{-t_0 (a_{\sigma,p}^+ b_{\sigma,p} + b_{\sigma,p}^+ a_{\sigma,p} + c_{\sigma,p}^+ d_{\sigma,p} + d_{\sigma,p}^+ c_{\sigma,p})$$
$$-t_2[a_{\sigma,p}^+ b_{\sigma,p} \exp(ip_y) + b_{\sigma,p}^+ a_{\sigma,p} \exp(-ip_y) + c_{\sigma,p}^+ d_{\sigma,p} \exp(-ip_y)$$
$$+ d_{\sigma,p}^+ c_{\sigma,p} \exp(ip_y)]$$
$$-t_1[a_{\sigma,p}^+ d_{\sigma,p} \exp(-ip_y - i\sqrt{3}p_x) + d_{\sigma,p}^+ a_{\sigma,p} \exp(ip_y + i\sqrt{3}p_x) +$$
$$+ a_{\sigma,p}^+ d_{\sigma,p} \exp(-i\sqrt{3}p_x) + d_{\sigma,p}^+ a_{\sigma,p} \exp(i\sqrt{3}p_x)]$$
$$-t_3 \cdot [b_{\sigma,p}^+ c_{\sigma,p} \exp(-ip_y) + b_{\sigma,p}^+ c_{\sigma,p} + c_{\sigma,p}^+ b_{\sigma,p} \exp(ip_y) + c_{\sigma,p}^+ b_{\sigma,p}]\}$$

The secular equation for the spectrum E_p of elementary excitations then is

$$\left| -E_p \delta_{ij} + H_{ij} \right| = 0, \qquad (2)$$

where the matrix elements, $H_{ij} = \rangle \Psi_i | H | \Psi_j \langle$, of the interdimer transfer Hamiltonian H are expressed via the Bloch functions Ψ_i of the energy state of the i-th dimer pair in a unit cell, i.e.,

$$\begin{vmatrix} -E_p & A & 0 & B \\ A^* & -E_p & C & 0 \\ 0 & C^* & -E_p & A^* \\ B^* & 0 & A & -E_p \end{vmatrix} = 0,$$

where

$$A = -t_0 - t_2 \exp(-ip_y),$$
$$B = -t_1 \exp(-ip_y - i\sqrt{3}p_x) - t_1 \exp(-i\sqrt{3}p_x),$$
$$C = -t_3 - t_3 \exp(-ip_y).$$

It should be stressed that in [22] this secular equation has been derived in terms of the bonding-antibonding orbital basis. From (2), one obtains the carrier energy dispersion relations for an ideal triangular lattice with interdimer hopping integrals,

$$E_p^{\pm} = \pm\{t_0^2 + t_2^2 + 2t_0 t_2 \cos p_y + 2\cos^2\frac{p_y}{2}(t_1^2+t_3^2) \pm 2\cos\frac{p_y}{2}$$
$$[(t_0^2+t_2^2+2t_0t_2\cos p_y)\cdot(t_1^2+t_3^2+2t_1t_3\cos\sqrt{3}p_x)+(t_1^2-t_3^2)^2]^{\frac{1}{2}}\}^{\frac{1}{2}}$$

In the limit of completely isotropic hopping ($t_{1-3} = t$) and $t_0 \gg t$ the carrier energy dispersions take the form $E_p^{\pm} = \pm t_0 \pm \varepsilon_p$, where

$$\frac{\varepsilon_p}{t} = \varepsilon_p^{\pm} = \cos p_y \pm 2\cos\frac{p_y}{2}\cos\frac{\sqrt{3}p_x}{2}. \quad (3)$$

This derivation of the electron antibonding/bonding (owing to dimerization) branches, $E_p^{\pm}$ from (2) is justified if the energy difference, $2t_0$, between E_p^+ and E_p^- pairs exceeds the energy splitting between bonding or antibonding branches (owing to two dimers per unit cell). Around Γ-point of the Brillouin zone (Bz) the energy dispersion relations of Eq. (3) acquire the form $\varepsilon_p^+ = 3 - 3p^2/4$ and $\varepsilon_p^- = -1 + (3p_x^2 - p_y^2)/4$. The lower band has isoenergetic lines $p_y = \pm\sqrt{3}p_x$ and saddle points are at lower branches of $E_p^{\pm}$ bands of Eq. (3).

At the Bz points (p_x, π)-(0, 0) the energy dispersion branch ε_p^+, in Eq.(3), achieves the energy boundaries $-1 \le \varepsilon_p^+ \le 3$ and the lower branch, ε_p^-, is located in the energy range $-3/2 \le \varepsilon_p^- \le 1$ (for the carrier momenta in the interval $(0, 2\pi/3) - (\pi/\sqrt{3}, 0)$). Energy dispersion surfaces of an antibonding band ε_p^+, Eqs. (3), are plotted in Fig. 2.

If there is a difference in nonazimuthal interdimer hopping integrals $\Delta t \equiv |t_1 - t_3|$ one can extend the expression for ε_p to

$$\varepsilon_p = t_2 \cos p_y \pm \cos\frac{p_y}{2}\sqrt{t_1^2+t_3^2+2t_1t_3\cos(\sqrt{3}p_x)} \quad (4)$$

(t_2 is an azimuthal hopping integral).

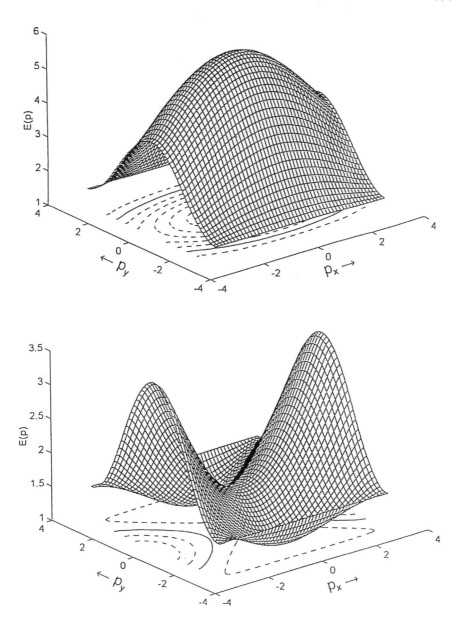

Figure 2. Energy dispersion surfaces of an antibonding band given by Eq. (3)..
Momenta over ET$_2$-plane are given in reduced units $|p_x| = |p_y| = \pi$ on the respective
Bz boundaries. The Fermi plane intersects both surfaces providing open and closed
sections of the Fermi surface in the inset of Fig. 6. For the lower energy surface E^-
(p_x,p_y) clearly seen the saddle point $\Gamma(0,0)$. The κ-ET$_2$X Fermi surface is illustrated
by the solid contours in the p_x - p_y plane.

According to (4) the energy dispersions and Fermi surface are plotted in Fig. 3 and in the inset of Fig. 6a. It is noteworthy that an upper and lower energy dispersion surfaces on Fig. 2 are responsible for open and closed Fermi surface portions, respectively, shown on Fig. 6a.

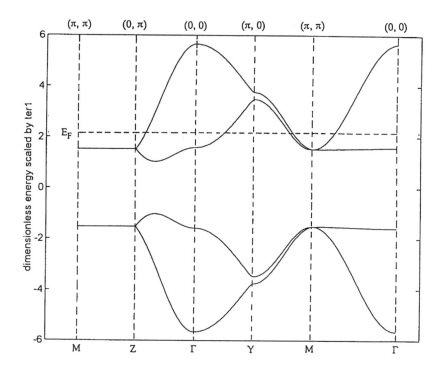

Figure 3. The tight-binding energy dispersions given by Eq. (4) in the Bz symmetry points with the parameters: $t_0 = 0.257$eV, azimuthal interdimer hopping $t_2 = 0.105$eV, nonazimuthal one $t_1 = 0.100$eV and a difference of nonazimuthal hoppings $\Delta t = 0.014$ eV. Momenta over ET_2-plane are given in reduced units $|p_x| = |p_y| = \pi$ on the respective Bz boundaries. The Fermi level is inside the antibonding band.

The above consideration suggests that the conduction band being constructed of the HOMOs of ET molecules, has no dispersion along the interlayer direction c. To consider a finite value of an energy dispersion along the c-direction we need to take into account essential ET_2 layers shown on Fig. 4.

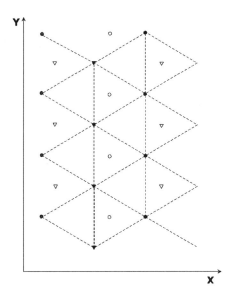

Figure 4. Closed circles and triangles denote the dimers in the ET2 plane. Open circles and triangles illustrate ET dimers of the neighbouring dimerized ET2 plane.

The corresponding multi-ET$_2$ layer Hamiltonian (here for simplicity we assume that $t_{1,2,3} \equiv t$) reads as:

$$\begin{aligned}
H = \sum_{i,j,k} [&\frac{t}{2}(\alpha^+_{i,j,k}\alpha_{i,j-1,k} + \beta^+_{i,j,k}\beta_{i,j-1,k} + \chi^+_{i,j,k}\chi_{i,j-1,k} + \delta^+_{i,j,k}\cdot\delta_{i,j-1,k}) \\
+&\frac{t}{2}(\alpha^+_{i,j,k}\cdot\beta_{i,j,k} + \alpha^+_{i,j,k}\cdot\beta_{i,j-1,k} + \alpha^+_{i,j,k}\cdot\beta_{i-1,j-1,k} + \alpha^+_{i,j,k}\cdot\beta_{i-1,j,k}) \\
+&\frac{t}{2}(\chi^+_{i,j,k}\cdot\delta_{i,j,k} + \chi^+_{i,j,k}\cdot\delta_{i,j-1,k} + \chi^+_{i,j,k}\cdot\delta_{i-1,j-1,k} + \chi^+_{i,j,k}\cdot\delta_{i-1,j,k}) \\
+&\frac{\tau}{2}(\alpha^+_{i,j,k}\cdot\chi_{i,j,k} + \alpha^+_{i,j,k}\cdot\chi_{i,j,k-1} + \beta^+_{i,j,k}\cdot\delta_{i,j,k} + \beta^+_{i,j,k}\cdot\delta_{i,j,k-1}) \\
+&\frac{\tau}{2}(\alpha^+_{i,j,k}\cdot\delta_{i-1,j,k} + \alpha^+_{i,j,k}\cdot\delta_{i-1,j-1,k} + \alpha^+_{i,j,k}\cdot\delta_{i-1,j,k-1} + \alpha^+_{i,j,k}\cdot\delta_{i-1,j-1,k-1}) \\
+&\frac{\tau}{2}(\beta^+_{i,j,k}\cdot\chi_{i,j,k} + \beta^+_{i,j,k}\cdot\chi_{i,j-1,k} + \beta^+_{i,j,k}\cdot\chi_{i,j,k-1} + \beta^+_{i,j,k}\cdot\chi_{i,j-1,k-1})],
\end{aligned} \quad (5)$$

in terms of the bonding-antibonding orbital basis, where $\alpha = \frac{1}{\sqrt{2}}(a_1 - b_1), \chi = \frac{1}{\sqrt{2}}(a_2 - b_2)$,

$$\alpha = \frac{1}{\sqrt{2}}(a_1 - b_1), \chi = \frac{1}{\sqrt{2}}(a_2 - b_2), \beta = \frac{1}{\sqrt{2}}(c_1 - d_1), \text{ and } \delta = \frac{1}{\sqrt{2}}(c_2 - d_2)$$

are antibonding molecular orbitals of dimers ET_2 with the lattice vector $\mathbf{R} = i \cdot \mathbf{x} + j \cdot \mathbf{y} + k \cdot \mathbf{z}$. Taking into account a small interlayer carrier hopping τ along the axis c perpendicular to ET_2 layers from (2) we get the general energy dispersion relations,

$$\omega_p^{1,2} = \varepsilon_p^+ \pm \frac{\tau}{t} \cos\frac{p_z c}{2} \sqrt{3 + 2\varepsilon_p^+}$$
$$\omega_p^{3,4} = \varepsilon_p^- \pm \frac{\tau}{t} \cos\frac{p_z c}{2} \sqrt{3 + 2\varepsilon_p^-}$$
(6)

These carrier energy dispersions are different from the cited ones, $\varepsilon = \varepsilon_p^\pm + t_\perp \cos p_z c$ .. It is easily seen from (6) that the effective carrier hopping,

$$t_\perp = \tau \frac{\cos\frac{p_z c}{2}}{\cos p_z c}, \qquad (7)$$

increases with the increase of the interlayer separation c, in agreement with the experimental visualization [24]. Related to spectra branches (6) the Fermi surface has a corrugated topology contrary to the conventional two-dimensional Fermi surface plotted in the inset of Fig. 6a, which has been derived on the basis of the energy dispersion relations (3) with the neglect of the interlayer hopping τ.

3. The electronic density of states (DOS) in normal phase

The electronic DOS per spin, $\rho(\varepsilon) = \sum_{p_x, p_y}[\delta(\varepsilon - \varepsilon_p^+) + \delta(\varepsilon - \varepsilon_p^-)]$, for the dimensionless energies (3) is defined for the unit cell volume by the equation

$$\rho_\pm(\varepsilon) = \frac{2}{\pi^2}\int_0^\pi dp_y \int_0^{\frac{\pi}{2}} dp_x \delta(\cos p_y \pm 2\cos\frac{p_y}{2}\cos p_x - \varepsilon).$$

Hereafter the momenta over ET_2-plane are given in the reduced units $|p_x| = |p_y| = \pi$ on the respective Bz boundaries. As a result of integrations one can obtain the partial DOS explicitly, as follows

$$\rho_+(-1 \le \varepsilon^+ \le 1) = \frac{2}{\pi^2\sqrt{u}} F\left(\arcsin\frac{1}{q}\sqrt{\frac{u+1}{2u}}; q\right),$$

$$\rho_+(1 \leq \varepsilon^+ \leq 3) = \frac{2}{\pi^2 \sqrt{t}} K(q), \qquad (8)$$

and

$$\rho_-(-3/2 \leq \varepsilon^- \leq -1) = \frac{2}{\pi^2 q \sqrt{t}} K(\frac{1}{q}),$$

$$\rho_-(-1 \leq \varepsilon^- \leq 1) = \frac{2}{\pi^2 \sqrt{t}} F\left(\arcsin\frac{1}{2q}\sqrt{\frac{(t+1)(5-t^2)}{2}}; q\right), \qquad (9)$$

where F and K denote, respectively, the elliptic integral and the complete elliptic integral of the 1st kind in the Legendre normal form with $q = \sqrt{\frac{(1+t)^3(3-t)}{16t}}$ dependent on the new energy variable $u = \sqrt{2\varepsilon + 3}$. The partial DOS are plotted in Fig. 5 [22]. They have logarithmic divergence along the lines $\varepsilon = -1$ of the Bz (c. f. Eq.(3)).

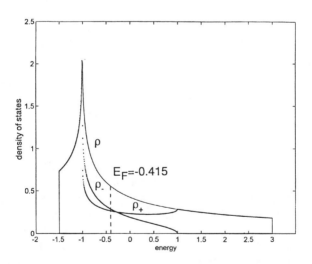

Figure 5.. The partial DOS ρ_+ and ρ_- calculated from (8) and (9) and the total DOS ρ for the κ-ET$_2$X tight-binding band structure resulting from Eqs. (10)-(11).

The spectrum branches (3) overlap in the energy range $-1 \leq \varepsilon \leq 1$ and the total DOS is defined in the range $-3/2 \leq \varepsilon \leq 3$. Taking into account Eqs. (8) and (9) the DOS per spin and per ET molecule is expressed by the following relations:

$$\rho(-1 \leq \varepsilon \leq 3) = \rho_-(-1 \leq \varepsilon^- \leq 1) + \rho_+(-1 \leq \varepsilon^+ \leq 1),$$

$$\rho_+(1 \leq \varepsilon^+ \leq 3) = \frac{2}{\pi^2 \sqrt{t}} K(q), \qquad (10)$$

$$\rho(-3/2 \leq \varepsilon \leq -1) = \frac{2}{\pi^2 q \sqrt{u}} K\left(\frac{1}{q}\right), \qquad (11)$$

which diverge near $\varepsilon = -1$ as $\rho(\varepsilon \approx -1) \approx \frac{3}{\pi^2} \ln \frac{1}{\varepsilon+1}$ inside the energy range $-3/2 \leq \varepsilon \leq 3$ [12]. Contrary to the conventional DOS for the two-dimensional and one-dimensional lattices the obtained κ-ET$_2$X DOS (see Eqs. (10) and (11)) have an energy-dependent pre-factor. The total DOS given by Eqs.(8) and (9)) is plotted in Fig. 5. The logarithmic divergence in Eq. (11) along the Bz edges are close to the upper dispersion surface on Fig. 2. Those along diagonals are close to the lower dispersion surface with a saddle point.

As seen from Fig. 2 the total DOS has a van Hove divergence of a logarithmic type at an energy $\varepsilon = \varepsilon_p / t = -1$, that is along the Bz boundaries M(π,π) -Z(0,π) – (-π,π) ((-π,-π)- (π,-π)) for the upper (lower) branch and along the Bz diagonal lines M(π,π) -Γ(0,0) – (-π,-π) ((-π,π), (π,-π)) for lower (upper) branch of the band E_p^+ (E_p^-) given by (3). To show a DOS shape with account of the influence of distinct nonazimuthal hopping integrals t_1 and t_3 (cf. Eq. (4)) numerical calculations have been fulfilled on the basis of the energy histogram over all meshed Bz provided $t_1=t_2$. The numerical results are shown on Fig. 6.

They demonstrate that at small t_3-t_1 magnitudes two divergencies $\varepsilon = -1 - (t_1 - t_3)^2 / 8t_3^2$ and $\varepsilon = -1 - (t_1 + t_3) / t_3$ [22] coincide very closely with $\varepsilon=-1$ (that is with -t_2) (see Fig. 6a). With an increase of the t_3-t_1 magnitude a splitting of logarithmic divergence manifests itself (see Fig. 6b). Small perturbations of the DOS shape on Fig. 6 are due to an applied numerical procedure. One can see that Eqs. (2), (3), (10), and (11) and Figs. 3, 5, and 6 reproduce the main results of the first-principles calculation [25] and tight-binding method [20,26,27] for the κ-ET$_2$X band structure (see also earlier results obtained, e.g., in [28]). Taking into account the electron concentration, three electrons per dimer ET_2^+ in κ-ET$_2$X materials, one can find that the Fermi level is located inside the conducting antibonding band (the Fermi level is shown on Figs. 3, 5, and 6).

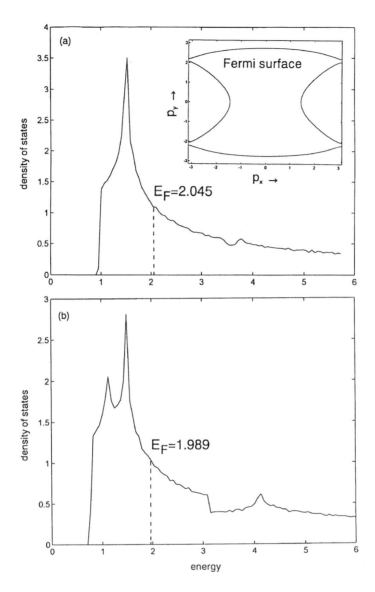

Figure 6. Numerical calculations of the DOS of the antibonding band. The Bz was divided into 201x201 mesh, and the numbers of these states which come into the energy range $E+\Delta E$ ($\Delta E = E_{max}/100$) were counted. (a) The top panel shows the DOS calculated by this energy histogram using the realistic transfer integrals, $t_2 = 0.105$eV, $t_1 = 0.100$eV with a difference of nonazimuthal hoppings $\Delta t = 0.014$ eV. The inset shows the Fermi surface. The closed and open sections of the Fermi surface are splitted at the zone boundary. (b) The bottom panel shows the DOS using the same magnitudes of transfer integrals but with $\Delta t = 0.05$ eV. The divergent peak in DOS is splitted.

So, a half-filling of an antibonding band E_p^+ by two of six $2\,\mathrm{ET}_2^+$ electrons leads to the metallic nature of κ-ET$_2$X compounds contrary to the existing κ-ET$_2$X semiconductors [29, 30]. The band structure of κ-ET$_2$X insulators with polymerised anion sheets such as the complexes with X ≡ β''-I Br Cl, β''-Cl$_2$, or 7,7,8,8-tetracyanoquinodimethane (triclinic), or κ'-Cu$_2$(CN)$_3$ and Cu[N(CN)]$_2$Cl [31] should be different from the similar band structure presented by Eq. (3) and (4). The reason for this contradiction is that the Slater-Koster scheme or any band calculation operates outside the electron-electron correlation scale.

4. The κ-ET$_2$X insulating model

When two electrons occupy one dimer ET$_2$ with strong intramolecular repulsion U_{ET} the increase of the dimer energy is ~ $2t_0$ (i. e., the loss of the intradimer kinetic energy). This magnitude is supposed to be the effective Anderson-Hubbard energy per dimer/site, $U_{ET_2} \approx 2t_0$ ~0.5eV [19].

We employ the X-operator machinery of Refs. [12,19,32-33] for the generalised Hubbard] -Okubo [34, 35] operators, $X_A^B = |B\rangle\langle A|$, projecting multielectron A states of the crystal cell to B states. After a standard transformation from the Heisenberg to the interaction representation, all disconnected diagrams in the simplest temperature Green's function, $D_{\alpha\beta}^0(\mathbf{r}t,\mathbf{r}'t') = -\langle \hat{T}\tilde{X}_\alpha(\mathbf{r}t)\tilde{X}_{-\beta}(\mathbf{r}'t')\rangle$, should be cancelled. Then the local ($\mathbf{r} = \mathbf{r}'$) Green's function take the form $D_{\alpha\beta}^0(\mathbf{r}t,\mathbf{r}'t') = \delta_{\mathbf{r},\mathbf{r}'} G_\alpha^0(\mathbf{r}t,\mathbf{r}'t')\langle [X_\alpha X_{-\alpha}]_\pm \rangle_0$. In Matsubara's ω-representation it can be rewritten as

$$D_{\alpha(BA)}^0(\omega_n) = f_\alpha G_{\alpha(BA)}^0(\omega_n) = \frac{f_\alpha}{-i\omega_n + \varepsilon_A - \varepsilon_B}, \quad (12)$$

where the subscripts $\alpha(BA)$ numerate the transitions $A\to B$ in the discrete spectrum of the non-pertubative Hamiltonian H_0 and the correlation factor [32-33,36-38], $f_\alpha = \rangle [X_\alpha X_{-\alpha}]_\pm \langle_0 = \rangle X_A^A \langle \pm \rangle X_B^B \langle = n_A \pm n_B$, is determined by the Boltzmann populations, $n_{A,B}$, of the H_0 eigenstates A and B. Expansions of the Fermi operators given by

$$a_\sigma = \sum_{\alpha(BA)} g_\alpha X_\alpha, \quad (13)$$

$$a_\sigma^+ = \sum_{\bar\alpha} g_{\bar\alpha} X_{\bar\alpha} \quad (\bar\alpha \equiv \alpha(AB)) \quad (14)$$

project the perturbative tunnel Hamiltonian, H_{int}, onto the X-operator basis.

The κ-ET$_2$X insulating problem can be described in the frame of the half-filled Hubbard model, $H=H_0+H_{int}$, with two ET_2^+ sites per unit cell on an ET$_2$- lattice. Its dimer Hamiltonian, H_0, has the lowest energy level ε_{GS} as the eigenvalue of the ground-state hole singlet $a_{\uparrow\downarrow}^+|0\rangle = |\pm\rangle$. The chosen chemical potential provides the equal energies, $\varepsilon_P= 0$ of the nearest excited zero-particle and two-particle polar (P) states, $|0\rangle$ and $|2\rangle = a_\uparrow^+ a_\downarrow^+ |0\rangle$, measured from the one-particle ground-state (GS) energy $\varepsilon_{GS} = -\varepsilon = -U_{ET_2}/2$ of the dimer ET_2^+. For the present problem the travelling carrier feels resonant energies $-\varepsilon$ and $+\varepsilon$ as poles of the dimer, or site Green's functions

$$D_{\alpha(0\uparrow)}^0(\omega_n) = \frac{n_{GS}+n_p}{-i\omega_n+\varepsilon_{GS}-\varepsilon_P} = \frac{1/2}{-i\omega_n-\varepsilon},$$

$$D_{\alpha(\downarrow 2)}^0(\omega_n) = \frac{1/2}{-i\omega_n+\varepsilon}. \tag{15}$$

Then according to the tight-binding approach for correlated electrons [33-34] the band spectra should be derived from the following equation

$$\sum_\alpha g_\alpha^2 D_\alpha^0(\omega_n) = \frac{\sum_{\alpha(PGS)} g_\alpha^2 f_\alpha}{-i\omega_n-\varepsilon} + \frac{\sum_{\alpha(GSP)} g_\alpha^2 f_\alpha}{-i\omega_n+\varepsilon} = -\frac{1}{t\varepsilon_p^\pm} \tag{16}$$

by an analytical continuation $i\omega_n \to \xi+i\delta$, where α numerate all possible transitions GS↔P between eigenlevels of the dimer Hamiltonian H_0, $\varepsilon_p^\pm$ denote the conventional tight-binding energy dispersions of uncorrelated electrons (cf. Eqs. (3) and (6)) and g_α are the expansion coefficients in (13) and (14). For the Hubbard model of κ-ET$_2$X salts the expansion (13) reduces to $a_\sigma = X_\sigma^0 + \sigma X_2^{\bar\sigma}$ with coefficients $g_\alpha = \pm 1$. In Eqs. (15) and (16) the correlation factor $f_{\alpha(0\uparrow)} = n_{P("0")} + n_{GS("\uparrow")} = 1 - n_{P("2")} - n_{GS("\downarrow")} = 1/2$ for the non-magnetic state reflects the filling of energy bands $\xi^\pm$ by carriers with carrier concentration of a hole per dimer ET_2^+, n=1. Then one can get from (16) the following correlated antibonding branches:

$$\xi_p^\pm = \frac{t_2}{2}\left[\varepsilon_p^\pm \pm \sqrt{(\varepsilon_p^\pm)^2 + (\frac{2\varepsilon}{t_2})^2}\right] \tag{17}$$

In Eq. (17) $\varepsilon_p^{\pm}$ are defined by.(2) and $\varepsilon = U_{ET_2}/2$. The energy branches, $\xi_p^{\pm}$, should replace the uncorrelated dispersion relations, $\varepsilon_p = t_2 \varepsilon_p^{\pm}$, in Eq. (3). The correlated energy dispersions of the antibonding band given by (3) and (17) are shown in Fig. 7.

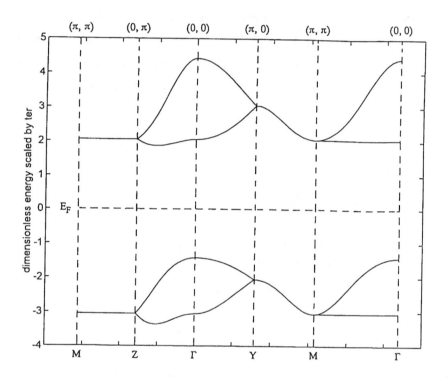

Figure 7.. The correlated splitting of energy dispersions of the antibonding band given by Eq. (17) with parameters $t_2 = 0.1$ eV and $U_{ET_2} = 0.5$ eV ~ 2 t_0. The Fermi level is inside of the band gap providing an insulating state of the κ-ET$_2$X model.

Two holes of the unit cell $2\,ET_2^+$ occupy the lower pair of antibonding bands, ξ_p^-. Thus, the insulating state of the κ-ET$_2$X system is provided by intradimer electron interactions. A pair of nearest antibonding energy dispersion surfaces are plotted in Fig. 8.

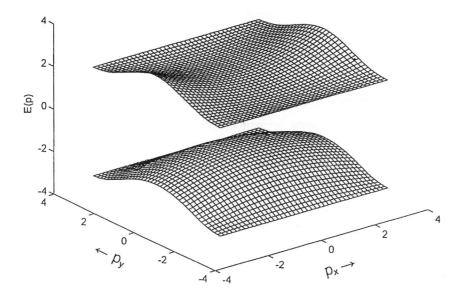

Figure 8. The closest pair of energy dispersion surfaces corresponding to the antibonding band given by Eqs. (3) and (17). A lower surface is completely occupied by electrons and an upper saddle energy surface is empty.

Below the Fermi plane, $E_F=0$, there is an energy dispersion surface with a positive curvature. Electron correlations manifest themselves in the inversion of surfaces with and without a saddle point (cf. Fig. 2 and 8). If the intradimer electron-electron interaction decreases the energy splitting of plotted dispersion surfaces in Fig. 8 also decreases, the topology of both surfaces increases and the upper unoccupied energy dispersion surface with a saddle point becomes closer to the lower surface. So, at the intermediate values of electron interactions a saddle point and logarithmic divergence in the DOS at the Fermi surface are likely to occur. At small values of the energy splitting the electron scattering on spin and charge degrees of freedom is essential and the insulator-metal transition occurs [19] (see Sec. 5).

From the tight-binding correlated bands (17) it follows that intradimer electron interactions are responsible for the insulating gap, $\Delta = \xi^+(\min \varepsilon_p^-) - \xi^-(\max \varepsilon_p^+)$. Taking into account the energy ranges of uncorrelated carries in Eq. (3), $-1 < \varepsilon_p^+ < 3$ and $-3/2 < \varepsilon_p^- < 1$, and dispersion relations (17), one can evaluate the bandgap as

$$\Delta = \frac{1}{2}\left[\sqrt{9+\left(\frac{2\varepsilon}{t_2}\right)^2} + \sqrt{\left(\frac{3}{2}\right)^2+\left(\frac{2\varepsilon}{t_2}\right)^2} - \frac{9}{2}\right]t_2 \qquad (18)$$

With the assumptions $\varepsilon = U_{ET_2}/2 \sim t_0 \sim 0.2\text{eV}$ and $t_2 \sim 0.1$ eV this magnitude is in agreement with the measured activation energy $E_g = \Delta/2 \sim 10^2\text{meV}$ in κ-ET$_2$X semiconductors [29,30].

5. The insulator-metal phase transition

As the correlated bands (17) widen, the electron scattering on spin and charge degrees of freedom leads to an insulator-metal phase transition of the Mott-Hubbard kind. The strong electron-electron interactions suppress charge fluctuations not only inside dimer but also in correlated bands of κ-ET$_2$X near the top of the band filling, and, therefore, the long-range interactions of electrons with interdimer lattice phonons are screened. This leads to a reduction of the electron-phonon coupling and we will neglect the role of phonons in the problem of insulator-metal transitions in the κ-ET$_2$X family at low temperatures.

In the general case of k-fold degeneracy of the GS in the unperturbed Hamiltonian H_0 the Mott-Hubbard phase transition is governed by singularities of the two-particle vertex Γ_{ab} for small momentum transfer,

$$\Gamma_{ab} = A\sum_{p,a',b'} t_p^{a,a'} \Gamma_{a'b'} t_p^{b,b'},$$

$$\frac{A}{k^2} = \sum_{\alpha,\beta} g_\alpha D_\alpha^0 g_{-\alpha} C_{\alpha\beta}^{(1)} g_\beta D_\beta^0 g_{-\beta} + \sum_{\alpha,\beta} g_\alpha D_\alpha^0 g_{-\alpha} C_{\alpha\bar\beta}^{(2)} g_{\bar\beta} D_{\bar\beta}^0 g_{-\bar\beta},$$

(19)

where the summation is over the momentum p in the ET$_2$-plane and over the transitions GS$\leftrightarrow$P in Eq. (13) with the Green's functions (15) at $\omega \to 0$. In Eq. (19) α and g_α ($\bar\alpha$ and $g_{\bar\alpha}$) denote intracell transitions and expansion coefficients in Eq. (14) for the spin projection $\uparrow(\downarrow)$.

The charge fluctuations are defined by the first kernel of Eq. (19) as $C_{\alpha\beta}^{(1)} = \overline{\delta\{X_\alpha X_{-\alpha}\}\delta\{X_\beta X_{-\beta}\}}$, $\delta\{X_\alpha X_{-\alpha}\} = \{X_\alpha X_{-\alpha}\} - \overline{\{X_\alpha X_{-\alpha}\}}$ with matrix elements $-1/k^2$ and $1/k - 1/k^2$. The second kernel in Eq. (19), the four-by-four matrix $\left\|C_{\alpha\beta}^{(2)}\right\|$ with elements $1/k$ and 0, determines the electron scattering on spin fluctuations. The dimension of these matrices, $\left\|C_{\alpha\beta}^{(1)}\right\|$ and $\left\|C_{\alpha\beta}^{(2)}\right\|$, equals to the number of transitions GS$\leftrightarrow$P or P$\leftrightarrow$GS in the discrete spectrum of the dimer Hamiltonian H_0.

For a general case of a k-fold degenerate GS with the total spin S the relation $A=(S+1)/S\epsilon^2$ for Eq. (19) is useful [39] In the Hubbard model employed for κ-ET$_2$X insulators, one hole per dimer on site ET_2^+ provides the spin value $S = 1/2$ of the one-particle GS.

Then the system of homogeneous equations (19) can be reduced to

$$\left[\frac{3(t_2)^2}{\epsilon^2}\sum_p (\epsilon_p^{+/-})^2 - 1\right]\Gamma_{++/--} = 0 \qquad (20)$$

and define vertices in the occupied correlated antibonding band $\xi_p^\pm(\epsilon_p^\pm)$ of Eq. (17). The solution of (20) yields singularities of the two-particle vertices in the complete inhomogeneous Bethe-Salpeter equation. After integrations over the ET$_2$-plane momentum, for a triangular lattice one can define the critical point of the insulator-metal phase transition as follows [40]

$$\left(\frac{t_2}{\epsilon}\right)_{crit} = \left(\frac{t_2}{U_{ET_2}/2}\right)_{crit} = \frac{4\pi}{\sqrt[4]{3}\sqrt{15\pi^2 + 64}} = 0.66 \qquad (21)$$

(noteworthy, for a square ET$_2$ - lattice the phase transition critical point is 0.43 [19]).

It is known that the empirical ratio $W_U/\Delta E = 1.1 - 1.2$ of the conducting bandwidth, W_U, to a dimer band splitting, $\Delta E = 2t_2$[31], separates a family of the κ-ET$_2$X insulators from the metallic κ-ET$_2$X compounds. Taking into account the antibonding bandwidth 4.5 t_2 (c. f. Eqs.(8) and (9)) the empirical relation can be considered as $t_2/t_0 = 0.48 - 0.54$. This value agrees fairly well with the calculated phase critical point. (21): $(t_2/t_0)_{crit} = 0.66$ ($U_{ET_2} \approx 2t_0$). We can conclude, e.g., that a κ-ET$_2$Cu[N(CN)$_2$]Cl paramagnet with the characteristic ratio $t_2/t_0 = 0.32 < (t_2/t_0)_{crit}$ [31] should be an insulator near the insulator-metal phase boundary. Recently it was reported that κ-ET$_2$Cu[N(CN)$_2$]Cl insulators manifest the phase transition to a metal under a moderate hydrostatic pressure of 30 MPa [41]. Substituting the semiconducting gap Δ [29, 30] into Eq.(18), one can extract from it the ratio of the material parameters as $2t_2/U_{ET_2} = t_2/t_0 = 0.56$, which is below the critical value (21) in the insulating phase. It is believed that all known κ-ET$_2$X salts are close to a calculated phase boundary (21). It is worth noting that if a Mott-Hubbard phase transition is not accompanied by crystallographic or magnetic changes the resulting metallic state will have to be the superconducting state [42]. This is the κ-ET$_2$X case because we have started from the insulating paramagnetic state (Sec. 5) without any order parameter.

6. Metallic phase

Intradimer hopping, t_0, is small compared with the intraET electron-electron correlations and in accordance with (21) a metallic state is accompanied by a decreased value of t_0. So, a metallic dimerized ET_2 layer in κ-ET_2X can be represented by a lattice of sites $ET_2 = ET_aET_b$ with degenerated energy levels of orbitals a and b, namely, described by the doubly degenerate Hubbard model with a hole concentration, n, of around unity [12]. Since one hole spreads over a dimer, ET_2^+ with three electrons it is easier to deal with the hole representation than with the electron representation. Let us denote eigenstates in the dimer $ET_2 = ET_aET_b$, as $\langle ab \rangle$. Then the one-particle GS is formed by four-fold degenerate states, $\langle \uparrow 0 \rangle, \langle \downarrow 0 \rangle, \langle 0\uparrow \rangle, \langle 0\downarrow \rangle$, each with a population, n_1, defined by the spin projection and hole position in $(ET_aET_b)^+$. The nearest excited states are the $\langle 00 \rangle$ state (the neutral dimer) with a population n_0 and the triplet states. Owing to the many-body interactions, the two-particle triplet states, $\langle \sigma\sigma \rangle$, and $T = (a_\uparrow b_\downarrow + a_\downarrow b_\uparrow)/\sqrt{2}$ the next excited singlet state is $S = (a_\uparrow b_\downarrow - a_\downarrow b_\uparrow)/\sqrt{2}$, and the singlet orbital Hubbard doublets are $\langle 20 \rangle = \sigma a_\sigma^+ a_{\bar\sigma}^+$ and $\langle 02 \rangle = \sigma b_\sigma^+ b_{\bar\sigma}^+$, which have higher energies than the empty hole state $\langle 00 \rangle$ [32,33,38]. Here for simplicity we assume $U_{ET} \gg t_{1,2,3}$. For the problem under consideration the expansion (13) is reduced to

$$a_\sigma/b_\sigma = \sum_\alpha g_\alpha^{a/b} X_\alpha = X_{\sigma 0/0\sigma}^{00} + \frac{1}{\sqrt{2}}(X_T^{0\bar\sigma/\bar\sigma 0} + \sigma X_S^{0\bar\sigma/\bar\sigma 0})$$
$$\pm X_{\sigma\sigma}^{0\sigma/\sigma 0} + \sigma X_{20/02}^{\bar\sigma 0/0\bar\sigma} \qquad (22)$$

with the populations of the GS and P, $\rangle X_{00}^{00} \langle = n_0$ and $\rangle X_{\sigma 0}^{\sigma 0}\langle = \rangle X_{0\sigma}^{0\sigma}\langle = n_1$.

For a set of the 16 dimer eigenstates, a completeness relation for X-operators yields $1 = \{a_\sigma^+ a_\sigma\} = \{b_\sigma^+ b_\sigma\} = \sum_A \langle X_A^A \rangle$ (which results from the orthogonality of the X-operator basis). Applying it to Eq. (22) and taking into account the four-fold degeneracy of the ground state, one can get the following system of equations for the site occupations n_1 and n_0,

$$\langle\{a_\sigma^+ a_\sigma\}\rangle = \langle\{b_\sigma^+ b_\sigma\}\rangle = \sum_A \langle X_A^A \rangle = 1, \quad 4n_1 + n_0 = 1,$$
$$\langle n_\sigma \rangle = \langle a_\sigma^+ a_\sigma \rangle = \langle b_\sigma^+ b_\sigma \rangle, \qquad n_1 \approx \frac{n}{4}. \qquad (23)$$

The perturbative tunnelling Hamiltonian is represented by all possible products of the X-operators for the neighbouring dimers, and the first-order perturbative Green's function for the metallic ET_2 layer is expressed as

$$G_{\sigma 0}^{00}(\mathbf{p},\omega_n) = \frac{f}{-i\omega_n - \mu + f\varepsilon_p} = G_{0\sigma}^{00}(\mathbf{p},\omega_n), \quad (24)$$

where the correlation factor $f = \langle\{X_{00}^{\sigma 0} X_{\sigma 0}^{00}\}\rangle = \langle X_{00}^{00}\rangle + \langle X_{\sigma 0}^{\sigma 0}\rangle = n_0 + n_1 = (4-3n)/4$ reflects the filling of the correlated energy band by carrier number n per a dimer. Similar to Eq. (16), the desired spectrum of the one-particle excitations follows from the pole of this Green's function:

$$\xi_p^{a,b} = f\varepsilon_p - \mu = \frac{4-3n}{4}\varepsilon_p - \mu, \quad (25)$$

where ε_p is related to the dispersions in Eqs. (3), (4), and (6).

Assuming a flat weighted DOS with a half-bandwidth, w, for non-interacting carriers from the equation for the hole concentration one can get

$$n_\sigma = \frac{n}{2} = f\sum_{p,\omega_n} \exp(i\delta)\left[G_{\sigma 0}^{00}(\mathbf{p},\omega_n) + G_{0\sigma}^{00}(\mathbf{p},\omega_n)\right] = 2f\sum_p \theta(-\xi_p^{a,b}), \quad (26)$$

where a dependence of the chemical potential, μ, on the carrier concentration is given by

$$\mu = \frac{5n-4}{4}w. \quad (27)$$

Eq. (27) reflects the correlated narrowing of the energy band: at $n=1$ (the κ-ET$_2$X salt case) we conclude that $\mu = w/4$ (quarter band filling, because the chemical potential origin is at the bottom of the energy band). The correlated reduction of the energy band may be involved in the difference between the optical [26] and cyclotron [23] effective masses.

7. Superconducting pairing in the ET$_2$-layer model

As it is seen from (3), (4), and (25), the band structure effects are important when the Fermi surface is near the Bz boundary (this is the case for the κ-ET$_2$X family) where the influence of the crystal potential is strong. In the tight-binding approach with neglect of retardation effects the symmetry of the Bloch functions, $\Psi(\mathbf{R})$, and the symmetry of the Fermi surface manifest in the symmetry of the superconducting order parameter $\Delta(\mathbf{p})$ [43-45], which represents the correlation function of the Cooper pairs as $\Delta(\mathbf{p}) \propto \langle c_{\sigma,p} c_{\bar{\sigma},-p} \rangle$. In the tight binding approach we have

$$\Delta_{pp'} = \sum_{r,r'} \exp i(\mathbf{p'r'} - \mathbf{pr}) \int \Psi^+(\mathbf{r}-\mathbf{R})\Delta(\mathbf{R})\Psi(\mathbf{r'}-\mathbf{R})d\mathbf{R} \approx$$
$$\approx \sum_{l,r} \exp i\mathbf{lp} \int \Psi^+(\mathbf{r}-\mathbf{R})\Delta(\mathbf{R})\Psi(\mathbf{r}+\mathbf{l}-\mathbf{R})d\mathbf{R} = \Delta(\mathbf{p}). \qquad (28)$$

From this relation it is easily seen that the energy gap $\Delta(\mathbf{p})$ has the same symmetry as the Fermi surface, but the dispersions in Eqs. (3)-(4) and (25) yield an anisotropic Fermi surface unlike the sphere in the Bardeen-Cooper-Schrieffer (BCS) theory. A complicated Fermi surface does not support conventional isotropic s-wave pairing, giving rise either to a constant pairing interaction or to a constant energy gap, $\Delta(\mathbf{R})$, as was mentioned earlier [46], and the symmetry of $\Delta(\mathbf{p})$ is lower than that of the Fermi surface.

Experiments imply that, in the κ-ET$_2$X superconductors, anisotropic singlet with a d-type pairing and nodes of the order parameter given by $\Delta_d(\mathbf{p}) \propto \cos p_x - \cos p_y$ (i.e. so-called $d_{x^2-y^2}$ pairing in the reference system of the square lattice) or other one (d_{xy}) occurs at the Fermi surface. They are also consistent with anisotropic singlet s^*-pairing with nodes of the $\Delta_s(\mathbf{p}) \propto \cos p_x + \cos p_y$ without a sign change or minima of the gap on the Fermi surface but in the same direction in the Bz as in the case of d-pairing. To determine the type of Cooper pairing we start from correlated Green's functions (24) (see comments in [42]) to apply the formalism [47] and [48].

Here we assume a triangular ET$_2$-lattice and express Fermi-operators via X-operators in the effective pairing Hamiltonian. Its unspecified interaction depends on momenta difference of pairing fermions, namely $V(\mathbf{p}-\mathbf{p'})$, and for the Cooper channel in the reference system of colliding fermions the superconducting order parameter at finite temperatures follows the standard BCS equation

$$\Delta(\mathbf{p}) = \sum_{\mathbf{p'},\alpha=\pm} V(\mathbf{p}-\mathbf{p'}) \cdot \Delta(\mathbf{p'}) \cdot \frac{\tanh \frac{\sqrt{(\xi_p^\alpha(\mathbf{p'}))^2 + \Delta^2(\mathbf{p}©)}}{2T}}{2\sqrt{(\xi_p^\alpha(\mathbf{p'}))^2 + \Delta^2(\mathbf{p'})}}, \qquad (29)$$

where the energy dispersions of correlated carriers ξ_p^α are determined by Eq. (25). In Refs. [12-13] the studies of superconductivity take into account only effective attraction between nearest electrons on a square lattice modelling ET$_2$-layer. For a particular triangular lattice with a single point basis of κ-ET$_2$X model (Fig. 1) the general form of attractive interaction between fermions, namely

$$V(\mathbf{p}-\mathbf{p}') = 2\Big[V_1 \cos(p_y - p_y^\odot) + V_2 \cos\frac{\sqrt{3}(p_x - p_x^\odot) + p_y - p_y^\odot}{2}$$
$$+ V_3 \cos\frac{\sqrt{3}(p_x - p_x^\odot) - (p_y - p_y^\odot)}{2}\Big], \quad (30)$$

conserves a symmetry of elementary excitation dispersions (3)-(4) irrespectively of the superconducting pairing mechanism. For an isotropic case, $V_{1,2,3} = V$, its expansion over the basis functions of irreducible representations of point symmetry group of the triangular lattice is given by:

$$V(\mathbf{p}-\mathbf{p}') = 2V\sum_{i=1}^{6} \eta_i(\mathbf{p})\eta_i(\mathbf{p}'), \quad (31)$$

where

$$\eta_1(\mathbf{p}) = \frac{1}{\sqrt{3}}\left(\cos p_y + 2\cos\frac{p_y}{2}\cos\frac{\sqrt{3}p_x}{2}\right),$$

$$\eta_2(\mathbf{p}) = \frac{2}{\sqrt{6}}\left(\cos p_y - \cos\frac{p_y}{2}\cos\frac{\sqrt{3}p_x}{2}\right),$$

$$\eta_3(\mathbf{p}) = \sqrt{2}\sin\frac{p_y}{2}\sin\frac{\sqrt{3}p_x}{2},$$

$$\eta_4(\mathbf{p}) = \frac{1}{\sqrt{3}}\left(\sin p_y + 2\sin\frac{p_y}{2}\cos\frac{\sqrt{3}p_x}{2}\right), \quad (32)$$

$$\eta_5(\mathbf{p}) = \frac{2}{\sqrt{6}}\left(\sin p_y - \sin\frac{p_y}{2}\cos\frac{\sqrt{3}p_x}{2}\right),$$

$$\eta_6(\mathbf{p}) = \sqrt{2}\cos\frac{p_y}{2}\sin\frac{\sqrt{3}p_x}{2}.$$

Here the basis functions $\eta_1(\mathbf{p})$, $\eta_2(\mathbf{p})$ and $\eta_3(\mathbf{p})$ describe anisotropic singlet s^*-pairing ("extended" s-pairing), $d_{x^2-y^2}$-pairing and d_{xy}-pairing, respectively, in a given reference system (see Fig. 1). The basis functions $\eta_{4,5,6}(\mathbf{p})$ are linear combinations of basis functions for two-dimensional representation corresponding to the triplet p-pairing.

In order to solve Eq. (29) we substitute it into expansions of an attractive interaction, (31), and an order parameter, $\Delta(\mathbf{p}) = \sum_{i=1}^{6} \Delta_i \eta_i(\mathbf{p})$, over basis functions of irreducible representations in our reference system (Sec. 2).

Then Eq. (29) is reduced to the following system of algebraic equations for T_c in terms of orthogonality of the basis functions $\{\eta_i(\mathbf{p})\}$

$$\Delta_i = 2V \sum_{\mathbf{p}; j=1; \alpha=\pm}^{6} \frac{\tanh \frac{\xi_p^\alpha}{2T_c}}{2\xi_p^\alpha} \Delta_j \eta_i(\mathbf{p}) \eta_j(\mathbf{p}), \tag{33}$$

where summation with respect to momenta p is completed over the ET$_2$-plane. In Eqs. (29) and (33) no account has been taken for the mutual transfers of Cooper pairs between Fermi surface sections of different curvatures. The reason is that electron sections of the Fermi surface (ξ^+) and hole portions (ξ^-) are separated by a small gap $\xi_p^+ - \xi_p^- \cong |t_1 - t_3|/4 \approx 4$ meV [49] due to weak anisotropy of interdimer hopping. In the general case the two-band model of superconductivity is worth consideration according to Refs. [50, 51].

The system of equations (33) has solutions for T_c when the following six-by-six determinant vanishes :

$$\left| \delta_{ij} - 2V \sum_{\mathbf{p}, \alpha = \pm} \frac{\tanh \frac{\xi_p^\alpha}{2T_c}}{2\xi_p^\alpha} \Delta_j \eta_i(\mathbf{p}) \eta_j(\mathbf{p}) \right| = 0. \tag{34}$$

Because of the oddness of integrals in (34) with respect to momenta, p_y, superconducting pairing of symmetries $\eta_1(\mathbf{p})$ with $\eta_2(\mathbf{p})$ and $\eta_4(\mathbf{p})$ with $\eta_5(\mathbf{p})$ can be mixed only. In Eq. (34) anisotropic singlet pairings of the d- and s^*- tips break down to one-dimensional d_{xy}-pairing and mixed $s^* + d_{x^2-y^2}$ - pairing with respect to the basis functions $\eta_3(\mathbf{p})$ and $\eta_1(\mathbf{p}) + \eta_2(\mathbf{p})$. For different values of effective attraction parameters, $V_1 \neq V_3 \neq V_2$ (cf. (30)), all singlet pairings are separated from the triplet p-wave pairing. The Knight shift measurements [9,10] indicate only a singlet form of an electron pairing in κ-ET$_2$X superconductors.

8. The symmetry of superconducting order parameter.

The essential part of this secular equation elements is

$$\sum_{\mathbf{p};\alpha=\pm} \frac{\tanh\frac{\xi_p^\alpha}{2T_c}}{2\xi_p^\alpha} \eta_i(\mathbf{p})\eta_j(\mathbf{p}) = \int_{-\infty}^{\infty} d\varepsilon \frac{\tanh\frac{\xi_p^\alpha}{2T_c}}{2\xi_p^\alpha} F_{ij}(\varepsilon), \tag{35}$$

where

$$F_{ij}(\varepsilon) = \sum_{\mathbf{p}} \delta(\varepsilon - \varepsilon_p^+)\eta_i(\mathbf{p})\eta_j(\mathbf{p}) + \sum_{\mathbf{p}} \delta(\varepsilon - \varepsilon_p^-)\eta_i(\mathbf{p})\eta_j(\mathbf{p}) =$$

$$\frac{\sqrt{3}}{4\pi^2} \cdot \int_{-\pi}^{\pi} dp_y \int_{-\frac{\pi}{\sqrt{3}}}^{\frac{\pi}{\sqrt{3}}} dp_x \delta\left(\varepsilon - \cos p_y - 2\cos\frac{p_y}{2}\cos\frac{\sqrt{3}p_x}{2}\right)\eta_i(\mathbf{p})\eta_j(\mathbf{p}) + \tag{36}$$

$$+ \frac{\sqrt{3}}{4\pi^2} \cdot \int_{-\pi}^{\pi} dp_y \int_{-\frac{\pi}{\sqrt{3}}}^{\frac{\pi}{\sqrt{3}}} dp_x \delta\left(\varepsilon - \cos p_y + 2\cos\frac{p_y}{2}\cos\frac{\sqrt{3}p_x}{2}\right)\eta_i(\mathbf{p})\eta_j(\mathbf{p}),$$

in dimensionless energy units of Eq. (25), $\xi_p/2ft$ as $\varepsilon_p^\pm$.

The $F_{ij}(u)$ functions depending on new variable $u = \sqrt{2\varepsilon + 3}$ are given by

$$F_{11}(u) = \frac{1}{3\pi^2}\left(4I_2 - 2(1+u^2)I_1 + \left(\frac{(u^2-1)^2}{4} - 2 - u^2\right)I_0\right)$$

$$F_{22}(u) = \frac{1}{6\pi^2}\left(I_2 + (u^2-5)I_1 + \left(\frac{(u^2-1)^2}{4} + 1 - u^2\right)I_0\right)$$

$$F_{33}(u) = \frac{1}{2\pi^2}\left(I_2 - \frac{(u^2-1)^2}{2}I_{-1} - (3+u^2)I_1 + \left(\frac{(u^2-1)^2}{4} - 1 + u^2\right)I_0\right) \tag{37}$$

$$F_{12}(u) = \frac{\sqrt{2}}{3\pi^2}\left(I_2 - \frac{u^2-1}{4}I_1 - \left(\frac{(u^2-1)^2}{8} + 2 - \frac{u^2}{2}\right)I_0\right)$$

for all kinds of superconducting singlet pairing. In Eqs. (37) the following notations have been introduced, namely

$$G(y) = (y+1)\left(y - \frac{t^2 - 1 - 2t}{2}\right)\left(\frac{t^2 - 1 + 2t}{2} - y\right)(1-y),$$

$$\begin{aligned}
I_0 &= \int_{-1}^{1} \frac{dy}{\sqrt{G(y)}}, \\
I_1 &= \int_{-1}^{1} \frac{y\,dy}{\sqrt{G(y)}}, \\
I_{-1} &= \int_{-1}^{1} \frac{dy}{(y+1)\sqrt{G(y)}}, \\
I_2 &= \int_{-1}^{1} \frac{(y+1)^2 dy}{\sqrt{G(y)}} = \frac{t^2+3}{2}(I_0 + I_1) - \left(\frac{t^2-1}{2}\right)^2 I_{-1} + \frac{\sqrt{G(y)}}{y+1},
\end{aligned} \tag{38}$$

which can be expressed via elliptic integrals over the whole energy regions of electron dispersions (3). For dimensionless energies in the region $-3/2 < \varepsilon < -1$ of the lower antibonding band ε_p^-, that is $0 < u < 1$, the integrals (36) are reduced to squarings

$$\begin{aligned}
I_0 &= \frac{4}{\sqrt{(3-u)(1+u)^3}} \cdot K(k), \\
I_1 &= \frac{4}{\sqrt{(3-u)(1+u)^3}} \cdot \left[-K(k) + \frac{(u-1)^2}{2}\Pi\left(\frac{\pi}{2}, \frac{4u}{(1+u)^2}, k\right)\right], \\
I_{-1} &= \frac{4}{(u-1)^2\sqrt{(3-u)(1+u)^3}} \cdot \left[\frac{(3-u)(u+1)}{2}E(k) + \frac{(u-1)^2}{2}K(k)\right],
\end{aligned} \tag{39}$$

where $k^2 = \dfrac{16u}{(3-u)(1+u)^3}$ is the modulus k squared.

For next energy region $-1 < \varepsilon < 3$ of the upper antibonding band, that is $1 < u < 3$, the integrals (37) have the following explicit form

$$I_0 = \frac{1}{\sqrt{u}} K(q)$$

$$I_1 = \frac{1}{\sqrt{u}} \left[-K(q) + \frac{(u-1)^2}{2} \Pi\left(\frac{\pi}{2}, \frac{(3-u)(u+1)}{4}, q\right) \right], \quad (40)$$

$$I_{-1} = \frac{4}{(u-1)^2(u+1)^2 \sqrt{u}} \cdot \left[2uE(q) + \frac{(u-1)^2}{2} K(q) \right]$$

with modulus $q = \frac{1}{k}$.

In formulae (39) and (40) the $K(k)$, $E(k)$ and $\Pi(\pi/2,n,k)$ denote the complete elliptic integrals of the first, second and third kind, respectively, in the normal Legendre form.

The symmetry of the representation with the largest T_c agrees with the BCS type description of superconductivity in [52] and [53] and the developed approach to κ-ET$_2$X superconductors [12,13,54]. Rearrangement of the matrix elements (35) of secular equation. (34) gives:

$$\int_{-\infty}^{\infty} d\varepsilon \cdot \frac{\tanh\frac{\xi_p}{2T_c}}{2\xi_p} F_{ij}(\varepsilon) = \int_{-\omega_c}^{\omega_c} d\varepsilon \cdot \frac{\tanh\frac{\xi_p}{2T_c}}{2\xi_p} F_{ij}(\varepsilon) =$$

$$= \int_{\xi(-\omega_c)}^{\xi(\omega_c)} d\xi \frac{\tanh\frac{\xi}{2T_c}}{2\xi f} F_{ij}(\varepsilon(\xi)) \quad (41)$$

with a cut-off energy parameter ω_c of the pairing interaction (30) in Eqs. (33) and (34). In Eq. (41) the divergence energy point $\xi=0$ can be converted to $\varepsilon = \frac{\mu}{f}$, where the chemical potential magnitude $\mu / f = -0.415$ has been found in [22] and its value corresponds to the realistic filling of energy bands by carriers in our κ-ET$_2$X salt model.

Applying the logarithmic approximation to matrix elements Eq. (34), one can get that for d_{xy} - pairing the superconducting critical temperature T_c satisfies the explicit equation

$$1 = \frac{2V}{f} F_{33} \ln \frac{\omega_c f}{2T_c}. \quad (42)$$

Recalling the correlation factor $f = 1/4$ and Eq. (37) from Eq. (42) we immediately get the corresponding coupling constant

$$\lambda_{xy} = 8VF_{33}(1.47). \qquad (43)$$

The superconducting critical temperature for the order parameter of the mixed symmetry $s^* + d_{x^2-y^2}$ is specified by the quadratic equation,

$$\begin{vmatrix} 1 - 8VF_{11} \ln\dfrac{\omega_c}{8T_c} & -8VF_{12}\ln\dfrac{\omega_c}{8T_c} \\ -8VF_{12}\ln\dfrac{\omega_c}{8T_c} & 1 - 8VF_{22}\ln\dfrac{\omega_c}{8T_c} \end{vmatrix} = 0. \qquad (44)$$

To calculate the parameters F_{ij} according to relations (37)-(40) it is useful to find the numerical values of elliptic integrals such as $K(0.99)=3.35$; $E(0.99)=1.03$; $\Pi\left(\dfrac{\pi}{2};0.94;0.99\right) = 24.65$. Substituting these values into Eq. (37) we obtain after lenghly calculations: $F_{11}(1.47) = \dfrac{1.60}{\pi^2}$; $F_{22}(1.47) = \dfrac{0.36}{\pi^2}$; $F_{33}(1.47) = \dfrac{2.09}{\pi^2}$; $F_{12}(1.47) = -\dfrac{0.48}{\pi^2}$. Then the mixed symmetry $s^* + d_{x^2-y^2}$ coupling constants are evaluated from Eq. (44) as follows:

$$\lambda_{d_{x^2-y^2}+s^*} = 8VF_{1,2} \qquad (45)$$

with $F_1 = \dfrac{1.76}{\pi^2}$ and $F_2 = \dfrac{0.20}{\pi^2}$ .I.e., both coupling constants (45) are smaller in magnitude than the coupling constant in Eq. (43) for the superconducting d_{xy}-pairing. Hence, it follows that the superconducting order parameter of the d_{xy}-symmetry is more preferable for the κ-ET$_2$X superconductor model under consideration.

For the amplitude of the effective pairing interaction $V = 0.022$ made dimensionless by interdimer hopping integral t, the ratio of superconducting critical temperatures of interest is $T_c^{d_{xy}} / T_c^{s^* + d_{x^2-y^2}} = 1.65$.

In view of the same V value and assuming that the cut-off energy parameter, ω_c, equals to the interdimer hopping integral, $\omega_c = t \sim 0.1$ eV, in the logarithmic approximation we can estimate from Eq. (42) the superconducting transition temperature T_c, 10 K, for the derived d_{xy}-wave pairing. This value is of a reasonable magnitude for κ-ET$_2$X superconductors.

For final conclusion about pairing, however, it is useful to compare not only T_c but also energies of different pairing symmetries [52,55]. Also we can not exclude that in κ-ET$_2$X materials with distorted lattice in ET$_2$-layers the electron interdimer hopping asymmetry can lead to contributions of other components in their pairing symmetry. Similar admixture of different pairing components was observed for high- T_c cuprates in [56-58].

The presented in Sects. 7 and 8 analysis may favour some kind of nonphonon superconducting pairing mechanism due to short coherence length and cut-off energy parameter which is assumed to the interdimer hopping integral amplitude, namely $\omega_c = t$. Really, in the Introduction we mentioned the absence of the isotopic effect according to the isotopic shift, ΔT_c, measurements in ^{34}S- and ^{13}C=^{13}C - substituted ET-superconductors. This implication is also consistent with the absence of the Hebel-Slichter peak in the ^{13}C NMR relaxation rate near T_c as reviewed in [9,10].

9. Electronic pair in an empty dimer ET$_2$ lattice

As the two fermions with the angular quantum number $l = 0$ scatters in the Cooper s-channel, the positive phase shift $\delta_{l=0}(E_F) > 0$ caused by scattering at finite Fermi energy (Cooper superconducting instability) in plane is a sufficient condition for a positive phase shift in an empty lattice $\delta_{l=0}(0) > 0$ (binding two-particle state) [59, 60]. In preceding Section it was shown that the anisotropic d_{xy}-pairing is likely to take place in the κ-ET$_2$X model. Contrary to the s-pairing in the case of d- or p-wave pairing the Cooper pair formation and the occurrence of the pair electron binding state in an empty lattice are different problems, and from the first problem it is not inferred the second one. For the proposed model of κ-ET$_2$X superconductors the Cooper pair can be formed due to a interelectron attraction smaller than it is necessary for the appearance of the binding pair in an empty lattice. To calculate the critical value of an electron attraction, forming the binding state of an electronic pair in an empty triangular ET$_2$-lattice, one have to start from the Bethe-Salpeter equation for the scattering T-matrix, that is

$$T_{\mathbf{p}\mathbf{p}'}(E) = V_{\mathbf{p}\mathbf{p}'} + \frac{\Omega}{(2\pi)^2} \iint \frac{V_{\mathbf{p}\mathbf{p}'} T_{\mathbf{p}'\mathbf{p}''}(E)}{E - 2\varepsilon_{\mathbf{p}}^-} d\mathbf{p}'', \qquad (46)$$

where ε_p^- is the one-particle energy (3), $\Omega = \sqrt{3}$ is a unit cell volume (the lattice constant $a=1$). For the derived anisotropic d_{xy}-wave pairing the pairing interaction is

$V_{\mathbf{p}\mathbf{p}'} = -2V \sin\frac{p_x}{2} \sin\frac{p_y}{2} \sin\frac{p_x'}{2} \sin\frac{p_y'}{2}$. Assuming the same symmetry of the scattering T-matrix and putting it, namely

$$T_{pp'}(E) = C \sin\frac{p_x}{2} \sin\frac{p_y}{2} \sin\frac{p_x'}{2} \sin\frac{p_y'}{2},$$ into Eq. (46) we obtain we get

$$C = -2V - C\frac{V}{t}\frac{\Omega}{(2\pi)^2} \iint dy dz \frac{\sqrt{4\cos^2\frac{y}{2} - (\cos y - z)^2} \left(\sin\frac{y}{2}\right)^2}{2\cos\frac{y}{2}(\varepsilon - z)}, \qquad (47)$$

for arbitrary values of C and T. Here the integration have to be completed over the region of variables y and z, where the integrand in in Eq. (47) is real. It is worthy to note that the binding energy ε of an electronic pair in an empty lattice is below the band bottom $\varepsilon_{min}^-/t = -\frac{3}{2}$. To evaluate the integral in Eq. (47), let us take the prefactor outside the square root at a point where the integrand has a maximum. After this operation it is easy to find the critical value for the interaction amplitude where the scattering matrix changes its sign and the binding state of two electrons appears,

$$V^{crit} \approx \frac{2\pi}{3} t, \qquad (48)$$

where the numerical coefficient reflects a triangular lattice symmetry. It should be mentioned that for the binding pair in the d_{xy}-channel the calculated attraction threshold (48) in a triangular ET_2 lattice is smaller than for a square lattice in the Hubbard model [61]. However, it is somewhat more larger than the critical attraction for the t-J model for the square lattice [62]. Earlier, in the mean-field approximation for the Hubbard model it has been shown that the next nearest neighbours on the square lattice decrease the attraction threshold for the s-pairing [63-65].

10. The superconducting density of electronic states

With the quasiparticle spectra expected for a superconducting state with the derived symmetry a quasiparticle can be excited with an infinitesimal small energy. This comes about because of the zero crossings of the gap giving no nodes or lines of nodes on the Fermi surface where $\Delta_{d_{xy}} = 0$. Consequently, the number of quasiparticles have a power law dependence on the temperature and will depend on the topology of both the Fermi suraface and the superconducting order parameter. So, the quantities of the superconducting condensate which depend on density of states (specific heat, relaxation rate, penetration depth of a weak magnetic field, scattering rate etc.) will have a different dependence compared with a conventional BCS picture.

The nodes obtained in Sec. 8 for the anisotropic superconducting order parameter $\Delta_{d_{xy}}$ manifest themselves in the behaviour of the superconducting density of electronic states both on open (electronic) and closed (hole) sections of the Fermi surface. In its turn this behaviour is reasonable for the description of measurable quantities in low-energy experiments of κ-ET$_2$X superconductors.

For an energy range $E > 0$ the density of electronic states in the superconducting phase is defined by

$$\rho_s^{\pm}(E) = \frac{\sqrt{3}}{4\pi^2} \int_{-\pi}^{\pi} dp_y \int_{-\pi/\sqrt{3}}^{\pi/\sqrt{3}} \delta\left(E - \sqrt{(\xi_p^{\pm})^2 + \Delta_p^2}\right) dp_x, \qquad (49)$$

with the order parameter $\Delta_p = \Delta_0 \sin\frac{p_y}{2} \sin\frac{\sqrt{3}p_x}{2}$ and one-particle energies

$$\xi_p^{\pm} = ft\left(\cos p_y \pm 2\cos\frac{p_y}{2}\cos\frac{\sqrt{3}p_x}{2}\right) - \mu \quad \text{(cf. Eq. (25))}$$

for correlated carriers. Deriving (49) we have taken into account that the coherence factors $1 \pm \xi_p/\sqrt{\xi_p^2 + \Delta_p^2}$ vanishes owing to oddness of second terms with respect to correlated energy ξ_p.

The value Δ_p is small in the neighbourhood of four nodes on the Fermi surface inside the first Bz near the straight lines $p_x = 0$, $p_y = 0$. In Eq. (48) we expand ξ_p and Δ_p in terms of variations from their values at the nodes of the order parameter $\Delta_p=0$ on the Fermi surface $\mu(p_x,p_y)$. As a result, one can find that close to the node $p_x = 0, p_y = 2\arccos\left(\sqrt{\frac{3}{4} + \frac{\mu}{2ft} - \frac{1}{2}}\right)$ on the electronic section $\xi_p^+ = 0$ of the Fermi surface the density of electronic states is given by

$$\rho_s^+(E) = \frac{E}{2\pi\Delta_0 f t \sin^2\frac{p_y}{2}\left(2\cos\frac{p_y}{2} + 1\right)} = \beta_+ E. \qquad (50)$$

In a similar manner close to the node $p_y = 0$, $p_x = \frac{2}{\sqrt{3}}\arccos\left(\frac{1}{2} - \frac{\mu}{2ft}\right)$ on the hole section of the Fermi surface the density of electronic states is as follows:

$$\rho_s^-(E) = \frac{E}{2\pi\Delta_0 ft \sin^2 \frac{\sqrt{3}p_x}{2}} = \beta_- E. \qquad (51)$$

Provided the conventional isotropic superconducting gap is of the s-type the density of electronic states equals to zero. As one would expect, Eqs. (50) and (51) show that for an anisotropic d_{xy}-order parameter the superconducting density of electronic states is gapless and linearly proportional to the energy near the order parameter nodes on the Fermi surface.

It is significant that the contributions of the Fermi surface portions with different curvatures to the linear dependencies in Eqs. (50) and (51) are characterised by different coefficients. Taking into account the calculated chemical potential $\mu/tf = -0.415$ [22] one can evaluate the ratio of the coefficients,

$$\frac{\beta_-}{\beta_+} \cong 3. \qquad (52)$$

This ratio is essential for description of superconducting phase characteristics of layered organic materials like κ-ET$_2$X superconductors.

11. Characteristics of the anisotropic superconducting phase

Thanks to the short coherence length in κ-ET$_2$X superconductors the London penetration depth $\lambda_{\alpha\alpha}$ can be calculated in the local electrodynamic approximation [66, 67],

$$\frac{c^2}{4\pi ne^2}\lambda_{\alpha\alpha}^{-2} = \sum_p \left(\frac{\partial \xi_p}{\partial p_\alpha}\right)^2 \left\{\left[-\frac{\partial N_F(\xi_p)}{\partial \xi_p}\right] - \left[-\frac{\partial N_F(E_p)}{\partial E_p}\right]\right\}, \qquad (53)$$

where n is the carrier concentration, α is the screening superconducting current orientation with respect to the coordinate axes (cf. Fig. 1), N_F denotes the Fermi-Dirac function, ξ_p the is normal one-particle electron energy (25) with respect to the Fermi level μ, and $E_p = \sqrt{\xi_p^2 + \Delta_p^2}$ is the Bogoljubov quasiparticle energy. To calculate the penetration depth for the magnetic field in the direction perpendicular to the ET$_2$-layer one needs to use an energy relation s ξ_p^+ and ξ_p^- in Eq. (53). At low temperatures the second term in Eq. (53) vanishes and the Fermi-Dirac function converts to the Dirac δ-function. As a result of a cumbersome procedure we obtain the penetration depth tensor in the form,

$$\lambda_{xx}^{-2} \propto \frac{3}{4\pi^2} f^2 \left\{ -I_2^x + (2\mu+4)I_1^x + (3-\mu^2)I_0^x \right\},$$

$$\lambda_{yy}^{-2} \propto \frac{1}{\pi^2} f^2 \left\{ \left(\frac{3-\mu^2}{4}\right)I_0^y - \frac{\mu}{2}I_1^y + \frac{(\mu+1)^2}{2}I_{-1}^y - \frac{1}{4}I_2^y \right\}.$$

(54)

Here the functions $I_{0,-1,1,2}^i$, where

$$I_2^i = (\mu/ft+3)I_1^i + (\mu/ft+3)I_0^i + (\mu/ft+1)^2 I_{-1}^i + \frac{\sqrt{G(y)}}{y+1},$$

are defined by the elliptic integrals after integrating in the limits from -1 to μ for λ_{xx} and from μ to 1 for λ_{yy}. Electron-electron correlations give rise to the correlation factor $f = 1/4$ which renormalise the chemical potential μ influencing λ_{xx}, λ_{yy} via incomplete elliptic integrals.

After calculation of $I_{0,-1,1,2}^i$ in Eq. (54) one can get the closed to each other penetration depths along the chosen coordinate axes x and y on the triangular lattice of dimers, as follows:

$$\frac{\lambda_{xx}}{\lambda_{yy}} \approx 1.1. \quad (55)$$

The superconducting density of states obtained in the previous Sec. makes it possible to derive the temperature dependence of the electronic specific heat and spin-lattice relaxation time affected by conduction electrons in the ET$_2$-plane. These values are defined by the density of states averaged over both portions of the Fermi surface. The linear dependencies (50) and (51) of the density of electronic states on energy in the superconducting condensate leads to the quadratic temperature relation for the electronic specific heat, namely

$$C_s = 2\sum_{p\alpha} \xi_p^\alpha \frac{\partial N_F}{\partial T} = 2\int_0^\infty (\beta_+ + \beta_-)E^2 \left(\frac{\partial N_F}{\partial T}\right) dE =$$

$$= 6(\beta_+ + \beta_-)T^2 \int_0^\infty \frac{x^2 dx}{\exp x + 1} = 9(\beta_+ + \beta_-)\varsigma(3)T^2 = 10.8(\beta_+ + \beta_-)T^2$$

(56)

for a unit cell of the ET$_2$ layer. Here $\zeta(3)$ is the Riemann ζ-function. Notice that according to (52) the carriers on closed, hole portions of the Fermi surface ($\xi_p^- = 0$) contribute significantly to the superconducting specific heat. Substituting into Eq. (56)

the parameters t=0.12 eV, Δ_0=(2.5÷3.5)T_c [68, 69], T_c = 10K for κ-ET$_2$X salts we get the superconducting specific heat per κ-(ET)$_2$X mole,

$$C_m = \alpha T^2, \qquad (57)$$

where the coefficient $\alpha = 10.8 N_A (\beta_+ + \beta_-) k_B^3 / 2$ (N_A is the Avogadro's number and k_B is the Boltzman's constant) varies between 1.59 and 2.23 mJ/K^{3}mole. Such α values are in agreement with results of measurements. [70] where the experimental value of α has been estimated as 1.58-1.85 mJ/K^{3}mole and 3.53 mJ/K^{3}mole for superconductors κ-(ET)$_2$Cu[N(CN)$_2$]Br with T_c =11.6K and κ-(ET)$_2$Cu(NCS)$_2$ with T_c =10K, respectively. The more accurate measurements of the same authors gave the value α = 2.2 mJ/K^{3}mole for the κ-(ET)$_2$Cu[N(CN)$_2$]Br superconductor [71] and a somewhat less value for κ-(ET)$_2$Cu(NCS)$_2$.

In the ET molecule central fragment the carbon ^{13}C nuclear magnetic momentum damps out through the conduction electrons during NMR conditions. The corresponding spin-lattice relaxation rate $R=1/T_1$ is defined in the superconducting phase by [66]

$$R_s \approx \sum_p \sum_{p'} \cdot \frac{E_p E_{p'} + \Delta_p \Delta_{p'} + \xi_p \xi_{p'}}{2 E_p E_{p'}} \\ \cdot \frac{N_F(E_p) - N_F(E_{p'})}{\exp((E_p - E_{p'})/T) - 1} \delta(E_{p'} - E_p - \nu) \qquad (58)$$

where ν is the frequency of an oscillating magnetic field. In. (58) the region of the integration is defined by the sign changes of Δ_p, $\Delta_{p'}$, ξ_p, and $\xi_{p'}$. Near the Δ_p nodes we have $\Delta_{p'} = \pm\Delta_p \left(1 + \frac{\nu}{E_p}\right)$ and $\xi_{p'}^\alpha = \pm\xi_p^\alpha \left(1 + \frac{\nu}{E_p}\right)$, which yields. the ratio of the relaxation rates for superconducting and normal phases,

$$\frac{R_s}{R_n} = \frac{2}{T} \int_0^w \frac{(\beta_+ + \beta_-)^2 E^2}{[\rho(\mu)]^2} \frac{\exp(E/T)}{(\exp(E/T)+1) \cdot (\exp((E+\nu)/T)+1)} dE. \qquad (59)$$

In Eq. (59) the parameters β_+ and β_- are defined in (50) and (51), ρ denotes the normal density of states at the Fermi energy and w is the cut-off energy parameter for linearly dependent on energy the superconducting density of states. For the radio-frequency ν~10MHz~10^{-4}K, at low temperatures $T<<w$ Eq. (59) gives

$$\frac{R_s}{R_n} = \frac{2T^2\beta^2\zeta(2)}{\rho^2(\mu)} \cdot \left(1 - \frac{\nu \ln 2}{T\,\zeta(2)}\right). \tag{60}$$

For the normal phase $R_n \sim T$ in accordance with the Korringa law and, as a consequence, we find that for the superconducting phase the spin-lattice relaxation rate has a cubic low-temperature dependence, $R_s \sim T^3$.

Electron-electron correlations influence the coefficients β_+ and β_- in Eqs. (50), (51), (56) and (59) via the factor $f = \frac{1}{4}$ and the renormalized chemical potential. The derived temperature dependence (60) differs from an activated dependence $R_s \sim \exp(-\Delta/T)$ at low temperatures in superconductors with conventional s-wave pairing and it has been observed in κ-ET$_2$X salts [72-74] at $T \ll T_c$.

12. Concluding remarks.

It is well known that an analytical calculation of the electronic structure in one-dimensional organic compounds have allowed to analyse various observed phenomena and to predict several new effects. The layered κ-ET$_2$X salts may change molecule arrangements, which is accompanied by carrier hopping changes. The present description of the κ-ET$_2$X electronic structure gives an opportunity to study κ-ET$_2$X energy bands, the Fermi surface topology and a variety of properties in erms of the measurable characteristic energies, $U_{ET} > t_0 > t_{1,2,3} > \tau$. The energy branches (3) of elementary excitations provide the anisotropic Fermi surface with the saddle open (electronic) and saddle closed (hole) portions. The interlayer carrier hopping τ leads to weakly three-dimensional energy dispersion relations (6) and to a corrugation of the initial Fermi surface, derived for an effective dimer Hamiltonian (1) of ET$_2$-layer. These energy dispersions are different from the ones in [75] postulated without a microscopic origin. The plotted energy dispersions and the normal densities of electronic states with logarithmic divergencies reproduce the main features of the first principles calculation for 10K class κ-ET$_2$Cu(NCS)$_2$ salt [25] or the tight-binding method ([18, 20, 26]. In [25, 28] slight asymmetry between bonding and antibonding dispersions is caused by inserting of an additional small hopping integral between nonconjugated molecules of neighbouring ET$_2$, which we did not take into account. The characteristic energy scale provides the correlated splitting of an uncorrelated tight-binding band and the energy gap Δ (18) due to strong electron interaction U_{ET}, in satisfactory agreement with the measured activation energy $E_g = \Delta/2$ for κ-ET$_2$X semiconductors [29, 30] (Sec. 4).

Variations of the ET donor arrangement at a fixed carrier concentration lead to a variation of an energy scale. When the energy gap (18) decreases the insulator-metal phase transition of the Mott-Hubbard type happens due to the electron scattering on spin and charge fluctuations [19, 40, 54]. The all known κ-ET$_2$X semiconductors and metals are positioned in the vicinity of the calculated phase point (21) of the insulator-metal phase transition. The decreased value of the $t/U_{ET_2} \propto t/t_0$ in (21) indicates that the insulator-metal phase transition in κ-ET$_2$X organics can be realised under the lower applied pressure than, e. g., in transition metal oxide inorganics. Recently, the insulator-superconducting metal phase transition was obtained in the paramagnetic κ-ET$_2$Cu[N(CN)$_2$]Cl single-crystals under a moderate hydrostatic pressure [41]. The κ-ET$_2$X materials that are metals at ambient pressure have smaller intradimer hopping, t_0, values in agreement with. (21) and remarks of Ref. [18]. So, the κ-ET$_2$X salts can be viewed as a correlated electron system with doubly degenerate sites in the triangular ET$_2$ layer, where ET$_2$ dimers are considered as entities with two degenerate energy levels. Strong electron correlations renormalise the hopping integrals and the chemical potential due to the correlation factor $f = \dfrac{4-3n}{4}$ in Eq. (25) and lead to correlated narrowing of the carrier energy band in the paramagnetic state. Also from Eq. (25) it follows that the magnetic breakdown gap between the closed and open portions of the Fermi surface (cf. Figs. 3 and inset in Fig. 6a) is

$$\xi_p^+ - \xi_p^- = 2f \cos\frac{p_{y0}}{2}|t_1 - t_3| \cong |t_1 - t_3|/4 \approx 4\text{meV}$$ (at the Fermi momentum $p_{y0} \approx 2\pi/3$) for a realistic difference of nonazimuthal hopping integrals $t_{1,3}$. This magnetic breakdown gap value is in agreement with its experimental findings. [76-79].

The symmetry of an electronic structure has had an immediate impact on a symmetry of an effective pairing attraction (30) of an unspecified nature beyond the details of the pairing aspect of the microscopic theory. For singlet pairing in a clean limit of the proposed model it is obtained that the superconducting condensate in κ-ET$_2$X salts has the singlet order parameter, $\Delta_p = \Delta_0 \sin\dfrac{p_y}{2}\sin\dfrac{\sqrt{3}p_x}{2}$ (see Sec. 8). Electron correlations via factor f = 1/4 dramatically enhance the tendency to the d_{xy}-wave symmetry of the superconducting condensate, which is also supported by the topology of the Fermi surface. According to Sects. 7-9 the superconducting coupling constant λ_{xy}) is provided by an effective attraction $V < V^{crit}$, which is still lacking for binding state formation in the same singlet d_{xy}-channel in an empty triangular lattice of ET$_2$ dimers. The absence of the Knight shift in κ-ET$_2$X superconductors [9,10] also favours the singlet superconductivity.

Close to the nodes of the superconducting order parameter on the Fermi sections the superconducting density of electronic states (50) and (51) is directly proportional to an excitation energy. As a result, the number of elementary excitations has power dependence on temperature, contrary to the conventional s-wave superconductors. The superconducting specific heat is therefore quadratic with respect to temperature (56) and the spin-lattice ^{13}C relaxation rate is cubic in temperature (60) at low temperature.

It is worthy to note that for three-dimensional anisotropic superconductors the DOS is quadratic in energy near the nodes of the d-wave superconducting order parameter on the spherical Fermi surface [80, 81]. Up to now in the studies of an anisotropic order parameter in systems with the isotropic Fermi surface and quadratic energy dispersions for electrons are popular, contrary to the Anderson remarks [46]. Only long ago [82] the influence of the anisotropic Fermi surface topology on the superconducting condensate has been studied for an isotropic s-wave order parameter. It should be emphasized that the expansions like Eq. (31) restores in terms of basis functions the earlier ideas of Refs. [83] and [84] about factorable kernels in the BCS integral equation.

In the present model the Fermi surface has the pair of saddle sections even near the Γ-point [22] of the Bz when filling of the conducting band is small.. Therefore, the Fermi sphere is then inappropriate as the κ-ET$_2$X Fermi surface for studies of an order parameter symmetry and related properties (cf. [85]). In this work we examined the interplay of the Fermi surface topology and properties of the κ-ET$_2$X superconductors. According to (50) and (51) on the hole saddle portion of the Fermi surface the superconducting DOS is by a factor 3 larger than on the electronic portion of the Fermi surface. It is manifested both in the superconducting specific heat,. and in the spin-lattice relaxation rate.. Therefore, in layered ET based superconductors without saddle sections of the Fermi surface the decreased contributions $\propto T^2$ should be expected in specific heat. The α-ET$_2$NH$_4$Hg(SCN)$_4$ superconductor seems to fall into this cathegory having significant contributions $\propto T^2$ to the superconducting specific heat according to the measurements [86, 87].

The calculated cubic temperature dependence of the spin-lattice relaxation rate (60) due to conduction electrons is in agreement with experimental findings, such as $R_s(^{13}C) \propto T^3$ [72-74] for the nuclear magnetic spins of central carbons isotopes in ET at low temperatures $T \ll T_c$. The absence of the Hebel-Slichter peak at $T \leq T_c$ can be explained by strong electron interactions, $U_{ET} \gg t_{1,2,3}$ [88]..

Noteworthy, the quadratic temperature dependence of the superconducting specific heat has been measured earlier in 1-2-3 HTSC [89], where the cubic law temperature dependence of spin-lattice relaxation rate $R_s \propto T^3$ [90, 91] and the absence of the Knight shiftwere also observed. These experimental data have been interpreted in the frame of superconducting pairing with $d_{x^2-y^2}$-wave symmetry for CuO$_2$ layers (see [92, 93] and review [94]). This pairing symmetry corresponds to d_{xy}-wave symmetry in our reference system.

The correlation factor $f = \dfrac{4-3n}{4}$ ($n = 1$ is the hole number per dimer) affects superconducting condensate properties. In Ref. [95] according to ESR copper signals the Cu^{2+} concentration change has been measured in the anion layer of κ-ET$_2$Cu$_2$(CN)$_3$ = $\kappa - ET_2^{1-x} Cu_{2-x}^+ Cu_x^{2+}(CN^-)_3$. The increase of paramagnetic ions Cu^{2+} decreases the concentration 1-x of hole carriers in cation layer ET$_2$ and leads to increase of correlation factor $f = \dfrac{1+3x}{4}$ resulting in the T_c decrease, according to Eqs. (42)-(43), which agrees with the previous results [95].

In the normal phase f-factor renormalises dispersion relations for correlated carriers. This suggests the four-fold narrowing of an energy band with possible difference of optical and cyclotron electron masses [23, 26, 96] and decrease of cross-section of hole orbits in κ-ET$_2$XCu[N(CN)$_2$]Br [97].

The baric experiments, in the present state of art, allow to act along fixed crystal directions of κ-ET$_2$X salts using external uniaxial pressure [98]. From this point of view it can be useful to present an analytical formulation of the electron structure in terms of characteristic energy parameters.

From Eqs. (56) and (60) for the κ-ET$_2$X superconducting phase it is follows a relation between the spin-lattice relaxation rate and electronic specific heat,

$$\frac{R_s C_n^2}{R_n C_s^2} = \frac{4\zeta^3(2)}{81\zeta^2(3)} = 0.29, \tag{61}$$

where $C_n = 2\zeta(2)\rho(\mu)T$. It is interesting to check relation (61) experimentally, which may shed light on additional mechanisms of ^{13}C nuclear spin damping.

The derived d$_{xy}$-wave superconducting order parameter affects the temperature dependence of the nuclear relaxation rate below T$_c$ differently than in the conventional BCS s-wave pairing model. Averaging over the Fermi surface in Eq. (58) leads to the softening of the singularity in superconducting DOS and to a disappearance of the Hebel-Slichter coherence peak for R$_s$, according to Maleyev scenario [99].

The developed approach can be applied to other layered organic systems, e. g., to recently discovered p-d organics such as (BETS)$_2$MX$_4$ molecular conductors (BETS = bis(ethylenedithio)tetraselenafulvalene) [100].

Acknowledgments

This work is supported by Grant N 96149 from the Russian State Program "Fullerenes and atomic clusters" and Russian Foundation for Basic Research.

References

1. Akamatsu, H. and Inokuchi, H. (1950) On the electrical conductivity of violanthrone, iso-violanthrone, and pyranthrone, *J. Chem. Phys.* **18,** 810 - 813.
2. Inokuchi, H. (1954) Photoconductivity of the condensed polynuclear aromatic compounds, *Bull. Chem. Soc. Jpn.*, **27,** 22-28.
3. Akamatsu, H., Inokuchi, H., and Matsunaga, Y. (1954) Electrical conductivity of the perylene-bromine complex, *Nature* **173,** 168-169.
4. Jerome, D., Mazaud, A., Ribault, M., and Bechgaard, K. (1980) A new ambient pressure organic superconductor based on BEDT-TTF with T$_c$ higher than 10K (T$_c$ = 10.4K), *J. Phys. Lett.* **41,** L95-L97.
5. Ishiguro, T. and Yamaji, K. (1990) *Organic Superconductors,* Springer-Verlag, Berlin.
6. Yagubskii, E.B., Shchegolev, I.F., Laukhin, V.N., and Kononovich, P.A. (1984) Normal-pressure superconductivity in an organic metal (BEDT-TTF)$_2$I$_3$, [bis(ethylenedithiolo) tetrathiofulvalene triiodide], *JETP Lett.* **39,** 12-16; Kaminskii, V.F., Prokhorova, T.G., Shibaeva, R.P., and Yagubskii, E.B. (1984) Crystal structure of organic superconductor (BEDT-TTF)$_2$I$_3$, *Sov. Phys.:JETP Lett.* **39,** 17-23.
7. Kartsovnik, M.V., Laukhin, V.N., Nizhankovskii, V.I., Ignat'ev, A.A. (1988) Transverse magnetoresistance and Shubnikov-de Haas oscillations in the organic superconductor β-(ET)$_2$IBr$_2$, *Sov. Phys.:JETP Lett.* **47,** 363-367; Kartsovnik, M.V., Kononovich, P.A., Laukhin, V.N., and Shchegolev, I.F. (1988) Anisotropy of magnetoresistance and the Shubnikov-de Haas oscillations in the organic metal β-(ET)$_2$IBr$_2$, *Sov. Phys.: JETP Lett.* **48,** 541-545.

8. Murata, K., Toyota, N., Honda, Y., Sasaki, T., Tokumoto, H., Bando, H., Anzai, H., Muto, Y., and Ishiguro, T. (1988) Magnetoresistance in β-(BEDT-TTF)$_2$I$_3$ and β-(BEDT-TTF)$_2$IBr$_2$: Shubnikov - de Haas effect, *J. Phys. Soc. Jpn.* **57,** 1540-1544.
9. Jerome, D. (1994) Organic superconductors: from (TMTSF)$_2$PF$_6$ to fullerenes, in J.-P. Farges (ed.), *Organic Conductors,* Marcel Dekker Inc.,pp. 405-494.
10. Ishiguro, T., Ito, H., Yamauchi, Y., Ohmichi, E., Kubota, M., Yamochi, H., Saito, G., Kartsovnik, M.V., Tanatar, M.A., Sushko Yu.V., and Logvenov, G.Yu. (1997) Electronic phase diagram and Fermi surfaces of κ-(ET)$_2$X, the high T$_c$ organic superconductors, *Synthetic Metals* **85,** 1471 - 1478.
11. Brooks, J.S., Uji, S., Aoki, H., Kato, R., Sawa, H., Clark, R.G., McKenzie, R., Athas, G.J., Sandhu, P., Valfells, S., Campos, C.E., Perenboom, J.A.A.J., van Bentum J., Tokumoto, M., Kinoshita, N., Kinoshita, T., Tanaka, Y., Anzai, H., Fisk, Z., and Sarrao, J. (1996) Molecular conductors: the Mesopotamia of low-dimensional physics, in Z. Fisk et al. (eds), *Physical Properties at High Magnetic Fields - II* ., World Scientific Press, pp.249-263.
12. Ivanov, V. and Kanoda, K.(1996) Strong electron correlations in κ-(BEDT-TTF)$_2$X salts, *Physica* **C 268,** 205-216.
13. Ivanov, V. and Kanoda, K. (1996) κ-(BEDT-TTF) salts as layered correlated electron systems, *Molec. Cryst. Liq. Cryst.* **285,** 211-216.
14. Okuno, Y. and Fukutome, H. (1997) A model for the electronic structures of β- and κ-(BEDT-TTF)$_2$X, *Solid State Commun.* **101,** 355-360.
15. Fortunelli, A. and Painelli, A. (1997) Ab initio estimate of Hubbard model parameters: a simple procedure applied to BEDT-TTF salts, *Phys. Rev.* **B55,** 16088-16095.
16. Bulaevskii, L.N. (1988) Organic layered superconductors, *Adv. in Physics* **37,** 443-470.
17. Toyota, N., Fenton, E.W., Sasaki, T., and Tachiki, M. (1989) Evidence of many-body renormalizations in some organic conductors, *Solid State Commun.* **72,** 859-862.
18. Kino, H. and Fukuyama, H. (1996) Phase diagram of two-dimensional organic conductors: (BEDT-TTF)$_2$X, *J. Phys. Soc. Jpn.* **65,** 2158-2169.
19. Ivanov, V.A. (1996) Insulator-metal phase transition in the κ-ET$_2$X salts, where "ET" is the bis(ethylenedithio)tetrathiafulvalene (BEDT-TTF) molecule, *Physica* **C271,** 127-132.
20. Tamura, M., Tajima, H., Yakushi, K., Kuroda, H., Kobayashi, A., Kato, R., and Kobayashi, H. (1994) Reflectance spectra of κ-(BEDT-TTF)$_2$I$_3$, *J. Phys. Soc. Jpn.* **60,** 3861-3873.
21. Oshima, K., Mori, T., Inokuchi, H., Urayama, H., Yamochi, H., and Saito, G.(1988) Shubnikov-de Haas effect and the Fermi surface in an ambient pressure organic superconductor [bis(ethylenedithiolo)tetrathiafulvalene]$_2$Cu(NCS)$_2$, *Phys. Rev.* **B38,** 938-947.
22. Ivanov, V., Yakushi, K., and Ugolkova, E. (1997) Electronic structure of κ-ET$_2$X salts where "ET" is the bis(ethylenedithio)tetrathiafulvalene (BEDT-TTF) molecule, *Physica* **C275,** 26-36.
23. Caulfield, J., Lubczynski, W., Pratt, F.L., Singletone, J., Ko, D.Y.K., Hayes, W., Kurmoo, M., and Day, P. (1994) Magnetotransport studies of the organic superconductor κ-(BEDT-TTF)$_2$Cu$_2$(NCS)$_2$ under pressure: the relationship between carrier effective mass and critical temperature, *J. Phys.: Condensed Matter.* **6,** 2911-2924.
24. Wosnitza, J., Goll, G., Beckmann, D., Wanka, S., Schweitzer, D., and Strunz, W. (1996) The Fermi surfaces of β-(BEDT-TTF)$_2$X, *J. Phys. I France* **6,** 1597-1608.
25. Xu, Y.-N., Ching, W.Y., Jean, Y.C., and Lou, Y. (1995) First-principles calculation of the electronic and optical properties of the organic superconductor κ-(BEDT-TTF)$_2$Cu(NCS)$_2$, *Phys. Rev.* **B52,** 12946-12950.
26. Ugawa, A., Ojima, G., Yakushi, K., and Kuroda, H.(1988) Optical and electric properties of di[bis(ethylenedithio)tetrathiafulvalenium] dithiocyano cuprate (I), organic superconductor, (BEDT-TTF)$_2$[Cu(SCN)$_2$], *Phys. Rev.* **B38,** 5122-5125.
27. Kino, H. and Fukuyama, H. (1995) Electronic states of conducting organic κ-(BEDT-TTF)$_2$X, *J. Phys. Soc. Jpn.* **64,** 2726-2729.
28. Jung, D., Evain, M., Novoa ,J.J., Whangbo, M.H., Beno, M.A., Kini, A.M., Schultz, A.J., Williams, J.M., and Nigrey, P.J. (1989) Similarity and differences in the structural and electronic properties of κ-phases organic conducting and superconducting salts, *Inorg. Chem.* **28,** 4516-4524.
29. Williams, J.M., Kini, A.M. , Wang, H.H., Carlson, K.D., Geiser, U., Montgomery, L.K., Pyrka, G.J., Watkins, D.M., Kommers, J.M., Boryschuk, S.J., Crouch, A.V.S., and Kwok, W.K. (1990) From semiconductor-semiconductor transition (42 K) to the highest-T$_c$ organic superconductors κ-(ET)$_2$Cu[N(CN)$_2$]Cl (T$_c$=12.5 K), *Inorg. Chem.* **29,** 3272-3274.

30. Sato, H., Sasaki, T., and Toyota, N. (1991) Electrical resistance and superconducting transitions in non-deuterated and deuterated κ-(BEDT-TTF)$_2$Cu[N(CN)$_2$]Br, *Physica* **C185-189**, 2679-2680.
31. Saito, G., Otsuka, A., and Zahidov, A.A. (1996) Overview of 10K class organic superconductors κ-(BDT-TTF)$_2$X (X=Cu(NCS)$_2$, Cu(CN)[N(CN)$_2$], Cu[N(CN)$_2$]X' (X'=Cl, Br)) and a search for superconductivity in alkali doped C$_{60}$ complexes, *Molec. Cryst. Liq. Cryst.* **3**, 284-292.
32. Ivanov, V.A. (1993) The tight-binding method for correlated electrons with particular attention to high-T$_c$ superconductors, in A. Narlikar (ed), *Studies on High-Tc Superconductors*, vol. 11,: Nova Science, New York, pp. 331-351.
33. Ivanov, V.A. (1994) The tight-binding approach to the corrundum structured compounds, *J. Phys.: Condensed Matter* **6**, 2065-2076.
34. Hubbard, J. (1965) Electron correlations in narrow energy bands: IV The atomic representation, *Proc. Roy. Soc.* **A285**, 512-560.
35. Okubo, S. (1962) Note on unitary symmetry in strong interactions, *Prog. Theor. Phys.* **27**, 949-966.
36. Nakajima, S., Sato, M., and Murayama, Y., (1991) Mean field theory of the holon-spinon system, *J. Phys. Soc. Jpn.* **60**, 2333-2340.
37. Murayama, Y. and Nakajima, S. (1991) The mechanism of hole-doped high-T$_c$ superconductivity, *J. Phys. Soc. Jpn.* **60**,4265-4279.
38. Murayama, Y., Ivanov, V.A., and Nakajima, S. (1996), Superconducting ruthenate viewed from the kinematic interaction mechanism, *Phys. Lett.* **A211**, 181-183.
39. Zaitsev, R.O. and Ivanov, V.A. (1985) Mott transition in a system of d-electrons with a half-filled band, *Sov. Phys.: Solid State* **27**, 2147-2152.
40. Ivanov, V. and Murayama, Y. (1997) Electronic structure, insulator-metal transition and superconductivity in κ-(BEDT-TTF)$_2$CuX salts, in S. Nakajima, M. Murakami (eds.), *Advances in Superconductivity* IX , Springer-Verlag: Tokyo, pp. 315-320.
41. Ito, H., Ishiguro, T., Kubota, M., and Saito, G. (1996) Metal-nonmetal transition and superconductivity localization in two-dimensional conductor κ-(BDT-TTF)$_2$Cu[N(CN)$_2$]Cl under pressure, *J. Phys. Soc. Jpn.* **65**, 2987-2993.
42. Eliashberg, G.M. and von der Linden, W. (1994) The Mott transition and superconductivity, *Sov. Phys.: JETP Lett.* **59**, 441-445.
43. Geilikman, B.T. and Kresin, V.Z. (1961) Effect of anisotropy on the properties of superconductors, *Sov. Phys. : JETP* **13**, 677-678.
44. Geilikman, B.T. and Kresin, V.Z. (1964) Effect of anisotropy on the properties of superconductors, *Sov. Phys.: Solid State* **5**, 2605-2611.
45. Pokrovskii, V.L. (1961) Thermodynamics of anisotropic superconductors, *Sov. Phys.: JETP* **13**, 447-450.
46. Anderson, P.W. (1959a) Theory of dirty superconductors, *J. Phys. Chem. Solids* **11**, 26-30; (1959b) Knight shift in superconductors, *Phys. Rev. Lett.* **3**, 325-326.
47. Gor'kov, L.P. (1958) On the energy spectrum of superconductors, *Sov. Phys. JETP* **7**, 505-509.
48. Nambu, Y. (1958) Quasi-particles and gauge invariance in the theory of superconductivity, *Phys. Rev.* **117**, 648-663.
49. Ivanov, V., Ugolkova, E., and Zhuravlev, M. (1998) Band structure and superconductivity in κ-(BEDT-TTF)$_2$X salts, *Sov. Phys.: JETP*, in press.
50. Suhl, H., Mattias, B., and Wolker, L.(1958) Bardeen-Cooper-Schrieffer theory of superconductivity in the case of overlapping bands, *Phys. Rev. Lett.* **3**, 552-556.
51. Moskalenko, V.A. (1959) Superconductivity of metals with taking into account of an energy bands overlapping, *Fiz. Met. Metalloved. (in Russian)* **8**, 503-513.
52. Miyake, K., Matsuura, T., Jichu, H., and Nagaoka, Y. (1984) A model for Cooper pairing in heavy fermion superconductor, *Prog. Theor. Phys.* **72**, 1063-1080.
53. Fehrenbacher, R. and Norman, M.R. (1995) Phenomenological BCS theory of the high-T$_c$ cuprates, *Phys. Rev. Lett.* **74**, 3884-3887.
54. Ivanov, V. (1997) κ-(BEDT-TTF)$_2$X organics (BEDT-TTF= bis(ethylenedithio)tetrathiafulvalene), as seen for hubbardists, *Phil. Mag.* **B76,** 697-713.
55. Szabo, Zs. and Gulacsi, Zs. (1997) Superconducting phase diagram in the extended Hubbard model, to be published.
56. Valls, O.T. and Beal-Monod, M. T. (1995) Effect of interaction anisotropy on the superconducting transition temperature, *Phys. Rev.* **B51**, 8438-8445.
57. Maki, K. and Beal-Monod, M.T. (1995) Anisotropic d+s wave superconductivity, *Phys. Lett.* **A208**, 365-368.

58. Kim, H. and Nicol, E.J. (1995) Effect of impurity scattering on a (d+s)-wave superconductor, *Phys. Rev.* **B52**, 13576-13584.
59. Miyake, K. (1983) Fermi liquid theory of dilute submonolayer ^{3}He on thin ^{4}He II film, *Prog. Theor. Phys.* **69**, 1794-1797.
60. Randeria, M., Duan, J..M., and Shieh, L.Y. (1989) Bound states, Cooper pairing and Bose condensation in two dimensions, *Phys. Rev. Lett.* **62**, 981-984.
61. Ivanov, V.A., Kornilovich, P.E., and Bobryshev, V.V. (1994) Possible enhancement of the phonon pairing mechanism by a staggered magnetic field, *Physica* **C235-240**, 2369-2370.
62. Kagan, M.Yu. and Rice, T.M. (1994) Superconductivity in the two-dimensional t-J model at low electron density, *J. Phys.: Condensed Matter* **6**, 3771-3780.
63. Micnas, R., Ranninger, J., and Robaszkiewicz, S. (1988) An extended Hubbard model with inter-site attraction in two-dimensions and high-temperature superconductivity, *J. Phys.* **C21**, L 145-L151.
64. Okhawa, F.J. and Fukuyama, H. (1984) Anisotropic superconductivity in the Kondo lattice, *J. Phys. Soc. Jpn.* **53**, 4344-4352.
65. Lacroix, C., Bastide, C., and da Rosa Simoes, A. (1990) s-wave superconductivity in the presence of a strong Coulomb repulsion, *Physica* **B163**, 124-126.
66. Parks, R. D. (ed.), (1981). *Superconductivity*, Plenum Press, NY, L.
67. Frick, M. and Schneider, T., (1990) On the theory of layered high-temperature superconductors: Finite temperature properties, *Z. Phys.* **B78**, 159-168.
68. Dressel, M., Klein, O., Gruner, G., Karlson, K.D., Wang, H.H., and Williams, J.M. (1994) Electrodynamics of the organic superconductors κ-(BEDT-TTF)$_2$Cu(NCS)$_2$ and κ-(BEDT-TTF)$_2$Cu[N(CN)$_2$]Br, *Phys. Rev.* **B50**, 13603-13615.
69. Lang, M., Toyota, N., Sasaki, T., and Sato, H.(1992) Magnetic penetration depth of κ-(BEDT-TTF)$_2$Cu(NCS)$_2$: strong evidence for conventional Cooper pairing, *Phys. Rev. Lett.* **69**, 1443-1446.
70. Nakazawa, Y.and Kanoda, K., (1997) *Physica* **C 282-287**, 1897-1898.
71. Nakazawa, Y.and Kanoda, K.(1997) Low-temperature specific heat of κ-(BEDT-TTF)$_2$Cu[N(CN)$_2$]Br, *Phys. Rev.* **B55**, 8670-8680.
72. De Soto, S.M., Slichter, C.P., Kini, A.M., Wang, H.H., Geiser, U., and Williams, J.M. (1995) ^{13}C NMR studies of the normal and superconducting states of the organic superconductor κ-(BEDT-TTF)$_2$Cu[N(CN)$_2$]Br, *Phys. Rev.* **B52**, 10364-10368.
73. Mayaffre, H., Wzietek, P., Jerome, D., Lenoir, C., and Batail, P. (1995) Superconducting state of κ-(ET)$_2$Cu[N(CN)$_2$]Br studied by ^{13}C NMR: evidence for vortex-core-induced nuclear relaxation and unconventional pairing, *Phys. Rev. Lett.* **75**, 4122-4125.
74. Kanoda, K., Miyagawa, K., Kawamoto, A., and Nakazawa, Y. (1996) NMR relaxation rate in the superconducting state of the organic conductor, κ-(BDT-TTF)$_2$Cu[N(CN)$_2$]Br, *Phys. Rev.* **B54**, 76-79.
75. Yago, R. and Iye, Y. (1994) On the peak inversion of the angular dependent magnetoresistance oscillation of cylindrical Fermi surfaces with different corrugation symmetries, *Solid State Commun.* **89**, 275-281.
76. Sasaki, T., Sato, H., and Toyota, N.(1991) On the magnetic breakdown oscillations in organic superconductors Realization of superconductivity at ambient pressure by band-filling control in κ-(BEDT-TTF)$_2$Cu(NCS)$_2$, *Physica* **C185-189**, 2687-2688.
77. Heidman, C.-P., Muller, H., Biberacher, K., Neumaier, K., Probst, Ch., Andres, K., Jansen, A., G., M., and Joss, W. (1991) Shubnikov - de Haas effect and the Fermi surface of κ-(BEDT-TTF)$_2$Cu(NCS)$_2$, *Synth. Met.* **41-43**, 2029-2032.
78. Wosnitza, J., Crabtree, G. W., Wang, H. H., Geiser, U., Williams, J. M., and Carlson, K. D. (1992) de Haas - van Alphen studies of the organic superconductors α-(ET)$_2$(NH$_4$)Hg(SCN)$_4$ and κ-(ET)$_2$Cu(NCS)$_2$ [with ET= bis(ethylenedithio)-tetrathiafulvalene], *Phys. Rev.* **B45**, 3018-3025.
79. Caulfield, J., Singleton, J., Pratt, F.L., Doporto, M., Lubczynski, W., Hayes, W., Kurmoo, M., Day, P., Hendriks, P.T.J., and Perenboom J.A.A.J. (1993) The effects of open section of the Fermi surface on the physical properties of 2D organic molecular metals, *Synth. Met.* **61**, 63-67.
80. Volovik, G.E.and Gor'kov, L.P. (1985) Unusual superconductivity in UBe$_{13}$, *Sov. Phys.:JETP Lett.* **61**, 843-847.
81. Hasegawa, Y. (1996) Density of states and NMR relaxation rate in anisotropic superconductivity with intersecting line nodes, *J. Phys. Soc. Jpn.* **65**, 3131-3133.
82. Geilikman, B.T. (1966) On the electron mechanism of superconductivity, *Sov. Phys.: Uspekhi* **9**, 142-160.

83. Bogolubov, N.N., Tolmachev, V.V., and Shirkov, D.V. (1958) A new method in the theory of superconductivity, *Fortschritte der Physik* **6**, 605-682.
84. Nakajima, S. (1964) Superconductivity of the Falicov-Cohen model, *Progress Theor. Physics* **32**, 871-884.
85. Ichinomiya, T. and Yamada, K. (1996) Damping effect on spin-lattice relaxation rate in organic superconductors, *J. Phys. Soc. Jpn.* **65**, 1764-1768.
86. Andraka, B., Stewart, G.R., Carlson, K.D., Wang, H.H., Vashon, M.D., and Williams, J.M. (1990) Specific heat in zero and applied magnetic fields of the organic superconductor α-di[bis(ethylenedithio)tetrathiafulvalene]-ammonium-tetra(thiocyanato)mercurate [α-(ET)$_2$(NH$_4$)Hg(SCN)$_4$], *Phys. Rev.* **B42,** 9963-9966.
87. Nakazawa, Y., Kawamoto, A., and Kanoda, K., (1995) Characterization of low-temperature electronic states of the organic conductors α-(BEDT-TTF)$_2$MHg(SCN)$_4$ (M=K, Rb and NH$_4$), *Phys. Rev.* **B52**, 12890-12894.
88. Kopaev, Yu.V. and Tagirov, L.R. (1989) Peculiarities of nuclear relaxation in superconductor with strong interelectron correlations, *Sov. Phys.: JETP Lett.* **49,** 499-453.
89. Moler, K.A., Baar, D.J., Urbach, J.C., Liang, R.X., Hardy, W.N., and Kapitulnik, A. (1994) Neutron diffraction Studies of Flowing and Pinned Magnetic Flux Lattices in 2H-NbSe$_2$, *Phys. Rev. Lett.* **73**, 2744-2747.
90. Imai, T., Shimizu, T., Yasuoka, H., Ueda, Y., and Kosuge, K. (1988) Anomalous temperature dependence of Cu nuclear spin-lattice relaxation YBa$_2$Cu$_3$O$_{6.91}$, *J. Phys. Soc. Jpn.* **57**, 2280-2283.
91. Kitaoka, Y., Hiramatsu, S.,. Konori, Y., Ishida, K., Kondo, T., Shiba, H., Asayama, K., Takagi, H., Uchida, S., Iwabuchi, H., and Tanaka, S. (1988) Nuclear relaxation and Knight shift studies of ^{63}Cu in 90K- and 60K- class YBa$_2$Cu$_3$O$_{7-y}$, *Physica* **C153-155**, 83-86.
92. Bulut, N. and Scalapino, J. D. (1991) Calculation of the transverse nuclear relaxation rate for YBa$_2$Cu$_3$O$_7$ in the superconducting state *Phys. Rev. Lett.* **67**, 2898-2901.
93. Thelen, D., Pines, D., and Lu, J.P. (1993) Evidence for $d_{x^2-y^2}$ pairing from nuclear-magnetic experiments in the superconducting state of YBa$_2$Cu$_3$O$_7$, *Phys. Rev.* **B47**, 9151 - 9154.
94. Schrieffer, J.R. (1994) Symmetry of the order parameter in high-temperature superconductors, *Solid State Commun.* **92**, 129-139.
95. Komatsu, T.M., Matsukawa, N., Inoue, T., and Saito, G., (1996) Realization of superconductivity at ambient pressure by band-filling control in κ-(BDT-TTF)$_2$Cu$_2$(CN)$_3$, *J. Phys. Soc. Jpn.* **65**, 1340-1354.
96. Pratt, F.L., Singleton, J., Kurmoto, M., Spermon, S.J.R.M., Hayes, W., and Day., P. (1990) Fermi surface and band structure of κ-(BEDT-TTF)$_2$Cu(NCS)$_2$, in: G. Saito, S. Kagoshima (eds.), *The Physics and Chemistry of Organic Superconductors*, Springer-Verlag Berlin, Heidelberg,, pp. 200-203.
97. Kartsovnik, M.V., Logvenov, G.Yu., Ito H., Ishiguro, T., and Saito, G. (1995) Shubnikov - de Haas oscillations in the organic superconductor κ-(BDT-TTF)$_2$Cu[N(CN)$_2$]Br, where BEDT-TTF is bis(ethylenedithio)tetrathiafulvalene, *Phys. Rev.* **B52**, 15715-15718.
98. Campos, C.E., Sandhu, P.S., Brooks, J.S., and Ziman, T. (1996) An extended Huckel-tight binding study of the effect of pressure and uniaxial stress on the electronic structure of α - (BEDT-TTF)$_2$KHg(SCN)$_4$ and κ-(BEDT-TTF)$_2$Cu(SCN)$_2$, *Phys. Rev.* **B53**, 12725-12732.
99. Maleyev, S.V., Yashenkin, A.G., and Aristov, D.N. (1994) Nuclear relaxation rate in layered superconductors with unconventional pairing, *Phys. Rev.* **B50,** 13825- 13 828.
100. Kobayashi, H., Tomito, H., Naito, T., Kobayashi, A., Sakai, F., Watanabe, T., and Cassoux, P. (1996) New BETS conductors with magnetic anions (BETS = bis(ethylenedithio)tetraselenafulvalene), *J. Am. Chem. Soc.* **118**, 368-377.

THE MULTIBAND ANALYSIS OF THE ELECTRON DISPERSION AT THE TOP OF THE VALENCE BAND IN UNDOPED CUPRATES

S.G. OVCHINNIKOV
*L.V.Kirensky Institute of Physics and Theoretical Physics Department,
Physical Faculty of Krasnoyarsk State University,
Krasnoyarsk, 660036 Russia*

Abstract. The dispersion of a hole in the antiferromagnetic background is calculated in the framework of the 6-band p-d model and compared with results of the t-J model and ARPES measurements on $Sr_2CuO_2Cl_2$. In $(0,0)$-(π,π) and $(0,\pi)$-$(\pi,0)$ directions our calculations show a better agreement vs t-J model (in $(0,0)$-$(\pi,0)$ directions), where the largest discrepancy of t-J and ARPES results takes place, we get a much smaller dispersion. The new ARPES data on $Sr_2CuO_2Cl_2$ are in good agreement with our calculation in the $(0,0)$-$(\pi,0)$ direction. The implication for the pairing mechanism and the nature of the local boson (spin exciton) resulted from the multiband strongly correlated electron structure analysis are discussed.

The 3-band p-d model [1, 2] is the simplest model of copper oxides that takes into account the effects of strong electron correlations on Cu sites and the charge transfer nature of the insulator state of the undoped parent oxides. The extra hole induced by p-type doping and the hole localized on the Cu form the Zhang-Rice singlet [3]. The corresponding triplet state is known to be well separated with the energy difference $\Delta\varepsilon = \varepsilon_t - \varepsilon_s \approx 2 \div 4$ eV and so is irrelevant in the low energy physics. A lot of theoretical works have been devoted to the 3-band p-d model, its reduction to the one-band Hubbard model and to the t-J model [4-6].

There are some theoretical and experimental indications of the importance of the other states beyond the 3-band p-d model. Polarized x-ray absorption spectroscopy [7] (XAS) and electron-energy-loss spectroscopy [8] (EELS) show a sizable (~(10/15)%) occupancy of Cu $d_{3z^2-r^2}$ orbitals in all the investigated superconducting copper oxides. To account for observed states the 3-band model must be further enlarged by including the $d_{3z^2-r^2}$ orbitals of copper [9, 10]. The first-principles calculations of the electronic structures of CuO_4 and CuO_6 clusters by the multiconfiguration self-consistent field method with configuration interaction have shown that the decreasing of the apical oxygen-Cu distance results in the stabilization of the $^3B_{1g}$ triplet vs $^1A_{1g}$ singlet [12]. The perturbative study of the two-hole spectrum

[13] of the CuO_4 cluster with the model parameters determined from the CuO XPS data results in the $^3B_1$ level lying *1.5 eV* above the $^1A_1$ singlet. This value appears to be sensitive to the crystal field energy and goes to zero or a negative value when the apical oxygen was added [14]. The small energy separation of the singlet and triplet states changes the low energy spectrum of the Fermi-type quasiparticles [15]. The strong mixing of the singlet and triplet states is shown to occur away from the Ã point [16].

Two problems are studied in this paper. The first one is, whether the Zhang-Rice singlet is well-separated from the higher energy triplet $^3B_{1g}$ and whether the triplet-singlet separation is small. The second is the effect of triplet-singlet mixing on the electronic band structure of CuO_2 layer and on the possible mechanism of pairing.

To answer the first question we have performed the exact diagonalization of the multiband p-d model Hamiltonian for the CuO_4 and CuO_6 clusters. In contrast to the previons papers [12-14] we have studied the dependence of the two-hole terms on the crystal field parameter $\Delta_d = \varepsilon_{3z^2-r^2} - \varepsilon_{x^2-y^2}$. In the limit $\Delta_d \to \infty$ our model is equivalent to the 3-band p-d model and we obtain large singlet-triplet separation $\Delta\varepsilon > 2$ eV, in agreement with [3]. The realistic value is $\Delta_d \sim 1$ eV. Decreasing the value Δ_d we find the increasing contribution of the $^3B_{1g}$ state to the triplet state and decreasing singlet-triplet separation. $\Delta\varepsilon$ is rather small, depending on the set of the model parameters, $\Delta\varepsilon \approx 0.1 \div 0.5$ *eV*. It means that for realistic copper oxides

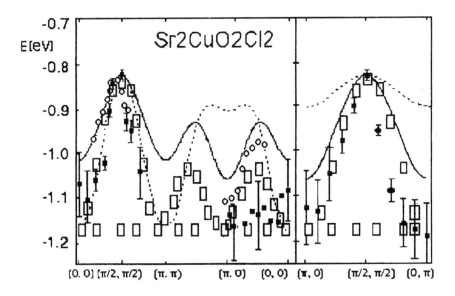

Figure 1. The quasiparticle spectrum of $Sr_2CuO_2Cl_2$. The filled sguares show the ARPES data [18], empty ovals-polarized ARPES data [20] dotted and solid lines show the results [19] of the *t-J* and *t-t'-J* models. The empty rectangulars show our result.

parameters the singlet is not well separated from the triplet. This fact makes limits to the reduction of the multiband *p-d* model to the one-band Hubbard model and to the *t-J* model.

To answer the second question we have performed the quasiparticle band structure calculation using the cluster perturbation method [17]. To combine the small cluster exact diagonalization and band structure effect of the infinite lattice we cover the CuO_2 lattice by a set of linear *O-Cu-O* clusters. Every copper and oxygen atom belongs to some CuO_2 cluster and after the exact diagonalization of the cluster the intercluster hopping results in the quasiparticle band structure. We have considered the top of the valence band and have shown that the triplet-singlet mixing away from the Ã point explains the disagreement of the experimental dispersion of the valence band in $Sr_2CuO_2Cl_2$ measured by ARPES [18] and calculated in the framework of *t-J* model [19].

Our results are shown in the Fig.1. by empty rectangulars as well as ARPES data by two groups [18, 20] and *t-J, t-t'-J* model results. The most controversial region of the $\vec{k}$-space is the $(0,0)-(\pi,0)$ line, where no clear structure in ARPES data [18] was found. In general, our results are in a better agreement for the whole Brillouin zone then the *t-J* model predicts. We have also shown that the triplet-singlet mixing due to intercluster hopping can be written as the fermion-boson interaction and may result in the superconducting pairing. Here boson is the spin exciton, the singlet-triplet excitation.

References

1. Emery, V.J (1987) Theory of high-T_c superconductivity in oxides, *Phys. Rev. Lett.* **58**, 2794-2798.
2. Varma, C.M, Schmitt-Rink, S., and Ruckenstein, A.E. (1987) Charge transfer excitations and superconductivity in ionic metals, *Solid State Commun.* **62**, 681-685.
3. Zhang, F.C. and .Rice,T.M. (1988) Effective Hamiltonian for superconductivity Cu oxides, *Phys. Rev.* **B37**, 3759-3762.
4. Dagotto, E. (1994) Correlated electrons in high-temperature superconductors, *Rev. Mod. Phys.* **66**, 763-840.
5. Kampf, A. (1994) Magnetic correlations in cuprate superconductors, *Phys. Rep.* **249**, 219-361.
6. Brenig, W. (1995) Aspects of electrons in the cuprate superconductors, *Phys. Rep.* **251**, 153-266.
7. Pompa, M., Bianconi, A., Castellano, A.C., Della Londa, S., Flank, A.M., Lagarde, P., and Udron, D. (1991) Polarized X-ray absorption spectra of cuprate superconductors, *Physica* **C184**, 51-57.
8. Romberg, H., Alexander, M.A., Nucker, N., Adelmann, P., and Fink, J. (1990) Electronic structure of the system $La_{2-x}Sr_xCuO_{4+x}$, , *Phys. Rev.* **B42**, 8768-8771.

9. Gaididei, Yu.B. and Loktev,V.M. (1988) On a theory of electronic spectrum and magnetic properties of high-T_C superconductors, *Phys. Stat. Sol.* **b147,** 307-316.
10. Weber,W.(1988) A Cu d-d excitation model for the pairing in the high-T_C cuprates, *Z. Phys.* **B70,** 323-329.
11. Grilli, M., Castellani, C., and Di Castro,V.(1990) Renormalized band structure of CuO_2 layers in superconducting compounds: A mean field approach, *Phys. Rev.* **B42,** 6233-6241.
12. Kamimura, H. and Eto, M.(1990) $^1A_{1g}$ to $^3B_{1g}$ conversion at the onset of superconductivity in $La_{2-x}Sr_xCuO_4$ due to the apical oxygen effect, *J. Phys. Soc. Jap.* **59**, 3053-3056.
13. Eskes, H. and .Sawatzky,G.A. (1991) Single, triple or multiple-band Hubbard model, *Phys. Rev.* **B44,** 9656-9660.
14. Eskes,H., Tjeng, L.H., and Sawatzky, G.A.(1990) Cluster model calculation of the electronic structure of CuO, *Phys. Rev.* **B41**, 288-292.
15. Ovchinnikov, S.G.(1994) Density of hole states in strongly correlated electron system of copper oxides, *Phys. Rev.* **B49,** 9891-9897.
16. Emery, V.J. and Reiter, G.(1988) Validity of the *t-J* model, *Phys. Rev.* **B38,** 11938-1141.
17. Ovchinnikov, S.G. and .Sandalov, I.S (1989) The band structure of strongly correlated electrons in $La_{2-x}Sr_xCuO_4$ and $YBa_2Cu_3O_{7-y}$, *Physica* **C 161**, 607-617.
18. Wells, B.O., Shen, Z.-X., Matsuura, A., King,D.M., Kastner, M.A., Greven, and M., Birgeneau, R.J. (1995) E vs k relations and many body effects in the model insulating copper oxide $Sr_2CuO_2Cl_2$, ,*Phys. Rev. Lett.* **74**, 964-967.
19. Nazarenko, A., Vos, K.J.E., Haas, S., Dagotto, E., and Gooding, R.J. (1995) Quasiparticle dispersion of $Sr_2CuO_2Cl_2$, *J.Supercond.* **8**, 671-673.
20. Grioni, M., Berger, H., Larosa, S., Vobornik, I., Zwick, F., Margaritondo, G., Kelley, R., Ma, J., and Onellion, M. (1997, Doping effects on the electronic structure of CuO_2 planes, *Physica B* (in press).

SPIN-PEIERLS MAGNET $CuGeO_3$

G.A. PETRAKOVSKII
*Institute of Physics SB RAS,
Krasnoyarsk, 660036, Russia.*

Abstract. The study of the magnetic and resonance properties of the single crystal $CuGeO_3$ were initiated by Petrakovskii and co-authors who first discovered the sharp decrease of the magnetic susceptibility for three main axes of the crystal at the temperature below 14 K. Later Hase and co-authors also observed the similar effect and interpreted it as a spin-Peierls transition. The effect was interpreted in terms of the $S=1/2$ uniform antiferromagnetic chain which is settled down in the 3D crystal below some critical temperature, and a second order phase transition to a dimerizated singlet state with the energy gap spectrum of magnetic excitations is likely to arise. The main investigations proving that the $CuGeO_3$ crystal is the spin-Peierls magnet are discussed in this review with the special emphasis on the following facts: the singletization of the ground magnetic state, the presence of the structural phase transition corresponding to the lattice doubling along the axis of the chain, the energy gap character of the magnetic excitation spectrum, the anomaly of specific heat at the transition temperature, and the specification of the magnetic phase diagram. The magnetoelastic properties of the $CuGeO_3$ crystal arealso discussed here.

1. Introduction

The copper-oxides magnets related to high-T_C superconductors have attracted much interest because they show rich varieties of the magnetic properties. In particular, it is important to understand the mechanisms of the formation of the singlet ground magnetic state in the copper oxide compositions. $CuGeO_3$ is of the special interest in connection with its unusual magnetic properties at low temperatures.

The magnetic and resonance properties of the orthorhombic crystal $CuGeO_3$ (D_{2h}^5) were studied first in our paper [1]. We have found a sharp decrease of the magnetic susceptibility at a temperature below 14 K (Fig.1). It was also shown that the neutron scattering investigation did not indicate the long magnetic ordering at 4 K.

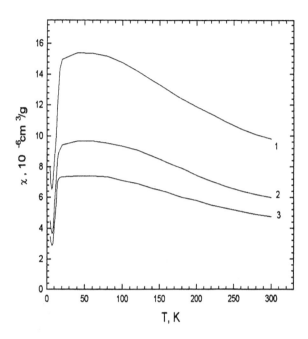

Figure 1. The temperature dependence of the magnetic susceptibility of CuGeO$_3$ for the crystal axes a,b,c (curves 1,2,3).

The sharp decrease of the susceptibility of the crystal CuGeO$_3$ was also observed later in [2] where it was interpreted as a spin-Peierls (SP) phase transition to a singlet state in a magnet with a chain of S=1/2 spins. To clarify this intensive investigations of the CuGeO$_3$ have been performed by different research groups.

2. Theory

According to the basic concept of the SP transition the S=1/2 uniform antiferromagnetic chain settled down in the 3D crystal can undergo a second order phase transition to a dimerized singlet state [3,4] below some critical temperature. The ground state is then a singlet separated from the excited triplet state by the energy gap which depends on the degree of the lattice dimerization and temperature.

The theory of the SP transition is based on the Hamiltonian of noninteracting chains with the Heisenberg interaction of the nearest spins and the spin-phonon interaction:

$$H = H_S + H_I = \sum_n \sum_{l=1}^{N} J_n(l,l+1) (S_{n,l}S_{n,l+1} - \frac{1}{4}) + H_I, \qquad (1)$$

where $S_{n,l}$ is the operator of the first spin in the n-th chain, N is the number of spins in the chain, $J(l,l+1)$ is the exchange integral which depends linearly on the magnetic ions displacement $u_n(l)$:

$$J_n(l,l+1) = J + [u_n(l) - u_n(l+1)]\nabla_1 J(l,l+1), \qquad (2)$$

$$H_I = \sum_{n,l,n',l'} K_{n,l,n',l'} u_{n,l} u_{n',l'} + \frac{1}{2}\sum_{n,l} M(\dot{u}_{n,l})^2, \qquad (3)$$

Here H_I includes the elastic energy of the ions displacement and the kinetic energy. Analyzing the Hamiltonian (1) in the Hartree-Fock approximation Pytte [3] has showed that the antiferromagnetic spin chain with the spin-phonon interaction like in the one dimensional electron-phonon system with the Peierls transition is unstable due to the lattice doubling and, as a consequence, there appears the static lattice displacement of ions, below some critical temperature T_C. The phase transition temperature T_C is given by

$$T_C = 2.28 pJe^{-\frac{1}{\lambda}}, \quad \lambda = \frac{4g^2 p^2 N(0)}{\omega_0^2(Q)}, \quad N(0) = \frac{1}{4\pi pJ},$$

$$g(\lambda q) = \frac{e(lq)}{MN^{1/2}} \nabla_1 J(l,l+1), \quad g = 2ig(\lambda q)p\sin(ka), \quad q = 2k_F, \qquad (4)$$

where $N(0)$ is the density of states of electrons with the Fermi momentum k_F, λ is the spin-phonon coupling constant, $\omega_0(Q)$ is the double lattice phonon energy, $e(lq)$ is the polarization vector, $p(T)$ depends slightly on the temperature and equals 1,64 at T=0. The static displacement of ions u increases when the temperature decreases. The temperature dependence of $u_0(T)$ is similar to the energy gap temperature dependence in the BCS model. The energy gap in the triplet excitations spectrum increases together with u_0: $\Delta = u_0 \omega_0 [\frac{\lambda M}{4N(0)}]$, where M is the mass of the magnetic ion. As a result, the magnetic susceptibility decreases exponentially with temperature decreasing. It is obvious that the temperature T_C may be large depending on the spin-phonon coupling constant λ. The analysis [5] shows that for the usual quasi-1D magnets with the exchange parameter J about 100 K the spin-Peierls phase transition is possible only in the case of the preliminary strong softening of phonons related to the lattice doubling. This fact is the reason of the rarity of spin-Peierls systems because the usual appearance of the long range magnetic order stabilizes the magnetic chain below T_C. In the more precise theory [5] the temperature T_C is given by:

$$T_c = 0.8\lambda J \quad , \tag{5}$$

Bulaevskii [6] has investigated the 1D Heisenberg S=1/2 chain with the alternating exchange:

$$H = J_1 \sum_{i=1}^{i=N/2} (S_{2i} S_{2i+1} + \alpha S_{2i} S_{2i+1}) \quad , \tag{6}$$

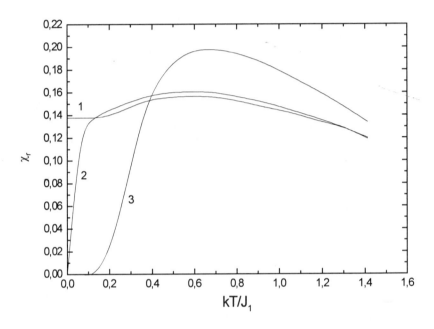

Figure 2. The temperature dependence of the magnetic susceptibility $\chi = \chi_0 J_1 / Ng^2 m^2$ [6]. The curves 1,2,and 3 correspond to $\alpha = 1.0; 0.95;$ and 0.00, respectively.

The magnetic susceptibility for the cases $\alpha=1; 0,95;$ and 0 is shown in Fig.2. At T=0 it is finite for the uniform chain ($\alpha=1$) and decreases exponentially for the alternating chain. The exact spin-wave spectrum for the antiferromagnetic chain at T=0 is [7]

$$\omega_k = \frac{1}{2} J\pi \left| \sin\frac{ka}{2} \right| \quad , \tag{7}$$

The lowest-lying spin-wave excitation degenerates in the ground state. For the alternating lattice [6] there are the singlet-triplet bonds with the energy gap $\Delta \neq 0$ between the ground state and the magnetic excitations band. If the alternating exchange at the dimerizated lattice is

$$J_{1,2} = J[1 \pm \delta(T)] , \qquad (8)$$

then the energy gap in magnetic excitations spectrum equals

$$\Delta(T) = 1.64 J \delta(T) , \qquad (9)$$

The energy gap Δ depends on the pressure and magnetic field. The critical temperature T_C depends also on magnetic field as H^2 and decreases when magnetic field increases.

Thus, for an unambiguous identification of the spin-Peierls phase transition the following facts are important: the present or the absence of the structure phase transition with the lattice doubling along the chain, existance of the singlet magnetic ground state, and the energy gap type of the magnetic excitations spectrum. It is necessary to note that the singlet ground state with the energy gap character of the magnetic excitations can appear without dimerization of the lattice, e.g., in the model of the resonant valence bonds [8], or the competition exchange bonds [9].

3. Experiment

Now let us outline the main experimental data proving that the spin-Peierls phase transition takes place in the compounds $CuGeO_3$. $CuGeO_3$ is the orthorhombic crystal (space group D_{2h}^5) with the lattice parameters a=4,801 A , b=8,473 A , c=2,942 A at T =296 K. The elementary lattice cell consists of two formula units. The fragment of the crystal lattice $CuGeO_3$ is shown in Fig.3. The Cu^{2+} ions are localized in the octahedral sites formed by the oxygen ions. The oxygen octahedrons are connected with each other in the chains. All copper ions are equivalent. The Cu-Cu bonds along the c-axis of the crystal are the most short. So this fact gives the possibility to qualify $CuGeO_3$ crystal as 1D magnet.

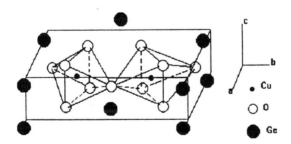

Figure 3. The crystal structure of $CuGeO_3$.

As it was mentioned the magnetic susceptibility of the $CuGeO_3$ crystal sharply decreases below 14 K [1,2]. It is necessary to note that we found the growth of the susceptibility below 7 K [1]. Investigations [10,11] have shown that there are two types of the $CuGeO_3$ crystals with different color: blue and green, depending on the synthesis conditions. The blue crystals are characterized by the monotonous decrease of the susceptibility, and the green crystals show the low temperature susceptibility growth. The different behavior of the blue and green crystals is due to the different content of oxygen. The blue crystals are characterized by the most stoichiometric composition. The isotropic exponential decrease of the magnetic susceptibility in these crystals points to the complete singletization of the magnetic subsystem ground state. The measurement of the magnetic resonance confirms this conclusion. The EPR technique is extremely sensitive to the dynamics of the low dimensional spin systems and it was used widely for studying 1D organic spin-Peierls systems [4]. The EPR intensity has to decrease sharply at low temperatures because of the energy gap arising in the spectrum of the singlet-triplet states. The experimental results [10] confirm this theoretical result (Fig.4).

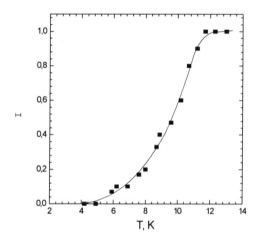

Figure 4. The temperature dependence of the integral intensity (arbitrary units) of the magnetic resonance in $CuGeO_3$ at the frequency 42,6 GHz.

We treat the experimental data basing on Bulaevskii's theory [6] and using Eqs. (8) and (9), which gives $\delta(0)=0,17$ (J /J =0,71) and $\Delta(0)=24$ K [2]. From (5) we get $\lambda =0,2$. Similar values can be obtained from the EPR measurements taking into account the exponential dependence of the resonance intensity in Bulaevskii's theory [12].

It is necessary to note that the temperature dependence of the magnetic susceptibility at $T>T_C$ does not agree with the theory [13] which takes into account only the nn exchange interchain interactions. As it was recently shown [14] it is necessary to introduce the nnn interactions which are only by a factor 3 weaker than the nn couplings. The experimental investigation of the temperature dependence of the specific heat of $CuGeO_3$ [15] also confirms the existence of the second order phase transition at the temperature 14 K.

From the point of view of the identification of the spin-Peierls phase transition it is important to study the influence of the magnetic field and temperature on the magnetic state of $CuGeO_3$. Such investigations have been made in [2,16,18]. In [2] it was found that the temperature T_C decreases with the increase of the magnetic field according to the theory [19]:

$$1 - \frac{T_C(H)}{T_C(0)} = 0.46 \frac{\mu_B H}{k_B T_C(0)} \quad , \tag{10}$$

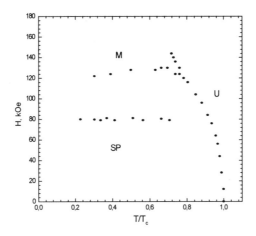

Figure 5. The magnetic phase diagram of $CuGeO_3$. SP, U, M are spin-Peierls, uniform and magnetic phase, respectively.

The magnetic field effect have no explanation within the conventional theory of structural phase transitions. The phase diagram of the magnetic state of $CuGeO_3$ [16,17] is shown in Fig.5. The phase boundary spin-Peierls-uniform(U) states are confirmed by the theory [19]. In the magnetic field $H > H_C$ (H_C is some critical field) the new magnetic phase (M) appears. Up to now the magnetic structure of this phase is not understood.. Recently it was found [20] that strong magnetic fields induce the transition to the phase with the nonzero magnetic susceptibility. The wave vector of the high-field phase depends on the field and it is incommensurate with the crystal lattice. The transition takes place in the magnetic field 12,95 T and is the phase transition of the first order.

However, there are contradictory opinions [16] based on the theory of the soliton structure of the M phase [21]. The typical magnetization curve for the $CuGeO_3$ single crystal was measured in [18] and is shown in Fig.6. The critical fields which destruct the spin-Peierls state are 12,8; 12,5; and 13,5 T along a, b, and c axes, respectively. The additional line on the phase diagram inside the SP-phase was also found by the ultrasonic study (Fig.5) [17]. The nature of this line is yet unknown.

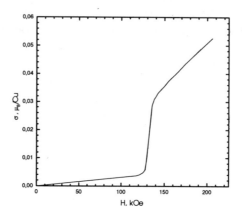

Figure 6. The magnetization curve of the crystal $CuGeO_3$ in the magnetic field H along the crystal axis a at the temperature 4,2 K.

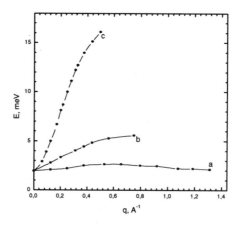

Figure 7. The dispersion dependencies of the spin excitations along axes a,b, and c, respectively.

Thus, the magnetic and resonance measurements show that the magnetic ground state of $CuGeO_3$ at a temperature below 14,2 K is the singlet state. Additional evidence of the existence of the spin-Peierls phase transition has been obtained by the neutron scattering measurements. Nishi et. al. [22] measured the energy gap $\Delta = 2,1$ meV (T=0 K) and the magnetic excitation spectrum (Fig.7) by the inelastic neutron scattering. Fig.7 shows that there is large enough dispersion of the excitations not only along the c-axis but also along the b-direction. So, this system is not exactly the 1D magnet. Authors found the exchange parameters $J_c = 10.4$ meV, $J_b = 0.1 J_c$, $J_a = 0.01 J_c$ and the temperature dependence of the energy gap:

$$\Delta(T) = \Delta(0)[1 - \frac{T}{T_C}]^{0.093} \qquad , \qquad (11)$$

It is necessary to note that the energy gap $\Delta(T)$ measured by the recent neutron scattering investigations has a nonzero value $\Delta(T) = (1/2)\Delta(0)$ [23] at the temperature $T=T_C$, as it follows from (11). The energy gap dependencies on the pressure and magnetic field were investigated by neutron scattering [24,25]. However, the lattice dimerization was not found neither in [22] nor in the other investigations [26]. However, Lorenzo et. al. [26] found unusual softening of the longitudinal acoustic phonons accompanied by the spontaneous distortion along the b- axis [28]. Then Pouget et.al. [29] by the X-ray method and Kanimura et. al. [30]. by the electron diffraction method found the superlattice reflections at the temperature below T_C. Finally, Hiroto et. al. [31] carried out the detailed neutron scattering study of the superlattice reflections and got the lattice dimerization picture shown in Fig.8.

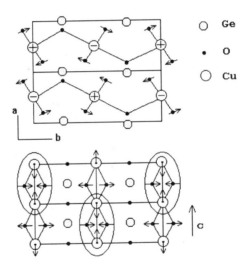

Figure 8. The schematic representation of the low-temperature structure of $CuGeO_3$ in the spin-Peierls state. The cell becomes doubled in the a and c directions below T_C. The arrows and signs indicate the directions of displacements

The picture shows the displacements of the oxygen atoms in ab-plane of about 0,01A which gives approximately the same displacements for the copper ions along the chain axis. The dimerization picture is very close to the simplest model for this compound. It is interesting to note that there are no anomalies in the temperature dependence of the lattice parameter c, in spite of the lattice dimerization along the c-axis.

The important information about the energy structure of $CuGeO_3$ in spin-Peierls state can be obtained from the EPR measurements in the mm and submm range. For the first time such investigations of the $CuGeO_3$ crystal were made in our paper [1]. The EPR measurements at 9,4 GHz showed that the resonance magnetic field depends slightly on the temperature and the g-values are 2,19; 2,266, and 2,083 for the magnetic field directions along a,b, and c-axes, respectively. Submm investigations in strong magnetic fields up to 14T [32] gave the possibility to observe the different resonances in the SP and the high magnetic field M phases. The most detailed EPR investigation in the range 160-1627 GHz in the magnetic fields up to 14T [33] gave the possibility to find the structure of the electron excitations.

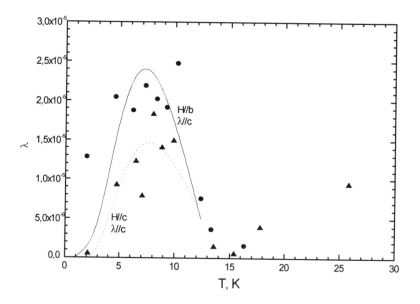

Figure 9. The temperature dependence of the magnetostriction λ of $CuGeO_3$ in the magnetic field 12 kOe. Solid and dotted lines are calculated from (13) and (14) with the fitting parameters: $A(H=0) = 47,6 \cdot 10^{-4}$ K, $A(H=12\ kOe) = -0.8 \cdot 10^{-4}$ K, $\Delta_0 = 18$ K, $\Delta_1 = 5,7$ K for H along the b-axis; and with $A(H=0)=28,9 \cdot 10^{-4}$ K, $A(H=12\ kOe)=3,0 \cdot 10^{-4}$ K; $\Delta=18$ K, $\Delta=13,7$ K for H along the c-axis; closed circles and triangles are experimental results for the magnetic field along b and c axes, respectively.

The magnetoelastic interaction is a very important characteristic of the SP magnet. The investigation of this interaction has been made in our paper [34]. We have measured the magnetostriction by the tenzometric method in the magnetic fields up to 12 T in the temperature range 2 - 25 K. . The result of the measurements is shown in Fig,9. The temperature dependence of the magnetostriction along the c-axis of the crystal has a maximum at the temperature about 7 - 8 K for the magnetic field parallel to the c- and b - axes of the crystal. The magnitostriction disappears at low temperatures. The magnetic field dependence of the magnitostriction shows its growth with the increase of the magnetic field. The rate of increasing has a maximum at the temperature about 8 K.

The results of our measurements of the magnetostriction of $CuGeO_3$ can be understood on the basis of a simple model where the SP state is described by the two-levels singlet-triplet system. At T = 0K such a system is nonmagnetic and the spin subsystem of the crystal does not influence the lattice. The temperature dependence of this effect is related to the change of the triplet state population. At high temperatures when the populations of triplet and singlet states are approximately equal the influence of the spin subsystem on the lattice is small and is mainly caused by the deformational dependence of the exchange interactions. One may expect that at a temperature T_C the influence of the magnetic subsystem will be maximal.

In our model for the intrinsic energy of the SP system we have

$$U = U_{lat} + U_{mag} = Eu^2 + \frac{\alpha \Delta_e N}{1+\exp[\frac{\Delta_e}{T}]} , \qquad (12)$$

where E is Young's modulus, u is the lattice deformation, α is the parameter of the theory, Δ_e is the energy of magnetic excitations, N is the number of Cu^{2+} ions in 1 cm^3 of the crystal. The energy gap Δ_e depends on the temperature, magnetic field, and deformations. From (12) for the equilibrium deformation we have

$$u_0 = A\{ \Delta_e^{-1} - \frac{1}{T}\frac{\exp[\frac{\Delta_e}{T}]}{1+\exp[\frac{\Delta_e}{T}]} \}\frac{1}{1+\exp[\frac{\Delta_e}{T}]} , \qquad (13)$$

where $A = -\frac{N\Delta_e \alpha}{2E}\frac{\partial \Delta_e}{\partial u}$.

The magnetostriction of the crystal in our model is the spontaneous change of the lattice deformation u_0 under the varying magnetic field because the energy gap Δ decreases. If the energy gap in the magnetic fields H=0 and H=$H_1 \neq 0$ equals Δ_0 and Δ_1, respectively, the magnetostriction is

$$\lambda = u_0(\Delta_1) - u_0(\Delta_0) , \qquad (14)$$

The temperature dependence of the magnetostriction is in a qualitative agreement with the experimental results.

4. The exchange interactions

It is important to discuss the formation of the exchange interactions for understanding of the anomalous magnetic properties of $CuGeO_3$. It is found experimentally that the antiferromagnetic exchange between Cu^{2+} ions along the chain (c-axis of the crystal) is dominating in the spin system of $CuGeO_3$. Cu^{2+} ions (electronic configuration d^9, spin $S=1/2$) are localized in the centers of the oxygen octahedrons pulling out along the b-direction. The exchange interaction of the nn ions Cu^{2+} inside the chain takes place across the oxygen bridges Cu^{2+} - O - Cu^{2+}. The bond configuration is almost 90^0. According to the well-known Goodenough-Kanamori-Anderson rules the 90^0-exchange between two half-filled orbits is ferromagnetic in contradiction with the experiment. Geertsma and homskii [35] have shown that it is possible to remove this contradiction by taking into account the presence of side-groups Cu^{2+} - O - Ge^{4+}. In [35] it was also shown that the exchange interaction of the nnn Cu^{2+} along the chain is important and it equals approximately 1/4 of the main exchange[36].

Baiukov [37] proposed the alternative mechanism for the appearance of the antiferromagnetic exchange among nn Cu^{2+} ions. This mechanism is based on the assumption that the ground state of the 3d-hole of Cu ion ($d_{x^2-y^2}$ orbital) and the first excited state (d_{XY} orbital) are closed to each other because of the crystal field configuration peculiarity. In this case the energy distance between these levels is less than exchange energy for the pair of spins of the nn Cu ions where one hole is on the $d_{x^2-y^2}$ orbital and the other one is on the d_{XY} orbital. This distribution of the holes is possible because the total energy of the pair decreases due to the exchange interactions. It should be also mentioned that the microscopic mechanism for the crystal $CuGeO_3$ dimerization have been also discussed by Mattheiss [38].

5. Conclusion

The existing experimental evidence for the magnetic ground state singletization, the crystal lattice dimerization, the T,H- magnetic phase diagram, the gap energy character of the magnetic excitations, the temperature dependencies of the magnetostriction, lattice expansion, and the spin magnetic resonance prove that the $CuGeO_3$ system has the second order spin-Peierls phase transition. It should be mentioned that other mechanisms of the ground state singletization of the spin $S=1/2$ antiferromagnetic chain can also take place in the $CuGeO_3$ crystal. E.g., it was pointed out [36] that the nnn exchange interactions may result in an interplay of the exchange interactions and lead to an alternative mechanisms of the singletization. Another mechanism of the singletization accompanied by the appearance of the energy gap in the magnetic excitation spectrum is the four-spin exchange interaction [39]. So, the properties of the spin-Peierls magnet $CuGeO_3$ should be quantitatively undersood within a more advanced theory basing on a more realistic Hamiltonian which include, e.g., the interchain exchange interactions, the exchange anisotropy, and the exchange nnn interactions in the chain.

References

1. Petrakovskii, G.A., Sablina, K.A., Vorotinov, A.M., Kruglik, A.I., Klimenko, A.G., Balayev, A.D., and Aplesnin, S.S. (1990) The magnetic and resonance properties of the crystal and amorphous $CuGeO_3$, *Sov. Phys.: JETP* **71**, 772-780.
2. Hase, M., Terasaki, I., Uchinokura, K. (1993) Observation of the Spin-Peierls transition in linear Cu^{2+} (spin - 1/2) chains in an inorganic compound $CuGeO_3$, *Phys. Rev. Lett.* **70**, 3651-3655.
3. Pytte, E. (1974) Peierls instability in Heisenberg chains, *Phys. Rev.* **B10**, 4637-4642.
4. Jacobs, I.S., Bray, J.W., Hart, H.R., Interrante, L.V., Kasper, J.S., Watkins, G.D., Prober, D.E., and Bonner, J.C. (1976) Spin-Peierls transitions in magnetic donor-acceptor compound of tetrathiafulvalene (TTF) with bisdithiolene metal complexes, *Phys. Rev.* **B14**, 3036-3051.
5. Buzdin, A.I., Bulaevskii, L.N. (1980) Spin-Peierls transition in the quasi one-dimentional crystals, *Uspekhi Fiz. Nauk*, **131**, 495-510 (in Russian).
6. Bulaevskii, L.N. (1963) On the theory of the nonuniform antiferromagnetic chain of spin, *Sov. Phys.: JETP* **44**, 1008-1012.
7. Cloizeaux, J., Pearson ,J.J. (1962) Spin-wave spectrum of the antiferromagnetic linear chain, *Phys. Rev.* **128**, 2131-2139.
8. Anderson, P.W. (1973) Resonanting valence bonds: a new kind of insulator ?,*Mater. Res. Bull.* **8**, 153-156.
9. Majumdar, C.K., Ghosh, D.K. (1969) On next-nearest-neighbour interaction in linear chain, *J.Math.Phys.* **10**, 1388-1395.
10. Petrakovskii, G.A., Pankrats, A.I., Sablina, K.A., Vorotinov, A.M., Velikanov, D.A., Szymszak, H., and Kolesnik, S. (1996) The influence of the termal treatment on the magnetic and resonance properties of $CuGeO_3$, *Fizika Tverdogo Tela* **38**, 1857-1867 (in Russian).
11. Petrakovskii, G.A., Vorotinov, A.M., Sablina, K.A., Pankrats, A.I., and Velikanov, D.A. (1996) The influence of the diamagnetic dilution by the Li^{1+} and Ga^{3+} on the magnetic and resonance properties of $CuGeO_3$, *Fizika Tverdogo Tela* **38**, 3430-3438 (in Russian).
12. Bulaevskii, L.N. (1960) Magnetic susceptibility of antiferromagnetic chain of spins, *Fizika Tverdogo Tela* **11**, 1132-1140 (in Russian).
13. Bonner, J. C. and Fisher, M. E. (1964) Linear magnetic chain with anisotropic chain, *Phys.Rev.* **135**, 640-649.
14. Riera, J. and Dobry, A. (1995) Magnetic susceptibility in the spin-Peierls system $CuGeO_3$, *Phys. Rev.* **B51**, 16098-16102.
15. Lui, X., Wosnitza, J., Lohneysen, H.V., and Kremer, R.K. (1995) Specific heat of the spin-Peierls compound $CuGeO_3$, *Z. Phys.* **B98**, 163-165.
16. Hase, M., Terasaki, I., Ushinokura, K., Tokunaga, M., Miura, N., Obara, H. (1993) Magnetic phase diagram of the spin-Peierls cuprate $CuGeO_3$, *Phys. Rev.* **B48**, 9616-9619.
17. Poirier, M., Castonguay, M., Revcolevschi, A., and Dhallenne, G. (1995) Ultrasonic study of the magnetic phase diagram of the spin-Peierls sysrem $CuGeO_3$, *Phys. Rev.* **B51**, 6147-6150.
18. Ohta, H., Imagawa, S., Ushiroyama, H., Motokawa, M., Fujita, O., and Akimitsu, J. (1994) Electron spin resonance of spin-Peierls material $CuGeO_3$. *J. Phys. Soc. Japan* **63**, 2870-2873.
19. Cross, M. C. (1979) Effect of magnetic field on a spin-Peiers transition , *Phys. Rev.* **B20**, 4606-4611.
20. Kiryukhin, V. and Keimer, B. (1995) Incommensurate lattice modulation in the spin-Peierls system $CuGeO_3$, *Phys. Rev.* **B52**, 704-709.
21. Bonner, J.C., Northby, J.A., Jacobs, I.S., and Interrante, L.V. (1987) Hign field specific - heat susceptibility measurements : relevance to the spin-Peierls phase diagram and the validity of a soliton picture, *Phys. Rev.* **B35**, 1791-1795.
22. Nishi, M., Fujita, O., and Akimitsu, J. (1994) Neutron scattering study on the spin-Peierls transition in a quasi- one- dimentional magnet $CuGeO_3$, *Phys. Rev.* **B50**, 6508-6512.
23. Regnault, L.P., Ain, M., Hennion, B., Dhalenne, G., and Revcolevschi, A. (1996) Inelastic neutron scattering of the spin-Peierls system $CuGeO_3$, *Phys. Rev.* **B53**, 5579-5597.

24. Nishi, M., Fujita, O., Akimitsu, J., Kakurai, K., and Fujii, Y. (1995) High-pressure effects on the spin-Peierls compound $CuGeO_3$, *Phys. Rev.* **B52**, 6959-6963.
25. Fujita, O., Nishi, M., Akimitsu, J., Okumura, H., Kakurai, K., and Fujii, Y. (1995) Characterization of the spin-Peierls gap in $CuGeO_3$ by means of inelastic neutron scattering, *Physica* **B213-214**, 281-283.
26. Roessli, B., Fischer, P., Schefer, J., Buhrer, W., Furrer, A., Vogt, T., Petrakovskii, G., and Sablina, K., (1994) Elastic and inelastic neutron study of $CuGeO_3$, *J. Phys.: Condens. Matter* **6**, 8469-8477.
27. Lorenzo, J., Hirota, K., Shirane, G., Tranguada, J., Hase, M., Uchinokura, K., Kojima, H., Tanaka, I., and Shibuya, Y. (1994) Soft longitudinal modes in spin-singlet $CuGeO_3$, *Phys. Rev.* **B50**, 1278-1281.
28. Harris, Q., Feng, Q., Birgeneau, R., Hirota, K., Kakurai, K., Lorenzo, J., Shirane, G., Hase, M., and Uchinokura, K. (1994) Thermal contraction of the spin-Peierls transition in $CuGeO_3$, *Phys. Rev.* **B50**, 12606-12610.
29. Pouget, J., Rengault, L. P., Ain, M., Hennion, B., Renard, J. P., Viellet, P., Dhalenne, G., and Revcolevshi, A. (1994) Structural evidence for a spin -Peierls ground state in the quasi-one-dimensional compound $CuGeO_3$, *Phys. Rev. Lett.* **72**, 4037-4040.
30. Kanimura, O., Terauchi, M., Tanaka, M., Fujita, O., and Akimitsu, J. (1994) Electron diffraction study of an inorganic spin-Peierls system $CuGeO_3$, *J. Phys. Soc. Japan* **63**, 2467-2471.
31. Hirota, K., Cox, D., Lorenzo, J., Shirane, J., Tranguada, J., Hase, M., Uchinokura, K., and Kojima, H. (1994) Dimerization of $CuGeO_3$ in the spin-Peierls state, *Phys. Rev. Lett.* **73**, 736-740.
32. Ohta, H., Imagawa, S., Yamamoto, Y., Motokawa, M., Fujita ,O., and Akimitsu (1995) EPR study of high and low field phases of spin-Peierls system $CuGeO_3$, *JMMM* **140-144**, 1685-1686.
33. Brill, T., Boucher, J., Voiron, J., Dhalenne, G., Revcolevschi, A., and Renard, J. (1994) High-field electron spin resonance and magnetization in the dimerised phase of $CuGeO_3$,*Phys. Rev. Lett.* **73**, 1545-1548.
34. Petrakovskii, G., Sablina, K., Vorotinov, A.,Krynetskii, I., Bogdanov, A., Szymczak, H., and Gladczuk, L. (1997) The magnetostriction of $CuGeO_3$, *Sol. St. Commun.* **101**, 545-547.
35. Geertsma, W., Khomskii, D. (1996) Influence of side group on 90°superexchange: a modification of the Goodenough-Kanamori-Anderson rules, *Phys. Rev.* **B54**, 3011-3016.
36. Khomskii, D., Geersma, W., and Mostovoy, M. (1996) Elementary excitations, exchange interaction and spin-Peierls transition in $CuGeO_3$, *Czech. J. of Phys.* **46**, 32-39.
37. Bayukov, O. - private communication.
38. Mattheiss, L.P. (1994) Band picture of the spin-Peierls transition in the spin -1/2 linear - chaincuprate $CuGeO_3$, *Phys. Rev.* **B49**, 14050-14055.
39. Aplesnin, S.S. (1996) Dimerization of the antiferromagnetic chain witn four-spin interaction, *Fizika Tverdogo Tela* **38**, 1868-1875 (in Russian).

ELECTRON ACOUSTIC EFFECTS IN METALLIC MAGNETIC MULTILAYERS

V.I. OKULOV AND V.V.USTINOV
*Institute of Metal Physics of Urals Branch of the Academy of Sciences,
Ekaterinburg, 620219, Russia*
E.A.PAMYATNYKH AND V.V.SLOVIKOVSKAYA
*Urals State University,
Ekaterinburg, 620083, Russia*

1. Introduction

The main properties of metallic magnetic multilayers (superlattice) of Fe/Cr type such as the type of magnetic ordering, dependence of electrical resistance on the magnetic field are defined by itinerant electrons. However, sufficiently complete theory of electron states in these multilayered metallic systems has not been developed up to now. Therefore, there are considerable difficulties, in particular, in the rigorous treatment of the mechanisms of the giant magnetoresistance and the role of intermediate nonmagnetic layers in forming the antiparallel and noncollinear magnetic structures [1]. These difficulties are mainly due to the deficit of experimental data on electron parameters. In the present paper we notice new possibilities which can be of interest in studying the propagation of elastic waves in metallic multilayers. At low temperatures the ultrasonic absorption and dependence of acoustic parameters on the magnetic field are defined by itinerant electrons. We have developed quasi-classical kinetic theory of electron magnetoacoustic effects in multilayers in the frameworks of the model of imperfect boundaries proposed in [2]. The results obtained allow us to describe the dependencies of effective elastic moduli on the magnetization and magnetic field strength. These dependencies are defined both by the parameters of electron structure and the defects of interfaces. We can estimate what of these factors is dominating.

2. Electron acoustic parameters of a metal layer

In order to calculate the electron acoustic parameters of an individual metal layer we solve an equation for the elastic oscillations simultaneously with the kinetic equation for electrons and Maxwell equations. At the interfaces the distribution functions of electrons in the neighbor layers are connected through the boundary conditions. However, in our model of imperfect interfaces [2] it is believed that the electrons while going to the neighbor layer are scattered by defect boundaries. In this model the connection of the distribution functions of electrons in different layer is absent. Therefore, the boundary condition involves the effective diffuseness parameter of the reflection Π which describe scattering of an electron by defects of an interface both while reflecting into a layer and penetrating into a neighbor layer.

We consider the propagation of a transverse ultrasonic wave of a frequency ω perpendicular to the boundary of a layer (along the axis of the crystal symmetry) occupying the area $0 \leq z \leq L$. After the Fourier-cosine transformation for the elastic displacement $u(z)$ the system of equations for the amplitudes u_N takes the following form:

$$(\lambda q_N^2 - \rho_m \omega^2) u_N = F(0) - (-1)^N F(L) - q_N^2 \sum_{N'} \left(-i\omega \alpha_{NN'} u_{N'} + \beta_{NN'} E_{N'}\right) \quad (1)$$

$$\frac{c^2 q_N^2}{4\pi i \omega} E_N = \sum_{N'} \left(\sigma_{NN'} E_{N'} - i\omega \beta_{NN'} u_{N'}\right), \quad (2)$$

where ρ_m, λ are the density and elastic modulus of a lattice, respectively; $F(0)$ and $F(L)$ are the quantities characterizing the external stresses at the boundaries; E_N is the Fourier transform of the amplitude of electric field strength; $q_N = \pi N / L$, $N = 0,1,2...$; $\alpha_{NN'}$, $\beta_{NN'}$, $\sigma_{NN'}$ are the electron kinetic coefficients. Solving the system of equations (1) and (2), we get the equality

$$u(L) = T_L F(0) - R_L F(L), \quad (3)$$

which defines the coefficients of reflection R_L and penetration T_L. If the layer thickness L is small in comparison with the wave length, one may write the following expansions

$$T_L = \frac{1}{\rho_m \omega^2 L} + \frac{1}{6} \frac{L}{\lambda_r} ; \quad R_L = \frac{1}{\rho_m \omega^2 L} - \frac{1}{3} \frac{1}{\lambda_t}, \quad (4)$$

where λ_r and λ_t are the parameters characterizing the acoustic properties of a thin layer.

To calculate the electron kinetic coefficients we assume that the layer thickness L is small compared with the specific mean free path of electrons. In this approximation the formula for the conductivity σ is analogous to the one obtained in [2] and for the magnetic metal has the form:

$$\sigma_{00} = \rho^{-1} = e^2 \left[G_+ \left(\frac{1}{\Pi_+} - \frac{1}{2} \right) + G_- \left(\frac{1}{\Pi_-} - \frac{1}{2} \right) \right] L \ln 1/L, \quad (5)$$

where the indices define the projections of the electron spin on the direction of the magnetization, and coefficients $G_\pm$ are dependent on the shape of the Fermi surfaces for electrons with corresponding projections of spin.

Using the same approximation we calculate the electron contributions to the parameters λ_r, λ_t,

$$\lambda_r = \lambda - i\lambda_{re}; \quad \lambda_t = \lambda - i\lambda_{te}; \quad \lambda_{re}, \lambda_{te} \ll \lambda \quad (6)$$

$$\lambda_{re} = \omega \frac{6}{L^2} \left[\sum_{N,N'} \alpha_{NN'} \Big/ q_{N'}^2 - \rho \left(\sum_N \beta_{0N} \right)^2 \right] \quad (7)$$

$$\lambda_{te} = -\omega \frac{12}{L^2} \left[\sum_{N,N'} (-1)^N \alpha_{NN'} \Big/ q_{N'}^2 + \rho \left(\sum_N \beta_{0N} \right)^2 \right] \quad (8)$$

Formulae (7) and (8) define dynamical electron contributions to the acoustical parameters of a layer. Previously we have considered static electron terms defined by the magnetodeformational interaction of electrons with the lattice [3].

3. Effective elastic modulus of a multilayer

To consider the acoustic properties of magnetic multilayer we use the well-known approximation treating a multilayer as a set of thin layers characterized by an effective elastic modulus $\bar{\lambda}$. If the total thickness of a multilayered film D is small compared with the wave length, the corresponding coefficients T_D and R_D defined analogously to (3) can be written in the form:

$$T_D = \frac{1}{\rho_m \omega^2 D} + \frac{1}{6} \frac{D}{\bar{\lambda}}; \quad R_D = \frac{1}{\rho_m \omega^2 D} - \frac{1}{3} \frac{D}{\bar{\lambda}} \quad (9)$$

If a multilayer consists of alternating layers of two metals of thicknesses L_1 and L_2, respectively, $\bar{\lambda}$ is given by

$$\frac{1}{\bar{\lambda}} = \frac{L_1}{L_1+L_2}\frac{1}{3}\left(\frac{1}{\lambda_{t1}}+\frac{2}{\lambda_{r1}}\right) + \frac{L_2}{L_1+L_2}\frac{1}{3}\left(\frac{1}{\lambda_{t2}}+\frac{2}{\lambda_{r2}}\right) \qquad (10)$$

This formula together with the results given above is used to calculate the electron contribution to the effective elastic modulus, λ_e, according to the definition $\bar{\lambda} = \bar{\lambda}_0 - i\lambda_e$, where $\bar{\lambda}_0$ is the lattice term. Taking into account the ferromagnetic layers only, we have

$$\lambda_e = \omega \sum_\sigma K_\sigma \Pi^*_\sigma L \ln\frac{1}{L} , \qquad (11)$$

where $\Pi^*_\sigma = \Pi_\sigma/(1-\Pi_\sigma/2)$ and K_σ are the coefficients dependent on the parameters of electron energy spectrum only (in particular, on the averaged over Fermi surface deformational potential) and independent on the characteristics of scattering. The parameters Π^*_σ characterizing the scattering by interfaces depend on the angle θ between the directions of the magnetizations in neighbor layers or average magnetization $M = M_0 \cos a/2$, where M_0 is the modulus of the magnetization of a layer. The electron energy and hence the coefficients K_σ depend on the magnetization M as well. Thus the electron imaginary contribution to the effective elastic modulus is the sum of two quantities which depend on the magnetization in a different manner.

4. Electron magnetoacoustic effect in ultrasound attenuation

In the magnetic field H the angle θ between the directions of the magnetizations of the neighbor ferromagnetic layers changes and reaches a saturation, $\theta = 0$, when the field increases. Therefore, the electron elastic modulus is changed following the magnetization curve. This magnetoacoustic effect is analogous to the effect of magnetoresistance since the formulae (5) and (11) for ρ and λ_e have the similar form. Following the approach [2] we introduce the relative change of the parameter $\Pi_+(H)$ with the magnetic field:

$$\frac{\Pi_+(H)-\Pi_+(0)}{\Pi_+(0)} = \xi(H) = \xi_0 \frac{M^2(H)}{M_0^2} \qquad (12)$$

Then the relative change of $\lambda_e(H)$ can be written in the form:

$$\frac{\Delta\lambda_e}{\lambda_e} = \frac{\lambda_e(H)-\lambda_e(0)}{\lambda_e(0)} = -a_1\xi(H)\frac{a_2+\xi(H)}{1+a_3\xi(H)+a_4\xi^2(H)} \qquad (13)$$

This formula describes the magnetoacoustic effect which is an analogue of the giant magnetoresistance effect in multilayers of the Fe/Cr type. The function $\xi(H)$ describes the dependence of the scattering parameters on the field and the coefficients a_i expressed in terms of the energy parameters depend on the magnetic field through the spin splitting energy. Thus (13) describes the interplay of different factors defining the field dependencies of the magnetoacoustic effect. The correlation of the magnetoacoustic effect with magnetoresistance is essential as well. If the role of energy parameters is not pronounced and the spin splitting of the Fermi surface can be neglected, $\lambda_e(H)$ is linearly proportional to $\rho(H)$, and we have

$$\frac{\Delta \lambda_e}{\lambda_e} = \frac{\Delta \rho}{\rho} = -\frac{(\Pi_+ - \Pi_-)^2}{(\Pi_+ + \Pi_-)(\Pi_+ + \Pi_- - \Pi_+ \Pi_-)} \qquad (14)$$

Another magnetoacoustic effect considered in [3] is related to the real contribution of electrons to the elastic modulus $\overline{\lambda}$ due to the magnetodeformational interaction and spin splitting of electron energy.

5. Conclusions

We have derived the dynamic electron contribution to the acoustic parameters of a magnetic multilayer of the Fe/Cr type assuming that the interfaces are imperfect and the thickness of a multilayer is small compared with the wavelength. This electron contribution depends on the magnetization of a multilayer and, through it, on the magnetic field strength. Therefore, the magnetoacoustic effect is analogous to the giant magnetoresistance effect and its observation would provide a new insight into the properties of magnetic multilayers.

Acknowledgements

The research was supported by the Russian Foundation for Basic Researche (Grant No. 95-02-04813).

References

1. Camley, R.E. and Stamps, R.L. (1993) Magnetic multilayers: spin configurations exitations and giant magnetoresistasnce, *J.Phys: Condens. Matter* **5**, 3727-3786.
2. Kravtsov, E.A., Okulov, V.I., and Ustinov, V.V. (1996) Simple theory of giantmagnetoresistance effect in magnetic metallic superlattices with imperfect interfaces, *Solid State Commun.* **99**, 39-41.
3. Okulov, V.I., Pamyatnykh, E.A., Slovikovskaya, V.V., and Ustinov, V.V. (1996) Electronic mechanism of sound velocity variation induced by magnetic field in magnetically ordered metalsuperlattices, *Low Temp. Phys.* **22**, 639 - 640.

Author Index

Akbar, S. 193
Arzhnikov, A. K. 375
Aso, N. 43

Baranov, N. V. 337, 345
Barar, V. 229
Bernhoeft, N. 43
Bourges, Ph. 67
Brauneck, W. 229

Cywinski R. 15

Dakin S. J. 15
Delin A. 323
Dobysheva L. V. 375
Dolja S.N. 309
Dubenko I. S. 261

Endoh Y. 43
Eriksson O. 323

Fawcett E. 27

Gaidukova I. Yu. 261
Granovsky A. 353
Granovsky S. A. 261
Grechnev G. E. 323

Hennion B. 61
Hiess A. 43

Ivanov A. S. 67
Ivanov V. A. 391

Jankowska-Kisielinska J. 61
Johansson B. 323

Kakehashi Y. 193, 363
Kamenev K. 345

Kazantsev V. A. 243
Khan H. R. 353
Kimura N. 193
Komatsubara T. 43
Kubler J. 161

Lacroix C. 303
Lander G. H. 43
Levitin R. Z. 261, 285
Lubashevsky I. A. 285

Maleyev S. V. 67
Markosyan A. S. 261
Menshikov A. Z. 243
Mikke K. 61
Milczarek J. J. 61
Musaev G. M. 285

Nakamura H. 1, 309
Noakes D. R. 27

Okulov V. I. 451
Ovchinnikov S. G. 433

Pamyatnykh E. 151, 451
Panfilov A. S. 309, 323
Petitgrand D. 67
Petrakovskii G. A. 437
Petropavlovsky A.B. 261
Pirogov A. N. 337
Platonov V. V. 285
Podgornykh S. M. 243
Poltavets A. 151
Prudnikov V. 353
Prudnikova M. 353

Rainford B. D. 15
Ritter C. 337
Rodimin V. E. 261
Roessli B. 43

Sandratskii L. M. 161
Sato N. 43
Schweizer J. 337
Shabalin M. 151
Shiga M. 1, 309
Shimizu M. 123
Slovikovskaya V. V. 451
Snegirev V. V. 261
Solontsov A. 89
Stewart J. R. 15
Svechkarev I. V. 309, 323

Tatsenko O. M. 285

Uchida T. 363
Ugolkova E. A. 391
Uhl M. 161
Ustinov V. V. 451

Valiev E. Z. 243
Vasil'ev A. 89

Wagner D. 89, 229
Wills J. M. 323

Yermakov A. A. 331

Zemlyanski S. V. 339
Zhuravlev M. Ye. 395
Zverev V. M. 213
Zvezdin A. K. 195

Subject Index

actinides 43, 161, 323
amorphous magnetism 363
antiferromagnetic chain 285, 437
antiferromagnetic fluctuations 1, 15, 27, 43, 89

band structure 161, 391, 433
β-Mn-Al alloys 15

cerium compounds 303
chromium 27
cobalt 161
coherant potential approximation 375
competing interactions 193
composites 353
correlated electron systems 303, 433
Coulomb interactions 391
critical fluctuations 27

density functional theory 161
dimerization 285, 437
disordered alloy 375

effective medium approach 353
elastic moduli 269, 451
electrical resistivity 337, 345, 353
electron acoustic effects 451
energy dispersive X-ray analysis 353
energy gap 437
EPR 437
extraordinary Hall effect 353

Falicov-Kimball model 229
fcc-Fe 161, 193
fcc-Fe-Mn 143
fcc-Mn 143
Fe_2O_3 ...161
FeRh 161
Fermi liquid 89
frustration 1, 15, 89
functional integration method 193

Gaussian approximation 89, 123
giant magnetoresistance 451
Ginzburg-Landau model 89, 161, 229
granular alloys 353

heavy fermions 151, 303
heavy fermion superconductor 43
Hubbard model 229, 375, 391, 433

intermediate valence 309
Invar problem 89, 229, 243, 269
isotope effect 269
itinerant electron antiferromagnetism 1, 15, 27, 89
itinerant electron ferrimagnetism 285
itinerant electron magnetism 1, 15, 27, 43, 61, 89, 123, 161, 193, 229, 243, 261, 269, 307, 337, 345, 375
itinerant electron spin glass 363

Kondo lattice 303, 309

layerd structures 391, 451

magnetic anisotropy 61
magnetic phase diagram 1, 15, 27, 67, 89, 285, 303, 363, 375
magnetic structure 1, 15, 27, 67, 161, 193, 437
magnetoacoustic effects 451
magnetoelasticity 269
magnetoresistance 337, 345, 353
magnetostriction 1, 15, 243, 437
magnetovolume effect 15, 89, 275, 353
manganese 1
manganese alloys 61
Mn_3Sn 161
metallic magnetic multilayers 451

metamagnetic transitions 285
molecular dynamics approach 193
muon spin relaxation 15

neutron polarisation analysis 43
neutron scattering 1, 15, 43, 61, 67, 337
nickel 161
NMR 1
noncollinear magnetism 67, 161, 363

orbital magnetism 323
orbital susceptibility 123
organic superconductor 391

paramagnetic susceptibility 15, 89, 123, 151, 285
Peierls instability 437
photoemission 433
pressure effect 269, 309, 323

rare-earth compounds 1, 15, 261, 309, 337

screening depth 391
singlet state 285, 437
soft-mode spin fluctuations 89
specific heat 1,, 43, 89, 391

spin anharmonicity 89
spin density wave 27
spin dynamics in UPd_2Al_3 43
spin-flop transitions 285
spin fluctuations 1, 15, 27, 61, 89, 123, 151, 161, 229, 243, 375
spin-glass 1, 363
spin-lattice relaxation time 391
spin liquid 1
spin waves 27, 67, 437
structural disorder 363
Stoner excitations 375
Stoner model 89, 123
superconductivity 43, 391

thermal expansion 89, 243, 261
tricritical point 27

ultrasonic absorption 451
uranium compounds 43, 161

valence phase transition 309
volume magnetostriction 243, 261

YMn_2 1, 15, 261
zero-point spin fluctuations 1, 89